Progress in Computer Science and Applied Logic

Volume [illegible]

Editor

John [illegible], National Science Foundation

Associate Editors

Robert [illegible], Cornell University

Jean [illegible], University of Pennsylvania

Richard [illegible], Cornell University

[illegible], Carnegie Mellon University

Progress in Computer Science and Applied Logic
Volume 23

Coding, Cryptography and Combinatorics

Keqin Feng
Harald Niederreiter
Chaoping Xing
Editors

Springer Basel AG

Editors:

Keqin Feng
Department of Mathematical Sciences
Tsinghua University
Beijing 100084
China
kqfeng@math.hkbu.edu.hk
kfeng@math.tsinghua.edu.cn

Harald Niederreiter
Department of Mathematics
National University of Singapore
2 Science Drive 2
Singapore 117543
Republic of Singapore
nied@math.nus.edu.sg

Chaoping Xing
Department of Mathematics
National University of Singapore
2 Science Drive 2
Singapore 117543
Republic of Singapore
matxcp@nus.edu.sg

2000 Mathematics Subject Classification 11G18, 11G20, 11T71, 94A55, 94A60, 94A62, 94B05, 94B35, 94B60, 94B65

A CIP catalogue record for this book is available from the Library of Congress, Washington D.C., USA

Bibliographic information published by Die Deutsche Bibliothek
Die Deutsche Bibliothek lists this publication in the Deutsche Nationalbibliografie; detailed bibliographic data is available in the Internet at <http://dnb.ddb.de>.

ISBN 978-3-0348-9602-3

Originally published by Birkhäuser Verlag, Basel in 2004
Softcover reprint of the hardcover 1st edition 2004

Printed on acid-free paper produced of chlorine-free pulp. TCF ∞

ISBN 978-3-0348-9602-3 ISBN 978-3-0348-7865-4 (eBook)
DOI 10.1007/978-3-0348-7865-4

www.birkhasuer-science.com

9 8 7 6 5 4 3 2 1

Table of Contents

Preface vii

Invited Papers

Claude Carlet
On the Secondary Constructions of Resilient and Bent Functions 3

Tadao Kasami
Adaptive Recursive MLD Algorithm Based on Parallel Concatenation Decomposition for Binary Linear Codes 29

Wen-Ching Winnie Li
Modularity of Asymptotically Optimal Towers of Function Fields 51

Peizhong Lu and Lianzhen Huang
A New Correlation Attack on LFSR Sequences with High Error Tolerance 67

Amin Shokrollahi
LDPC Codes: An Introduction 85

Contributed Papers

Jintai Ding and Dieter Schmidt
The New Implementation Schemes of the TTM Cryptosystem Are Not Secure 113

Elona Erez and Meir Feder
The Capacity Region of Broadcast Networks with Two Receivers 129

Fang-Wei Fu, San Ling and Chaoping Xing
Constructions of Nonbinary Codes Correcting t-Symmetric Errors and Detecting All Unidirectional Errors: Magnitude Error Criterion 139

Aline Gouget
On the Propagation Criterion of Boolean Functions 153

Tor Helleseth, Jyrki Lahtonen and Petri Rosendahl
On Certain Equations over Finite Fields and Cross-Correlations of m-Sequences 169

Françoise Levy-dit-Vehel and Ludovic Perret
A Polly Cracker System Based on Satisfiability 177

Jing Li
Combinatorially Designed LDPC Codes Using Zech Logarithms and Congruential Sequences ... 193
Lei Li and Shoulun Long
New Constructions of Constant-Weight Codes ... 209
San Ling and Patrick Solé
Good Self-Dual Quasi-Cyclic Codes over $\mathbf{F}_q$, q Odd ... 223
Wilfried Meidl
Linear Complexity and k-Error Linear Complexity for p^n-Periodic Sequences ... 227
Jean-Francis Michon, Pierre Valarcher and Jean-Baptiste Yunès
HFE and BDDs: A Practical Attempt at Cryptanalysis ... 237
Harald Niederreiter
Digital Nets and Coding Theory ... 247
Harald Niederreiter and Ferruh Özbudak
Constructive Asymptotic Codes with an Improvement on the Tsfasman-Vlăduţ-Zink and Xing Bounds ... 259
Josef Pieprzyk and Huaxiong Wang
Malleability Attacks on Multi-Party Key Agreement Protocols ... 277
Charles C. Pinter
Combinatorial Tableaux in Isoperimetry ... 289
Yuansheng Tang
On the Error Exponents of Reliability-Order-Based Decoding Algorithms for Linear Block Codes ... 303
Xiaojian Tian and Cunsheng Ding
A Construction of Authentication Codes with Secrecy ... 319
Hitoshi Tokushige, Jun Asatani, Marc P.C. Fossorier and Tadao Kasami
Selection Method of Test Patterns in Soft-Input and Output Iterative Bounded Distance Decoding Algorithm ... 331
Yejing Wang, Luke McAven and Reihaneh Safavi-Naini
Deletion Correcting Using Generalised Reed-Solomon Codes ... 345
Arne Winterhof
A Note on the Linear Complexity Profile of the Discrete Logarithm in Finite Fields ... 359
Huapeng Wu, M. Anwar Hasan and Ian F. Blake
Speeding Up RSA and Elliptic Curve Systems by Choosing Suitable Moduli ... 369
Gang Yao, Feng Bao and Robert H. Deng
Security Analysis of Three Oblivious Transfer Protocols ... 385
Gang Yao, Guilin Wang and Yong Wang
An Improved Identification Scheme ... 397

Preface

It has long been recognized that there are fascinating connections between coding theory, cryptology, and combinatorics. Therefore it seemed desirable to us to organize a conference that brings together experts from these three areas for a fruitful exchange of ideas. We decided on a venue in the Huang Shan (Yellow Mountain) region, one of the most scenic areas of China, so as to provide the additional inducement of an attractive location. The conference was planned for June 2003 with the official title Workshop on Coding, Cryptography and Combinatorics (CCC 2003). Those who are familiar with events in East Asia in the first half of 2003 can guess what happened in the end, namely the conference had to be cancelled in the interest of the health of the participants. The SARS epidemic posed too serious a threat.

At the time of the cancellation, the organization of the conference was at an advanced stage: all invited speakers had been selected and all abstracts of contributed talks had been screened by the program committee. Thus, it was decided to call on all invited speakers and presenters of accepted contributed talks to submit their manuscripts for publication in the present volume. Altogether, 39 submissions were received and subjected to another round of refereeing. After careful scrutiny, 28 papers were accepted for publication. The selected papers cover a wide range of topics from coding theory, cryptology, and combinatorics and they contain significant advances in these areas as well as very useful surveys.

We extend our cordial thanks to the international program committee consisting of A.R. Calderbank (USA), C. Carlet (France), C.S. Ding (Hong Kong), K.Q. Feng (China, co-chair), T. Helleseth (Norway), H. Imai (Japan), D. Jungnickel (Germany), A. Klapper (USA), P.V. Kumar (USA), S. Ling (Singapore), S.L. Ma (Singapore), J.L. Massey (Denmark/Sweden), H. Niederreiter (Singapore, co-chair), T. Okamoto (Japan), D.Y. Pei (China), J.Y. Shao (China), Z.X. Wan (China), and G.Z. Xiao (China). We are grateful to all referees of the manuscripts for their conscientious work and their valuable advice. We acknowledge with gratitude the organizational and financial support that was provided by the University of Science and Technology of China in Hefei and the National Science Foundation of China. Special thanks go to Shoulun Long of USTC in Hefei.

Finally, we express our thanks to Birkhäuser Verlag, and especially to Dr. Thomas Hempfling, for agreeing to publish this volume and for the advice and help we have received.

January 2004

Keqin Feng
Harald Niederreiter
Chaoping Xing

Invited Papers

Progress in Computer Science and Applied Logic, Vol. 23, 3–28

On the Secondary Constructions of Resilient and Bent Functions

Claude Carlet

Abstract. We first give a survey of the known secondary constructions of Boolean functions, permitting to obtain resilient functions achieving the best possible trade-offs between resiliency order, algebraic degree and nonlinearity (that is, achieving Siegenthaler's bound and Sarkar et al.'s bound). We introduce then, and we study, a general secondary construction of Boolean functions. This construction includes as particular cases the known secondary constructions previously recalled. We apply this construction to design more numerous functions achieving optimum trade-offs between the three characteristics (and additionally having no linear structure). We conclude the paper by indicating generalizations of our construction to Boolean and vectorial functions, and by relating it to a known secondary construction of bent functions.

Keywords. Stream ciphers, Boolean function, correlation immunity, resiliency, nonlinearity, algebraic degree.

1. Introduction

Boolean functions are extensively used in stream cipher systems. Important necessary properties of Boolean functions used in these systems are balancedness, high-order correlation immunity, high algebraic degree and high nonlinearity. An n-variable Boolean function $f : F_2^n \mapsto F_2$ is called *balanced* if its output is uniformly distributed over $\{0,1\}$. It is called mth order correlation immune if the distribution probability of its output is unaltered when any m of its input bits are kept constant. A balanced mth order correlation immune function is called *m-resilient.* The *algebraic degree* of an n-variable Boolean function equals the degree of its algebraic normal form (see Section 2), and its *nonlinearity* equals its Hamming distance to the set of all n-variable affine functions.

Bounds exist, showing the limits inside which lie necessarily these characteristics for all Boolean functions:

– Siegenthaler showed in [29] that any n-variable mth order correlation immune function ($0 \leq m < n$) has algebraic degree smaller than or equal to $n - m$, and that any n-variable m-resilient function ($0 \leq m < n$) has algebraic degree smaller than or equal to $n - m - 1$ if $m < n - 1$ and equal to 1 if $m = n - 1$.

– Sarkar and Maitra showed in [28] a divisibility bound on the Walsh transform values of an n-variable, mth order correlation immune (resp. m-resilient) function, with $m \leq n-2$: these values are divisible by 2^{m+1} (resp. by 2^{m+2}). This provided a nontrivial upper bound on the nonlinearity of resilient functions (and also of correlation immune functions, but non-balanced functions present less cryptographic interest), independently obtained by Tarannikov [31] and by Zheng and Zhang [35]: the nonlinearity of any n-variable, m-resilient function is upper bounded by $2^{n-1} - 2^{m+1}$. Tarannikov showed that resilient functions achieving this bound must have degree $n - m - 1$ (that is, achieve Siegenthaler's bound); thus, they achieve best possible trade-offs between resiliency order, degree and nonlinearity. Moreover, they must be plateaued (see Section 2), which gives them a better chance of resisting the attack of Leveiller et al. [17]. For $m \leq \frac{n}{2} - 2$, the upper bound $2^{n-1} - 2^{m+1}$ on the nonlinearity of m-resilient functions cannot be tight, since we know that the nonlinearity of any balanced n-variable function is strictly upper bounded by $2^{n-1} - 2^{\frac{n}{2}-1}$. But thanks to the divisibility property due to Sarkar and Maitra, there exists then a better upper bound: if n is even and if $m \leq \frac{n}{2} - 2$, then the nonlinearity of any m-resilient function is upper bounded by $2^{n-1} - 2^{\frac{n}{2}-1} - 2^{m+1}$. If n is odd, the bound is more complex (see [28]), but a potentially better upper bound can be given, whatever is the evenness of n: Sarkar-Maitra's divisibility bound shows that $\widehat{\mathrm{f}}(a) = \varphi(a) \cdot 2^{m+2}$ where $\varphi(a)$ is integer-valued. But Parseval's relation $\sum_{a\in F_2^n} \widehat{\mathrm{f}^2}(a) = 2^{2n}$ and the fact that $\widehat{\mathrm{f}}(a)$ is null for every word a of weight $\leq m$ implies

$$\sum_{a;\ w_H(a)>m} \varphi^2(a) = 2^{2n-2m-4}$$

and thus

$$\max_{a\in F_2^n} |\varphi(a)| \geq \sqrt{\frac{2^{2n-2m-4}}{2^n - \sum_{i=0}^m \binom{n}{i}}} = \frac{2^{n-m-2}}{\sqrt{2^n - \sum_{i=0}^m \binom{n}{i}}}.$$

Thus we have

$$\max_{a\in F_2^n} |\varphi(a)| \geq \left\lceil \frac{2^{n-m-2}}{\sqrt{2^n - \sum_{i=0}^m \binom{n}{i}}} \right\rceil$$

(where $\lceil \lambda \rceil$ denotes the smallest integer greater than or equal to λ) and this implies that the nonlinearity of f is upper bounded by

$$2^{n-1} - 2^{m+1} \left\lceil \frac{2^{n-m-2}}{\sqrt{2^n - \sum_{i=0}^m \binom{n}{i}}} \right\rceil.$$

We shall call *Sarkar et al.'s bound* the collection of all these upper bounds on the nonlinearities of m-resilient functions.

More recently, it has been shown in [5] that the Walsh transform values of n-variable, m-resilient, degree d functions are divisible by $2^{m+2+\lfloor\frac{n-m-2}{d}\rfloor}$. This provided more precise upper bounds on the nonlinearities of resilient functions (see also a further improvement in [8]).

Constructions of Boolean functions possessing a good combination of all characteristics have been proposed in [6, 7, 19, 20, 24, 27, 28, 31, 32]. But the knowledge in this matter is still insufficient. For given n – and even for low values of n – we know very few (if any) n-variable Boolean functions achieving Siegenthaler's and Sarkar et al.'s bounds. Knowing numerous such functions gives more chances to find, among them, functions satisfying additional constraints needed for applications. Obviously, the known "good" Boolean functions on n variables will always be an insignificant proportion of the total number of functions; but the number of n-variable Boolean functions being huge, the number of known good functions can however become sufficiently large for a better use in applications.

Functions achieving good characteristics can be obtained by using two kinds of constructions: the primary ones [1, 4, 6, 7], which directly give functions whose characteristics can be calculated (or at least can be lower bounded); and the secondary constructions [19, 20, 24, 27, 28, 31, 32] which build n-variable resilient functions from n'-variable ones (with $n' < n$ in general). The primary constructions could seem preferable, since they lead to potentially more numerous functions. Unfortunately, the known primary constructions do not permit (except in extreme cases, which do not present a real cryptographic interest, see [6]), to build, alone, resilient functions in any numbers of variables, achieving Siegenthaler's and Sarkar et al.'s bounds. They have been used, however, to design optimum functions in small numbers of variables, and they could also be modified (see, *e.g.*, [27, 28]) to lead to functions in larger numbers of variables, achieving good characteristics. But this often needed computer help and it is hardly generalizable. Fortunately, secondary constructions have been used successfully to design optimum functions. Elementary constructions have been combined into a nice construction, introduced by Tarannikov [31] and later studied and slightly modified by Pasalic, Maitra, Johansson and Sarkar [24], which permits to build an infinite sequence of such functions, *cf.* [31, 24, 19, 28, 32].

As we wrote above, these constructions lead, for given n, to very few n-variable functions achieving the bounds. In order to produce, for every n, more numerous such n-variable functions, we have either to find better primary constructions (and this is an open problem), or to find more functions obtained from secondary constructions. The aim of the present paper is to give a general secondary construction, including as particular cases the constructions cited above, and leading to many more functions achieving the bounds.

The paper is organized as follows. In Section 2, we recall the necessary background on Boolean functions. In Sections 3 and 4, we give a survey of the known elementary constructions, including the efficient combination of elementary constructions introduced by Tarannikov and modified by Pasalic et al. (we call it

Tarannikov et al.'s construction). In Section 5, we give a generalization of Tarannikov et al.'s construction, which gives an explanation why this construction works so well; our generalized construction is simpler to understand, thanks to a nice symmetry property, and leads to a multiple branching infinite tree of functions, whereas Tarannikov et al.'s construction leads only to an infinite sequence. As Tarannikov et al.'s construction, our general construction uses pairs of functions achieving Siegenthaler's and Sarkar et al.'s bounds, and whose Walsh spectra are disjoint. In Section 6, we study in detail the problem of generating such pairs. We give a complete description of these pairs for high resiliency orders ($m \geq n-3$). For the remaining resiliency orders, we give, for every d, examples of such pairs of degree d, for all but finitely many cases. In Section 7, we indicate generalizations of our construction to Boolean and vectorial functions. In Section 8, we study how our construction permits to design bent functions, and we relate it to a previous construction of bent functions.

2. Preliminaries

In this section we introduce a few basic concepts and results. By F_2 we denote the finite field $GF(2)$. The (Hamming) distance $d(f,g)$ between two Boolean functions f and g on F_2^n (two n-variable Boolean functions) equals the size of the set $\{x \in F_2^n / \, f(x) \neq g(x)\}$. The (Hamming) weight of f is its distance to the null function, that is the size of its support $\{x \in F_2^n; \; f(x) = 1\}$. It is denoted by $wt(f)$. An n-variable Boolean function f is *balanced* if $wt(f) = 2^{n-1}$. The function f can be represented uniquely by a multivariate polynomial over F_2 of the form $f(x) = \sum_{I \subseteq \{1,\ldots,n\}} a_I \left(\prod_{i \in I} x_i\right)$, called its *algebraic normal form.* The degree of this polynomial is called the *algebraic degree* or simply *degree* of f, and it is denoted by $d^\circ f$. The degree of any cryptographic function must be high (see [2, 14, 15, 16, 34]).

The functions of degrees at most one are called affine functions. The set of all n-variable affine functions is denoted by $A(n)$. Affine functions have constant derivatives $D_a f(x) = f(x) + f(x+a)$. On the contrary, cryptographic functions have preferably no constant derivative $D_a f$ ($a \neq 0$), that is no nonzero *linear structure.* This is a strong requirement in block ciphers (see [13]); in the case of stream ciphers, the existence of nonzero linear structures for a combining function or a filtering one, in pseudorandom generators, is a potential weakness (even if no attack using it has been found so far) that can preferably be avoided.

The *nonlinearity* N_f of an n-variable function f is defined as

$$N_f = \min_{g \in A(n)} (d(f,g)),$$

i.e., N_f is the distance between f and the set of all n-variable affine functions. It must be high (*cf.* [2, 21, 34]). An important tool for the analysis of Boolean functions is the *Walsh transform,* which we define next. The Walsh transform of an n-variable function $f(x_1, \ldots, x_n)$ is the real-valued function over F_2^n whose value at every $a \in F_2^n$ is defined as

$$\widehat{\mathrm{f}}(a) = \sum_{x \in F_2^n} (-1)^{f(x)+a \cdot x},$$

where $a \cdot x = a_1x_1 + \cdots + a_nx_n$ is the usual inner product in F_2^n. We have

$$N_f = 2^{n-1} - \frac{1}{2} \max_{a \in F_2^n} |\widehat{\mathrm{f}}(a)|. \tag{2.1}$$

The nonlinearity of any Boolean function is upper bounded by $2^{n-1} - 2^{\frac{n}{2}-1}$ (we shall call this bound the *universal bound*), due to Parseval's relation $\sum_{a \in F_2^n} \widehat{\mathrm{f}}^2(a) = 2^{2n}$. An n-variable function f is called *bent* if it achieves nonlinearity $2^{n-1} - 2^{\frac{n}{2}-1}$ (this is possible only if n is even), which is equivalent to the fact that $\widehat{\mathrm{f}}(u) = \pm 2^{\frac{n}{2}}$ for all $u \in F_2^n$. These functions (whose degrees can be at most equal to $\frac{n}{2}$, see [25]) present the best possible theoretical resistance to linear attacks, and in the same time to differential attacks, since a Boolean function f is bent if and only if (see [25]) all of its derivatives $D_a f(x)$, $a \in F_2^{n*}$ are balanced (we say that f satisfies then the propagation criterion of degree n, $PC(n)$; more generally, f satisfies $PC(\ell)$ for some integer ℓ if $D_a f$ is balanced for every nonzero vector a of Hamming weight at most ℓ). But bent functions are never balanced, which makes them inappropriate for a cryptographic use. The class of bent functions is included in the class of *plateaued functions*, whose Walsh transform values all belong to a set of the form $\{0, \lambda, -\lambda\}$ for some positive value of λ, called the *amplitude* of the function.

Correlation immune functions were introduced by Siegenthaler [29, 30], to withstand a class of divide-and-conquer attacks on certain models of stream ciphers: we recall that a function $f(x_1, \cdots, x_n)$ is mth order correlation immune if the distribution probability of its output is unaltered when any m of its inputs are kept constant. Xiao and Massey [33] provided a spectral characterization of correlation immune functions. A function f is mth order correlation immune if and only if its Walsh transform $\widehat{\mathrm{f}}$ satisfies: $\widehat{\mathrm{f}}(u) = 0$, for $1 \leq wt(u) \leq m$, where $wt(u)$ denotes the Hamming weight of u. Notice that the two constant Boolean functions are mth order correlation immune, but they do not present interest from cryptographic point of view. Function f is balanced if and only if $\widehat{\mathrm{f}}(0) = 0$. A balanced mth order correlation immune function is called *m-resilient*. The nonlinearity bounds recalled in the introduction are direct consequences of Relation (2.1), of Sarkar's and Maitra's divisibility bound and of the universal bound. The fact that any m-resilient function with nonlinearity $2^{n-1} - 2^{m+1}$ has degree $n - m - 1$ can be deduced from the improved divisibility bound given in [5]; any such function must also be plateaued because $\widehat{\mathrm{f}}(a)$ being divisible by 2^{m+2} and having magnitude upper bounded by this same number, according to Relation (2.1), it must equal 0 or $\pm 2^{m+2}$.

By an (n, m, d, N) function we mean an n-variable, m-resilient function having degree d and nonlinearity N. In the above notation, we may replace some component by $-$ if we do not want to specify it.

3. The Known Elementary Constructions

In this section, we gather the known results on secondary constructions, which have appeared in scattered ways in the literature.

3.1. Direct Sums of Functions

3.1.1. Adding a variable. Let f be an r-variable t-resilient function. The Boolean function on F_2^{r+1}:

$$h(x_1, \ldots, x_r, x_{r+1}) = f(x_1, \ldots, x_r) + x_{r+1}$$

(the addition being obviously computed in F_2) is $(t+1)$-resilient [29]. If f is an $(r, t, r-t-1, 2^{r-1} - 2^{t+1})$ function, then h is an $(r+1, t+1, r-t-1, 2^r - 2^{t+2})$ function, and thus achieves Siegenthaler's and Sarkar et al.'s bounds. This construction does not permit to increase the degree. Also, h has the linear structure $(0, \ldots, 0, 1)$.

3.1.2. Generalization. If f is an r-variable t-resilient function and if g is an s-variable m-resilient function, then the function:

$$h(x_1, \ldots, x_r, x_{r+1}, \ldots, x_{r+s}) = f(x_1, \ldots, x_r) + g(x_{r+1}, \ldots, x_{r+s})$$

is $(t+m+1)$-resilient. This construction has been first introduced by Rothaus in [25], for generating bent functions. The resiliency property of h comes from the easily provable relation $\widehat{\mathrm{h}}(a, b) = \widehat{\mathrm{f}}(a) \times \widehat{\mathrm{g}}(b)$, $\forall a \in F_2^r$, $b \in F_2^s$. We have also $d^\circ h = \max(d^\circ f, d^\circ g)$ and, thanks to Relation (2.1), $N_h = 2^{r+s-1} - \frac{1}{2}(2^r - 2N_f)(2^s - 2N_g) = 2^r N_g + 2^s N_f - 2N_f N_g$. Such function does not give full satisfaction (J. Dillon already explained in [12] that such *decomposable* functions have weaknesses). For instance, h has low degree, in general. And if $N_f = 2^{r-1} - 2^{t+1}$ and $N_g = 2^{s-1} - 2^{m+1}$, then $N_h = 2^{r+s-1} - 2^{t+m+3}$ and h does not achieve Sarkar's and Maitra's bound (note that this is not in contradiction with the properties of the construction recalled at Subsection 3.1.1, since the function $g(x_{r+1}) = x_{r+1}$ is 0-resilient, that is, balanced, but has nonlinearity greater than $2^0 - 2^1$).

Function h has no nonzero linear structure if and only if f and g both have no nonzero linear structure.

3.2. Siegenthaler's Construction

Let f and g be two Boolean functions on F_2^r. Consider the function

$$h(x_1, \cdots, x_r, x_{r+1}) = (x_{r+1} + 1) f(x_1, \cdots, x_r) + x_{r+1} g(x_1, \cdots, x_r)$$

on F_2^{r+1}. Note that the truth-table of h can be obtained by concatenating the truth-tables of f and g. We have:

$$\widehat{\mathrm{h}}(a_1, \ldots, a_r, a_{r+1}) = \widehat{\mathrm{f}}(a_1, \ldots, a_r) + (-1)^{a_{r+1}} \widehat{\mathrm{g}}(a_1, \ldots, a_r).$$

Thus:

1. If f and g are m-resilient, then h is m-resilient [29] (since the vector $(a_1,\ldots,a_r)$ has Hamming weight smaller than or equal to the vector $(a_1,\ldots,a_{r+1})$); moreover, if for every $a \in F_2^r$ of Hamming weight $wt(a) = m+1$, we have $\widehat{f}(a)+\widehat{g}(a) = 0$, then h is $(m+1)$-resilient. Note that the construction recalled at Subsection 3.1.1 corresponds to $g = f+1$ and satisfies this condition. Another possible choice of a function g satisfying this condition is $g(x) = f(x_1+1,\ldots,x_r+1)+\epsilon$, where $\epsilon = m \bmod 2$ (it was first pointed out in [1]), since $\widehat{g}(a) = \sum_{x\in F_2^r}(-1)^{f(x)+\epsilon+(x+(1,\ldots,1))\cdot a} = (-1)^{\epsilon+wt(a)}\widehat{f}(a)$. It leads to a function h having also a nonzero linear structure (namely, the all-one vector);
2. The number $\max_{a_1,\ldots,a_{r+1}\in F_2} |\widehat{h}(a_1,\ldots,a_r,a_{r+1})|$ is clearly upper bounded by $\max_{a_1,\ldots,a_r\in F_2} |\widehat{f}(a_1,\ldots,a_r)| + \max_{a_1,\ldots,a_r\in F_2} |\widehat{g}(a_1,\ldots,a_r)|$; this implies the inequality $2^{r+1}-2N_h \leq 2^{r+1}-2N_f-2N_g$, that is $N_h \geq N_f+N_g$;
   a. if f and g achieve nonlinearity $2^{r-1}-2^{m+1}$ and if h is $(m+1)$-resilient, then the nonlinearity 2^r-2^{m+2} of h is the best possible;
   b. if f and g are such that, for every word a, at least one of the numbers $\widehat{f}(a)$, $\widehat{g}(a)$ is null (in other words, if the supports of the Walsh transforms of f and g are disjoint), then the number
   $$\max_{a_1,\ldots,a_{r+1}\in F_2} |\widehat{h}(a_1,\ldots,a_r,a_{r+1})|$$
   is equal to
   $$\max\left(\max_{a_1,\ldots,a_r\in F_2} |\widehat{f}(a_1,\ldots,a_r)|;\ \max_{a_1,\ldots,a_r\in F_2} |\widehat{g}(a_1,\ldots,a_r)|\right).$$
   Hence, we have $2^{r+1}-2N_h = 2^r-2\min(N_f,N_g)$ and N_h equals therefore $2^{r-1}+\min(N_f,N_g)$; thus, if f and g achieve nonlinearity $2^{r-1}-2^{m+1}$ then h achieves best possible nonlinearity 2^r-2^{m+1};
3. If the monomials of highest degree in the algebraic normal forms of f and g are not all the same, then $d^\circ h = 1+\max(d^\circ f, d^\circ g)$. Note that this condition is not satisfied in the two cases indicated above in 1, where h is $(m+1)$-resilient.
4. For every $a = (a_1,\ldots,a_r) \in F_2^r$ and every $a_{r+1} \in F_2$, we have, denoting $(x_1,\ldots,x_r)$ by x:
   $$D_{(a,a_{r+1})}h(x,x_{r+1})$$
   $$= D_af(x)+a_{r+1}(f+g)(x)+x_{r+1}D_a(f+g)(x)+a_{r+1}D_a(f+g)(x).$$
   If $d^\circ(f+g) \geq d^\circ f$ and if there does not exist $a \neq 0$ such that D_af and D_ag are constant and equal to each other, then h admits no nonzero linear structure (this is in fact a particular case of Corollary 5.2 below).

This construction permits to obtain:

- from any two m-resilient functions f and g having disjoint Walsh spectra, achieving nonlinearity $2^{r-1}-2^{m+1}$ and such that $d^\circ(f+g) = r-m-1$, an

m-resilient function h having degree $r-m$ and having nonlinearity $2^r - 2^{m+1}$, that is, achieving Siegenthaler's and Sarkar et al.'s bounds; note that this construction increases (by 1) the degree;

- from any m-resilient function f achieving degree $r-m-1$ and nonlinearity $2^{r-1} - 2^{m+1}$, a function h having resiliency order $m+1$ and nonlinearity $2^r - 2^{m+2}$, that is, achieving Siegenthaler's and Sarkar et al.'s bounds and having same degree as f (but having nonzero linear structures).

So it permits, when combining these two methods, to keep best trade-offs between resiliency order, degree and nonlinearity, and to increase by 1 the degree and the resiliency order.

3.3. Tarannikov's Elementary Construction

Let f be any Boolean function on ${F_2}^r$. Define the Boolean function h on ${F_2}^{r+1}$ by $h(x_1,\dots,x_r,x_{r+1}) = x_{r+1} + f(x_1,\dots,x_{r-1},x_r+x_{r+1})$. For every $(a_1,\dots,a_{r+1}) \in F_2^{r+1}$, if we denote $(a_1,\dots,a_{r-1})$ by a and $(x_1,\dots,x_{r-1})$ by x, then $\widehat{\mathrm{h}}(a_1,\dots,a_{r+1})$ is equal to

$$\sum_{x_1,\dots,x_{r+1}\in F_2} (-1)^{a\cdot x+f(x_1,\dots,x_r)+a_r(x_r+x_{r+1})+(a_{r+1}+1)x_{r+1}} =$$

$$\sum_{x_1,\cdots,x_{r+1}\in F_2} (-1)^{a\cdot x+f(x_1,\dots,x_r)+a_rx_r+(a_r+a_{r+1}+1)x_{r+1}};$$

it is null if $a_{r+1} = a_r$ and it equals $2\,\widehat{\mathrm{f}}(a)$ if $a_{r+1} = a_r + 1$. Thus:

1. $N_h = 2\,N_f$;
2. If f is t-resilient, then h is t-resilient (since the vector $(a_1,\dots,a_r)$ has Hamming weight smaller than or equal to the vector $(a_1,\dots,a_{r+1})$). If, additionally, $\widehat{\mathrm{f}}(a_1,\dots,a_{r-1},1)$ is null for every vector $(a_1,\dots,a_{r-1})$ of weight at most t, then h is $(t+1)$-resilient (since the only case where $\widehat{\mathrm{h}}(a_1,\dots,a_{r+1})$ may be nonzero is when $a_{r+1} = a_r + 1 = 0$); note that, in such case, if f has nonlinearity $2^{r-1} - 2^{t+1}$ then the nonlinearity of h, which equals $2^r - 2^{t+2}$ achieves then Sarkar et al.'s bound (and, hence, Siegenthaler's bound). The condition that $\widehat{\mathrm{f}}(a_1,\dots,a_{r-1},1)$ is null for every vector $(a_1,\dots,a_{r-1})$ of weight at most t is achieved if f does not actually depend on its last input bit; but the construction is then a particular case of the construction recalled at Subsection 3.1.1. The condition is also achieved if f is obtained from two t-resilient functions, by using Siegenthaler's Construction (recalled at Subsection 3.2).
3. $d^\circ h = d^\circ f$ if $d^\circ f \geq 1$.
4. h has the nonzero linear structure $(0,\dots,0,1,1)$.

Tarannikov combined in [31] this construction with the constructions recalled at Subsections 3.1 and 3.2, to build a more complex secondary construction, which permits to increase in the same time the resiliency order and the degree of the functions and which leads to an infinite sequence of functions achieving Siegenthaler's and Sarkar et al.'s bounds. Increasing then, by using the construction recalled at

Subsection 3.1.1, the set of ordered pairs (n, m) for which such functions can be constructed, he deduced the existence of n-variable m-resilient functions achieving Siegenthaler's and Sarkar et al.'s bounds for any number of variables n and any resiliency order m such that $m \geq \frac{2n-7}{3}$ and $m > \frac{n}{2} - 2$. Pasalic et al. slightly modified this more complex Tarannikov's construction in [24], into a construction that we call *Tarannikov et al.'s construction*, which permitted, when iterating it together with the construction recalled at Subsection 3.1.1, to relax slightly the condition on m into $m \geq \frac{2n-10}{3}$ and $m > \frac{n}{2} - 2$ (the use of Construction 3.1.1 gives then functions with nonzero linear structures). We describe precisely this construction at Section 4.

3.4. Maiorana-McFarland's Construction

We use, at Section 6, a primary construction of resilient functions called Maiorana-McFarland's construction, introduced in [1] (and later studied in [4, 6, 10, 11]): let m, s_1 and s_2 be positive integers such that $s_1 > m$; let g be any Boolean function on $F_2^{s_2}$ and ϕ a mapping from $F_2^{s_2}$ to $F_2^{s_1}$ such that every element in $\phi(F_2^{s_2})$ has Hamming weight strictly greater than m. Then the function:

$$f_{\phi,g}(x, y) = x \cdot \phi(y) + g(y),\ x \in F_2^{s_1},\ y \in F_2^{s_2} \tag{3.1}$$

is m-resilient. Indeed, for every $a \in F_2^{s_1}$ and every $b \in F_2^{s_2}$, we have

$$\widehat{\mathrm{f}_{\phi,\mathrm{g}}}(a, b) = 2^{s_1} \sum_{y \in \phi^{-1}(a)} (-1)^{g(y)+b \cdot y}, \tag{3.2}$$

since every (affine) restriction of f to a coset of $F_2^{s_1}$: $x \mapsto f_{\phi,g}(x, y) + a \cdot x + b \cdot y$ either is constant or is balanced on $F_2^{s_1}$, in which last case it contributes for 0 in the sum $\sum_{x \in F_2^{s_1}, y \in F_2^{s_2}} (-1)^{f_{\phi,g}(x,y)+x \cdot a+y \cdot b}$. Note that if ϕ is injective, then $f_{\phi,g}$ has nonlinearity $2^{s_1+s_2-1} - 2^{s_1-1}$, and that two such functions corresponding to two mappings ϕ_1 and ϕ_2 such that $\phi_1(F_2^{s_2}) \cap \phi_2(F_2^{s_2}) = \emptyset$ have disjoint Walsh supports.

This construction of resilient Maiorana-McFarland's functions is an adaptation of a construction of bent functions (see [12]): if n is even, if π is a permutation of $F_2^{n/2}$ and if g is a Boolean function on $F_2^{n/2}$, then the function $f(x, y) = x \cdot \pi(y) + g(y)$, $x, y \in F_2^{n/2}$, is bent.

4. Tarannikov et al.'s Construction

In this section, we consider the construction introduced in [31], and modified in [24]. We call it *Tarannikov et al.'s construction.* Let us first present it as this has been done in [24]. It uses two $(n-1, t, d-1, 2^{n-2} - 2^{t+1})$ functions f_1 and f_2 to design an $(n+3, t+2, d+1, 2^{n+2} - 2^{t+3})$ function h, assuming that $f_1 + f_2$ has also degree $d-1$ and that the supports of the Walsh transforms of f_1 and f_2 are disjoint. The two restrictions $h_1(x_1, \ldots, x_{n+2}) = h(x_1, \ldots, x_{n+2}, 0)$ and $h_2(x_1, \ldots, x_{n+2}) = h(x_1, \ldots, x_{n+2}, 1)$ have then also disjoint Walsh supports, and these two functions can then be used in the places of f_1 and f_2 (all these properties will be proved again below). This permits to generate an infinite sequence of functions achieving Sarkar et al.'s and Siegenthaler's bounds.

Remark 4.1. As observed in [19, 24], the assumption that the supports of the Walsh transforms of f_1 and f_2 are disjoint is equivalent to the assumption that the function $(1+x_n)f_1 + x_n f_2$ has nonlinearity $2^{n-1} - 2^{t+1}$ (if they are not, then $(1+x_n)f_1 + x_n f_2$ has nonlinearity $2^{n-1} - 2^{t+2}$).

Also, the assumption that $f_1 + f_2$ has degree $d-1$ is equivalent to the assumption that this same function $(1+x_n)f_1 + x_n f_2$ has degree d. ◇

The function h is defined by the relations

$$f(x_1,\dots,x_n) = (1+x_n)f_1(x_1,\dots,x_{n-1}) + x_n f_2(x_1,\dots,x_{n-1}),$$

$$F(x_1,\dots,x_{n+2}) = x_{n+2} + x_{n+1} + f(x_1,\dots,x_n),$$

$$\begin{aligned}G(x_1,\dots,x_{n+2}) &= (1+x_{n+2}+x_{n+1})f_1(x_1,\dots,x_{n-1}) \\ &\quad + (x_{n+2}+x_{n+1})f_2(x_1,\dots,x_{n-1}) + x_{n+2} + x_n\end{aligned}$$

and

$$h(x_1,\dots,x_{n+3}) = (1+x_{n+3})F(x_1,\dots,x_{n+2}) + x_{n+3}G(x_1,\dots,x_{n+2}).$$

If we translate this definition of h into a single formula, we obtain that $h(x_1,\dots,x_{n+3})$ equals:

$$\begin{aligned}(g(x_n,\dots,x_{n+3})+1)\, f_1(x_1,\dots,x_{n-1}) + g(x_n,\dots,x_{n+3})\, f_2(x_1,\dots,x_{n-1}) \\ + g'(x_n,\dots,x_{n+3}),\end{aligned}$$

where the functions g and g' are defined by $g(x) = x_1x_4 + x_2x_4 + x_3x_4 + x_1$ and $g'(x) = x_1x_4 + x_2x_4 + x_2 + x_3$. This can be checked by a direct calculation. It can also be deduced by considering the truth-table of h. By definition, this truth-table can obtained by concatenating the truth-tables of the functions

$$f_1, f_2, \overline{f_1}, \overline{f_2}, \overline{f_1}, \overline{f_2}, f_1, f_2, f_1, \overline{f_1}, f_2, \overline{f_2}, \overline{f_2}, f_2, \overline{f_1} \text{ and } f_1.$$

This implies that the function $h(x_1,\dots,x_{n+3})$ is equal to the function

$$\begin{aligned}&(g(x_n,\dots,x_{n+3})+1)\, f_1(x_1,\dots,x_{n-1}) \\ &+ g(x_n,\dots,x_{n+3})\, f_2(x_1,\dots,x_{n-1}) + g'(x_n,\dots,x_{n+3}),\end{aligned}$$

where the two 4-variable Boolean functions g and g' have truth-tables given in Table 1.

The ANF of g can easily be deduced from its truth-table and equals $x_1x_4 + x_2x_4 + x_3x_4 + x_1$; the ANF of g' equals $x_1x_4 + x_2x_4 + x_2 + x_3$. Thus we have $h(x_1,\dots,x_{n+3}) = (1 + x_nx_{n+3} + x_{n+1}x_{n+3} + x_{n+2}x_{n+3} + x_n)\, f_1(x_1,\dots,x_{n-1}) + (x_nx_{n+3} + x_{n+1}x_{n+3} + x_{n+2}x_{n+3} + x_n)\, f_2(x_1,\dots,x_{n-1}) + x_nx_{n+3} + x_{n+1}x_{n+3} + x_{n+1} + x_{n+2}$.

x_1	x_2	x_3	x_4	$g(x)$	$g'(x)$
0	0	0	0	0	0
1	0	0	0	1	0
0	1	0	0	0	1
1	1	0	0	1	1
0	0	1	0	0	1
1	0	1	0	1	1
0	1	1	0	0	0
1	1	1	0	1	0
0	0	0	1	0	0
1	0	0	1	0	1
0	1	0	1	1	0
1	1	0	1	1	1
0	0	1	1	1	1
1	0	1	1	1	0
0	1	1	1	0	1
1	1	1	1	0	0

TABLE 1

In order to find an explanation of the nice properties of Tarannikov et al.'s construction, let us first calculate the Walsh transforms of g, g' and $g'' = g + g'$. They are given in Table 2.

We observe that g is balanced but that it is not 1-resilient. On the contrary, we see that g' and g'' are both 1-resilient. Moreover, the supports of their Walsh transforms are disjoint and they achieve Siegenthaler's and Sarkar et al.'s bounds. We also observe that the functions $|\widehat{g'}(x)|$ and $|\widehat{g''}(x)|$ are invariant under the mapping $x \mapsto x + (0,0,0,1)$ (that is, they do not depend in fact on x_4).

Let us calculate the Walsh transform of h. To simplify the notation, we shall denote $(x_1, \dots, x_{n-1})$ by x and $(x_n, \dots, x_{n+3})$ by y. Thus we have $h(x,y) = (g(y)+1)\, f_1(x) + g(y)\, f_2(x) + g'(y)$, $x \in F_2^{n-1}, y \in F_2^4$. The Walsh transform of h equals:

$$\widehat{h}(a,b) = \sum_{y \in F_2^4 /\, g(y)=0} (-1)^{g'(y)+b\cdot y} \left(\sum_{x \in F_2^{n-1}} (-1)^{f_1(x)+a\cdot x} \right)$$

$$+ \sum_{y \in F_2^4 /\, g(y)=1} (-1)^{g'(y)+b\cdot y} \left(\sum_{x \in F_2^{n-1}} (-1)^{f_2(x)+a\cdot x} \right)$$

$$= \widehat{f_1}(a) \sum_{y \in F_2^4 /\, g(y)=0} (-1)^{g'(y)+b\cdot y} + \widehat{f_2}(a) \sum_{y \in F_2^4 /\, g(y)=1} (-1)^{g'(y)+b\cdot y}$$

x_1	x_2	x_3	x_4	$\widehat{g}(x)$	$\widehat{g'}(x)$	$\widehat{g''}(x)$
0	0	0	0	0	0	0
1	0	0	0	8	0	0
0	1	0	0	0	0	0
1	1	0	0	0	0	8
0	0	1	0	0	0	0
1	0	1	0	0	8	0
0	1	1	0	8	8	0
1	1	1	0	0	0	8
0	0	0	1	0	0	0
1	0	0	1	8	0	0
0	1	0	1	0	0	0
1	1	0	1	0	0	-8
0	0	1	1	0	0	0
1	0	1	1	0	-8	0
0	1	1	1	-8	8	0
1	1	1	1	0	0	8

TABLE 2

$$= \widehat{f_1}(a) \sum_{y \in F_2^4} (-1)^{g'(y)+b\cdot y} \left(\frac{1+(-1)^{g(y)}}{2}\right) + \widehat{f_2}(a) \sum_{y \in F_2^4} (-1)^{g'(y)+b\cdot y} \left(\frac{1-(-1)^{g(y)}}{2}\right)$$

$$= \frac{1}{2}\widehat{f_1}(a)\left[\widehat{g'}(b) + \widehat{g''}(b)\right] + \frac{1}{2}\widehat{f_2}(a)\left[\widehat{g'}(b) - \widehat{g''}(b)\right].$$

The functions g' and g'' being 1-resilient and the functions f_1 and f_2 being t-resilient, we can see that h is $(t+2)$-resilient (indeed, if $(a,b) \in F_2^{n-1} \times F_2^4$ has weight at most $t+2$ then either a has weight at most t or b has weight at most 1). The degree of h clearly equals $d+1$ since f_1+f_2 has degree $d-1$ and since g has degree 2, and we have $\max_{(a,b)\in F_2^{n-1}\times F_2^4} |\widehat{h}(a,b)| = 4\left(\max_{i\in\{1,2\}} \max_{a\in F_2^{n-1}} |\widehat{f_i}(a)|\right)$, since the supports of the Walsh transforms of g' and g'' are disjoint, as well as those of f_1 and f_2, and since $|\widehat{g'}(b)|$ and $|\widehat{g''}(b)|$ equal 0 or 8, for every $b \in F_2^4$. Notice that g does not play a direct role above (except for its degree): only g' and g'' play roles. Moreover, if we denote g' and g'' by g_1 and g_2 (g becomes then g_1+g_2), a nice symmetry between (f_1, f_2) and (g_1, g_2) appears, and this leads to Theorem 5.1. This theorem and Proposition 5.1, by exhibiting a simple and more general framework for Tarannikov et al.'s construction, give an explanation of the nice properties of this construction.

5. A Generalization of Tarannikov et al.'s Construction

Theorem 5.1. Let r, s, t and m be positive integers such that $t < r$ and $m < s$. Let f_1 and f_2 be two r-variable t-resilient functions. Let g_1 and g_2 be two s-variable m-resilient functions. Then the function

$$h(x,y) = f_1(x) + g_1(y) + (f_1 + f_2)(x)\,(g_1 + g_2)(y),\; x \in F_2^r, y \in F_2^s \tag{5.1}$$

is an $(r+s)$-variable $(t+m+1)$-resilient function. If $f_1 + f_2$ and $g_1 + g_2$ are non-constant, then the algebraic degree of h equals $\max(d^\circ f_1, d^\circ g_1, d^\circ(f_1+f_2) + d^\circ(g_1+g_2))$. The value of the Walsh transform of h at $(a,b) \in F_2^r \times F_2^s$ equals

$$\widehat{\mathrm{h}}(a,b) = \frac{1}{2}\widehat{\mathrm{f}_1}(a)\,[\widehat{\mathrm{g}_1}(b) + \widehat{\mathrm{g}_2}(b)] + \frac{1}{2}\widehat{\mathrm{f}_2}(a)\,[\widehat{\mathrm{g}_1}(b) - \widehat{\mathrm{g}_2}(b)]\,. \tag{5.2}$$

This implies

$$N_h \geq -2^{r+s-1} + 2^s(N_{f_1} + N_{f_2}) + 2^r(N_{g_1} + N_{g_2}) - (N_{f_1} + N_{f_2})(N_{g_1} + N_{g_2}). \tag{5.3}$$

Moreover, if the Walsh transforms of g_1 and g_2 have disjoint supports, then, denoting by f the function $f(x, x_{r+1}) = (1 + x_{r+1})f_1(x) + x_{r+1}f_2(x)$, we have

$$N_h \geq 2^{s-1}N_f + (2^r - N_f)\min_{i\in\{1,2\}} N_{g_i}. \tag{5.4}$$

If, additionally, the Walsh transforms of f_1 and f_2 have disjoint supports, then

$$N_h = \min_{i,j\in\{1,2\}} \left(2^{r+s-2} + 2^{r-1}N_{g_j} + 2^{s-1}N_{f_i} - N_{f_i}N_{g_j}\right). \tag{5.5}$$

Proof. We have:

$$\begin{aligned}
\widehat{\mathrm{h}}(a,b) = & \sum_{y\in F_2^s /\, g_1+g_2(y)=0} \left(\sum_{x\in F_2^r} (-1)^{f_1(x)+a\cdot x}\right)(-1)^{g_1(y)+b\cdot y} \\
& + \sum_{y\in F_2^s /\, g_1+g_2(y)=1} \left(\sum_{x\in F_2^r} (-1)^{f_2(x)+a\cdot x}\right)(-1)^{g_1(y)+b\cdot y} \\
= \widehat{\mathrm{f}_1}(a) & \sum_{y\in F_2^s /\, g_1+g_2(y)=0} (-1)^{g_1(y)+b\cdot y} + \widehat{\mathrm{f}_2}(a) \sum_{y\in F_2^s /\, g_1+g_2(y)=1} (-1)^{g_1(y)+b\cdot y} \\
& = \widehat{\mathrm{f}_1}(a) \sum_{y\in F_2^s} (-1)^{g_1(y)+b\cdot y}\left(\frac{1 + (-1)^{(g_1+g_2)(y)}}{2}\right) \\
& + \widehat{\mathrm{f}_2}(a) \sum_{y\in F_2^s} (-1)^{g_1(y)+b\cdot y}\left(\frac{1 - (-1)^{(g_1+g_2)(y)}}{2}\right).
\end{aligned}$$

We deduce $\widehat{\mathrm{h}}(a,b) = \frac{1}{2}\widehat{\mathrm{f}_1}(a)\,[\widehat{\mathrm{g}_1}(b) + \widehat{\mathrm{g}_2}(b)] + \frac{1}{2}\widehat{\mathrm{f}_2}(a)\,[\widehat{\mathrm{g}_1}(b) - \widehat{\mathrm{g}_2}(b)]$, that is, Relation (5.2).

If (a,b) has weight at most $t+m+1$ then a has weight at most t or b has weight at most m; hence we have $\widehat{\mathrm{h}}(a,b) = 0$. Thus, h is $t+m+1$-resilient.

If f_1+f_2 and g_1+g_2 are non-constant, then the algebraic degree of h equals $\max(d^\circ f_1, d^\circ g_1, d^\circ(f_1+f_2)+d^\circ(g_1+g_2))$ because the terms of highest degrees in $(g_1+g_2)(y)\,(f_1+f_2)(x)$, in $f_1(x)$ and in $g_1(y)$ cannot cancel each others. We deduce from Relation (5.2):

$$\max_{(a,b)\in F_2^r\times F_2^s}|\widehat{\mathrm{h}}(a,b)| \le \frac{1}{2}\max_{a\in F_2^r}|\widehat{\mathrm{f}_1}(a)|\left(\max_{b\in F_2^s}|\widehat{\mathrm{g}_1}(b)|+\max_{b\in F_2^s}|\widehat{\mathrm{g}_2}(b)|\right)$$
$$+\frac{1}{2}\max_{a\in F_2^r}|\widehat{\mathrm{f}_2}(a)|\left(\max_{b\in F_2^s}|\widehat{\mathrm{g}_1}(b)|+\max_{b\in F_2^s}|\widehat{\mathrm{g}_2}(b)|\right),$$

that is, using Relation (2.1):

$$2^{r+s}-2N_h \le \frac{1}{2}(2^r-2N_{f_1})\left((2^s-2N_{g_1})+(2^s-2N_{g_2})\right)$$
$$+\frac{1}{2}(2^r-2N_{f_2})\left((2^s-2N_{g_1})+(2^s-2N_{g_2})\right),$$

or equivalently Relation (5.3). If the Walsh transforms of g_1 and g_2 have disjoint supports, then Relation (5.2), which can be rewritten

$$\widehat{\mathrm{h}}(a,b) = \frac{1}{2}\widehat{\mathrm{g}_1}(b)\left[\widehat{\mathrm{f}_1}(a)+\widehat{\mathrm{f}_2}(a)\right]+\frac{1}{2}\widehat{\mathrm{g}_2}(b)\left[\widehat{\mathrm{f}_1}(a)-\widehat{\mathrm{f}_2}(a)\right]$$
$$=\frac{1}{2}\widehat{\mathrm{g}_1}(b)\left[\widehat{\mathrm{f}}(a,0)\right]+\frac{1}{2}\widehat{\mathrm{g}_2}(b)\left[\widehat{\mathrm{f}}(a,1)\right],$$

implies:

$$\max_{(a,b)\in F_2^r\times F_2^s}|\widehat{\mathrm{h}}(a,b)| \le \frac{1}{2}\max_{u\in F_2^{r+1}}|\widehat{\mathrm{f}}(u)|\times\max_{i\in\{1,2\}}\left(\max_{b\in F_2^s}|\widehat{\mathrm{g}_i}(b)|\right),$$

that is, using Relation (2.1):

$$2^{r+s}-2N_h \le \frac{1}{2}(2^{r+1}-2N_f)\times\max_{i\in\{1,2\}}(2^s-2N_{g_i}),$$

or equivalently, Relation (5.4). If the supports of the Walsh transforms of f_1 and f_2 are disjoint, as well as those of g_1 and g_2, we deduce also from Relation (5.2) that

$$\max_{(a,b)\in F_2^r\times F_2^s}|\widehat{\mathrm{h}}(a,b)| = \frac{1}{2}\max_{i,j\in\{1,2\}}\left(\max_{a\in F_2^r}|\widehat{\mathrm{f}_i}(a)|\max_{b\in F_2^s}|\widehat{\mathrm{g}_j}(b)|\right). \tag{5.6}$$

Using Relation (2.1), we deduce

$$2^{r+s}-2N_h = \frac{1}{2}\max_{i,j\in\{1,2\}}\left((2^r-2N_{f_i})(2^s-2N_{g_j})\right),$$

which is equivalent to Relation (5.5). □

Relations (2.1) and (5.6) (as well as Relations (5.5) or (5.4)), imply:

Corollary 5.1. Let f_1 and f_2 be two $(r,t,-,2^{r-1}-2^{t+1})$ functions with disjoint Walsh supports and such that f_1+f_2 has degree $r-t-1$. Let g_1 and g_2 be two $(s,m,-,2^{s-1}-2^{m+1})$ functions with disjoint Walsh supports and such that g_1+g_2 has degree $s-m-1$. Then the function $h(x,y)=f_1(x)+g_1(y)+(f_1+f_2)(x)\,(g_1+$

$g_2)(y)$ is an $(r+s, t+m+1, r+s-t-m-2, 2^{r+s-1}-2^{t+m+2})$ function. Hence, it achieves Siegenthaler's and Sarkar et al.'s bounds.

Remark 5.1. The elementary secondary constructions, recalled in Section 3, are particular cases of our construction; Rothaus' construction (see Subsection 3.1.2) corresponds to $f_1 = f_2$ or $g_1 = g_2$, Siegenthaler's construction corresponds to $s = 1$, $g_1 = 0$ and $g_2(y_1) = y_1$; Tarannikov's construction does not seem to enter in our framework, in general; but it does in the particular situations in which it is actually applied by Tarannikov.

Remark 5.2. If $t \leq \frac{r}{2} - 2$ (r even) and $m = \frac{s}{2} - 1$ (s even), if f_1 and f_2 are two $(r, t, -, 2^{r-1} - 2^{\frac{r}{2}-1} - 2^{t+1})$ functions (achieving Sarkar et al.'s bound) with disjoint Walsh supports and if g_1 and g_2 are two $(s, m, -, 2^{s-1} - 2^{m+1})$ functions with disjoint Walsh supports, then h is an $(r+s, t+m+1, -, 2^{r+s-1} - 2^{\frac{r+s}{2}-1} - 2^{m+t+2})$ function (achieving Sarkar et al.'s bound if $t+m+1 \leq \frac{r+s}{2} - 2$), according to Relations (2.1) and (5.6) and to the equality $\frac{1}{2}\left(2^{\frac{r}{2}} + 2^{t+2}\right) 2^{m+2} = 2^{\frac{r+s}{2}} + 2^{m+t+3}$.

Proposition 5.1. Under the hypothesis of Theorem 5.1, let us assume that g_1 and g_2 are plateaued with the same amplitude (this is the case if g_1 and g_2 are $(s, m, s-m-1, 2^{s-1}-2^{m+1})$ functions), and that the Walsh transforms of f_1 and f_2 have disjoint supports, as well as g_1 and g_2. If the union of the supports of $\widehat{g_1}$ and $\widehat{g_2}$ is invariant under the mapping $y \mapsto y + (0, \ldots, 0, 1)$, then the Walsh transforms of the restrictions $h_1(x, y) = h(x, y_1, \ldots, y_{s-1}, 0)$ and $h_2(x, y) = h(x, y_1, \ldots, y_{s-1}, 1)$ of h have disjoint supports.

Proof. For every $i \in \{1, 2\}$, we have:

$$\widehat{\mathrm{h_i}}(a, b_1, \ldots, b_{s-1}) = \frac{1}{2}\left(\widehat{\mathrm{h}}(a, b_1, \ldots, b_{s-1}, 0) - (-1)^i \widehat{\mathrm{h}}(a, b_1, \ldots, b_{s-1}, 1)\right), \quad (5.7)$$

and we know, according to Relation (5.2), that the numbers $\widehat{\mathrm{h}}(a, b_1, \ldots, b_{s-1}, 0)$ and $\widehat{\mathrm{h}}(a, b_1, \ldots, b_{s-1}, 1)$ are then either equal to each other or opposite. Thus, at most one of the values $\widehat{\mathrm{h_1}}(a, b_1, \ldots, b_{s-1})$ and $\widehat{\mathrm{h_2}}(a, b_1, \ldots, b_{s-1})$ can be nonzero. Hence, the supports of the Walsh transforms of h_1 and h_2 are disjoint. □

Note that these two restrictions of h are related to the corresponding restrictions of g_1 and g_2 through Relation (5.1), just as h is related to g_1 and g_2. We deduce that, *if h has been obtained by Corollary* 5.1, *then h_1 and h_2 satisfy the hypothesis for f_1 and f_2 in this same corollary,* with $r+s-1$ in the place of r, and $t+m+1$ in the place of t. Corollary 5.1 can then be applied again, with h_1 and h_2 instead of f_1 and f_2. This leads to infinite sequences of functions. This has been observed by Pasalic et al. in [24], in the particular case of their construction. But what is new in the present paper is the symmetry between f_1, f_2 and g_1, g_2. Starting, at least, with two pairs $\{f_1, f_2\}$ and $\{g_1, g_2\}$ of functions satisfying the hypothesis of Corollary 5.1 and such that the union of the supports of $\widehat{f_1}$ and $\widehat{f_2}$ is invariant under the mapping $x \mapsto x + (0, \ldots, 0, 1)$ and the union of the supports of $\widehat{g_1}$ and $\widehat{g_2}$ is invariant under the mapping $y \mapsto y + (0, \ldots, 0, 1)$, we get a whole infinite multiple branching tree of functions, instead of a simple infinite sequence. This gives more values of n and t for which $(n, t, n-t-1, 2^{n-1}-2^{t+1})$ functions (having no nonzero linear structure, see below) can be obtained. It gives also, for every value of n and t such that $t \geq \frac{2r-7}{3}$ and $t > \frac{r}{2} - 2$, many more $(n, t, n-t-1, 2^{n-1}-2^{t+1})$ functions obtained thanks to the combination with the elementary construction "adding a variable", than Tarannikov et al.'s construction, combined with this same construction, does (see [31]).

Moreover, as we prove in the next proposition, the functions obtained with the generalized construction can have no nonzero linear structure.

Proposition 5.2. Let f_1 and f_2 be two r-variable Boolean functions. Let g_1 and g_2 be two s-variable Boolean functions.

1. If f_1+f_2 or g_1+g_2 is constant and if f_1, f_2, g_1 and g_2 admit no nonzero linear structure, then the function $h(x,y) = f_1(x)+g_1(y)+(f_1+f_2)(x)\,(g_1+g_2)(y)$ has no nonzero linear structure.
2. If f_1+f_2 and g_1+g_2 are non-constant and if the two following conditions are satisfied, then h has no nonzero linear structure:
   a. for every a, the function $f_1+f_2+D_af_1$ is non-constant and for every b, the function $g_1+g_2+D_bg_1$ is non-constant;
   b. There does not exist $a \neq 0$ such that D_af_1 and D_af_2 are constant and equal to each other; there does not exist $b \neq 0$ such that D_bg_1 and D_bg_2 are constant and equal to each other.

Proof. For every nonzero $(a,b) \in F_2^r \times F_2^s$, we have $D_{(a,b)}h(x,y) = D_af_1(x) + D_bg_1(y) + (g_1+g_2)(y)D_a(f_1+f_2)(x) + (f_1+f_2)(x)D_b(g_1+g_2)(y) + D_a(f_1+f_2)(x)D_b(g_1+g_2)(y)$.

1. If f_1+f_2 is null then $D_{(a,b)}h(x,y) = D_af_1(x) + D_bg_1(y)$ and if f_1+f_2 equals the constant function 1, then $D_{(a,b)}h(x,y) = D_af_1(x) + D_bg_2(y)$; in both cases, f_1, f_2, g_1 and g_2 admitting no nonzero linear structure, $D_{(a,b)}h$ is not constant. Similarly, if g_1+g_2 is constant, then $D_{(a,b)}h(x,y)$ is not constant.

2. The terms of highest degree with respect to x in $(f_1+f_2)(x)D_b(g_1+g_2)(y)$, and the terms of highest degree with respect to y in $D_a(f_1+f_2)(x)(g_1+g_2)(y)$, can be all cancelled in $D_{(a,b)}h(x,y)$ only if $D_a(f_1+f_2)$ and $D_b(g_1+g_2)$ are constant – say $D_a(f_1+f_2) = \epsilon$ and $D_b(g_1+g_2) = \eta$ – in which case $D_{(a,b)}h(x,y)$ equals $[D_af_1(x)+\eta(f_1+f_2)(x)] + [D_bg_1(y)+\epsilon(g_1+g_2)(y)] + \epsilon\eta$. If $\epsilon = 1$ or $\eta = 1$, then Condition a completes the proof. Otherwise, $\epsilon = \eta = 0$ and Condition b completes the proof. □

Since, if f_1+f_2 is non-constant and has degree at least equal to $d^\circ f_1$, the function $f_1+f_2+D_af_1$ cannot be constant (D_af_1 having at most degree $d^\circ f_1 - 1$), we deduce:

Corollary 5.2. If $d^\circ(f_1+f_2) \geq d^\circ f_1 \geq 1$ and $d^\circ(g_1+g_2) \geq d^\circ g_1 \geq 1$, and if f_1 and f_2 have not a same nonzero linear structure, as well as g_1 and g_2, then h admits no nonzero linear structure.

Remark 5.3. We do not have to assume in the second alinea of Proposition 5.2 or in Corollary 5.2 that the functions f_1, f_2, g_1 and g_2 have no nonzero linear structure. This is why Tarannikov et al.'s construction, which is a combination of elementary constructions producing functions having nonzero linear structures, permits to design functions having no nonzero linear structure.

6. Examples of Pairs of Functions Satisfying the Hypothesis of Corollary 5.1 and the Additional Condition of Proposition 5.1

We are looking for pairs of functions (which can indifferently play the role of $\{f_1, f_2\}$ or the role of $\{g_1, g_2\}$) satisfying the hypothesis of Corollary 5.1, (and having preferably no nonzero linear structure). We want them to lead to infinite classes of functions, and thus, to satisfy the condition of Proposition 5.1. So we seek pairs $\{g_1, g_2\}$ of $(s, m, s-m-1, 2^{s-1}-2^{m+1})$ functions with disjoint Walsh supports, such that g_1+g_2 has degree $s-m-1$ and such that the union of the supports of $\widehat{g_1}$ and $\widehat{g_2}$ is invariant under the mapping $y \mapsto y+(0,\ldots,0,1)$. We reduce ourselves to the cases $m \leq s-2$ to avoid the degenerate situation in which g_1+g_2 is constant. *We shall be able to give a complete description of all pairs for* $m = s-2$ *and for* $m = s-3$. For $m \leq s-4$, we shall give examples valid for all but finitely many cases, but the complete classification is open.

6.1. The Case $m = s-2$

For every $s \geq 2$, two $(s, s-2, 1, 0)$ functions g_1 and g_2 such that g_1+g_2 has degree 1 and with disjoint Walsh spectra are two affine functions $g_1(x) = u\cdot x+\epsilon$ and $g_2(x) = v\cdot x+\eta$ where u and v are distinct and have weights at least $s-1$. This leads to the secondary construction $(f_1, f_2) \mapsto f_1(x)+u\cdot y+\epsilon+(f_1+f_2)(x)((u+v)\cdot y+\epsilon+\eta)$. It is possible to have additionally that the union of the supports of $\widehat{g_1}$ and $\widehat{g_2}$ is invariant under the mapping $y \mapsto y+(0,\ldots,0,1)$: the supports of these Walsh transforms are $\{u\}$ and $\{v\}$; so we just have to take $u+v = (0,\ldots,0,1)$. But, we have then essentially Siegenthaler's construction.

Clearly, the functions g_1 and g_2 do not satisfy the conditions of Proposition 5.2, and indeed, h admits as linear structure any vector $(0, b)$ such that $b \in (u+v)^{\perp}$, since we have $D_{(0,b)}h(x, y) = u\cdot b+(f_1+f_2)(x)((u+v)\cdot b)$.

So taking $m = s-2$ is not completely satisfactory.

6.2. The Case $m = s-3$

We determine now the pairs of $(s, s-3, 2, 2^{s-1}-2^{s-2})$ functions g_1 and g_2 such that g_1+g_2 has degree 2 and with disjoint Walsh spectra: for every $i = 1, 2$, g_i having nonlinearity 2^{s-2}, there exists an affine function ℓ_i such that $g_i+\ell_i$ has weight 2^{s-2} and we know (cf. [18]) that, being quadratic, $g_i+\ell_i$ is then the indicator of an $(s-2)$-dimensional flat. Without loss of generality, set $g_i(x) = (a_i\cdot x)(b_i\cdot x)+c_i\cdot x+\epsilon_i$, where a_i and b_i are linearly independent ($i = 1, 2$). We have then (see for instance [7]) $\widehat{g_i}(u) = 0$ if a_i, b_i and $u+c_i$ are linearly independent; and we have $|\widehat{g_i}(u)| = 2^{s-1}$ otherwise (that is, if $u \in c_i+\langle a_i, b_i\rangle$ where $\langle a_i, b_i\rangle$ is the vector space spanned by a_i and b_i). The supports of $\widehat{g_1}$ and $\widehat{g_2}$ are disjoint if and only if $(c_1 + \langle a_1, b_1\rangle) \cap (c_2 + \langle a_2, b_2\rangle) = \emptyset$ (which is equivalent to saying that the function $(1 + x_{s+1})g_1(x) + x_{s+1}g_2(x)$ belongs to the class $\mathcal{Q}_1$ introduced in [7]). And g_1 and g_2 are $(s-3)$-resilient if and only if the flats $c_1 + \langle a_1, b_1\rangle$ and $c_2 + \langle a_2, b_2\rangle$ have minimum weights at least $s - 2$.

There are only two situations in which the condition of Proposition 5.1 (i.e., the union of the supports of $\widehat{g_1}$ and $\widehat{g_2}$ is invariant under the mapping $y \mapsto y + (0, \ldots, 0, 1)$) is also satisfied: either we have $\langle a_1, b_1\rangle = \langle a_2, b_2\rangle$ (so that the set $(c_1 + \langle a_1, b_1\rangle) \cup (c_2 + \langle a_2, b_2\rangle)$ is a flat) and $(0, \ldots, 0, 1) \in \langle c_1 + c_2, a_1, b_1\rangle$ (since $\langle c_1 + c_2, a_1, b_1\rangle$ is the direction of this flat), or the vector $(0, \ldots, 0, 1)$ belongs to $\langle a_1, b_1\rangle$ and to $\langle a_2, b_2\rangle$.

- If $\langle a_1, b_1\rangle = \langle a_2, b_2\rangle$, then $g_1 + g_2$ is affine. We may assume without loss of generality that $a_1 = a_2$ and $b_1 = b_2$. Function $h(x, y)$ has the form $f_1(x) + (a_1 \cdot y)(b_1 \cdot y) + c_1 \cdot y + \epsilon + (d \cdot y + \eta)(f_1 + f_2)(x)$, where $d = c_1 + c_2$ and $\eta = \epsilon_1 + \epsilon_2$. Then $D_{(0,b)}h(x, y)$ has the form $e \cdot y + (a_1 \cdot b)(b_1 \cdot b) + c_1 \cdot b + (d \cdot b)(f_1 + f_2)(x)$, where $e = (b_1 \cdot b)\, a_1 + (a_1 \cdot b)\, b_1$. Thus, for every $b \in \{a_1, b_1, d\}^\perp$, the vector $(0, b)$ is a linear structure for h, and h admits therefore nonzero linear structures, if $s \geq 4$.
- If $(0, \ldots, 0, 1)$ belongs to $\langle a_1, b_1\rangle$ and to $\langle a_2, b_2\rangle$, then we may assume without loss of generality that $b_1 = b_2 = (0, \ldots, 0, 1)$. We still have to find a_1, c_1, a_2 and c_2, whose last coordinates can be taken, without loss of generality, equal to 0, and such that $c_1, c_1 + a_1$, c_2 and $c_2 + a_2$ are distinct and have weights at least $s - 2$. The set of vectors of length s, with last coordinates equal to 0, and with weights at least $s - 2$ has size $\binom{s-1}{s-2} + \binom{s-1}{s-1} = s$. Thus we need $s \geq 4$. The choice of any two disjoint pairs of vectors (that is, of disjoint lines) $\{c_1, c_1 + a_1\}$ and $\{c_2, c_2 + a_2\}$ in this set leads then to a pair of functions with the desired properties.

If $s = 4$, then such choice is a partition and all choices lead in fact to the same pair of functions g_1 and g_2, up to permutation of the coordinates. Indeed, in such a partition, one line (and one only) must contain the vector $(1, 1, 1, 0)$ and the other vector of the same line is one of the 3 vectors of weight 2 covered by this vector of weight 3. So, the construction due to Tarannikov et al. is the only possible one, up to permutation of the coordinates.

Remark 6.1. Let us see, out of curiosity, what partition corresponds to the $(4, 1, 2, 4)$ functions in the construction of Tarannikov et al.: we have $g_1(x) = x_1x_4 + x_2x_4 + x_2 + x_3 = (a_1 \cdot x)(b_1 \cdot x) + c_1 \cdot x$ and $g_2(x) = x_3x_4 + x_1 + x_2 + x_3 = (a_2 \cdot x)(b_2 \cdot x) + c_2 \cdot x$, with $a_1 = (1, 1, 0, 0), b_1 = (0, 0, 0, 1), c_1 = (0, 1, 1, 0), a_2 = (0, 0, 1, 0), b_2 = (0, 0, 0, 1)$ and $c_2 = (1, 1, 1, 0)$. Thus the disjoint lines are $\{c_1, c_1 + a_1\} = \{(0, 1, 1, 0), (1, 0, 1, 0)\}$ and $\{c_2, c_2 + a_2\} = \{(1, 1, 1, 0), (1, 1, 0, 0)\}$. Note that g_1 and g_2 belong to Maiorana-McFarland's class, since $g_1(x) = (x_1, x_2, x_3) \cdot (x_4, 1 + x_4, 1)$ and $g_2(x) = (x_1, x_2, x_3) \cdot (1, 1, 1 + x_4)$. The mappings $\phi_1 : x_4 \mapsto (x_4, 1 + x_4, 1)$ and $\phi_1 : x_4 \mapsto (1, 1, 1 + x_4)$ are injective, all the elements of their images have weights at least 2, and $\phi_1(F_2)$ and $\phi_2(F_2)$ are disjoint, so that the Walsh supports of g_1 and g_2 are disjoint. According to Proposition 5.2, if $d^\circ(f_1+f_2) = d^\circ f_1$ and if f_1 and f_2 have not a same nonzero linear structure, then h has no nonzero linear structure, because for every $b \neq 0$, the function $g_1 + g_2 + D_b g_1$ is non-constant and there does not exist $b \neq 0$ such that $D_b g_1$ and $D_b g_2$ are constant and equal each other (g_1 and g_2 share here the nonzero linear structure $b = (1,1,0,0)$, but, for this value of b, the derivative $D_b g_1$ equals 1 and is different from the derivative $D_b g_2$, which equals 0). ◇

If $s \geq 5$, then we have, up to permutation of the coordinates, two possible choices of the lines $\{c_1, c_1 + a_1\}$ and $\{c_2, c_2 + a_2\}$: one in which the word $(1, \ldots, 1, 0)$ appears in one of the lines, and one in which it does not. The first choice gives, up to permutation of the coordinates:

$$c_1 + \langle a_1, b_1 \rangle$$
$$= \{(0,1,1,1,\ldots,1,0),(1,0,1,1,\ldots 1,0),(0,1,1,1,\ldots,1,1),(1,0,1,1,\ldots 1,1)\};$$
$$c_2 + \langle a_2, b_2 \rangle$$
$$= \{(1,1,1,1,\ldots,1,0),(1,1,0,1,\ldots 1,0),(1,1,1,1,\ldots,1,1),(1,1,0,1,\ldots 1,1)\};$$

which leads to the following functions: $g_1(x) = (x_1 + x_2)x_s + \sum_{i=2}^{s-1} x_i$, $g_2(x) = x_3 x_s + \sum_{i=1}^{s-1} x_i$ and thus to the secondary construction: $h(x,y) = f_1(x) + (y_1 + y_2)y_s + \sum_{i=2}^{s-1} y_i + (f_1 + f_2)(x)((y_1 + y_2 + y_3)y_s + y_1)$. The second choice gives, up to permutation of the coordinates:

$$c_1 + \langle a_1, b_1 \rangle$$
$$= \{(0,1,1,1,\ldots,1,0),(1,0,1,1,\ldots 1,0),(0,1,1,1,\ldots,1,1),(1,0,1,1,\ldots 1,1)\};$$
$$c_2 + \langle a_2, b_2 \rangle$$
$$= \{(1,1,1,0,1,\ldots,1,0),(1,1,0,1,\ldots 1,0),(1,1,1,0,1,\ldots,1,1),(1,1,0,1,\ldots 1,1)\};$$

which leads to the following functions: $g_1(x) = (x_1 + x_2)x_s + \sum_{i=2}^{s} x_i$, $g_2(x) = (x_3 + x_4)x_s + \sum_{i=1}^{s} x_i + x_3$ and thus to the secondary construction: $h(x,y) = f_1(x) + (y_1 + y_2)y_s + \sum_{i=2}^{s-1} y_i + (f_1 + f_2)(x)((y_1 + y_2 + y_3 + y_4)y_s + y_1 + y_3)$.

In both cases, from two $(r, t, r-t-1, 2^{r-1} - 2^{t+1})$ functions f_1 and f_2 with disjoint Walsh supports and such that $f_1 + f_2$ has degree $r - t - 1$, we construct an $(r+s, t+s+1, r+s-t-m-2, 2^{r+s-1} - 2^{t+s+2})$ function h, and this construction has exactly the same nice properties as Tarannikov et al.'s construction, except that h has the nonzero linear structures $(0, b)$ with $b_1 = b_2 = b_3 = b_s = 0$ and $b_4 = 1$ in the first case; $b_1 = b_2 = b_3 = b_4 = 1$ and $b_s = 0$ in the second one).

6.3. The Case $m = s - 4$

We wish now to obtain two $(s, s-4, 3, 2^{s-1} - 2^{s-3})$ functions g_1 and g_2 whose sum has degree 3 and with disjoint Walsh spectra. Note that, according to Remark 4.1, the pair obtained at Subsection 6.2 of $(s, s-3, 2, 2^{s-1} - 2^{s-2})$ functions ($s \geq 4$) whose sum has degree 2 and with disjoint Walsh spectra, gives an $(s+1, s-3, 3, 2^s - 2^{s-2})$ function. So, applying this observation to $s - 1 \geq 4$ gives an $(s, s-$

$4, 3, 2^{s-1} - 2^{s-3})$ function for every $s \geq 5$. Such function can also easily be obtained by Maiorana-McFarland's construction, as observed in [28]. But we seek a pair of such functions, we want their sum to have degree 3 and their Walsh spectra to be disjoint. According to Proposition 5.1, from every pair of functions obtained at Subsection 6.1, and every pair obtained at Subsection 6.2, we can construct the desired pair by using our construction, if the union of the supports of the Walsh transforms of the functions in one of the pairs is invariant under the mapping $y \mapsto y + (0, \ldots, 0, 1)$. Also, according to Remark 4.1, seeking the desired pair is equivalent to seek an $(s+1, s-4, 4, 2^s - 2^{s-3})$ function, that is, denoting $s+1$ by s', an $(s', s'-5, 4, 2^{s'-1} - 2^{s'-4})$ function. The universal bound and the fact that bent functions are not balanced implies that $s'-4 > \frac{s'}{2} - 1$, that is, $s' > 6$, which means that we can hope obtaining such pair for $s \geq 6$ only. A $(7, 2, 4, 56)$ function with no nonzero linear structure has been exhibited in [24]. The construction "adding a variable" permits then to obtain $(s', s'-5, 4, 2^{s'-1} - 2^{s'-4})$ functions for every $s' \geq 7$ and to deduce, for every $s \geq 6$, pairs (g_1, g_2) of $(s, s-4, 3, 2^{s-1} - 2^{s-3})$ functions g_1 and g_2 whose sum has degree 3, and with disjoint Walsh spectra.

6.4. The Cases $m \leq s - 5$

Here again, according to Proposition 5.1, from every desired pair of s-variable m-resilient functions such that $m = s - k$, and every desired pair of s'-variable m'-resilient functions such that $m' = s' - k$, we can construct a desired pair of s''-variable m''-resilient functions such that $m'' = s'' - k - k' + 1$, if the union of the supports of the Walsh transforms of the functions in one of the pairs is invariant under the mapping $y \mapsto y + (0, \ldots, 0, 1)$.

Also, for every positive integer k and for every $s \geq k - 2 + 2^{k-2}$, there exist $(s, s-k, k-1, 2^{s-1} - 2^{s-k+1})$ functions. Indeed, set $s_2 = k-2$, and for every s, set $s_1 = s - s_2$; there exists an injective mapping $\phi : F_2^{s_2} \mapsto \{y \in F_2^{s_1};\, wt(y) \geq s-k+1\}$ if and only if $\sum_{i=s-k+1}^{s_1} \binom{s_1}{i} = s-k+2 \geq 2^{s_2}$. We deduce the existence of an $(s-k)$-resilient Maiorana-McFarland's function for every s such that $s - k + 2 \geq 2^{k-2}$.

Consequently, according to Remark 4.1, for every positive integer k and for every $s \geq k - 3 + 2^{k-2}$, there exist pairs of $(s, s-k+1, k-2, 2^{s-1} - 2^{s-k+2})$ functions whose sum has degree $k-2$ and with disjoint Walsh spectra.

Remark 6.2. Another example of pair of functions (g_1, g_2) achieving Siegenthaler's and Sarkar et al.'s bounds, such that $g_1 + g_2$ has same degree as g_1 and g_2 and with disjoint Walsh spectra can be found in the literature: it is a pair of $(7, 1, 5, 52)$ functions coming from a $(8, 1, 6, 116)$ function in [19]. This pair is not of the same kind as the others, because the resiliency order 1 is upper bounded by $\frac{s}{2} - 2$, where s is the number of variables.

7. Generalizations of the Construction

7.1. For Boolean Functions

The construction of Theorem 5.1 can be generalized into

$$h(x^1,\dots,x^k)=\prod_{i=1}^{k}(f_1^i+f_2^i)(x^i)+\sum_{i=1}^{k}f_1^i(x^i),\ \text{where } x^i\in F_2^{r_i},\forall i=1,\dots,k.$$

It is a simple matter to see that the Walsh transform of the function $\sum_{i=1}^k f_1^i(x^i)$ is the function $(a^1,\dots,a^k)\mapsto \prod_{i=1}^k \widehat{\mathrm{f}_1^i}(a^i)$. Since the product $\prod_{i=1}^k (f_1^i+f_2^i)(x^i)$ equals 1 if and only if $(f_1^i+f_2^i)(x^i)$ equals 1 for every i, we deduce:

$$\widehat{\mathrm{h}}(a^1,\dots,a^k)=\prod_{i=1}^{k}\widehat{\mathrm{f}_1^i}(a^i)-2\prod_{i=1}^{k}\left(\sum_{x^i\in F_2^{r_i}\,/\,f_1^i(x^i)+f_2^i(x^i)=1}(-1)^{f_1^i(x^i)+x^i\cdot a^i}\right)$$

$$=\prod_{i=1}^{k}\widehat{\mathrm{f}_1^i}(a^i)-2\prod_{i=1}^{k}\left(\sum_{x^i\in F_2^{r_i}}(-1)^{f_1^i(x^i)+x^i\cdot a^i}\left(\frac{1-(-1)^{f_1^i(x^i)+f_2^i(x^i)}}{2}\right)\right)$$

$$=\prod_{i=1}^{k}\widehat{\mathrm{f}_1^i}(a^i)-\frac{1}{2^{k-1}}\sum_{I\subseteq\{1,\dots,k\}}(-1)^{|I|}\prod_{i\in I}\widehat{\mathrm{f}_2^i}(a^i)\prod_{i\in\{1,\dots,k\}\setminus I}\widehat{\mathrm{f}_1^i}(a^i).$$

7.2. For Vectorial Functions

Our construction can be also generalized to vectorial functions. Let F_1 and F_2 be mappings from F_2^r to a field F_{2^k}, and let G_1 and G_2 be mappings from F_2^s to F_{2^k}. Define $H(x,y)=F_1(x)+G_1(y)+(F_1+F_2)(x)\times(G_1+G_2)(y)$, where "$\times$" is the multiplication in the field F_{2^k}. Recall that, in the case of a vectorial function H valued in such a field, the Walsh transform is applied to the Boolean functions $tr(\alpha H)$, where tr is the trace function from F_{2^k} to F_2, and where α is any element of F_{2^k}. The value at $(a,b)\in F_2^r\times F_2^s$ of the Walsh transform of $tr(\alpha H)$ equals:

$$\sum_{\beta\in F_{2^k}}\ \sum_{\substack{y\in F_2^s/\\(G_1+G_2)(y)=\beta}}\ \sum_{x\in F_2^r}(-1)^{tr[\alpha F_1(x)+\alpha G_1(y)+\alpha\beta(F_1+F_2)(x)]+a\cdot x+b\cdot y}=$$

$$\frac{1}{2^k}\sum_{\beta\in F_{2^k}}\sum_{\gamma\in F_{2^k}}\sum_{y\in F_2^s}\sum_{x\in F_2^r}(-1)^{tr[\alpha F_1(x)+\alpha G_1(y)+\alpha\beta(F_1+F_2)(x)+\gamma((G_1+G_2)(y)+\beta)]+a\cdot x+b\cdot y}$$

$$=\frac{1}{2^k}\sum_{\beta\in F_{2^k}}\sum_{\gamma\in F_{2^k}}(-1)^{tr[\gamma\beta]}\left(\sum_{y\in F_2^s}(-1)^{tr[\alpha G_1(y)+\gamma(G_1+G_2)(y)]+b\cdot y}\right)$$

$$\left(\sum_{x\in F_2^r}(-1)^{tr[\alpha F_1(x)+\alpha\beta(F_1+F_2)(x)]+a\cdot x}\right)$$

$$= \frac{1}{2^k} \sum_{\beta \in F_{2^k}} \sum_{\gamma \in F_{2^k}} (-1)^{tr[\gamma\beta/\alpha]} \left(\sum_{y \in F_2^s} (-1)^{tr[\alpha G_1(y) + \gamma(G_1+G_2)(y)] + b \cdot y} \right)$$
$$\left(\sum_{x \in F_2^r} (-1)^{tr[\alpha F_1(x) + \beta(F_1+F_2)(x)] + a \cdot x} \right).$$

This leads to bounds on the nonlinearity of F (the minimum nonlinearity of all the functions $tr(\alpha H)$, $\alpha \neq 0$; see, *e.g.*, [9]) that we cannot include here.

8. A Secondary Construction of Bent Functions Revisited

We consider now the same secondary construction as in the previous sections, but applied to bent functions instead of resilient functions. Recall that any n-variable bent function f (n even), admits a dual $\widetilde{f}$ defined by $\widehat{\mathrm{f}}(u) = 2^{n/2}(-1)^{\widetilde{f}(u)}$, $\forall u \in F_2^n$. Let f_1 and f_2 be two r-variable bent functions (r even) and let g_1 and g_2 be two s-variable bent functions (s even). Let us denote their duals by $\widetilde{f_1}, \widetilde{f_2}, \widetilde{g}_1$ and $\widetilde{g}_2$. Define again $h(x,y) = f_1(x) + g_1(y) + (f_1 + f_2)(x)\,(g_1 + g_2)(y)$, $x \in F_2^r, y \in F_2^s$. According to Relation (5.2), we have

$$\widehat{\mathrm{h}}(a,b)$$
$$= 2^{\frac{r+s}{2}-1} \left[(-1)^{\widetilde{f}_1(a)+\widetilde{g}_1(b)} + (-1)^{\widetilde{f}_1(a)+\widetilde{g}_2(b)} + (-1)^{\widetilde{f}_2(a)+\widetilde{g}_1(b)} - (-1)^{\widetilde{f}_2(a)+\widetilde{g}_2(b)} \right]$$
$$= 2^{\frac{r+s}{2}} (-1)^{\widetilde{f}_1(a)+\widetilde{g}_1(b)+(\widetilde{f}_1+\widetilde{f}_2)(a))(\widetilde{g}_1+\widetilde{g}_2)(b))}.$$

This last equality can be easily checked in each of the 16 cases corresponding to the 16 possible values of $(\widetilde{f_1}(a), \widetilde{f_2}(a), \widetilde{g}_1(b), \widetilde{g}_2(b))$. We see that h is then bent and that its dual $\widetilde{h}$ can be obtained from $\widetilde{f_1}, \widetilde{f_2}, \widetilde{g}_1$ and $\widetilde{g}_2$ exactly in the same manner as h can be obtained from f_1, f_2, g_1 and g_2. Note that this construction generalizes the classical construction $f(x) + g(y)$ indicated by J. Dillon. But it has also the interest of generating functions which are not "decomposable". Moreover, if $f_1 + f_2$ has maximum possible degree $\frac{r}{2}$ and if $g_1 + g_2$ has maximum possible degree $\frac{s}{2}$, then h has maximum possible degree $\frac{r+s}{2}$.

This secondary construction leads to new bent functions. For instance, if we take f_1, f_2, g_1 and g_2 in Maiorana-McFarland's class, we see that for every permutations π_1 and π_2 on $F_2^{r/2}$, for every permutations π_1' and π_2' on $F_2^{s/2}$, for every $r/2$-variable Boolean functions h_1 and h_2 and for every $s/2$-variable Boolean functions h_1' and h_2', the function $(x, y, x', y') \in F_2^{r/2} \times F_2^{r/2} \times F_2^{s/2} \times F_2^{s/2} \mapsto x \cdot \pi_1(y) + h_1(y) + x' \cdot \pi_1'(y') + h_1'(y') + (x \cdot \pi_1(y) + h_1(y) + x \cdot \pi_2(y) + h_2(y))(x' \cdot \pi_1'(y') + h_1'(y') + x' \cdot \pi_2'(y') + h_2'(y'))$ is bent. The function $x \cdot \pi_1(y) + h_1(y) + x' \cdot \pi_1'(y') + h_1'(y')$ belongs to Maiorana-McFarland's class, but not the global function above, in general.

Remark 8.1. Thus, if f_1 and f_2 satisfy $PC(r)$ and if g_1 and g_2 satisfy $PC(s)$, then h satisfies $PC(r + s)$. Note that if f_1 and f_2 satisfy only $PC(\ell)$ with $\ell < r$ or if

g_1 and g_2 satisfy only $PC(\ell')$ with $\ell' < s$ then we lose most of the strength of this result: take for instance $f_1 = f_2$, then $h(x,y) = f_1(x) + g_1(y)$; the derivative $D_{(a,b)}(x,y) = D_a f_1(x) + D_b g_1(y)$ is balanced if and only if $D_a f_1$ is balanced or if $D_b g_1$ is balanced; thus, if f_1 and f_2 satisfy $PC(\ell)$ and if g_1 and g_2 satisfy $PC(\ell')$, then h may satisfy only $PC(\min(\ell,\ell'))$.

Remark 8.2. The construction $(f_1, f_2, g_1, g_2) \mapsto h$ is a particular case of the general secondary construction given in [3], that we describe now: let h be a Boolean function on $F_2^{r+s} = F_2^r \times F_2^s$ such that, for any element y of F_2^s, the function on F_2^r:

$$h_y : x \to h(x,y)$$

is bent. Then h is bent if and only if, for any element u of F_2^r, the function

$$\varphi_u : y \to \widetilde{h_y}(u)$$

is bent on F_2^s. If this condition is satisfied, then the dual of h is the function $\widetilde{h}(u,v) = \widetilde{\varphi_u}(v)$.

Here, for every y, h_y equals f_1 plus a constant or f_2 plus a constant (depending on the values of y) and thus is bent; and φ_u equals g_1 plus a constant or g_2 plus a constant (depending on the values of u), and thus is bent too.

What is interesting in the particular case of this construction $(f_1, f_2, g_1, g_2) \mapsto h$ is that it only assumes the bentness of f_1, f_2, g_1, and g_2 for deducing the bentness of h; no extra condition is needed, contrary to the general construction recalled above.

9. Conclusion

We have given a general framework for the best known secondary construction of resilient functions achieving Siegenthaler's and Sarkar et al.'s bounds, and avoiding nonzero linear structures. This has led us to a generalization of this construction, leading to an infinite multiple branching tree (instead of an infinite sequence) of such functions. The original secondary construction permitted to build few functions achieving Siegenthaler's and Sarkar et al.'s bounds for any number of variables n and any resiliency order t such that $t \geq \frac{2n-10}{3}$ and $t > \frac{n}{2} - 2$. Our generalization permits to build many more such functions with the same number of variables and resiliency order. It also permits to design vectorial functions and bent functions.

References

[1] P. Camion, C. Carlet, P. Charpin, N. Sendrier. On correlation-immune functions. *Advances in Cryptology: Crypto '91, Proceedings, Lecture Notes in Computer Science*, V. 576, pp. 86–100, 1991.

[2] A. Canteaut and M. Trabbia. Improved fast correlation attacks using parity-check equations of weight 4 and 5, *Advanced in Cryptology-EUROCRYPT 2000. Lecture Notes in Computer Science* 1807, pp. 573–588, 2000.

[3] C. Carlet. A construction of bent functions. *Finite Fields and Applications, London Mathematical Society Lecture Series* 233, Cambridge University Press, pp. 47–58, 1996.

[4] C. Carlet. More correlation-immune and resilient functions over Galois fields and Galois rings. *Advances in Cryptology, EUROCRYPT' 97, Lecture Notes in Computer Science* 1233, Springer Verlag, pp. 422–433, 1997.

[5] C. Carlet. On the coset weight divisibility and nonlinearity of resilient and correlation-immune functions. *Proceedings of SETA'01* (Sequences and their Applications 2001), Discrete Mathematics and Theoretical Computer Science, Springer, pp. 131–144, 2001.

[6] C. Carlet. A larger class of cryptographic Boolean functions via a study of the Maiorana-McFarland construction. *Advances in Cryptology - CRYPTO 2002, Lecture Notes in Computer Science* 2442, pp. 549–564, 2002.

[7] C. Carlet and E. Prouff. On plateaued functions and their constructions. Proceedings of *Fast Software Encryption* 2003, *Advances in Cryptology, Lecture Notes in Computer Science* 2887, pp. 54–73, Springer 2003.

[8] C. Carlet and P. Sarkar. Spectral domain analysis of correlation immune and resilient Boolean functions. *Finite Fields and Applications* 8, pp. 120–130, 2002.

[9] F. Chabaud and S. Vaudenay (1995). Links between differential and linear cryptanalysis. *EUROCRYPT'94, Advances in Cryptology, Lecture Notes in Computer Science* 950, Springer Verlag, 356–365.

[10] S. Chee, S. Lee, K. Kim and D. Kim. Correlation immune functions with controlable nonlinearity. *ETRI Journal*, vol 19, no 4, pp. 389–401, 1997.

[11] S. Chee, S. Lee, D. Lee and S.H. Sung. On the correlation immune functions and their nonlinearity. *Proceedings of Asiacrypt'96*, LNCS 1163, pp. 232–243.

[12] J.F. Dillon. Elementary Hadamard Difference sets. Ph. D. Thesis, Univ. of Maryland, 1974.

[13] J.H. Evertse. Linear structures in block ciphers. In *Advances in Cryptology - EUROCRYPT' 87*, no. 304 in Lecture Notes in Computer Science, Springer Verlag, pp. 249–266, 1988.

[14] T. Jakobsen and L.R. Knudsen. The interpolation attack on block ciphers. *Fast Software Encryption'97*, Lecture Notes in Computer Science 1267, pp. 28–40, 1997.

[15] L.R. Knudsen. Truncated and higher order differentials. *Fast Software Encryption, Second International Workshop*, Lecture Notes in Computer Science, n 1008. pp. 196–211. - Springer Verlag, 1995.

[16] X. Lai. Higher order derivatives and differential cryptanalysis. *Proc. Symposium on Communication, Coding and Cryptography*, in honor of J.L. Massey on the occasion of his 60'th birthday. R. Blahut, editor. Kluwer Academic Publishers, 1994.

[17] S. Leveiller, G. Zemor, P. Guillot and J. Boutros. A new cryptanalytic attack for PN-generators filtered by a Boolean function. *Proceedings of Selected Areas of Cryptography 2002*, LNCS 2595, pp. 232–249 (2003).

[18] F.J. MacWilliams and N.J. Sloane. *The Theory of Error-Correcting Codes*, Amsterdam, North Holland, 1977.

[19] S. Maitra and E. Pasalic. Further constructions of resilient Boolean functions with very high nonlinearity. *IEEE Transactions on Information Theory*, vol. 48 (7), pp. 1825–1834, 2002.

[20] S. Maitra and P. Sarkar. Modifications of Patterson-Wiedemann functions for cryptographic applications. *IEEE Trans. Inform. Theory*, Vol. 48, pp. 278–284, 2002.

[21] M. Matsui. Linear cryptanalysis method for DES cipher. *Advances in Cryptology - EUROCRYPT'93, number 765 in Lecture Notes in Computer Science.* Springer Verlag, pp. 386–397, 1994.

[22] N.J. Patterson and D.H. Wiedemann. The covering radius of the $[2^{15}, 16]$ Reed-Muller code is at least 16276. *IEEE Trans. Inform. Theory*, IT-29, pp. 354–356, 1983.

[23] N.J. Patterson and D.H. Wiedemann. Correction to [22]. *IEEE Trans. Inform. Theory*, IT-36(2), pp. 443, 1990.

[24] E. Pasalic, S. Maitra, T. Johansson and P. Sarkar. New constructions of resilient functions and correlation immune Boolean functions achieving upper bound on nonlinearity. *Proceedings of the Workshop on Coding and Cryptography* 2001, pp. 425–434, 2001.

[25] O.S. Rothaus. On bent functions. *J. Comb. Theory*, 20A, 300–305, 1976.

[26] R.A. Rueppel. *Analysis and Design of Stream Ciphers*, Com. and Contr. Eng. Series, Springer, Berlin, 1986.

[27] P. Sarkar and S. Maitra. Construction of nonlinear Boolean functions with important cryptographic properties. *Advances in Cryptology - EUROCRYPT 2000*, number 1807 in Lecture Notes in Computer Science, pp. 485–506. Springer Verlag, 2000.

[28] P. Sarkar and S. Maitra. Nonlinearity bounds and constructions of resilient Boolean functions. *CRYPTO 2000, LNCS* Vol. 1880, ed. Mihir Bellare, pp. 515–532, 2000.

[29] T. Siegenthaler. Correlation-immunity of nonlinear combining functions for cryptographic applications. *IEEE Transactions on Information theory*, V. IT-30, No. 5, pp. 776–780, 1984.

[30] T. Siegenthaler. Decrypting a class of stream ciphers using ciphertext only. *IEEE Transactions on Computer, V. C-34*, No. 1, pp. 81–85, 1985.

[31] Y.V. Tarannikov. On resilient Boolean functions with maximum possible nonlinearity. *Proceedings of INDOCRYPT 2000*, Lecture Notes in Computer Science 1977, pp. 19–30, 2000.

[32] Y.V. Tarannikov. New constructions of resilient Boolean functions with maximum nonlinearity. *Proceedings of FSE 2001*, 8th International Workshop, FSE 2001, Lecture Notes in Computer Science, vol. 2355, pp. 66–77, 2001.

[33] G.-Z. Xiao and J.L. Massey. A spectral characterization of correlation-immune combining functions. *IEEE Trans. Inf. Theory*, Vol IT 34, n° 3, pp. 569–571, 1988.

[34] G.-Z. Xiao, C. Ding and W. Shan. *The Stability Theory of Stream Ciphers*, vol. LNCS 561, Springer Verlag, 1991.

[35] Y. Zheng, X.-M. Zhang. Improving upper bound on the nonlinearity of high order correlation immune functions. *Proceedings of Selected Areas in Cryptography 2000*, Lecture Notes in Computer Science 2012, pp. 262–274, 2001.

Claude Carlet
INRIA
Projet CODES
BP 105
F-78153 Le Chesnay Cedex, France

also with

GREYC (Caen) and the University of Paris 8
e-mail: `claude.carlet@inria.fr`

Progress in Computer Science and Applied Logic, Vol. 23, 29–50

Adaptive Recursive MLD Algorithm Based on Parallel Concatenation Decomposition for Binary Linear Codes

Tadao Kasami

Abstract. Based on the original recursive MLD algorithm (RMLD), "top-down RMLD" has been proposed to reduce the average decoding complexity by a lazy evaluation strategy. In this paper, a revised version of top-down RMLD, called adaptive RMLD, is surveyed. In the adaptive RMLD, the coarsest parallel concatenation decomposition is adopted as the basis of recursion, and a new sufficient condition that a currently best candidate is the optimum at the current level of recursion is used as an early termination condition of the recursion. Preliminary simulation results for the (128, 64) Reed-Muller code are presented.

Keywords. MLD, Recursive, Adaptive, Parallel concatenation decomposition.
This paper is partially based on [11].

1. Introduction

Two types of maximum likelihood decoding algorithms for linear block codes have been proposed. The decoding complexity of the first type (for example, Viterbi type algorithms) is almost independent of the signal to noise ratio, and the complexity of the second type (for example, iterative decoding algorithms [1, 2]) decreases considerably as the signal to noise ratio increases. Based on the original recursive maximum likelihood decoding algorithm [3] of the first type, abbreviated as "Bottom-up RMLD", recursive maximum likelihood decoding algorithm of the second type, called "Top-down RMLD [4]", has been introduced by a lazy evaluation strategy. In contrast with Viterbi type algorithms, RMLD has a parallel structure and bottom-up RMLD provides a reduced decoding complexity compared with an optimally sectionalized Viterbi algorithm by A. Lafourcade and A. Vardy [5]. Furthermore, RMLD has a homogeneous structure for binary-transitive-invariant codes [6], e.g., RM codes and EBCH (extended primitive BCH) codes.

In this paper, a revised version of top-down RMLD, called "Adaptive RMLD" is introduced. In Adaptive RMLD, the coarsest parallel concatenation decomposition is adopted as the basis of recursion, and a new sufficient condition that a currently best candidate is the optimum at the current level of recursion is used as an early termination condition of the recursion. Preliminary simulation results (Figures 6 and 7) [12] on the (128, 64, 16) RM code show that the numbers of addition equivalent operations (AEO) are reduced to about 10^{-2} for 1.0dB to 3.0dB and about 10^{-1} for 4.0dB of those by top-down RMLD [7, 8] which used ordered statistic information [9].

2. Notations

For $i \leq j$, $[i,j]$ denotes the set of integers from i to j, called a section. For a positive integer n, V^n denotes the set of binary n-tuples. For $\boldsymbol{u} = (u_1, u_2, \ldots, u_n) \in V^n$ and a subset $I = \{i_1, i_2, \ldots, i_m\}$ of $[1,n]$, $p_I\boldsymbol{u} \triangleq (u_{i_1}, u_{i_2}, \ldots, u_{i_m})$. For two subsets I and J of $[1,n]$, $p_J(p_I\boldsymbol{u}) \triangleq p_{I\cap J}\boldsymbol{u}$. For $U \subseteq V^n, p_IU \triangleq \{p_I\boldsymbol{u} : \boldsymbol{u} \in U\}$ and $s_IU \triangleq p_I\{\boldsymbol{u} \in U : \mathrm{sup}(\boldsymbol{u}) \subseteq I\}$, where $\mathrm{sup}(\boldsymbol{u})$ denotes the support of $\boldsymbol{u}$. For $J = \{j_1, j_2, \ldots, j_g\} \subseteq I$, $\boldsymbol{u} \in p_IV^n$ and $U \subseteq p_IV^n$, define $p_J\boldsymbol{u} \triangleq (u_{j_1}, u_{j_2}, \ldots, u_{j_g})$ and $p_JU \triangleq \{p_J\boldsymbol{u} : \boldsymbol{u} \in U\}$. For a matrix M with n columns, p_IM denotes the submatrix of M consisting of the i_1th, the i_2th, ..., the i_mth columns in this order.

We assume that a binary (N, K) linear block code C is used over an AWGN channel with BPSK signaling and each codeword is equally transmitted. For a received sequence $\boldsymbol{r} = (r_1, r_2, \ldots, r_N)$ [1] , let $\boldsymbol{z} = (z_1, z_2, \ldots, z_N)$ denote the binary hard-decision sequence for $\boldsymbol{r}$. For $I \subseteq [1, N]$ and $\boldsymbol{u} \in p_IV^N$, define

$$L_I(\boldsymbol{u}) = \sum_{\{i \in I \,:\, u_i \neq z_i\}} |r_i|. \tag{2.1}$$

$L_I(\boldsymbol{u})$ (or simply $L(\boldsymbol{u})$) is called the correlation discrepancy of $\boldsymbol{u}$. For a nonempty subset U of p_IV^N, define $L[U] \triangleq \min_{\boldsymbol{u}\in U} L(\boldsymbol{u})$ and for $\boldsymbol{u} \in U$ such that $L(\boldsymbol{u}) = L[U]$, we write $\boldsymbol{u} = v[U]$ and call it the best [2] (or the most likely) in U. Define $L[\emptyset] \triangleq \infty$ for the empty set $\emptyset$. For the most likely codeword $\boldsymbol{c}_{\mathrm{ML}}$ of $C, \boldsymbol{c}_{\mathrm{ML}} = v[C]$. For a family T of subsets of p_IV^N and $1 \leq i \leq |T|$, let $\boldsymbol{v}_T(i)$ denote the ith best in $\{v[D] : D \in T\}$.

For a binary linear block code A_1 and its linear subcode A_2, let A_1/A_2 denote the set of cosets of A_2 in A_1. For a linear subcode A_3 of A_2, a coset $B \in A_1/A_2$ can be represented as a union of $|A_2/A_3|$ cosets in A_1/A_3. Define $B/A_3 \triangleq \{D \in A_1/A_3 : D \subseteq B\}$. In this paper, a coset D in a set T of cosets is denoted by **its** ***id*-number**. For $D \in T$, a unique binary sequence, denoted $id_T(D)$, is assigned. Let Id_T denote $\{id_T(D) : D \in T\}$. For $\boldsymbol{u} \in D$, define $id_T(\boldsymbol{u}) \triangleq id_T(D)$. For $\boldsymbol{u}$ in a coset in T, let $T(\boldsymbol{u})$ or $T(id_T(\boldsymbol{u}))$ denote the coset which contains $\boldsymbol{u}$. For

[1] For simplicity, unquantized case will be considered.

[2] For $\boldsymbol{u} \neq \boldsymbol{u}'$ in U, the probability that $L(\boldsymbol{u}) = L(\boldsymbol{u}')$ is zero.

$T = A_1/A_2$, we choose a linear code, denoted $[T]$, with dimension $m = \log_2 |T|$ as the set of coset leaders of A_1/A_2. Let $\{\boldsymbol{g}_1, \boldsymbol{g}_2, \ldots, \boldsymbol{g}_m\}$ be a chosen basis of $[T]$. For a coset $D \in T$ whose coset leader is $\sum_{i=1}^{m} a_i \boldsymbol{g}_i$ with $a_i \in \{0, 1\}$,

$$id_T(D) \triangleq (a_1, a_2, \ldots, a_m). \tag{2.2}$$

For $id \in Id_{A_1/A_2}$, let $\mu_{A_1/A_2}(id)$ or simply $\mu(id)$ denote the coset leader of a coset whose id-number is id. That is,

$$(A_1/A_2)(id) = \mu_{A_1/A_2}(id) + A_2. \tag{2.3}$$

In case that $T = D/A_3$ with $D \in A_1/A_2$ and $id_1 \triangleq id_{A_1/A_2}(D)$, for $E \in D/A_3$, there is a unique $id_2 \in Id_{A_2/A_3}$ such that

$$E = \mu_{A_1/A_2}(id_1) + \mu_{A_2/A_3}(id_2) + A_3. \tag{2.4}$$

Consequently, we define

$$id_{D/A_3}(E) \triangleq id_1 \circ id_2, \tag{2.5}$$

$$\mu_{D/A_3}(id_1 \circ id_2) \triangleq \mu_{A_1/A_2}(id_1) + \mu_{A_2/A_3}(id_2), \tag{2.6}$$

where infix "$\circ$" denotes the concatenation operation. We use the following notations. For $\alpha = \beta \circ \gamma \in \{0, 1\}^*$, $\beta \backslash \alpha \triangleq \gamma$ and $\alpha/\gamma \triangleq \beta$. For $\beta, \gamma \in \{0, 1\}^*$ and $A \subseteq \{0, 1\}^*$, $\beta \backslash A \triangleq \{\beta \backslash \alpha : \alpha \in A\}$ and $A/\gamma \triangleq \{\alpha/\gamma : \alpha \in A\}$. From the definition of Id_T and (2.3) to (2.6),

$$Id_{A_2/A_3} = id_1 \backslash Id_{D/A_3}, \tag{2.7}$$

$$(D/A_3)(id) = \mu_{D/A_3}(id) + A_3, \text{ for } id \in Id_{D/A_3}. \tag{2.8}$$

3. Parallel Concatenation Decomposition of Coset Sets

In the original, top-down or adaptive RMLD, how to divide into sub-problems is specified by a binary sectionalization where each section is labelled I_α with index $\alpha \in \{0, 1\}^*$. The length of α, denoted $|\alpha|$, is called the level of section I_α. I_λ denotes $[1, N]$. I_α with $|I_\alpha| \geq 2$, called a nonleaf section, is partitioned into $I_{\alpha 0}$ and $I_{\alpha 1}$. A binary sectionalization can be represented by a binary tree like Fig. 1. For code length $N = 2^m$, a sectionalization such that $|I_\alpha| = 2^{m-|\alpha|}$ with $0 \leq |\alpha| \leq m$ is called the **uniform binary sectionalization**. We abbreviate the suffix I_α as α, e.g., p_{I_α} as p_α and s_{I_α} as s_α. For $U \subseteq V^{I_\alpha}$ and $\beta \in \{0, 1\}^*$, $p_{\alpha\beta}U$ and $s_{\alpha\beta}U$ are abbreviated as $p_\beta U$ and $s_\beta U$, respectively.

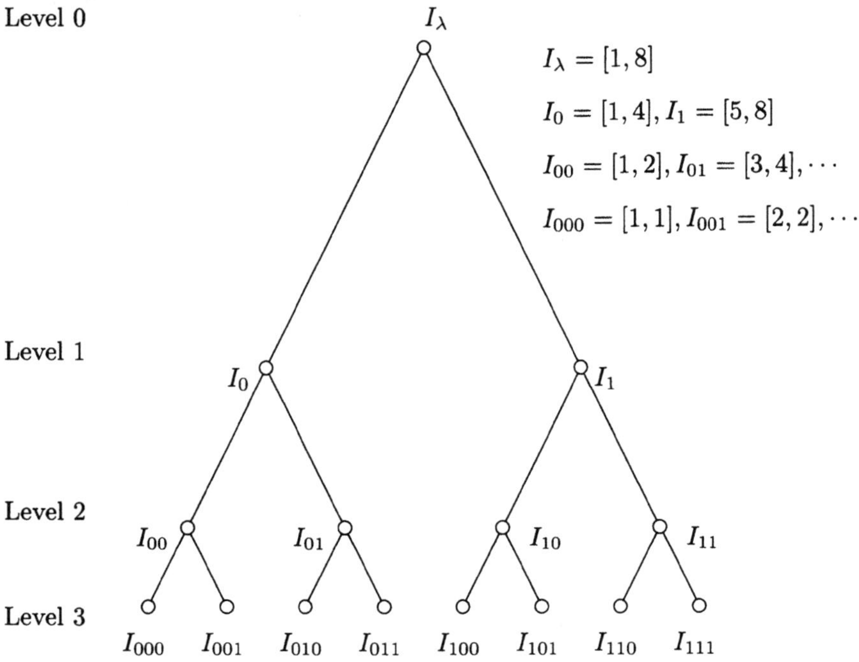

Figure 1: The uniform binary section tree with $N = 2^3$.

RMLD is based on decomposition techniques. For a nonleaf section I_α, let A and B be a binary linear code and its linear subcode over I_α, respectively. Since $s_0B \circ s_1B$ is a linear subcode of A, we have that

Parallel Concatenation Decomposition: A can be decomposed as

$$A = \bigcup_{D \in A/(s_0B \circ s_1B)} p_0D \circ p_1D. \tag{3.1}$$

△

For $\boldsymbol{u}$ and $\boldsymbol{u}'$ in $D \in A/(s_0B \circ s_1B)$, $p_b\boldsymbol{u} + p_b\boldsymbol{u}' \in s_bB$ with $b \in \{0,1\}$, that is,

$$p_bD \in p_bA/s_bB, \tag{3.2}$$

and if $s_bA = s_bB$ for $b \in \{0,1\}$, then there is a one-to-one correspondence between p_0A/s_0B and p_1A/s_1B. For $D \in A/(s_0B \circ s_1B)$, $p_bD \in p_bA/s_bB$ and the coset $p_{\bar{b}}D$ in $p_{\bar{b}}A/s_{\bar{b}}B$, where $\bar{0} \triangleq 1$ and $\bar{1} \triangleq 0$, $(p_bD, p_{\bar{b}}D)$ is called an adjacent pair, and $p_{\bar{b}}D$ is called the counter part of p_bD.

Concatenation Lemma: Let E_b be a linear code over $I_{\alpha b}$ with $b \in \{0,1\}$. Then,

$$E_0 \circ E_1 \subseteq B \Longleftrightarrow E_b \subseteq s_bB. \tag{3.3}$$

(3.3) follows from $\mathbf{0} \in E_b$ for $b \in \{0,1\}$. △

From the above lemma, the coarsest decomposition of type (3.1) is obtained by choosing B as A itself.

Coarsest Parallel Concatenation Decomposition:

$$A = \bigcup_{D \in A/(s_0 A \circ s_1 A)} p_0 D \circ p_1 D. \tag{3.4}$$

Any decomposition of type (3.1) is a refinement of the decomposition (3.4). Each coset in $A/(s_0A \circ s_1A)$ consists of $|s_0A/s_0B| \cdot |s_1A/s_1B|$ cosets of $A/(s_0B \circ s_1B)$. We consider $A/(s_0A \circ s_1A)$ in (3.4). We can choose a generator matrix G of A of the following form:

$$G = \begin{bmatrix} G_0 & 0 \\ 0 & G_1 \\ \multicolumn{2}{c}{G_{0,1}} \end{bmatrix}, \tag{3.5}$$

where 0 denotes a zero matrix, G_0 and G_1 are generator matrices of s_0A and s_1A, respectively, and $G_{0,1}$ is a generator matrix of the set of coset leaders of $A/(s_0A \circ s_1A)$. For a basis $B_a = \{\boldsymbol{g}_1, \boldsymbol{g}_2, \cdots, \boldsymbol{g}_m\}$ of $[A/(s_0A \circ s_1A)]$, p_bB_a with $b \in \{0,1\}$ is linearly independent. Otherwise, there is a linear sum $\boldsymbol{u}$ of rows of $G_{0,1}$ such that $p_0\boldsymbol{u}$(or $p_1\boldsymbol{u}$) $= \boldsymbol{0}$ and $p_1\boldsymbol{u}$(or $p_0\boldsymbol{u}$) $\neq \boldsymbol{0}$. Then, $p_1\boldsymbol{u}$(or $p_0\boldsymbol{u}$) is in s_1A(or s_0A), a contradiction.

We choose p_bB_a as a basis of $[p_bA/s_bA]$. Then, for $D \in A/(s_0A \circ s_1A)$ and $\sum_{i=1}^{m} a_i\boldsymbol{g}_i \in D$, it follows from (2.2) that for $b \in \{0,1\}$,

$$id_{A/(s_0A \circ s_1A)}(D) = id_{p_bA/s_bA}(p_bD) = \boldsymbol{a} \tag{3.6}$$

and

$$\begin{gathered}((p_0A/s_0A)(id), (p_1A/s_1A)(id)) \text{ is an adjacent pair,} \\ \text{and they are unique counter parts to each each.}\end{gathered} \tag{3.7}$$

This simplifies the specification of the adjacent coset. Eq.(3.4) can be rewritten as

$$A = \bigcup_{id \in Id_{A/(s_0A \circ s_1A)}} (p_0A/s_0A)(id) \circ (p_1A/s_1A)(id), \tag{3.8}$$

where $Id_{A/(s_0A \circ s_1A)} = Id_{p_bA/s_bA}$ with $b \in \{0,1\}$.

Let A' be a linear supercode of A. For a coset $D \in A'/A$, the following is a corollary of (3.8). Define $id_D \triangleq id_{A'/A}(D)$, $id_{p_0D} \triangleq id_{p_0A'/p_0A}(p_0D)$ and $id_{p_1D} \triangleq id_{p_1A'/p_1A}(p_1D)$. Since $(p_0A/s_0A)(id) = \mu_{p_0A/s_0A}(id) + s_0A$, $(p_1A/s_1A)(id) = \mu_{p_1A/s_1A}(id) + s_1A$ and $D = \mu_{A'/A}(id_D) + A$, it follows from (3.8) that

$$D = \bigcup_{id \in Id_{p_0A/s_0A} = Id_{p_1A/s_1A}} (p_0\mu_{A'/A}(id_D) + \mu_{p_0A/s_0A}(id) + s_0A) \circ$$
$$(p_1\mu_{A'/A}(id_D) + \mu_{p_1A/s_1A}(id) + s_1A).$$

Since $p_0D = p_0\mu_{A'/A}(id_D) + p_0A = \mu_{p_0A'/p_0A}(id_{p_0D}) + p_0A$, $p_0\mu_{A'/A}(id_D) = \mu_{p_0A'/p_0A}(id_{p_0D})$. From (2.4), $\mu_{p_0A'/p_0A}(id_{p_0D}) + \mu_{p_0A/s_0A}(id) + s_0A = (p_0D/s_0A)(id_{p_0D} \circ id)$. Consequently, we have (3.9):

$$D = \bigcup_{id \in Id_{A/(s_0A \circ s_1A)} = Id_{p_bA/s_bA}} (p_0D/s_0A)(id_{p_0D} \circ id) \circ (p_1D/s_1A)(id_{p_1D} \circ id), \tag{3.9}$$

where $((p_0D/s_0A)(id_{p_0D} \circ id), (p_1D/s_1A)(id_{p_1D} \circ id))$ is called an adjacent pair in D.

4. Recursive Maximum Likelihood Decoding

We briefly review recursive maximum likelihood decoding (RMLD) [3] and introduce new versions of RMLD, called top-down RMLD [4, 7, 8] and the revised version [11, 12], called adaptive RMLD, based on a "call by need" approach. For an index $\alpha \in \{0,1\}^*$, define $T_\alpha \triangleq p_\alpha C/s_\alpha C$.

Local Optimum [3]: For a coset D in T_α, there is a codeword $\boldsymbol{u} \in C$ such that $p_\alpha \boldsymbol{u} = v[D]$. Then, for any codeword $\boldsymbol{u}' \in C$ such that $p_\alpha \boldsymbol{u}' \in D$,

$$L_\alpha(\boldsymbol{u}) \le L_\alpha(\boldsymbol{u}'). \tag{4.1}$$

$\triangle$

The sub-codeword $v[D]$ is called **the most likely local (MLL) sub-codeword** in D or an MLL sub-codeword in T_α. Since $p_\lambda C/s_\lambda C = \{C\}$, an MLL sub-codeword in T_λ is the most likely codeword $\boldsymbol{c}_{\mathrm{ML}}$.

Let I_α be a nonleaf section. From the above lemma, the recursive search procedure for $\boldsymbol{c}_{\mathrm{ML}}$ in the original RMLD is based on the following decomposition, called MLL decomposition, which can be derived from (3.1) and (3.2) by substituting $p_\alpha C$ for A and $s_\alpha C$ for B:

$$p_\alpha C = \bigcup_{D \in PT_\alpha} p_0 D \circ p_1 D, \tag{4.2}$$

where

$$PT_\alpha \triangleq p_\alpha C/(s_{\alpha 0}C \circ s_{\alpha 1}C). \tag{4.3}$$

This type of decomposition is the same as the one which holds for the sets of label sequences between adjacent subsections of a trellis diagram [10].

Consider $\boldsymbol{v}_{T_\alpha}(i)$ with $1 \le i \le |T_\alpha|$, abbreviated as $\boldsymbol{v}_\alpha(i)$, which is the i-th best MLL sub-codeword in T_α, that is, with the smallest discrepancy in $T_\alpha \backslash \bigcup_{h=1}^{i-1}\{\boldsymbol{v}_\alpha(h) + s_\alpha C\}$. From (4.2), there is a pair $j_{\alpha 0}(i)$ and $j_{\alpha 1}(i)$ such that $1 \le j_{\alpha 0}(i) \le |T_{\alpha 0}|$, $1 \le j_{\alpha 1}(i) \le |T_{\alpha 1}|$ and

$$\boldsymbol{v}_\alpha(i) = \boldsymbol{v}_{\alpha 0}(j_{\alpha 0}(i)) \circ \boldsymbol{v}_{\alpha 1}(j_{\alpha 1}(i)). \tag{4.4}$$

The most likely codeword $\boldsymbol{v}_\lambda(1)$ is derived from $\boldsymbol{v}_0(j_0(1))$ and $\boldsymbol{v}_1(j_1(1))$ which can be obtained in turn recursively by (4.4).

In the original (bottom-up) RMLD, T_α-table whose entries are $(v[D], L[D])$ for every $D \in T_\alpha$ is constructed for every index α. For a leaf section I_α where $|I_\alpha| = 1$, $p_\alpha C \subseteq \{0,1\}$ and $s_\alpha C = \{0\}$ (if the minimum distance of C is 2 or greater), and therefore T_α-table $= \{(b, L_\alpha(b)) : b$ is the best in $p_\alpha C\}$. For a nonleaf section I_α, T_α-table is constructed recursively from $T_{\alpha b}$-tables with $b \in \{0,1\}$ by using (4.2) as follows: for $D \in T_\alpha$, $v[D] = v[\{D_0 \circ D_1 : D_b \in T_{\alpha b}, (D_0, D_1)$ is an adjacent pair in $D\}]$, $L[D] = \min(L[D_0] + L[D_1])$, where the minimum is taken over $(D_0, D_1) \in$ the set of adjacent pairs in D. For $\alpha = \lambda$, since $p_\lambda C/s_\lambda C = \{C\}$, T_λ-table contains the single entry $(\mathbf{c}_{\rm ML}, L(\mathbf{c}_{\rm ML}))$.

Example 1 [5]: Consider the l-th level section I_α of the uniform binary sectionalization for $\mathrm{RM}_{r,m}$ (the r-th order RM code of length 2^m) with $0 < l = |\alpha| < m$,

$$p_\alpha \mathrm{RM}_{r,m} = \mathrm{RM}_{\min\{r,m-l\},m-l}, \tag{4.5}$$

$$s_\alpha \mathrm{RM}_{r,m} = \begin{cases} \mathrm{RM}_{r-l,m-l}, & \text{for } r \geq l, \\ \{0\}, & \text{for } r < l. \end{cases} \tag{4.6}$$

Then,

$$\begin{aligned} \log_2 |T_\alpha| &= \log_2 |p_\alpha \mathrm{RM}_{r,m}| - \log_2 |s_\alpha \mathrm{RM}_{r,m}| \\ &= \sum_{i=\max\{r-l+1,0\}}^{\min\{r,m-l\}} \binom{m-l}{i}. \end{aligned} \tag{4.7}$$

△

This example shows that for RM codes with the uniform binary sectionalization, T_α are identical. This holds for a class of codes, including RM codes and EBCH codes, called binary transitive invariant. The definition of the class is given in Appendix A. For a binary transitive invariant code, the following lemma holds [6].

Transitive Invariant Lemma: Suppose that C of length 2^m is binary-transitive-invariant and the uniform binary sectionalization is used. Then $p_\alpha C$(or $s_\alpha C$) is the same for every section I_α of the same level $|\alpha|$. △

Now, we explain the main idea of top-down and adaptive RMLD.

(T) Introduction of call-by-need approach: Simulation results [4] show $j_{\alpha 0}(i) \ll |T_{\alpha 0}|$ or $j_{\alpha 1}(i) \ll |T_{\alpha 1}|$ in (4.4) for almost all cases of relatively small $|\alpha|$ and i, as the law of large numbers suggests. To make effective use of this fact, we reorganized the recursive search procedure in the original RMLD by a call by need approach. The simulation results [4] show that the computational complexity in terms of the number of addition equivalent operations can be remarkably reduced. In contrast to the original RMLD, the top-down RMLD requires only a very small portion of T_α-tables to be constructed in average.

(A) In the top-down RMLD [4], the MLL decomposition (4.2) was still used. We have noticed that those recursion levels whose complexities are dominant are a few higher levels with nonzero small $|\alpha|$'s. A weak point to use partition $T_\alpha = p_\alpha C/s_\alpha C$

is that the decreasing rate of block size $|s_\alpha C|$ as $|\alpha|$ increases is much greater than that of $|p_\alpha C|$, this is, the decreasing rate of $|T_\alpha|$ is smaller than that we expected.

For a top-down type algorithm based on call-by-need approach, it is essential to make use of an effective early termination condition. In general, such a condition is more effective for a partition with a relatively large block size.

(A1) From the above considerations, the adaptive RMLD presented in this paper adopts the **coarsest parallel concatenation** decomposition (3.4) as a basis of recursion. For a linear subcode of A of $p_\alpha C$,

$$A = \bigcup_{id \in Id_{A/(s_0 A \circ s_1 A)}} (p_0 A/s_0 A)(id) \circ (p_1 A/s_1 A)(id),$$

where $Id_{A/(s_0 A \circ s_1 A)} = Id_{p_b A/s_b A}$ with $b \in \{0,1\}$. Define

$$PF_\alpha \triangleq p_\alpha C/(s_0(p_\alpha C) \circ s_1(p_\alpha C)). \tag{4.8}$$

Except for $\alpha = \lambda$ where $PF_\alpha = PT_\alpha$, the number of blocks in the above decomposition where $A = p_\alpha C$, $|PF_\alpha|$, is considerably smaller than $|PT_\alpha|$, as shown in Table 1 for RM codes and several EBCH codes of length 128. Consequently, the worst case search spaces of recursively called subprocedures can be reduced effectively.

(A2) From the decomposition (3.8) or (3.9), once $\boldsymbol{v}_{\alpha 0}(j_{\alpha 0}(i))$ with $j_{\alpha 0}(i) < j_{\alpha 1}(i)$ in (4.4) has been found, the counterpart $\boldsymbol{v}_{\alpha 1}(j_{\alpha 1}(i))$ can be found in a very simple way.

(A3) In the adaptive RMLD, a new sufficient condition that a currently best candidate is the optimum at the current level of recursion is used as an early termination condition of the recursion (refer to Sec. 5).

(A4) Preliminary simulation results [12] for the (128, 64) RM code show the adaptive RMLD presented in this paper provides a considerably smaller average decoding complexity than the original RMLD [3, 6] and top-down RMLD [7, 8] in terms of the number of addition equivalent operations.

Table 1: The dimensions of PT_α and PF_α for several RM and EBCH codes of length 128

$\|\alpha\|$	RM(128, 64)		EBCH(128, 57)		EBCH(128, 64)		EBCH(128, 71)		EBCH(128, 79)	
	PT_α	PF_α	PT_α	PF_α	PT_α	PF_α	PT_α	PF_α	PT_α	PF_α
0	20	20	27	27	34	34	41	41	34	34
1	30	10	40	10	47	13	54	6	44	6
2	24	4	26	4	31	1	31	1	29	1
3	15	1	15	1	16	0	16	0	16	0
4	8	0	8	0	8	0	8	0	8	0

Example 2: For RM codes, we use binary polynomial representation. For $0 \le r \le m$, let $P_{r,m}$ denote the set of binary polynomials spanned by monomials of degree r with m binary variables. $P_{r,m}$ is the set of coset leaders of cosets in $\mathrm{RM}_{r,m}/\mathrm{RM}_{r-1,m}$. In (3.4), let $A = p_\alpha \mathrm{RM}_{r,m}$. It follows from (4.5) and (4.6) that

for $0 \le l \triangleq |\alpha| < m$ and $b \in \{0,1\}$,

$$p_{\alpha b}\mathrm{RM}_{r,m} = \mathrm{RM}_{\min\{r,m-l-1\},m-l-1}, \tag{4.9}$$

$$s_b(p_\alpha \mathrm{RM}_{r,m}) = \mathrm{RM}_{\min\{r,m-l\}-1,m-l-1}, \text{ for } b \in \{0,1\}. \tag{4.10}$$

Suppose $r < m - l$. Then, $\min\{r, m-l\} = \min\{r, m-l-1\} = r$ and the set of coset leaders of cosets in $p_{\alpha b}\mathrm{RM}_{r,m}/s_b(p_\alpha \mathrm{RM}_{r,m})$ is $P_{r,m-l-1}$, and Eq.(3.4) can be expressed as

$$p_\alpha \mathrm{RM}_{r,m} = \bigcup_{f \in P_{r,m-l-1}} \{f + s_0(p_\alpha \mathrm{RM}_{r,m})\} \circ \{f + s_1(p_\alpha \mathrm{RM}_{r,m})\}. \tag{4.11}$$

From the definition of PF_α for $C = \mathrm{RM}_{r,m}$, we have that

$$\log_2 |PF_\alpha| = \binom{m-l-1}{r}. \tag{4.12}$$

For $r \ge m - l$, $p_{\alpha b}\mathrm{RM}_{r,m} = s_b(p_\alpha \mathrm{RM}_{r,m}) = \mathrm{RM}_{m-l-1,m-l-1} = V^{2^{m-l-1}}$ and $\log_2 |PF_\alpha| = 0$.

For comparison, $|PT_\alpha| = |p_\alpha C/(s_{\alpha 0}(C) \circ s_{\alpha 1}(C))|$ for $C = \mathrm{RM}_{r,m}$ is

$$\begin{aligned}
&\log_2 |p_\alpha \mathrm{RM}_{r,m}|/|s_{\alpha 0}\mathrm{RM}_{r,m}|^2 \\
&= \sum_{i=0}^{\min\{r,m-l\}} \binom{m-l}{i} - \begin{cases} 2\sum_{i=0}^{r-l-1} \binom{m-l-1}{i}, & \text{for } r > l, \\ 0, & \text{for } r \le l. \end{cases} \\
&= \begin{cases} \sum_{i=r-l}^{\min\{r,m-l\}} \binom{m-l}{i} - \binom{m-l-1}{r-l-1}, & \text{for } r > l, \\ \sum_{i=0}^{\min\{r,m-l\}} \binom{m-l}{i}, & \text{for } r \le l. \end{cases}
\end{aligned} \tag{4.13}$$

△

5. Search Procedure for $v_T(i)$

Let $I_{\alpha b}$ with $b \in \{0,1\}$ be a nonleaf section. Let A, A' and B denote linear codes over I_α such that

$$s_\alpha C \subseteq B \subseteq A \subseteq A' \subseteq p_\alpha C. \tag{5.1}$$

For $D \in A'/A$, define

$$T \triangleq D/B. \tag{5.2}$$

We introduce procedure pick(T) which returns $v_T(i), L_\alpha(v_T(i))$ and $id_T(v_T(i))$ at the ith call with $1 \le i \le |T|$. $L_\alpha(v_T(i))$ and $id_T(v_T(i))$ are abbreviated as $L_T(i)$ and $id_T(i)$, respectively. Since $s_\alpha C \subseteq B$, $v_T(i)$ is an MLL sub-codeword in T. By definition, $id_T(i) \ne id_T(j)$ for $i \ne j$. D denotes the search space of pick(T), B specifies the search units (blocks) of pick(T), and $(|B/s_\alpha C|-1)$ MLL sub-codewords in each coset of T except for the best one can be ignored.

We present a recursive implementation of pick(T) based on the decomposition (3.9). From (3.9), we have that

$$D = \bigcup_{id \in Id_{A/(s_0 A \circ s_1 A)} = Id_{p_b A/s_b A}} (p_0 D/s_0 A)(id_{D_0} \circ id) \circ (p_1 D/s_1 D)(id_{D_1} \circ id), \quad (5.3)$$

where $id_{D_b} \triangleq id_{p_b A'/p_b A}(p_b D)$. We define

$$PF(D,A) \triangleq D/(s_0 A \circ s_1 A), \quad (5.4)$$

$$F_b(D,A) \triangleq p_b PF(D,A) = p_b D/s_b A, \text{ for } b \in \{0,1\}, \quad (5.5)$$

where parameter D may be replaced by $\boldsymbol{u} \in D$ or $id_{A'/A}(D)$, and if there is no possible confusion, then parameters D and A may be omitted, and $PF(A) \triangleq A/(s_0 A \circ s_1 A)$ and $F_b(A) \triangleq p_b A/s_b A$. Define $id_D \triangleq id_{A'/A}(D)$. From (5.3) to (5.5),

$$D = \bigcup_{id \in Id_{PF(A)}} PF(D,A)(id_D \circ id), \quad (5.6)$$

$$PF(D,A)(id_D \circ id) = F_0(D,A)(id_{D_0} \circ id) \circ F_1(D,A)(id_{D_1} \circ id), \quad (5.7)$$

$$F_b(D,A)(id_{D_b} \circ id) = \mu_{p_b A'/p_b A}(id_{D_b}) + \mu_{F_b(A)}(id) + s_b A. \quad (5.8)$$

If $D = A$, then $id_D = id_{D_b} = \lambda$.

5.1. Recursive Implementation of pick(T) Based on the Coarsest Parallel Concatenation Decomposition

There are two cases to be considered.

(Case I) $B = A, T = \{D\}$. For this case, T consists of a single coset in A'/A, and pick(D) is called only once. For example, let $\alpha = \lambda$ and $A' = A = B = C$. Then, $D = C \in T_\lambda = C/C$, and $PF = C/(s_0 C \circ s_1 C), F_b = p_b C/s_b C = T_b$ and pick(D) is an example of Case I. Another example is pick($T_b(id)$).

(Case II) $B \neq A, D \neq A$. For this case, $\alpha \neq \lambda$ from (5.1). Assume that

$$s_\alpha C \subseteq s_0 A \circ s_1 A. \quad (5.9)$$

Since $s_\alpha C \subseteq B$, the relation (5.9) holds if the following relation is true:

$$B \subseteq s_0 A \circ s_1 A. \quad (5.10)$$

Relation (5.10) with $A = p_\alpha C$ and $B = s_b(p_{\alpha/b} C)$, that is,

$$s_b(p_{\alpha/b} C) \subseteq s_0(p_\alpha C) \circ s_1(p_\alpha C), \quad (5.11)$$

holds for $C =$ a RM code or an EBCH code of length 128 and dimension 57, 64, 71 or 78, with the uniform binary sectionalization. For the EBCH codes, (5.11) is verified by constructing generator matrices for $s_b(p_\alpha C)$ and $s_b(p_{\alpha/b} C)$. A proof of (5.11) for $\mathrm{RM}_{r,m}$ is as follows: Since (5.11) holds for $r = 0$, assume that $r \geq$

1. Note that $p_\alpha \mathrm{RM}_{r-1,m} \subseteq p_{\alpha 0}\mathrm{RM}_{r-1,m} \circ p_{\alpha 1}\mathrm{RM}_{r-1,m}$. From (4.5) and (4.6), $p_\alpha \mathrm{RM}_{r-1,m} = s_b(p_{\alpha/b}\mathrm{RM}_{r,m})$ and $p_{\alpha b'}\mathrm{RM}_{r-1,m} = s_{b'}(p_\alpha \mathrm{RM}_{r,m})$ with $b' \in \{0,1\}$. Hence, (5.11) holds for $C = \mathrm{RM}_{r,m}$.

In contrast with Case I, pick(T) may be called two or more times in Case II. Hereafter, we consider Case II. Case I is a special case where $|T| = 1$. We consider the processing made by pick(T) at the hth call with $1 \le h \le |T|$. Suppose $\boldsymbol{v}_T(i)$, $L_T(i)$ and $id_T(i)$ with $1 \le i < h$ have been found and returned to the parent procedure. For $id \in Id_{PF(A)}$, define

$$\rho_h(id) \triangleq |\{\boldsymbol{v}_T(i) : \boldsymbol{v}_T(i) \in PF(id_D \circ id) \text{ and } 1 \le i < h\}|, \tag{5.12}$$

$$PF(id)_h \triangleq PF(id_D \circ id) \backslash \{\bigcup_{i=1}^{h-1} (\boldsymbol{v}_T(i) + B)\}. \tag{5.13}$$

In pick(T), subprocedure pick(F_b) with $b \in \{0,1\}$ is called by need which returns $\boldsymbol{v}_{F_b}(i)$, $L_{F_b}(i)$, $id_{F_b}(i)$ together with $id_{F_b'}(i)$ at the ith call, where F_b' is defined in 5.2.

(1) Suppose pick(F_b) has been called $\bar{i}_b$ times with $b \in \{0,1\}$ and $1 \le \bar{i}_b < |F_b|$, and $\boldsymbol{v}_{F_b}(i_b)$, $L_{F_b}(i_b)$, $id_{F_b}(i_b)$ and $id_{F_b'}(i_b)$ with $1 \le i_b \le \bar{i}_b$ have been found. Define

$$IP_F \triangleq \{id_{D_b} \backslash id_{F_b}(i_b) : 1 \le i_b \le \bar{i}_b \text{ and } b \in \{0,1\}\}. \tag{5.14}$$

From (2.7) and (3.6), $IP_F \subseteq Id_{F_b(A)} = Id_{PF(A)}$. Define cS_T and cS_T' as

$$cS_T \triangleq \bigcup_{id \in IP_F} PF(id)_h, \tag{5.15}$$

$$cS_T' \triangleq \bigcup_{id \in Id_{PF(A)} \backslash IP_F} PF(id)_h. \tag{5.16}$$

From (5.6),

$$D = \bigcup_{id \in Id_{PF(A)}} PF(id_D \circ id). \tag{5.17}$$

From (5.13) to (5.15),

$$D \backslash \{\bigcup_{i=1}^{h-1} T(\boldsymbol{v}_T(i))\} = cS_T \cup cS_T'. \tag{5.18}$$

Define

$$\boldsymbol{c}_{\text{best}} \triangleq v[cS_T] = \text{ the best of } \bigcup_{id \in IP_F} v[PF(id)_h]. \tag{5.19}$$

Hence, $\boldsymbol{v}_T(h)$ is either $\boldsymbol{c}_{\text{best}}$ or $v[cS_T']$.

Sufficient conditions that $\boldsymbol{v}_T(h) = \boldsymbol{c}_{\text{best}}$
If the following (TC_T-1) or (TC_T-2) holds, then $\boldsymbol{v}_T(h) = \boldsymbol{c}_{\text{best}}$.

$$(TC_T\text{-}1) \qquad L_{F_0}(\bar{i}_0) + L_{F_1}(\bar{i}_1) \ge L(\boldsymbol{c}_{\text{best}}), \tag{5.20}$$

$$(TC_T\text{-}2) \qquad |IP_F| = |PF(A)| = |A|/(|s_0A| \cdot |s_1A|). \tag{5.21}$$

Proof. (i) If $(TC_T\text{-}2)$ holds, then $v[cS'_T]$ is not defined.
(ii) Suppose $(TC_T\text{-}1)$ holds and $|IP_F| < |PF(A)|$. For $id \in Id_{PF(A)} = Id_{F_0(A)} = Id_{F_1(A)}$, it follows from (5.7) that

$$PF(id_D \circ id) = F_0(id_{D_0} \circ id) \circ F_1(id_{D_1} \circ id). \tag{5.22}$$

For $\boldsymbol{u} \in cS'_T$, there exists $id \in Id_{PF(A)} \backslash IP_F$ such that $p_b\boldsymbol{u} \in F_b(id_{D_b} \circ id)$ with $b \in \{0,1\}$. Since $L(\boldsymbol{v}) \geq L_{F_b}(\bar{i}_b)$ for any $\boldsymbol{v} \in F_b(id_{D_b} \circ id)$ with $id \in Id_{F_b} \backslash IP_F$,

$$L(p_b\boldsymbol{u}) \geq L_{F_b}(\bar{i}_b). \tag{5.23}$$

Hence, $L(\boldsymbol{u}) \geq L_{F_0}(\bar{i}_0) + L_{F_1}(\bar{i}_1) \geq L(\boldsymbol{c}_{\text{best}})$. △

If (5.10) holds, then the following $(TC_T\text{-}1')$ is a corollary of $(TC_T\text{-}1)$. Without loss of generality, suppose $\bar{i}_b$ is updated after $\bar{i}_{\bar{b}}$.

$$(TC_T\text{-}1') \qquad id_{F_b}(\bar{i}_b) = id_{D_b} \circ id \text{ with } id \in IP_F.$$

Proof. If $(TC_T\text{-}1')$ holds, then there is $i_{\bar{b}}$ such that $1 \leq i_{\bar{b}} \leq \bar{i}_{\bar{b}}$ and $id_{F_{\bar{b}}}(i_{\bar{b}}) = id_{D_{\bar{b}}} \circ id$. From (5.22), $\boldsymbol{v} \triangleq \boldsymbol{v}_{F_b}(\bar{i}_b) \circ \boldsymbol{v}_{F_{\bar{b}}}(i_{\bar{b}}) \in PF(id_D \circ id)$ and $\boldsymbol{v}$ is the best in $PF(id_D \circ id)$. Since any $\boldsymbol{u} \in D\backslash PF(id_D \circ id)$ belongs to some coset in D/B other than that containing $\boldsymbol{v}$ from (5.10), $\boldsymbol{v}$ may be output already as $\boldsymbol{v}_T(i)$ with $i < h$. Otherwise, since $L_{F_{\bar{b}}}(\bar{i}_{\bar{b}}) \geq L_{F_{\bar{b}}}(i_{\bar{b}})$, $L_{F_0}(\bar{i}_0) + L_{F_1}(\bar{i}_1) \geq L(\boldsymbol{v}) \geq L(\boldsymbol{c}_{\text{best}})$. △

(2) If $v[PF(id)_h]$ is found for every $id \in IP_F$, then $\boldsymbol{c}_{\text{best}}$ can be obtained from (5.19). We introduce pick$(PF(id)_h)$ with $id \in IP_F$ which returns the MLL subcodeword $\boldsymbol{u}$ with the smallest discrepancy in $PF(id)_h$ and its discrepancy. If $PF(id)_h = \emptyset$, then $\emptyset$ is returned.

Assume that the relation (5.10) holds, that is, $B \subseteq s_0A \circ s_1A$. Then pick$(PF(id)_h)$ can be reduced to pick$(PF'(id_D \circ id))$, where

$$PF'(id_D \circ id) \triangleq PF(id_D \circ id)/B. \tag{5.24}$$

Procedure pick$(PF'(id_D \circ id))$ returns $\boldsymbol{u} = v[PF(id)_h]$, its discrepancy and $id_{PF'(id_D \circ id)}(\boldsymbol{u})$ at the $(\rho_h(id)+1)$th call, if $PF(id)_h \neq \emptyset$. For a recursive implementation of pick$(PF'(id_D \circ id))$, a subprocedure of type pick$(p_bPF(id_D \circ id)/E_b)$, where $b \in \{0,1\}$ and E_b is a linear subcode of $p_bPF(id_D \circ id) = F_b(id_{D_b} \circ id)$, is called by need from pick$(PF'(id_D \circ id))$. We choose s_bB as E_b which is the maximal solution under the condition of $E_0 \circ E_1 \subseteq B$. Define

$$F'_b(id_{D_b} \circ id) \triangleq F_b(id_{D_b} \circ id)/s_bB. \tag{5.25}$$

As a result, pick$(PF'(id_D \circ id))$ makes use of the following relation based on (5.22). For $id \in Id_{PF(A)}$,

$$PF(id_D \circ id)/(s_0B \circ s_1B) = F'_0(id_{D_0} \circ id) \circ F'_1(id_{D_1} \circ id), \tag{5.26}$$

is a refinement of $PF'(id_D \circ id) = PF(id_D \circ id)/B$. For given $\boldsymbol{u}_0 \in F'_0(id_D \circ id)$ and $\boldsymbol{u}_1 \in F'_1(id_D \circ id)$, a simple procedure to decide if there is $\boldsymbol{v}_T(i) \in PF(id_D \circ id)$ with $1 \leq i < h$ such that $\boldsymbol{u}_0 \circ \boldsymbol{u}_1 + \boldsymbol{v}_T(i) \in B$ is shown in Appendix B.

Figure 2 shows the call-return relation among pick(T), pick(F_b) and pick$(PF(id)_h)$.

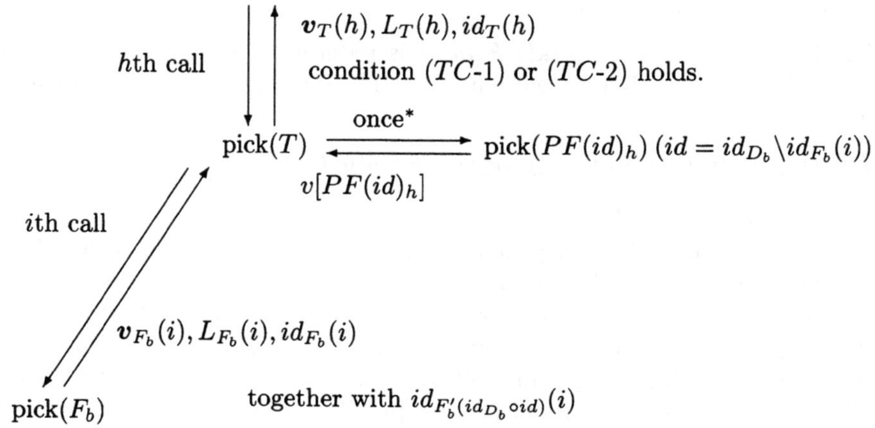

$$T = D/B, PF = D/(s_0A \circ s_1A), F_b = p_bD/s_bA$$

$$PF(id)_h = PF(id_D \circ id) \backslash \{ \bigcup_{i=1}^{h-1} (\boldsymbol{v}_T(i) + B) \}$$

* once for a return from pick$_{F_b}$ with $id_{F_b}(i) = id_D \circ id$

Figure 2: The call-return relation among pick(T), pick(F_b) and pick$(PF(id)_h)$.

Example 3: For $\alpha = \{0,1\}^*$, let $A' = A = D = p_\alpha C$. Then, $id_D = id_{D_0} = id_{D_1} = \lambda$.

(i) For $\alpha = \lambda$, $B = C$ and $T = C/C = \{C\}$. Then $PF(C) = C/(s_0C \circ s_1C)$, and $F_b(C) = p_bC/s_bC = T_b$ with $b \in \{0,1\}$. Since $C \supseteq s_0C \circ s_1C$, (5.9) does not hold in general.

(ii) We introduce the following new abbreviated notations for $\alpha \in \{0,1\}^*$:

$$PF_\alpha \triangleq PF(p_\alpha C) = p_\alpha C/(s_0(p_\alpha C) \circ s_1(p_\alpha C)), \tag{5.27}$$

$$F_{\alpha,b} \triangleq F_b(p_\alpha C) = p_{\alpha b}C/s_b(p_\alpha C), \text{ for } b \in \{0,1\}. \tag{5.28}$$

For $s_\alpha C \subsetneq B \subsetneq p_\alpha C$, pick$(p_\alpha C/B)$ can be implemented by calling pick$(F_{\alpha,b})$ and pick$(PF(id)_h)$, where $id \in Id_{PF_\alpha} = Id_{F_{\alpha,b}} = Id_{F_{\alpha,\bar{b}}}$ is a return value from pick$(F_{\alpha,b})$ at the i_b-th call.

(ii.1) If $\rho_h(id) = 0$, then $v[PF(id)_h] = v[PF_\alpha(id)]$ and $\boldsymbol{v}_{F_{\alpha,b}}(i_b) = v[F_{\alpha,b}(id)]$. From (5.7),

$$v[PF_\alpha(id)] = \boldsymbol{v}_{F_{\alpha,b}}(i_b) \circ v[F_{\alpha,\bar{b}}(id)], \tag{5.29}$$

where $F_{\alpha,\bar{b}}(id) = \mu_{F_{\alpha.\bar{b}}}(id) + s_{\bar{b}}(p_\alpha C)$. In (5.29), $v[F_{\alpha.\bar{b}}(id)]$ can be found by calling pick($F_{\alpha,\bar{b}}(id)$) once. For $\alpha = \lambda$ (Case I), $\rho_h(id) = 0$ and it is sufficient to consider pick($PF_\alpha(id)$) with $|\alpha| > 0$.

(ii.2) If $\rho_h(id) \geq 1$, then pick($PF(id)_h$) is reduced to $PF'(p_\alpha C)(id)$. Let $B = s_b(p_{\alpha/b}C)$ for $\alpha \in \{0,1\}^*b$ with $b \in \{0,1\}$. $PF'(p_\alpha C)(id)$ is abbreviated as $PF'_\alpha(id)$:

$$PF'_\alpha(id) \triangleq PF'(p_\alpha C)(id) = (\mu_{PF_\alpha}(id) + (s_0(p_\alpha C) \circ s_1(p_\alpha C)))/s_b(p_{\alpha/b}C). \quad (5.30)$$

On the complexity of pick(T)

The complexity of pick(T) depends on the average of $\bar{i}_0 + \bar{i}_1$ for which (TC_T-1) holds for given h. By definition, $\bar{i}_0 + \bar{i}_1 = |IP_F| +$ the number of occurrences of (TC_T-1$'$) to hold, which reduce the computational complexity. Furthermore, for $\alpha \in \{0,1\}^*$ and $b \in \{0,1\}$, h is not greater than the $\bar{i}_b$ of the parent procedure. As is shown in Table 2, the upper limit of $\bar{i}_b$, $|F_{\alpha.b}| = |p_{\alpha b}C/s_b(p_\alpha C)|$, is smaller than that of the MLL decomposition, $|T_{\alpha b}|(= |p_{\alpha b}C/s_{\alpha b}C|)$, except for $\alpha = \lambda$ where $T_{\alpha b} = F_{\alpha,b}$.

Table 2: The dimensions of $T_{\alpha b}$ and $F_{\alpha,b}$ for several RM and EBCH codes of length 128

$\|\alpha\|$	RM(128, 64)		EBCH(128, 57)		EBCH(128, 64)		EBCH(128, 71)		EBCH(128, 79)	
	$T_{\alpha b}$	$F_{\alpha,b}$	$T_{\alpha b}$	$F_{\alpha,b}$	$T_{\alpha b}$	$F_{\alpha,b}$	$T_{\alpha b}$	$F_{\alpha.b}$	$T_{\alpha b}$	$F_{\alpha,b}$
0	20	20	27	27	34	34	41	41	34	34
1	20	10	24	10	30	13	30	6	25	6
2	14	4	15	4	16	1	16	1	15	1
3	8	1	8	1	8	0	8	0	8	0
4	4	0	4	0	4	0	4	0	4	0

For a linear block code E, let $d_H(E)$ denote the minimum distance of E. Since $(p_0A) \circ (p_1A) \supseteq A \supseteq s_bA \circ \{\mathbf{0}\}$ with $b \in \{0,1\}$,

$$d_H(p_0A) + d_H(p_1A) \leq d_H(A) \leq d_H(s_bA). \quad (5.31)$$

Suppose A is binary transitive invariant and the uniform binary sectionalization is used. Then $p_0A = p_1A$, and therefore,

$$2d_H(p_bA) \leq d_H(s_bA). \quad (5.32)$$

Note that $d_H(p_bA)$ is the minimum distance between different cosets in F_b and $d_H(s_bA)$ is the minimum distance of each coset in F_b. For relatively small $|\alpha|, h$ and $\bar{i}_b < |F_b|$, the possibility that there exists better $\boldsymbol{u}$ in $p_bD \setminus \bigcup_{i=1}^{i=\bar{i}_b} F_b(\boldsymbol{v}_{F_b}(i))$ than $\boldsymbol{v}_{F_b}(\bar{i}_b)$ is small in average.

5.2. Outline of pick($PF'(id_D \circ id)$)

Note that pick($PF'(id_D \circ id)$) with $id \in IP_F$ is called in one of the following situations:

(i) id is just registered to IP_F and $v[PF(id)_h]$ is to be put in cS_T as the first representative from $PF(id_D \circ id)(= PF(id)_h)$.

(ii) pick(T) has returned $\boldsymbol{v}_{PF'(id_D \circ id)}(\rho_h(id))$ as $\boldsymbol{v}_T(h-1)$ and pick(T) is currently called to find $\boldsymbol{v}_T(h)$. Then, for finding a new $\boldsymbol{c}_{\text{best}}$ in (5.19), $\boldsymbol{v}_{PF'(id_D \circ id)}(\rho_h(id)+1)$ is required to make up $\boldsymbol{v}_T(h-1) = \boldsymbol{v}_{PF'(id_D \circ id)}(\rho_h(id))$.

Subprocedure pick($F'_b(id_{D_b} \circ id)$) is called by need to obtain $\boldsymbol{v}_{F'_b(id_{D_b} \circ id)}(j_b)$ at the j_bth call. When pick(F_b) at the ith call returns $\boldsymbol{v}_{F_b}(i)$, $L_{F_b}(i)$ and $id \triangleq id_{D_b} \backslash id_{F_b}(i)$ together with $id_{F'_b(id_{D_b} \circ id)}(i)$, that is, in the above situation (i), $\boldsymbol{v}_{F'_b(id_{D_b} \circ id)}(1) = \boldsymbol{v}_{F_b}(i)$. That is, the first call of pick($F'_b(id_{D_b} \circ id)$) simply refers to the return value of pick(F_b) at the ith call.

To make use of the structure (5.26), we introduce new notations.

Define a set of ordered pairs of positive integers $P \triangleq \{(i_0, i_1) : 1 \le i_b \le \nu_b \triangleq |s_b A|/|s_b B| \text{ for } b \in \{0,1\}\}$, $\boldsymbol{v}(i_0, i_1) \triangleq \boldsymbol{v}_{F'_0(id_{D_0} \circ id)}(i_0) \circ \boldsymbol{v}_{F'_1(id_{D_1} \circ id)}(i_1)$ for $(i_0, i_1) \in P$, $P_{\rho_h(id)} \triangleq \{(i_0, i_1) \in P : \boldsymbol{v}(i_0, i_1) = \boldsymbol{v}_{PF'(id_D \circ id)}(i) \text{ for } 1 \le i < \rho_h(id)\}$, and $\bar{P}_{\rho_h(id)} \triangleq P \backslash P_{\rho_h(id)}$.

We introduce the following partial order "$\le$" into P:

$$(i_0, i_1) \le (i'_0, i'_1) \iff i_0 \le i'_0 \text{ and } i_1 \le i'_1.$$

For $p = (i_0, i_1)$ and $p' = (i'_0, i'_1)$ in P, we write $p < p'$ iff $p \le p'$ and $p \ne p'$, and $p|p'$ iff $p \not\le p'$ and $p' \not\le p$. For $p \in P$, define $\kappa p \triangleq \{p' \in P : p \le p'\}$ and let $\partial \bar{P}_{\rho_h(id)}$ denote the set of minimal pairs in $\bar{P}_{\rho_h(id)}$. Then,

(∂1) for p and p' in $\partial \bar{P}_{\rho_h(id)}$, $p|p'$,

(∂2) $\bar{P}_{\rho_h(id)} = \bigcup_{p \in \partial \bar{P}_{\rho_h(id)}} \kappa p.$

Define $L(i_0, i_1) \triangleq L(\boldsymbol{v}(i_0, i_1))$. Then, for $p < p'$ in P, $Lp < Lp'$ [3]. Hence, $\boldsymbol{v}(i_0, i_1)$ is a candidate for $v[PF(id)_h]$ only if $(i_0, i_1) \in \partial \bar{P}_{\rho_h(id)}$. From ($\partial$1), we can number the pairs in $\partial \bar{P}_{\rho_h(id)}$ as follows:

(∂3) $\partial \bar{P}_{\rho_h(id)} = \{(i_0^{(j)}, i_1^{(j)}) : 1 \le j \le \delta \triangleq |\partial \bar{P}_{\rho_h(id)}|\}$, where

$$\nu_0 \ge i_0^{(1)} > i_0^{(2)} > \ldots > i_0^{(\delta)} \text{ and } i_1^{(1)} < i_1^{(2)} < \ldots < i_1^{(\delta)} \le \nu_1.$$

Here we assume that $\boldsymbol{v}_{F'_0(id_{D_0} \circ id)}(j_0)$ with $1 \le j_0 \le i_0^{(1)} = \bar{j}_0$ and $\boldsymbol{v}_{F'_1(id_{D_1} \circ id)}(j_1)$ with $1 \le j_1 \le i_1^{(\delta)} \triangleq \bar{j}_1$ have been computed by pick(F_b) or pick($F'_b(id_{D_b} \circ id)$) and $\{Lp : p \in \partial \bar{P}_{\rho_h(id)}\}$ is listed, e.g., by a priority queue. From (∂2), there exists unique j_m such that $1 \le j_m \le \delta$ and

(∂4) $\boldsymbol{v}(i_0^{(j_m)}, i_1^{(j_m)}) = v[\partial \bar{P}_{\rho_h(id)}]$.

There are two cases:

Case 1: If $id_{PF'(id_0 \circ id)}(\boldsymbol{v}(i_0^{(j_m)}, i_1^{(j_m)})) \notin \{id_{PF'(id_0 \circ id)}(\boldsymbol{v}_T(i)) : \boldsymbol{v}_T(i) \in PF'(id_0 \circ id) \text{ and } 1 \le i \le \rho_h(id)\}$, that is, $\boldsymbol{v}(i_0^{(j_m)}, i_1^{(j_m)}) \in PF(id)_h$ by using the decision algorithm in Appendix B, then output $\boldsymbol{v}(i_0^{(j_m)}, i_1^{(j_m)})$ as $v[PF(id)_h]$, delete $(i_0^{(j_m)}, i_1^{(j_m)})$ from $\partial P_{\rho_h(id)}$ and return.

---

[3] Refer to footnote 2.

Case 2: Otherwise, delete $(i_0^{(j_m)}, i_1^{(j_m)})$ from $\partial P_{\rho_h(id)}$ and update $\partial P_{\rho_h(id)}$ (if necessary).

For Case 1, $\partial \bar{P}_{\text{new } \rho_h(id)}$ is updated at the next call to pick$(PF'(id_D \circ id))$ with new $\rho_h(id) =$ current $\rho_h(id)+1$. For Case 2, $\partial \bar{P}_{\rho_h(id)}$ is to be updated, if necessary. Note that for $(i_0, i_1) \in P$,

$(\partial 5)$ $\kappa(i_0, i_1)\backslash\{(i_0, i_1)\} = \kappa(i_0 + 1, i_1)(\text{for } i_0 < \nu_0) \cup \kappa(i_0, i_1 + 1)(\text{for } i_1 < \nu_1)$.

Hence, in order to meet $(\partial 1)$ and $(\partial 2)$, each of the following pairs need to be added to $\partial \bar{P}_{\rho_h(id)} \setminus \{(i_0^{(j_m)}, i_1^{(j_m)})\}$ under the following specified conditions:
(i) $(i_0^{(j_m)} + 1, i_1^{(j_m)})$, if (a) either $j_m = 1$ and $i_0^{(1)} < \nu_0$ or (b) $j_m > 1$ and $i_0^{(j_m - 1)} - i_0^{(j_m)} \geq 2$,
(ii) $(i_0^{(j_m)}, i_1^{(j_m)} + 1)$, if (a) either $j_m = \delta$ and $i_1^{(\delta)} < \nu_1$ or (b) $j_m < \delta$ and $i_1^{(j_m+1)} - i_1^{(j_m)} \geq 2$.

For the case (a) only, pick$(F_b'(id_{D_b} \circ id))$ is called to find $\boldsymbol{v}_{F_b'(id_{D_b} \circ id)}(\bar{j}_b + 1)$.

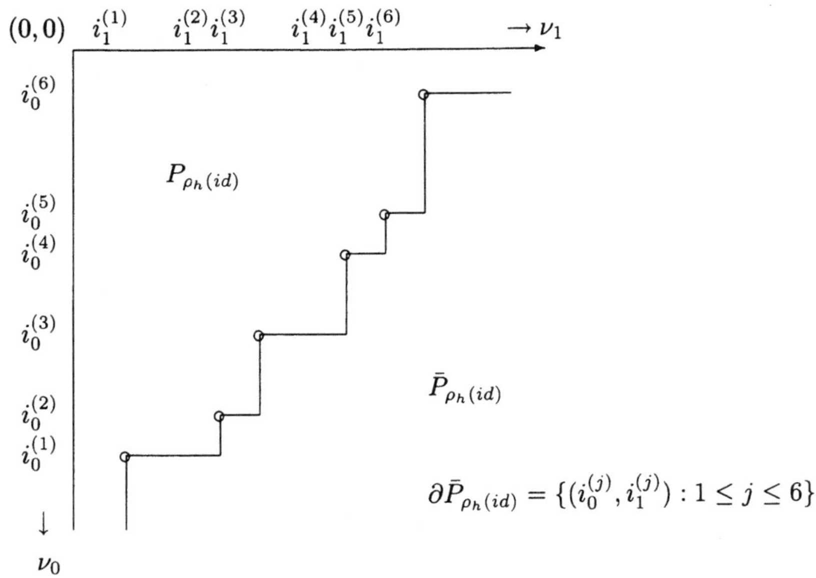

Figure 3: Illustration of $P, P_{\rho_h(id)}, \bar{P}_{\rho_h(id)}$ and $\partial \bar{P}_{\rho_h(id)}$.

In Figure 3, suppose $(i_0^{(2)}, i_1^{(2)})$ or $(i_0^{(5)}, i_1^{(5)})$ is deleted. Then no new pair is to be added. If $(i_0^{(3)}, i_1^{(3)})$ is deleted, then $(i_0^{(3)} + 1, i_1^{(3)})$ and $(i_0^{(3)}, i_1^{(3)} + 1)$ are to be added. If $(i_0^{(4)}, i_1^{(4)})$ is deleted, then $(i_0^{(4)} + 1, i_1^{(4)})$ is to be added.

Then, $|\partial \bar{P}_{\rho_h(id)}| - 1 \leq$ updated $\partial P_{\rho_h(id)}$ (current or updated) $\leq |\partial \bar{P}_{\rho_h(id)}| + 1$. If either $j_m = 1$ and $i_0^{(1)} < \nu_0$ or $j_m = \delta$ and $i_1^{(\delta)} < \nu_1$, then $\boldsymbol{v}_{F_0'(id_{D_0} \circ id)}(i_0^{(1)} + 1)$ with $i_0^{(1)} = \bar{j}_0$ or $\boldsymbol{v}_{F_1'(id_{D_1} \circ id)}(i_1^{(\delta)} + 1)$ with $i_1^{(\delta)} = \bar{j}_1$ is to be found, respectively.

Example 4: For $\alpha \in \{0,1\}^*b$ with $b \in \{0,1\}$, $A = p_\alpha C$ and $B = s_b(p_{\alpha/b}C)$.
(i) Subprocedure pick($PF'_\alpha(id^{(1)})$) with $id^{(1)} \in Id_{PF_\alpha} = Id_{F_{\alpha,b_1}}$ for $b_1 \in \{0,1\}$, calls pick($F'_{\alpha,b_1}(id^{(1)})$), by need, which is processed in turn by its subprocedures on descendant subsections. $F'_{\alpha,b_1}(id^{(1)})$ is the following abbreviation:

$$F'_{\alpha,b_1}(id^{(1)}) \triangleq (\mu_{F_{\alpha,b_1}}(id^{(1)}) + s_{b_1}(p_\alpha C))/s_{bb_1}(p_{\alpha/b}C). \tag{5.33}$$

Table 3 shows the dimensions of $PF'_\alpha(id)$(5.30) and $F'_{\alpha,b}(id)$(5.33) for several RM and EBCH codes of 128.

Table 3: The dimensions of $PF'_\alpha(id)$ and $F'_{\alpha,b}(id)$ for several RM and EBCH codes of length 128

$\|\alpha\|$	RM(128, 64)		EBCH(128, 57)		EBCH(128, 64)		EBCH(128, 71)		EBCH(128, 79)	
	PF'_α	$F'_{\alpha,b}$	PF'_α	$F'_{\alpha,b}$	PF'_α	$F'_{\alpha,b}$	PF'_α	$F'_{\alpha,b}$	PF'_α	$F'_{\alpha,b}$
1	10	10	17	15	21	17	35	24	28	19
2	6	6	6	6	12	10	5	5	5	5
3	3	3	3	3	1	1	1	1	1	1
4	1	1	1	1	0	0	0	0	0	0

(ii) For convenience, we extend the notations PF_α and $F_{\alpha,b}$ as follows:
For α and β in $\{0,1\}^*$ and $b \in \{0,1\}$ such that $I_{\alpha\beta}$ is a nonleaf subsection of I_α,

$$PF_{\alpha,\beta} \triangleq s_\beta(p_\alpha C)/(s_{\beta 0}(p_\alpha C) \circ s_{\beta 1}(p_\alpha C)), \tag{5.34}$$

$$F_{\alpha,\beta b} \triangleq s_{\beta b}(p_\alpha C)/s_{\beta b}(p_\alpha C). \tag{5.35}$$

Abbreviate $PF_{\alpha,\lambda}$ as PF_α. From (5.34) and (5.35),

$$p_b PF_{\alpha,\beta} = F_{\alpha,\beta b}. \tag{5.36}$$

(iii) Subprocedure pick($F'_{\alpha,b_1}(id^{(1)})$) can be implemented as the Case II in 5.1, where $A' = p_{\alpha b_1}C, A = s_{b_1}(p_\alpha C), D = p_{b_1}\mu_{PF_\alpha}(id^{(1)}) + A = \mu_{F_{\alpha,b_1}}(id^{(1)}) + A$ and $B = s_{bb_1}(p_{\alpha/b}C)$, by subprocedures pick($F_{\alpha,b_1b_2}(id^{(1)},-)$) with $b_2 \in \{0,1\}$ and pick($PF'_{\alpha,b_1}(id^{(1)},id^{(2)})$), where $id^{(2)} \in Id_{F_{\alpha,b_1b_2}(id^{(1)},-)}$ is one of return values from pick($F_{\alpha,b_1b_2}(id^{(1)},-)$),

$$F_{\alpha,b_1b_2}(id^{(1)},-) \triangleq (p_{b_2}\mu_{F_{\alpha,b_1}}(id^{(1)}) + p_{b_2}s_{b_1}(p_\alpha C))/s_{b_1b_2}(p_\alpha C), \tag{5.37}$$

$$\begin{aligned} PF'_{\alpha,b_1}(id^{(1)},id^{(2)}) \triangleq\ & (p_{b_1}\mu_{PF_\alpha}(id^{(1)}) + \mu_{PF_{\alpha,b_1}(id^{(1)})}(id^{(2)}) \\ & + s_{b_1 0}(p_\alpha C) \circ s_{b_1 1}(p_\alpha C))/s_{bb_1}(p_{\alpha/b}C). \end{aligned} \tag{5.38}$$

Refer to Figure 4, where

$$\begin{aligned} & F_{\alpha,b_1b_2}(id^{(1)},id^{(2)}) \\ & \triangleq p_{b_1b_2}\mu_{PF_\alpha}(id^{(1)}) + p_{b_2}\mu_{PF_{\alpha,b_1}}(id^{(2)}) + s_{b_1b_2}(p_\alpha C)/s_{bb_1b_2}(p_{\alpha/b}C). \end{aligned} \tag{5.39}$$

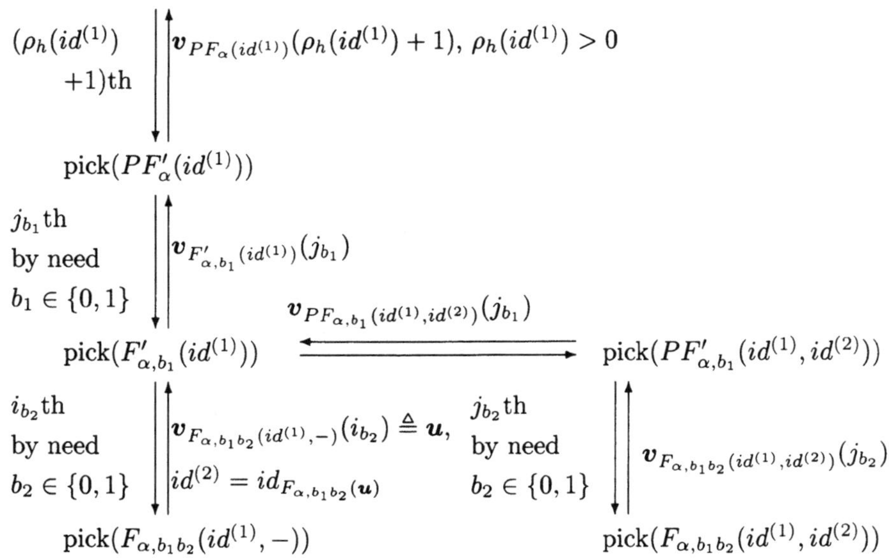

$PF'_\alpha(id^{(1)})$: (5.30),
$F'_{\alpha,b_1}(id^{(1)})$: (5.33),
$F_{\alpha,b_1b_2}(id^{(1)},-)$: (5.37),
$PF'_{\alpha,b_1}(id^{(1)},id^{(2)}))$: (5.38), $id^{(2)} \in Id_{PF'_{\alpha,b_1}(id^{(1)})} = Id_{F_{\alpha,b_1b_2(id^{(1)})}}$,
$F_{\alpha,b_1b_2}(id^{(1)},id^{(2)})$: (5.39).

Figure 4: The call-return relation among pick$(PF'_\alpha(id^{(1)}))$, pick$(F'_{\alpha,b_1}(id^{(1)}))$, pick$(PF'_{\alpha,b_1}(id^{(1)},id^{(2)}))$, pick$(F_{\alpha,b_1b_2}(id^{(1)},-))$ and pick$(F_{\alpha,b_1b_2}(id^{(1)},id^{(2)}))$.

6. Preliminary Simulation Results [12]

Figures 5 and 6 [12] show the simulation results of block error probabilities and average numbers of addition equivalent operations (AEO) for the (128, 64, 16) RM code, respectively. The number of AEO for the code by the standard Viterbi decoding with 128 sections is 16,897,966,073. As compared with the previously presented top-down RMLD for $\mathrm{RM}_{3,7}$ [7], [8], where the bit positions are permuted so that the left half 64 bits form the most reliable basis [9], the simulation range by the proposed algorithm has been extended to 0.0dB and the decoding complexity is significantly reduced. These are good indications of the effectiveness of the coarsest parallel concatenation decomposition technique to make use of the fine structure of the target code. Simulations for some EBCH codes and RM codes and the detailed analysis of decoding complexity are under study by coworkers.

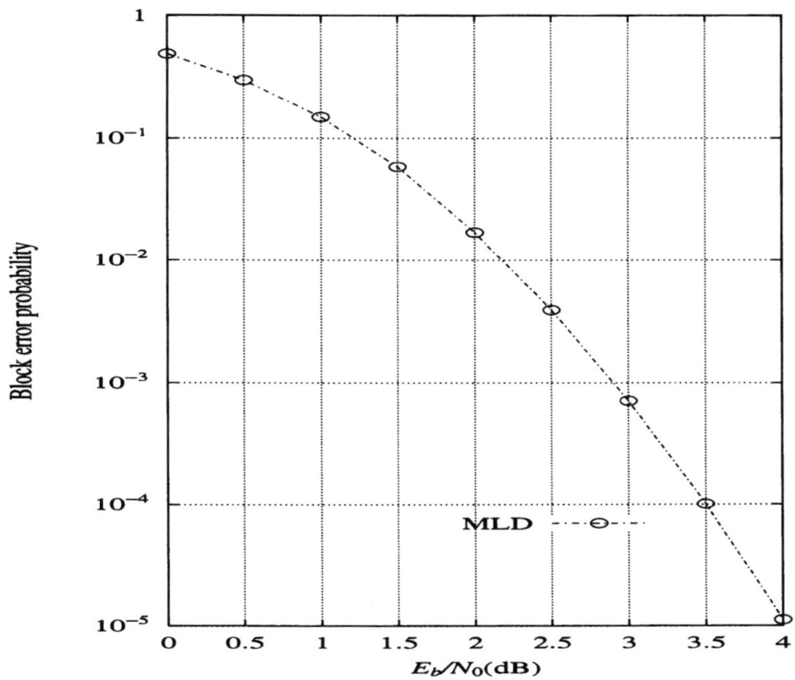

Figure 5: Block error probability for $RM_{3,7}$.

Appendix A: Binary Transitive Invariant Codes [6]

For a positive integer m and a nonnegative integer j less than 2^m, represent j in a binary expression as

$$j = j_1 2^{m-1} + j_2 2^{m-2} + \ldots + j_m, \qquad j_i \in \{0,1\} \qquad \text{for } 1 \le i \le m.$$

There is a one-to-one mapping φ_m from the set of binary polynomials with m variables, P_m, to V^{2^m} such that

$$\varphi_m(f) = (u_1, u_2, \ldots, u_{2^m}), \tag{A-1}$$

where $u_{j+1} \triangleq f(j_1, j_2, \ldots, j_m)$ for $0 \le j < 2^m$.

A binary block code B of length 2^m is **binary transitive invariant**, if and only if for any $f \in P_m$ and $(b_1, b_2, \ldots, b_m) \in V^m$, $\varphi_m(f(x_1, x_2, \ldots, x_m)) \in B \Leftrightarrow \varphi_m(f(x_1 + b_1, x_2 + b_2, \ldots, x_m + b_m)) \in B$. RM codes and EBCH codes are binary transitive invariant.

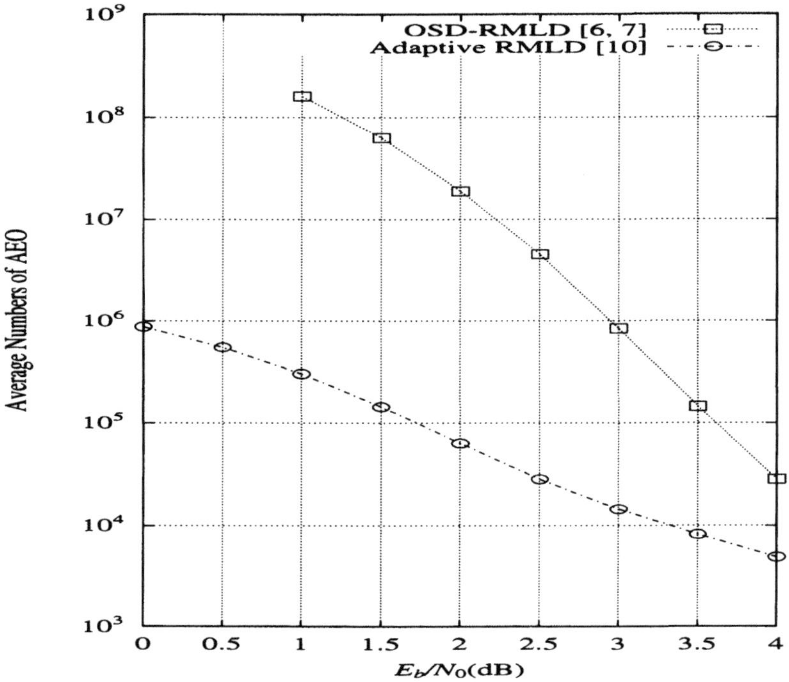

Figure 6: Average numbers of AEO.

Appendix B: A Decision Procedure Where $u + v \in B$ for u, $v \in s_0A \circ s_1A$ in (5.10)

We can choose a generator matrix G_B of B of the following form:

$$G_B = \begin{bmatrix} G_{B,0} & 0 \\ 0 & G_{B,1} \\ \multicolumn{2}{c}{G_{B,0,1}} \end{bmatrix}, \tag{B-1}$$

where $G_{B,b}$ with $b \in \{0, 1\}$ is a generator matrix of s_bB and $G_{B,0,1}$ is a generator matrix of $[B/(s_0B \circ s_1B)]$. As stated for (3.5), there is a one-to-one correspondence between the sets of rows in $p_0G_{B,0,1}$ and $p_1G_{B,0,1}$, respectively. From (B-1) and (5.10), we can choose a generator matrix G_b of s_bA of the form:

$$G_b = \begin{bmatrix} G_{B,b} \\ p_bG_{B,0,1} \\ G_b^{(1)} \end{bmatrix},$$

and submatrix

$$G_b' \triangleq \begin{bmatrix} p_bG_{B,0,1} \\ G_b^{(1)} \end{bmatrix}$$

is a generator matrix of $[s_bA/s_bB]$. For $id_b \in Id_{s_bA/s_bB}$, id_b can be partitioned into two subsections $id_{b,1}$ and $id_{b,2}$ corresponding to submatrices $p_bG_{B,0,1}$ and $G_b^{(1)}$, respectively.

Note that $G_{s_0A\circ s_1A} = \begin{bmatrix} G_{B,0} & 0 \\ 0 & G_{B,1} \\ p_0G_{B,0,1} & 0 \\ 0 & p_1G_{B,0,1} \\ G_0^{(1)} & 0 \\ 0 & G_1^{(1)} \end{bmatrix}$ is a generator matrix of $s_0A\circ s_1A$.

Define $G' \triangleq \begin{bmatrix} 0 & p_1G_{B,0,1} \\ G_0^{(1)} & 0 \\ 0 & G_1^{(1)} \end{bmatrix}$. Since $\begin{bmatrix} G_B \\ G' \end{bmatrix}$ is derived from $G_{s_0A\circ s_1A}$ by row operations, G' is a generator matrix of $[(s_0A\circ s_1A)/B]$. For $id \in Id_{(s_0A\circ s_1A)/B}$, id can be partitioned into three subsections id_1, id_2 and id_3 corresponding to submatrices of G', $[0, p_1G_{B,0,1}]$, $[G_0^{(1)}, 0]$ and $[0, G_1^{(1)}]$, respectively. It follows from (3.6) and the definitions of G'_b and G' that for $\boldsymbol{u}_b \in s_bA$, $id_1(\boldsymbol{u}_0\circ\boldsymbol{u}_1) = id_{1,1}(\boldsymbol{u}_1)$, $id_2(\boldsymbol{u}_0\circ\boldsymbol{u}_1) = id_{0,2}(\boldsymbol{u}_0)$, and $id_3(\boldsymbol{u}_0\circ\boldsymbol{u}_1) = id_{1,2}(\boldsymbol{u}_1)$. Consequently, $\boldsymbol{u} = \boldsymbol{u}_0\circ\boldsymbol{u}_1$ and $\boldsymbol{v} = \boldsymbol{v}_0\circ\boldsymbol{v}_1$, where $\boldsymbol{u}_b$ and $\boldsymbol{v}_b$ are in s_bA, are in the same coset in $(s_0A\circ s_1A)/B$, iff

$$id_{1,1}(\boldsymbol{u}_1) = id_{1,1}(\boldsymbol{v}_1), \tag{B-2}$$

$$id_{b,2}(\boldsymbol{u}_b) = id_{b,2}(\boldsymbol{v}_b). \tag{B-3}$$

Acknowledgment

The author is grateful to a reviewer and Drs. T. Fujiwara, Y. Kaji and T. Koumoto for their valuable suggestions to improve the manuscript, and Dr. H. Tokushige and Mr. I. Hisadomi for their help in preparing the manuscript.

References

[1] Y.S. Han, C.R.P. Hartmann, and C.-C. Chen, "Efficient priority first search maximum-likelihood soft-decision decoding of linear block codes," *IEEE Trans. Inform. Theory*, vol. 39, pp. 1514–1523, Sept. 1993.

[2] T. Kaneko, T. Nishijima, H. Inazumi, and S. Hirasawa, "An efficient maximum-likelihood decoding algorithm for linear block codes with algebraic decoder," *IEEE Trans. Inform. Theory*, vol. 40, pp. 320–327, Mar. 1994.

[3] T. Fujiwara, H. Yamamoto, T. Kasami and S. Lin, "A trellis-based recursive maximum likelihood decoding algorithm for linear codes," *IEEE Trans. Inform. Theory*, vol. 44, pp. 714–729, Mar. 1998.

[4] Y. Kaji, T. Fujiwara and T. Kasami, "An efficient call-by-need algorithm for the maximum likelihood decoding of a linear code," *2000 International Symposium on Information Theory and Its Applications*, pp. 335–338, Honolulu, HI, Nov. 2000.

[5] A. Lafourcade and A. Vardy, "Optimum sectionalization of a trellis," *IEEE Trans. Inform. Theory*, vol. 42, pp. 689–703, May 1996.

[6] T. Kasami, H. Tokushige, T. Fujiwara, H. Yamamoto and S. Lin, "A recursive maximum likelihood decoding algorithm for some transitive invariant binary block codes," *IEICE Trans. Fundamentals*, vol. E81-A, pp. 1916–1924, Sept. 1998.

[7] T. Koumoto, and T. Kasami, "Top-down recursive maximum likelihood decoding using ordered statistics information for half rate codes," Technical Report of IEICE, IT2002–29, The Institute of Electronics, Information and Communication Engineers, pp. 13–18, Japan, Sept. 2002.

[8] T. Koumoto and T. Kasami, "Top-down recursive maximum likelihood decoding using ordered statistics information," *Proc of the IEEE Inform. Theory Workshop*, pp. 202, Bangalore, India, Oct. 2002.

[9] M.P.C. Fossorier and S. Lin, "Soft-decision decoding of linear block codes based on ordered statistics," *IEEE Trans. Inform. Theory*, vol. 41, pp. 1379–1396, Sept. 1995.

[10] S. Lin, T. Kasami, T. Fujiwara and M. Fossorier, "Trellises and Trellis-Based Decoding Algorithms for Linear Block Codes," *Kluwer Academic Publishers*, Norwell, MA, 1998.

[11] T. Kasami, T. Fujiwara, Y. Kaji and T. Koumoto, "Adaptive recursive maximum likelihood decoding based on parallel concatenation decomposition," *Technical Report of IEICE*, IT2002–66, March, 2003.

[12] T. Koumoto, Y. Kaji, T. Fujiwara and T. Kasami, "Implementation and simulation results of adaptive recursive maximum likelihood decoding," *Technical Report of IEICE*, IT2003–28, July 2003.

Tadao Kasami
Ohata-cho 4-26
Nishinomiya-shi
Hyogo, 662-0836, Japan
e-mail: `kasami73@nifty.com`

Progress in Computer Science and Applied Logic, Vol. 23, 51–65

Modularity of Asymptotically Optimal Towers of Function Fields

Wen-Ching Winnie Li

Abstract. Elkies conjectures that all recursively defined asymptotically optimal towers of function fields over finite fields with square cardinality arise from elliptic modular curves, Shimura curves, or Drinfeld modular curves by appropriate reduction. In this paper we review the recursive asymptotically optimal towers constructed so far, discuss the reasons behind Elkies' conjecture, present numerical evidence of this conjecture, and sketch Elkies' proof of modularity of the new families.

Mathematics Subject Classification (2000). Primary 14G50, 14G05, 11R58.

Keywords. Modular curves, optimal towers.

1. Introduction

It is well known that for codes over a finite field F with square cardinality q at least 49, the algebraic geometry bound of the information rate is better than the Gilbert-Varshamov bound. This is achieved by exhibiting families, called asymptotically optimal family, of smooth curves defined over F such that the limit of the number of F-rational points over its genus approaches the optimal value $\sqrt{q}-1$. Appropriate reductions of elliptic modular curves, Shimura curves, and Drinfeld modular curves are shown to yield asymptotically optimal families. Families arising from such curves are called modular. For practical purposes, explicit constructions of asymptotically optimal families are desired. This was first done by Garcia and Stichtenoth in 1995, giving recursively constructed towers. To date, there are several known recursively defined asymptotically optimal towers, which are all proved by Elkies to be modular. Elkies further conjectures that all recursively defined asymptotically optimal towers over finite fields with square cardinality are modular.

This research is supported in part by the NSA grant MDA904-03-1-0069.

In this paper we review the recursive asymptotically optimal towers constructed so far, discuss the reasons behind Elkies' conjecture, present numerical evidence of this conjecture, and sketch Elkies' proof of modularity of the new families.

2. Algebraic Geometry Codes

Let X be a smooth projective curve of genus g defined over a finite field $\mathbb{F}$ of q elements. Choose n distinct $\mathbb{F}$-rational points $P_1, \ldots, P_n$ on X and an effective divisor G of X with support disjoint from the P_i's. Suppose $n > \deg G > g$. Denote by $\mathcal{L}(G)$ the finite-dimensional $\mathbb{F}$ vector space spanned by the nonzero $\mathbb{F}$-rational functions f on X such that $div f + G \geq 0$. Consider the $\mathbb{F}$-linear map ϕ from $\mathcal{L}(G)$ to $\mathbb{F}^n$ defined by

$$\phi(f) = (f(P_1), \ldots, f(P_n)).$$

This map is well defined since the poles of f lie in the support of G. Further, the condition $\text{div} f + G \geq 0$ implies that the total number of poles of a nonzero f, counting multiplicities, is at most $\deg G$, hence a nonzero f can have at most $\deg G$ zeros on X. As $n > \deg G$, we see that $\phi(f) \neq 0$ if $f \neq 0$. In other words, ϕ is an injection. The image of ϕ, denoted by $\mathcal{C} = \mathcal{C}(P_1, \ldots, P_n; G)$, is a linear code over $\mathbb{F}$ of length n, called an *algebraic geometry code.* Its dimension $k = \dim \mathcal{L}(G)$ is at least $\deg G - g + 1$ by the Riemann-Roch theorem. Its minimal distance d, which is the least number of nonzero components among nonzero codewords in $\mathcal{C}$, is at least $n - \deg G$, as explained above. From practical point of view, it would be desirable that $\mathcal{C}$ has large information rate $r(\mathcal{C}) := k/n$ so that it can transmit more messages. On the other hand, one would also desire that $\mathcal{C}$ has large error-correcting rate $\delta(\mathcal{C}) := d/n$ so that it can correct more errors. These two quantities apparently are opposite to each other. A code is said to be *good* if the sum of these two quantities is large. In our case, we have the following lower bound for an algebraic geometry code $\mathcal{C}$:

$$r(\mathcal{C}) + \delta(\mathcal{C}) \geq \frac{n-g+1}{n} = 1 + \frac{1}{n} - \frac{1}{n/g}.$$

Therefore, to construct good algebraic geometry codes, we seek curves X defined over $\mathbb{F}$ whose number of $\mathbb{F}$-rational points, $N_q(X)$, divided by its genus $g(X)$ is large. In fact, we'll need a family of curves $\{X_i\}$ defined over $\mathbb{F}$ such that the ratio $N_q(X_i)/g(X_i)$ is large as i approaches infinity. For this purpose, let $N_q(g)$ denote the maximal possible number of $\mathbb{F}$-rational points on a curve of genus g defined over $\mathbb{F}$, and consider the quantity

$$A(q) = \limsup_{g \to \infty} \frac{N_q(g)}{g}$$

first introduced by Ihara in 1981. He showed that [10]

$$A(q) \geq \sqrt{q} - 1 \qquad \text{if } q \text{ is a square.}$$

Then Drinfeld and Vladut [1] in 1983 proved the unconditional upper bound

$$A(q) \leq \sqrt{q} - 1.$$

Combining both, we conclude that

$$A(q) = \sqrt{q} - 1 \qquad \text{when } q \text{ is a square.}$$

To date, the precise value of $A(q)$ for a nonsquare q is unknown. However, there are various lower bounds for such $A(q)$. The reader is referred to [20], [16], [15], [18], and [12] for more information.

For the remainder of this paper, q is assumed to be a square. A sequence of curves $\{X_i\}$ defined over $\mathbb{F}$ with genus $g(X_i)$ approaching ∞ and $N_q(X_i)/g(X_i)$ approaching $A(q) = \sqrt{q} - 1$ as i increases to infinity is called an *asymptotically optimal* family of curves. Ihara obtained the aforementioned lower bound for $A(q)$ by exhibiting an asymptotically optimal family of Shimura curves. In [19] Tsfasman, Vladut, and Zink exhibited an asymptotically optimal family of modular curves. These families are not explicit in the sense that the defining equations of these curves are not explicit.

Instead of viewing a curve X geometrically, one may regard it algebraically by studying the associated field of $\mathbb{F}$-rational functions on X, called the function field $\mathbb{F}(X)$ attached to X. The $\mathbb{F}$-rational points on X correspond to degree one places of $\mathbb{F}(X)$, and the genus of $\mathbb{F}(X)$ is equal to the genus of X. Conversely, given a field F over $\mathbb{F}$ with transcendence degree one, there is a smooth projective curve X defined over $\mathbb{F}$, unique up to isomorphism, such that F is its function field. In the event that $\{X_i\}$ is a sequence of covers, the associated function fields $\{\mathbb{F}(X_i)\}$ form a tower under inclusion. Given a function field F over $\mathbb{F}$, we can compute the ratio of the number $N(F)$ of places of degree one versus its genus $g(F)$ and call a family of function fields $\{F_i\}$ over $\mathbb{F}$ *bad*, *good*, or *asymptotically optimal* if the limit of $N(F_i)/g(F_i)$, as i approaches infinity, is 0, nonzero, or $A(q)$, respectively. Most of the families are bad. Asymptotically optimal families are rare. In what follows, we shall discuss explicit asymptotically optimal towers which are defined recursively and Elkies' modularity conjecture.

3. Recursively Defined Towers

By a *recursively defined tower* over $\mathbb{F}$ we mean a strictly increasing tower $\mathcal{T}$ of function fields

$$F_1 \subset F_2 \subset F_3 \cdots \tag{3.1}$$

satisfying the following conditions:

1. Each F_i is a function field with field of constants $\mathbb{F}$;
2. F_{i+1} is a finite separable extension of F_i for all $i \geq 1$;
3. The genus $g(F_i)$ of F_i is greater than 1 for some i;

4. F_1 is the rational function field $\mathbb{F}(x_1)$, $F_{i+1} = F_i(x_{i+1})$ for $i \geq 1$, and there is a rational function $f(X,Y)$ in variables X and Y with coefficients in $\mathbb{F}$ such that $f(x_i, x_{i+1}) = 0$ for $i \geq 1$.

Clearly the fields F_i are defined by explicitly given equations. Moreover, the recursive defining equation facilitates the study of the splitting of degree one places in its immediate superfield, which provides a lower bound of the growth of the number of the places of degree one, and the study of the ramification in each consecutive extension, which, combined with the Hurwitz genus formula, describes the growth of the genus of fields. Recall that the limit as i goes to infinity of the quotient of the number $N(F_i)$ of places of degree one by the genus $g(F_i)$ for each F_i tells us how good the tower is.

Remark. For the sake of simplicity, we restrict ourselves to adding one variable and satisfying one recursive relation at each step. We shall see later an example of adding two variables and satisfying two relations. Obviously it extends to adding m variables and satisfying m conditions.

Exhibited below are a few examples of recursively defined asymptotically optimal towers whose field of constants $\mathbb{F}$ has square cardinality.

1. The first such tower was given by Garcia and Stichtenoth [6] in 1995 over $\mathbb{F}_{q^2}$ with the recursive polynomial
$$f(X,Y) = (YX)^q + YX - X^{q+1}.$$
2. In 1996, Garcia and Stichtenoth [7] found a subtower of the first tower, defined by the recursive relation
$$f(X,Y) = Y^q + Y - \frac{X^q}{X^{q-1}+1},$$
which is also asymptotically optimal. Elkies in [2] showed that the above two towers are in fact Drinfeld modular towers, that is, they arise from Drinfeld modular curves by reduction.
3. In [8] Garcia and Stichtenoth constructed two more towers over $\mathbb{F}_4$ and $\mathbb{F}_9$, respectively. They were shown by Elkies [3] to come from the reduction mod 2 of the elliptic modular curves $\{X_0(3^n)\}$ and reduction mod 3 of the modular curves $\{X_0(2^n)\}$, respectively. Solé in [17] gave a slightly different proof of this fact using Jacobi quartic identity.
4. In his Allerton conference paper [3] in 1997, Elkies listed six families of elliptic modular curves $\{X_0(\ell^n)\}$ for $\ell = 2,3,4,5,6$, and $\{X_0(3 \cdot 2^n)\}$, as well as two families of Shimura curves, showing that their induction mod p for primes p not dividing the level yield recursively defined asymptotically optimal towers over $\mathbb{F}_{p^2}$. We shall explain some of these families in the next section. It should be pointed out that the first two towers are wild towers, meaning that wild ramifications occur in the consecutive field extensions, while the towers in [4] are tame towers, that is, at most tame ramifications occurring in consecutive extensions.

5. With the help of computer search, in 2002 Li, Maharaj, and Stichtenoth [13] gave four new recursively defined asymptotically optimal towers over $\mathbb{F}_4, \mathbb{F}_9, \mathbb{F}_{25}, \mathbb{F}_{49}$, respectively. These are tame towers, and they are not subtowers of any previously known asymptotically optimal towers. Elkies showed that they are elliptic modular towers [4]. The recursive polynomials and the proofs will be discussed in later sections.
6. The most recent asymptotically optimal tower is the one constructed by Bezerra and Garcia in 2003. It is a subtower of the second tower with the recursive rational function
$$f(X,Y) = \frac{Y-1}{Y^q} - \frac{X^q-1}{X}.$$
The modularity of this tower is unknown.

The tower $\mathcal{T}$ can be seen from geometric point of view as follows. Denote by X_i the smooth irreducible curve defined over $\mathbb{F}$ whose function field is F_i. The increasing chain (3.1) means that geometrically we have a sequence of covering curves:

$$X_1 \longleftarrow X_2 \longleftarrow X_3 \cdots$$

with X_1 equal to the projective line $\mathbb{P}^1$ over $\mathbb{F}$, and X_2 is a curve in $X_1 \times X_1$ defined by $f(X,Y) = 0$. Inductively, we see that X_n is a curve in the product of n copies of X_1, namely, $X_1 \times \cdots \times X_1$, such that a point $(P_1, \dots, P_n)$ of the product $X_1 \times \cdots \times X_1$ lies in X_n if and only if (P_j, P_{j+1}) lies in X_2 for $j = 1, \dots, n-1$. In other words, X_n is obtained by iterating $n-1$ times the correspondence from X_1 to X_1 given by X_2.

4. Elkies' Conjecture

Before explaining the underlying philosophy of Elkies' conjecture, we recall some basic facts about elliptic modular curves. The group $\mathrm{SL}_2(\mathbb{Z})$ acts on the Poincaré upper-half plane $\mathfrak{H}$ by fractional linear transformations. Given a positive integer N, we are interested in the congruence subgroup

$$\Gamma_0(N) := \{\begin{pmatrix} a & b \\ c & d \end{pmatrix} \in \mathrm{SL}_2(\mathbb{Z}) : c \equiv 0 \pmod{N}\},$$

and its subgroup

$$\Gamma_1(N) := \{\begin{pmatrix} a & b \\ c & d \end{pmatrix} \in \mathrm{SL}_2(\mathbb{Z}) : a \equiv d \equiv 1 \pmod{N},\ c \equiv 0 \pmod{N}\}.$$

The quotients $Y_0(N) := \Gamma_0(N)\backslash\mathfrak{H}$ and $Y_1(N) = \Gamma_1(N)\backslash\mathfrak{H}$ are called modular curves; each has finitely many cusps. After adjoining the cusps, we obtain compactified modular curves $X_0(N)$ and $X_1(N)$, respectively. These are curves defined over $\mathbb{Q}$. Both $Y_0(N)$ and $Y_1(N)$ are moduli spaces, each parametrizes equivalence classes of elliptic curves defined over $\mathbb{C}$ with certain level N structure. We explain $Y_0(N)$ in more detail.

Consider the case $N = \ell^n$ for an integer $\ell > 1$ and integer $n \geq 0$. A point z in $Y_0(\ell^n)$ represents the equivalence class of an elliptic curve E together with a cyclic subgroup C_{ℓ^n} of order ℓ^n. We can also think of z as representing the isogeny from E to its quotient E/C_{ℓ^n}, which is again an elliptic curve. Being cyclic, the group C_{ℓ^n} contains a unique cyclic subgroup $C_{\ell^{n-1}}$. After $n-1$ iterations, we obtain a descending sequence

$$C_{\ell^n} \supset C_{\ell^{n-1}} \supset \cdots \supset C_\ell. \tag{4.1}$$

In terms of isogenies, this yields a chain

$$E = E_0 \longrightarrow E_1 \longrightarrow \cdots \longrightarrow E_n,$$

where $E_i = E/C_{\ell^i}$ for $i = 1, \ldots, n$.

On the other hand, the set of complex points on an elliptic curve over $\mathbb{C}$ may be identified with $\mathbb{C}$ divided by a rank two lattice. Another way to interpret z is to regard it as parametrizing the isogeny from the elliptic curve with lattice $\mathbb{Z} + z\mathbb{Z}$ to the the elliptic curve with lattice $\ell^{-n}\mathbb{Z} + z\mathbb{Z}$, which is equivalent to the lattice $\mathbb{Z} + \ell^n z\mathbb{Z}$.

The advantage of viewing a point z as the chain (4.1) is that we can break it into $n-1$ subchains of length 2 so that each subchain is a point in $Y_0(\ell^2)$, and consequently we obtain a map

$$\pi_n : \quad Y_0(\ell^n) \longrightarrow (Y_0(\ell^2))^{n-1}$$

by sending $E = E_0 \longrightarrow E_1 \longrightarrow \cdots \longrightarrow E_n$ to the point $(E_0 \to E_1 \to E_2,\ E_1 \to E_2 \to E_3, \cdots, E_{n-2} \to E_{n-1} \to E_n)$. In terms of points z in $\mathfrak{H}$, the map π_n sends z to the point $(z, \ell z, \ldots, \ell^{n-2}z)$ in $n-1$ copies of $Y_0(\ell^2)$. We extend π_n to a map from $X_0(\ell^n)$ to $(X_0(\ell^2))^{n-1}$.

The Atkin-Lehner involution w_{ℓ^n} acts on $X_0(\ell^n)$ by sending z to $\frac{-1}{\ell^n z}$, or equivalently, it maps $E_0 \to E_1 \to \cdots \to E_n$ to $E_n \to E_{n-1} \to \cdots \to E_0$. Here the map $E_j \to E_{j-1}$ is the dual isogeny of $E_{j-1} \to E_j$ for $j = 1, \ldots, n$. Note that there are two natural maps from $X_0(\ell^2)$ to $X_0(\ell)$: the first one starts with the involution w_{ℓ^2} on $X_0(\ell^2)$, then followed by the natural projection proj from $X_0(\ell^2)$ to $X_0(\ell)$, while the second one starts with proj from $X_0(\ell^2)$ to $X_0(\ell)$, then followed by the involution w_ℓ on $X_0(\ell)$. Comparison of these two maps yields a description of the image of π_n. More precisely, Elkies proved

Theorem 4.1. [Elkies [3]] *The map*

$$\pi_n : X_0(\ell^n) \longrightarrow (X_0(\ell^2))^{n-1}$$

given by

$$z \mapsto (z, \ell z, \ldots, \ell^{n-2}z)$$

is an injection. Its image consists of points $(P_1, \ldots, P_{n-1})$ in $(X_0(\ell^2))^{n-1}$ satisfying the relation

$$\mathrm{proj} \circ w_{\ell^2}(P_j) = w_\ell \circ \mathrm{proj}(P_{j+1}) \qquad \text{for } j = 1, \ldots, n-2. \tag{4.2}$$

When $X_0(\ell^2)$ has genus zero (and hence so does $X_0(\ell)$), we may parametrize the points on the curve by its Hauptmodul, that is, a generator of the function field of the curve. We compute the actions of w_{ℓ^2} and w_ℓ using the respective Hauptmodul, and further express the Hauptmodul of $X_0(\ell)$ in terms of the Hauptmodul x_1 of $X_0(\ell^2)$. In this way we obtain a recursive relation $f(X,Y)$ describing the relation (4.2) as $f(x_1(z), x_1(\ell z)) = 0$ for all $z \in \mathcal{H}$. This works for $\ell = 2,3,4,5$.

Example. Consider the case $\ell = 2$. The Hauptmodul for $X_0(4)$ is

$$h_4(z) = 1 + \frac{1}{8}(\frac{\eta(z)}{\eta(4z)})^8$$

and the Hauptmodul for $X_0(2)$ is

$$h_2(z) = (\frac{\eta(z)}{\eta(2z)})^{24} = 8\frac{(h_4(z)+1)^2}{h_4(z)-1}.$$

Here

$$\eta(z) = e^{2\pi i z/24}\prod_{m\geq 1}(1 - e^{2\pi i m z})$$

is a modular function of weight 1/2. The recursive rational function is

$$f(X,Y) = (X^2 - 1)((\frac{Y+3}{Y-1})^2 - 1) - 1.$$

A more interesting and complicated case is $\ell = 6$. The modular curve $X_0(36)$ has genus one. It is an elliptic curve with affine equation given by $y^2 = x^3 + 1$, hence points on $X_0(36)$ may be described by pairs (x, y) satisfying the defining equation. The curve $X_0(6)$ has genus 0. After going through the computations outlined above, Elkies [3] showed that the map π_n identifies the points in $X_0(6^n)$ with the points $((x_1, y_1), \ldots, (x_{n-1}, y_{n-1}))$ in $X_0(36)^{n-1}$ satisfying the conditions

$$(x_{j-1}^3 - 8)(z_j^3 - 8) = 72 \qquad \text{for } j = 2, \ldots, n-1, \tag{4.3}$$

where

$$z_j = (\frac{y_j + 3}{x_j - 2})^2 - x_j - 2$$

is the x-coordinate of the point (2, 3) – $(x_j,\ y_j)$ on $X_0(36)$. In conclusion, the tower of the fields over $\mathbb{F}_{p^2}$

$$F_1 \subset F_2 \subset F_3 \subset \cdots$$

obtained from $X_0(6^n)$ modulo a prime $p \neq 2, 3$ is constructed by adjoining two variables at each stage, namely, $F_j = F_{j-1}(x_j, y_j)$ for $j \geq 2$, which satisfy two relations

$$y_j^2 = x_j^3 + 1$$

and (4.3).

Based on the fact that all asymptotically optimal recursive towers known at the time were proved by him to arise from either elliptic modular curves, Shimura modular curves, or Drinfeld modular curves by reduction, Elkies conjectured in 1997 that this should be a general phenomenon.

Elkies' Modularity Conjecture [3]. Every asymptotically optimal recursively defined tower over a finite field with square cardinality is modular, that is, the fields in the tower are the function fields of either elliptic modular curves, Shimura modular curves, or Drinfeld modular curves by reduction.

After the paper [13], Elkies includes towers arising from compactification by adding cusps of $\mathfrak{H}/(\Delta \cap \Gamma_0(\ell^n))$, where Δ is some other congruence subgroup of $\mathrm{PGL}_2(\mathbb{Q})$, modulo primes coprime to ℓ. In his website [5], Elkies lists 15 (resp. 6) cases of elliptic modular towers arising from $\Delta \cap \Gamma_0(2)$ (resp. $\Delta \cap \Gamma_0(4)$) with the recursive relation $f(X,Y)$ a polynomial quadratic in X and Y.

5. Numerical Evidence of Elkies' Conjecture

Using a computer program called KASH, in a joint work with Maharaj and Stichtenoth [13], we performed an extensive search for polynomials $f(X,Y)$ of low degree over small finite fields which define good towers in general and asymptotically optimal towers over finite fields of square cardinality in particular. To achieve this goal, we considered only towers satisfying the three conditions in the following theorem, which provides an explicit lower bound of how good such towers are.

Theorem 5.1. [9] *Let $F_1 \subset F_2 \subset \cdots$ be a tower of function fields over $\mathbb{F}_q$ such that*

(i) *All consecutive extensions F_{n+1} over F_n are tame;*

(ii) *The set*

$$R = \{\textit{places } v \textit{ of } F_1 : v \textit{ is ramified in } F_n \textit{ for some } n \geq 2\}$$

is finite;

(iii) *The set*

$$S = \{\textit{places } v \textit{ of } F_1 : \deg v = 1 \textit{ and } v \textit{ splits completely in all } F_n\}$$

is nonempty.

Then

$$A(q) \geq \lim_{n\to\infty} \frac{N(F_n)}{g(F_n)} \geq \frac{2s}{2g(F_1) - 2 + r},$$

where s is the cardinality of S, and $r = \sum_{v \in R} \deg v$.

In the course of our search, all known asymptotically optimal recursive towers over small finite fields were recovered, but no good towers over a prime field were found. Based on this fact, we are tempted to make the following

Conjecture. No tame towers over prime fields satisfying conditions (i)–(iii) are recursively defined by a polynomial $f(X,Y)$ of degree 2 or 3.

In addition to the numerical evidence, there are some theoretical support to this conjecture. Lenstra in [11] proved that the construction of Garcia, Stichtenoth and Thomas [9] (for every finite field which is not prime) does not work over a prime field $\mathbb{F}_p$ and $f(X,Y) = Y^2 + aX^2 + bX$ over $\mathbb{F}_p$. In other words, such f confirms the conjecture above. Moreover, Maharaj, Stichtenoth and Wulftange [14] showed that the recursive polynomial $f(X,Y) = Y^2 + aX^2 + bX$ over $\mathbb{F}_p$ defines a tower over $\mathbb{F}_q$ satisfying the conditions (i)–(iii) above if and only if $p = 3, a = 1, b \neq 0$ and q is a square. Further, $f(X,Y) = Y^3 + aX^3 + bX^2 + cX$ over $\mathbb{F}_p$ defines a tower satisfying (i)–(iii) if and only if $p = 2, a = b = c = 1$ and q is a square. Hence they provide further support to the above assertion.

Of the conditions (i)–(iii) in Theorem 5.1, our experience indicates that the condition (iii) is more restrictive than (ii).

As a result of our computer search, 4 new asymptotically optimal towers were discovered in [13], which were proved by Elkies [4] to be modular as an appendix to [13]. This provides a numerical evidence of Elkies' modularity conjecture. We summarize the main results below.

Theorem 5.2. [Li-Maharaj-Stichtenoth [13] and Elkies [4]]

(1) *The polynomials*

$X^2Y^3 + (X^3 + X^2 + X)Y^2 + (X+1)Y + X^3 + X$ *over* $\mathbb{F}_4$

$2XY^2 + (X^2 + X + 1)Y + X^2 + X + 2$ *over* $\mathbb{F}_9$

$(4X+1)Y^2 + (X^2 + X + 2)Y + X + 3$ *over* $\mathbb{F}_{25}$

$(X^2+6)Y^2 + XY + X^2 + 4$ *over* $\mathbb{F}_{49}$

define recursive asymptotically optimal towers.

(2) *These towers are not subtowers of the known asymptotically optimal towers.*

(3) *These towers are new modular towers. More precisely, the nth curve in each tower is isomorphic with the elliptic modular curve associated with the following congruence subgroup of* $\mathrm{PSL}_2(\mathbb{Z})$*:*

$\Gamma_1(9) \cap \Gamma_0(3^{n+1})$ *over* $\mathbb{F}_4$

$\Gamma_1(5) \cap \Gamma_0(2^n)$ *over* $\mathbb{F}_9$

$\Gamma_1(12) \cap \Gamma_0(2^{n+1})$ *over* $\mathbb{F}_{25}$

$\Gamma_1(5) \cap \Gamma_0(2^n)$ *over* $\mathbb{F}_{25}$.

Several remarks are in order. The new towers, while they are not subtowers, are supertowers of previously known modular towers $X_0(3^{n+1})$ and $X_0(3 \cdot 2^{n+1})$, and of modular tower $X_0(5 \cdot 2^n)$, which can be obtained by known methods.

We note the following new features of the new towers:

(A) Every previous recursive tower of elliptic modular curves is either $\{X_0(\ell^n N_0)\}$ or a subtower of $\{X_0(\ell^n N_0)\}$; new towers require $\Gamma_0(\ell^n N_0) \cap \Gamma_1(N_0)$. Because of the involvement of Γ_1 groups, in the proof of modularity of new towers, we cannot use the usual models of these curves, in which rational functions have rational Fourier expansions at the cusp at infinity; instead, one has to use Igusa's model of the modular curve, which is a twist of the usual one.

(B) In previous modular towers, as shown in the previous section, the method is to find a modular function $x_1(\cdot)$ on the upper half-plane $\mathfrak{H}$ satisfying $f(x_1(z), x_1(\ell z)) = 0$ for all $z \in \mathfrak{H}$, leading to the parametrization of the point $(x_1, \ldots, x_n)$ by modular functions $(x_1(z), x_1(\ell z), \ldots, x_1(\ell^{n-1}z))$. In the new towers, the identity takes the form $f(x_1(z), \varepsilon(x_1(\ell z))) = 0$, where ε is a fractional linear transformation such that

$$f(X,Y) = 0 \qquad \text{if and only if} \qquad f(\varepsilon(X), \varepsilon(Y)) = 0.$$

Thus a point on a new tower has coordinates

$$(x_1(z), \varepsilon(x_1(\ell z)), \varepsilon^2(x_1(\ell^2 z)), \ldots, \varepsilon^{n-1}(x_1(\ell^{n-1}z))).$$

In each case the cyclic group generated by ε gives the action of $\Gamma_0(N_0)/\Gamma_1(N_0)$ (which is isomorphic to the abelian group $(\mathbb{Z}/N_0\mathbb{Z})^\times/\{\pm 1\}$) on the x_1-line of $X_1(N_0)$.

To give a flavor of Elkies' proof of the modularity of new towers, we demonstrate the case of the tower $\mathcal{F}$ over $\mathbb{F}_9$ defined by the recursive polynomial

$$f(X,Y) = 2XY^2 + (X^2 + X + 1)Y + X^2 + X + 2.$$

Note that each F_{n+1} is a quadratic extension of F_n for $n \geq 1$. The general strategy is to simplify the tower by successively dividing out symmetries until the tower becomes a recognizable modular tower.

The starting point is to find a symmetry on the curve X_2 defined by $f(X,Y) = 0$. We get some clue by looking at the set S of $\mathbb{F}$-rational points on the projective line which splits completely in all fields F_n in the tower. To find S, search for a maximal set of places of degree one in $F_1 = \mathbb{F}(X)$ which splits completely in F_2 such that the occurring degree one places in F_2 are the same as those in F_1 we started with. This then repeats itself as we go up through all extensions in the tower. Consequently the starting set is the set S we look for. Denote by ω a primitive root of $\mathbb{F}_9^\times$; it satisfies $\omega^2 - \omega - 1 = 0$. By straightforward computations, we find the following splitting information:

place in F_1	place in F_2
0	$\infty,\ 1$
1	$-1,\ 1$
∞	$\infty,\ -1$
-1	$-\omega,\ -\omega^3$
$-\omega$	$0,\ -\omega^3$
$-\omega^3$	$0,\ -\omega$

Therefore we obtain

$$S = \{0, 1, \infty, -1, -\omega, -\omega^3\}.$$

To S we attach a directed graph, called the graph of splitting points, with vertex set S and edge set given by the table above, that is, there is an out-edge from vertex u to vertices v and v' if and only if the place u of F_1 splits into places v and v' in F_2. Each vertex has two out-edges and two in-edges. Note that there is a loop at the vertex 1 and vertex ∞, respectively; a loop counts as an in-edge and an out-edge. This graph helps us visualize the following symmetry on S:

$$0 \leftrightarrow 0, \ -1 \leftrightarrow -1, \ 1 \leftrightarrow \infty, \ -\omega \leftrightarrow -\omega^3.$$

The fractional linear transformation

$$\varepsilon(X) = \frac{X}{X-1}$$

has order two and maps the symmetrical points to each other. Since an $\mathbb{F}$-rational involution on X_2 must preserve the symmetry on S, this suggests that ε is the desired involution on X_2, and inductively on all X_n. Indeed this can be verified by checking

$$f(X,Y) = 0 \qquad \text{if and only if} \qquad f(\varepsilon(X), \varepsilon(Y)) = 0.$$

Setting $U = X + \varepsilon(X)$ and $V = Y + \varepsilon(Y)$, we obtain a quotient tower $\mathcal{G}$ with recursive defining polynomial

$$g(U,V) = UV^2 - U^2V + (U+1)^2.$$

Proceed as before. To find a symmetry for the tower $\mathcal{G}$, we have to figure out its graph of splitting points, which arises from that of tower $\mathcal{F}$ under $U = X + \varepsilon(X)$. It has vertices $0, \infty, 1, -1$ and out-edges $1 \to -1, -1 \to -1, -1 \to 0, 0 \to \infty, \infty \to \infty, \infty \to 1$. Observe the symmetry

$$0 \leftrightarrow 1, \qquad \infty \leftrightarrow -1,$$

which suggests the involution $\mu(U) = \frac{1-U}{1+U}$. After verifying

$$g(U,V) = 0 \ \text{ if and only if } \ g(\mu(U), \mu(V)) = 0,$$

we conclude that μ is the desired involution on the $\mathcal{G}$ tower. Under $W = U + \mu(U)$ and $Z = V + \mu(V)$, we obtain a quotient tower $\mathcal{H}$ defined by the recursive polynomial

$$h(W,Z) = (W-1)Z^2 + (W - W^2)Z + W^2 + W.$$

So far, we have constructed three towers: tower $\mathcal{F}$ is a two-fold cover of tower $\mathcal{G}$, which is a two-fold cover of tower $\mathcal{H}$. We proceed to draw connection with modular towers from bottom up.

Theorem 5.3.

(i) *The $\mathcal{H}$ tower is isomorphic to the tower from the family of modular curves $\{X_0(5\cdot 2^n)/w_5\}$. Here w_5 is the Atkin-Lehner involution at 5.*

(ii) *The $\mathcal{G}$ tower is isomorphic to the tower from the family of modular curves $\{X_0(5\cdot 2^n)\}$ with isomorphism given by $(U,V)\mapsto(\alpha(U),\alpha(V))$ where*

$$\alpha(U)=\frac{U-I}{(I-1)U-1},\quad \text{and } I^2=-1 \quad \text{in } \mathbb{F}_9.$$

(iii) *The $\mathcal{F}$ tower is isomorphic to the tower from the family of curves $\{X_0(5\cdot 2^n)\times_{X_0(5)} X_1(5)\}$ with the isomorphism given by $(X,Y)\mapsto(\beta(X),\beta(Y))$ where*

$$\beta(X)=c\frac{X-a}{X-b},\qquad \beta(0)=-I,\quad \beta(-1)=I,$$

and a,b, being roots of $x^2-(I+1)x+I+1=0$, lie in a quadratic extension of $\mathbb{F}_9$.

Notice that the fiber product in case (iii) is nothing but the curve of the group $\Gamma_0(5\cdot 2^n)\cap\Gamma_1(5)=\Gamma_1(5)\cap\Gamma_0(2^n)$, as stated in Theorem 5.2. In each case the modular tower consists of the function fields over $\mathbb{F}_9$ of the reduction mod 3 of the corresponding modular curves.

It should be pointed out that the isomorphisms of the first two cases are over $\mathbb{F}_9$, while the last isomorphism is over a quadratic extension of $\mathbb{F}_9$ if the usual model on modular curves is used. However, if the Igusa model, which is a quadratic twist of the usual model, is used for the modular curves in case (iii), then the isomorphism in (iii) is again over $\mathbb{F}_9$.

We sketch the proof of Theorem 5.3. Start with the modular curve $X_0(10)$, which has genus zero and Hauptmodul

$$G(z)=\frac{\eta(2z)}{\eta(z)}\left(\frac{\eta(5z)}{\eta(10z)}\right)^5.$$

Its quotient $X_0(10)/w_5$ has genus zero and Hauptmodul

$$H(z)=\left(\frac{\eta(z)\eta(5z)}{\eta(2z)\eta(10z)}\right)^4=\frac{G^2-4G}{G+1}.$$

Write $G_i(z)$ for $G(2^iz)$ and $H_i(z)$ for $H(2^iz)$ for brevity. It is not hard to check that the G_i's and H_i's satisfy the following recursive relations respectively. In other words, the modular towers in (ii) and (i) are both recursive towers with recursive relation

$$\begin{aligned}G_{i+1}^2 &= G_i(G_iG_{i+1}-2G_{i+1}-4),\\ H_{i+1}^2 &= H_i(H_iH_{i+1}+8H_{i+1}+16),\end{aligned}$$

respectively. Compare the $\mathcal{H}$ tower with the tower from $\{X_0(5\cdot 2^n)/w_5\}$. Observe that an isomorphism of recursive towers should preserve fixed points of the recursive relations. This would give us a clue about the isomorphism. Indeed, solving

$h(W,W)=0$ against $H_i = H_{i+1}$, one is led to the isomorphism

$$(H_i, H_{i+1}) = (\frac{W}{1-W}, \frac{Z}{1-Z}).$$

A direct computation shows that this map brings the recursive relation on H_i, H_{i+1} to the recursive relation on W, Z. This proves (i).

For $\mathcal{G}$ tower, $G_i = G_{i+1}$ has 4 simple roots, while $g(U,U)=0$ has two double roots at $U=-1$ and $U=\infty$. Use the equivalent form of the $\mathcal{G}$ tower by applying the involution μ to only one variable. This yields an isomorphic tower $\mathcal{G}'$ with new recursive relation

$$g'(U,V) = (1-U^2)V^2 - (U^2+U+1)V + 1.$$

The $\mathcal{G}'$ tower can now be identified with the $\{X_0(5\cdot 2^n)\}$ tower by taking $(G_i, G_{i+1}) = (\alpha(U), \alpha(V))$ with

$$\alpha(U) = \frac{U-I}{(I-1)U-1}, \quad \text{where } I^2 = -1 \quad \text{in } \mathbb{F}_9.$$

This proves (ii).

Finally we prove (iii). The bottom curve of $\{X_0(5\cdot 2^n)\times_{X_0(5)} X_1(5)\}$ is $X_1(10)$, which has genus zero with Hauptmodul given by

$$G'(z) = e^{-2\pi i z}\prod_{n=1}^{\infty}(1-e^{2\pi i n z})^{c_n},$$

where

$$c_n = \begin{cases} -1 & \text{if } n \equiv \pm 1, \pm 2 \pmod{10}, \\ 1 & \text{if } n \equiv \pm 3, \pm 4 \pmod{10}, \\ 0 & \text{if } 5|n. \end{cases}$$

Further, $G = G' - \frac{1}{G'}$, or equivalently, $G'^2 - GG' - 1 = 0$. This implies that the double cover $X_1(10)$ over $X_0(10)$ is ramified at $G^2+4=0$. Reducing mod 3 and regard the reduced curves as over $\mathbb{F}_9$, the two ramified points are at $G=I$ and $G=-I$ in $\mathbb{F}_9$. Notice that $\alpha(0)=I$, $\alpha(1)=-I$, and $U=0$ and $U=1$ are the two branch points of the double cover of the U-line by the X-line given by $U = X + \varepsilon(X)$. One checks that the isomorphism $(X,Y)\mapsto(\beta(X),\beta(Y))$ as described in (iii) lifts the isomorphism $(U,V)\mapsto(\alpha(U),\alpha(V))$ given in (ii). This proves (iii).

Acknowledgment

The research of this work is supported in part by the NSA grant MDA904-03-1-0069.

References

[1] V.G. Drinfel'd and S.G. Vladut, *Number of points of an algebraic curve.* Funct. Anal. **17** (1983), 53–54.

[2] N.D. Elkies, *Explicit towers of Drinfeld modular curves.* Proceedings of the 3rd European Congress of Mathematics, Barcelona, 7/2000.

[3] N.D. Elkies, *Explicit modular towers,* Proceedings of the Thirty-Fifth Annual Allerton Conference on Communication, Control and Computing, T. Basar and A. Vardy, eds. (1997), 23–32.

[4] N.D. Elkies, Appendix to New optimal tame towers of function fields over small finite fields by W.-C. W. Li, H. Maharaj, and H. Stichtenoth, Lecture Notes in Computer Science **2369** C.Fieker and D.R.Kohel, eds. (2002), Springer-Verlag, Berlin, 384–389.

[5] N.D. Elkies, http://abel.math.harvard.edu/ elkies/compnt.html.

[6] A. Garcia and H. Stichtenoth, *A tower of Artin-Schreier extensions of function fields attaining the Drinfeld-Vladut bound,* Invent. Math. **121** (1995), 211–222.

[7] A. Garcia and H. Stichtenoth, *On the asymptotic behaviour of some towers of function fields over finite fields,* J. Number Theory **61** (1996), 248–273.

[8] A. Garcia and H. Stichtenoth, *Asymptotically good towers of function fields over finite fields,* C.R. Acad. Sci. Paris Sér. I Math. **322** (1996), 1067–1070.

[9] A. Garcia, H. Stichtenoth and M. Thomas, *On towers and composita of towers of function fields over finite fields.* Finite Fields Appl. **3** (1997), no. 3, 257–274.

[10] Y. Ihara, *Some remarks on the number of rational points of algebraic curves over finite fields,* J. Fac. Sci. Univ. Tokyo Sect. IA Math. **28** (1981), 721–724.

[11] H.W. Lenstra, Jr., *On a Problem of Garcia, Stichtenoth, and Thomas,* Finite Fields Appl. **8**, 1–5 (2001).

[12] W.-C.W. Li and H. Maharaj, *Coverings of curves with asymptotically many rational points,* J. Number Theory **96** (2002), 232–256.

[13] W.-C.W. Li, H. Maharaj, and H. Stichtenoth, *New optimal tame towers of function fields over small finite fields,* Lecture Notes in Computer Science **2369** C. Fieker and D.R. Kohel, eds. (2002), Springer-Verlag, Berlin, 372–389 .

[14] H. Maharaj, H. Stichtenoth, and J. Wulftange, *On a problem of Garcia, Stichtenoth, and Thomas II*, 2003, preprint.

[15] H. Niederreiter and C.P. Xing, *Towers of global function fields with asymptotically many rational places and an improvement on the Gilbert-Varshamov bound.* Math. Nachr. **195** (1998), 171–186.

[16] J.-P. Serre, *Rational Points on Curves over Finite Fields,* Lecture Notes, Harvard University, 1985.

[17] P. Solé, *Towers of function fields and iterated means,* IEEE Trans. Inform. Theory **46** (2000), 1532–1535.

[18] A. Temkine, *Hilbert class field towers of function fields over finite fields and lower bounds for $A(q)$,* J. Number Theory **87** (2001), 189–210.

[19] M.A. Tsfasman, S.G. Vladut and T. Zink, *Modular curves, Shimura curves and Goppa codes better than the Varshamov-Gilbert bound,* Math. Nachr. **109** (1982), 21–28.

[20] T. Zink, *Degeneration of Shimura surfaces and a problem in coding theory,* in Fundamentals of Computation Theory, L. Budach (ed.), Lecture Notes in Computer Science, Vol. **199**, Springer, Berlin, p. 503–511, 1985.

Wen-Ching Winnie Li
Department of Mathematics
Pennsylvania State University
University Park, PA 16802, USA
e-mail: wli@math.psu.edu

Progress in Computer Science and Applied Logic, Vol. 23, 67–83

A New Correlation Attack on LFSR Sequences with High Error Tolerance

Peizhong Lu and Lianzhen Huang

Abstract. Let $u = (u_1, u_2, \ldots, u_n)$ be N bits of a linear feedback shift register (LFSR) sequence with L the degree of the feedback polynomial. Let $z = (z_1, z_2, \ldots, z_N)$ be N bits of observed sequence such that $P(z_i = u_i) = 1/2 + \delta$ where $0 < \delta \leq \frac{1}{2}$. This paper presents a new efficient correlation attack on stream ciphers, which is equivalent to solve the problem of recovering the LFSR's initial state $(u_1, u_2, \ldots, u_L)$ from the observed output sequence z. We consider the problem as a decoding problem for a linear $[N, L]$ code. Our new approach has at least three advantages. Firstly, the new algorithm constructs much more independent parity check equations which results in significant decrease both of the decoding errors and of the required length N of the observed sequence. Secondly, by the combination of statistical test and repeatedly using of One-Step decoding algorithm, our novel scheme provides better performance and lower complexity than other reported methods. Thirdly, we find a new formula to describe the relationship between the tendency of attack performance, the weight w of parity check equations, the noise level δ, and N.

Mathematics Subject Classification (2000). Primary 94Z55; Secondary 94A60.

Keywords. Stream cipher, correlation attack, statistical test.

1. Introduction

In the design of stream cipher system, the initial states of some linear feedback shift registers (LFSR) are commonly used as secrete keys. The running keystreams are generated by some nonlinear combination of several LFSR sequences.

There are several classes of attacks against binary stream ciphers. One important class of attacks on LFSR-based stream ciphers is fast correlation attacks [1, 2, 3, 4, 5, 6, 7]. Siegenthaler [8] showed that it can happen that the observed output sequence is correlated to the output of a particular target LFSR. Thus it is reasonable to try to apply a so-called divide-and-conquer attack, i.e., try to restore the initial state of the target LFSR independently of the other unknown key bits.

The basic ideas of all reported fast correlation attacks consider the cryptographic problem as a suitable decoding one, namely one may consider the output of the target LFSR to have passed through an observation channel. The channel is modelled by the Binary Symmetric Channel (BSC), with some error probability $p = \frac{1}{2} - \delta$, for $\delta > 0$.

Let $u = (u_1, u_2, \dots, u_N)$ be N bits of a linear feedback shift register (LFSR) sequence with L the degree of the feedback polynomial $f(x)$. Then u is considered as a codeword of a binary linear $[N, L]$ block code. Let $z = (z_1, z_2, \dots, z_N)$ be N bits of observed sequence such that $P(z_i = u_i) = \frac{1}{2} + \delta$ where $0 < \delta \leq \frac{1}{2}$.

The correlation attack is a decoding problem of restoring the LFSR's initial state $u = (u_1, u_2, \dots, u_N)$ from the observed output sequence z.

Meier and Staffelbach [7] find a very efficient way of iteratively decoding the $[N, L]$ code when the feedback polynomial $f(x)$ has low weight.

Methods for fast correlation attacks for general feedback polynomials have been proposed [5]. Johansson and Jonsson [5][6] suggest a new fast correlation attack based on convolutional codes. They can be applied to arbitrary LFSR feedback polynomials. The Viterbi algorithm with memory orders $B \leq 18$ is used as the final decoding method. The performance of the algorithm is good. But the degree of the feedback polynomial should be less than 64 because of the limit of $B \leq 18$ in Viterbi algorithm.

Recently, there are some nice algorithms [1][3] for fast correlation attacks based on linear binary block codes, which can be applied to arbitrary LFSR feedback polynomials.

Mihaljevic, Fossorier and Imai [3] present two algorithms for the fast correlation attacks. These decoding procedures offer good trade-offs between the required sample length, overall complexity and performance. Chepyzhov, Johansson and Smeets [1] present a new simple algorithm for fast correlation attacks on stream ciphers. They associate with the target LFSR another binary linear $[n_2, k]$-code with $k < L$. The k information symbols of this code may coincide with the first k symbols of the initial state of the LFSR we want to recover. The codeword of this second code is considered to have passed through another BSC2 with a double "noise level" $p_2 = 2p(1-p) > p$. If the length of the new code can be chosen at least $n_2 = \lceil k/C(p_2) \rceil$, then the decoding of this code leads to the recovery of the first k symbols in the initial state of the LFSR. Since the new code has dimension k, the decoding complexity is decreased from $O(2^L \times L/C(P))$ to $O(2^k \times k/C(p_2))$, where $C(p) = 1 - H(p) = 1 - (-p \log_2 p - (1-p) \log_2(1-p)$ is the channel capacity of BSC.

In this paper, we present two new algorithms, Algorithm A and B, for fast correlation attacks. The two algorithms do not depend on the weight of the LFSR feedback polynomial. Although we are influenced by [1] and [3], our algorithms improve the construction of parity check sets such that the number of parity check equations we construct is $L - B$ times more than that in [1] and [3], which results in significant decrease both of the decoding errors and of the number of bits of the received degraded LFSR sequence.

We first define some random variables on the number of passed-parity-check equations. Then we propose a statistical test based on linear block codes which is a main step of our decoding algorithm. Our novel algorithm provides a remarkably better performance and lower complexity than other reported methods by repeatedly using One-Step decoding algorithm.

Our new approach is compared with recently proposed improved fast correlation attacks in [1] and [3] based on binary linear block codes. Plentiful experimental results show that our new algorithm yields better performance and lower complexity than the best algorithm reported up-to-now.

Some new interesting theoretical results are also derived in this paper. We find a new formula to describe the relationship between the tendency of attack performance, the weight w of parity check equations, the noise level p, and the required length N of the observed sequence, namely the performance of our correlation attack by using $(w+1)$-weight parity check equations is better than the one by only using w-weight equations if and only if

$$(1-2p)\sqrt{\frac{N}{w+1}} > 1.$$

The paper is organized as follows. Section 2 introduces some concepts used in correlation attack. Section 3 defines $L-B+1$ sets of parity check equations. Section 4 discusses some random variables of the number of passed-parity-check equations and their probability distributions. Section 5 presents our new fast correlation attacks. Comparisons between the recently reported best fast correlation attacks and our proposed algorithms are given in Section 6. Finally, the results of this paper are summarized in Section 7.

2. Concepts and Problem Descriptions

Let $z = (z_1, z_2, \dots, z_N)$ be the observed keystream sequence which is regarded as the received channel output. Let $u = (u_1, u_2, \dots, u_N)$ be the LFSR sequence which is considered as a codeword from an $[N, L]$ linear block code C. The code C is composed of all the 2^L sequences generated by an LFSR with a feedback polynomial of L degree. Due to the correlation between u_i and z_i, we can consider each z_i as the output of the binary symmetric channel, BSC, when u_i was transmitted. The correlation between u_i and z_i is described by the following probability:

$$P(z_i = u_i) = 1 - p = 1/2 + \varepsilon$$

where $p < 0.5$ and $\varepsilon > 0$.

The so-called fast correlation attack on a particular LFSR is to find the initial state $(u_1, u_2, \dots, u_L)$ of the LFSR sequence u by using z and the correlation probability $P(z_i = u_i) = 1 - p$ with complexity of order $O(2^{\alpha L})$ with respect to some $\alpha < 1$. Thus the problem of finding a fast correlation attack is equivalent to the problem of finding a fast decoding algorithm of the linear $[N, L]$ block code

C over a BSC with crossover probability p, where $(u_1, u_2, \ldots, u_L)$ is called the information word and $u_{L+1}, \ldots, u_N$ are the parity check symbols.

It is worth to notice that in the theory of correlation attack, the typical values of p are closed to $1/2$. For example, $p = 0.4$. However in the theory of error-correcting codes, the typical values of p are much smaller, for example, $p = 0.05$.

From the coding theory, we know that, to realize unique decoding for a codeword passing a BSC, the length N of codeword must be not less than $N_0 = \frac{L}{C(p)}$. Usually, fast correlation attacks perform better when $N \gg N_0$.

Similar as other algorithms reported for fast attacks, our algorithms have a precomputing procedure for constructing independent parity check equations in off-line. When we need to find the initial states of original LFSR sequence x after we received keystream sequence z, our new algorithm will decode z according to the parity check equations in disk in on-line.

3. Sets of Parity Check Equations

We define two types of sets of parity check equations. The first type has $L - B$ sets $\Omega_i, i = B+1, \ldots, L$, which correspond to the ith information symbol. The second type has one set Ω^*.

Let $G_{LFSR} = \begin{pmatrix} g_1 & g_2 & \cdots & g_N \end{pmatrix}$ be the generating matrix of the $[N, L]$ linear code C, where g_i is a L-dimensional column vector. Let $u = (u_1, u_2, \ldots, u_N)$ be a codeword of C. We can see that

$$u_i = U_0 g_i, i = 1, 2, \ldots, N, \tag{3.1}$$

where U_0 is the initial state of the LFSR for the sequence. Let $z = (z_1, z_2, \ldots, z_N)$ be N bits of observed sequence such that $P(z_i = u_i) = \frac{1}{2} + \delta = 1 - p$ where $0 < \delta \leq \frac{1}{2}$. We have the following parity-check equations corresponding to (3.1)

$$z_i \oplus Z_0 g_i, i = 1, 2, \ldots, N, \tag{3.2}$$

where $Z_0 = (z_1, z_2, \ldots, z_L)$. $\oplus$ is the sum of mod 2. If $z_i \oplus Z_0 g_i = 0$ for some i, we call it the *passed-parity-check equation.*

Definition 3.1. *For arbitrary $B < i \leq L$, and given a weight w, the set of parity check equations associated with the ith information symbol is the set Ω_i consisting of the following parity-check equations*

$$(z_{j_1} \oplus Z_0 g_{j_1}) \oplus (z_{j_2} \oplus Z_0 g_{j_2}) \oplus \cdots \oplus (z_{j_w} \oplus Z_0 g_{j_w})$$

where $1 \leq j_1, j_2, \ldots, j_w \leq N$ and $(g_{j_1} \oplus g_{j_2} \oplus \cdots \oplus g_{j_w})$ has arbitrary values in the first B coordinates, value one at the ith coordinate, and value zero in all the other $L - B - 1$ coordinates.

Definition 3.2. *The set Ω^* consists of the following parity-check equations*

$$(z_{j_1} \oplus Z_0 g_{j_1}) \oplus (z_{j_2} \oplus Z_0 g_{j_2}) \oplus \cdots \oplus (z_{j_w} \oplus Z_0 g_{j_w})$$

where $1 \leq j_1, j_2, \ldots, j_w \leq N$ and $(g_{j_1} \oplus g_{j_2} \oplus \cdots \oplus g_{j_w})$ has arbitrary values in the first B coordinates, value zero in all the other $L - B$ coordinates.

It is not difficult to see that $|\Omega^*| \approx |\Omega_i|$ for $i = B+1, \ldots, L$. Let $m = |\Omega^*|$. For $1 \leq j \leq m$, the jth parity check equation in Ω_i has the following relation:

$$\begin{aligned} c_{i_j} &= (z_{j_1} \oplus Z_0 g_{j_1}) \oplus (z_{j_2} \oplus Z_0 g_{j_2}) \oplus \cdots \oplus (z_{j_w} \oplus Z_0 g_{j_w}) \\ &= Z_0'(g_{j_1} \oplus g_{j_2} \oplus \cdots \oplus g_{j_w}) \oplus z_i \oplus \textstyle\sum_{k=1}^{w} z_{j_k} \\ &= Z_0'(g_{j_1} \oplus g_{j_2} \oplus \cdots \oplus g_{j_w}) \oplus b_{i_j} \end{aligned} \tag{3.3}$$

where $Z_0' = (z_1, z_2, \ldots, z_B, 0, \ldots, 0)$, $b_{i_j} = z_i \oplus \sum_{k=1}^{w} z_{j_k}$.

Similarly, by Definition 2, the value of the jth equation in Ω^* can be expressed as the following:

$$\begin{aligned} c_j &= (z_{j_1} \oplus Z_0 g_{j_1}) \oplus (z_{j_2} \oplus Z_0 g_{j_2}) \oplus \cdots \oplus (z_{j_w} \oplus Z_0 g_{j_w}) \\ &= Z_0'(g_{j_1} \oplus g_{j_2} \oplus \cdots \oplus g_{j_w}) \oplus \textstyle\sum_{k=1}^{w} z_{j_k} \\ &= Z_0'(g_{j_1} \oplus g_{j_2} \oplus \cdots \oplus g_{j_w}) \oplus b_j \end{aligned} \tag{3.4}$$

where $b_j = \sum_{k=1}^{w} z_{j_k}$.

We outline the precomputation of Ω^* and Ω_i in the following algorithm.

Precomputing Algorithm:
Input: Integers B, L, N, w and the generator matrix G_{LFSR}.
Processing steps: For arbitrary w columns $g_{j_1}, g_{j_2}, \ldots, g_{j_w}$ of G_{LFSR}, if $(g_{j_1} \oplus g_{j_2} \oplus \cdots \oplus g_{j_w})$ has arbitrary values in the first B coordinates, and value one at the ith coordinate, and value zero in all other coordinates, then the vector $(i_1, i_2, \ldots, i_w)$ and the vector of the first B coordinates of $(g_{j_1} \oplus g_{j_2} \oplus \cdots \oplus g_{j_w})$ are stored as a record in the set Ω_i.

If $(g_{j_1} \oplus g_{j_2} \oplus \cdots \oplus g_{j_w})$ has arbitrary values in the first B coordinates, and value zero in all the other coordinates, then the vector $(i_1, i_2, \ldots, i_w)$ and the vector of the first B coordinates of $(g_{j_1} \oplus g_{j_2} \oplus \cdots \oplus g_{j_w})$ are stored as a record in the set Ω^*.

Output: The sets of parity check equations Ω^* and Ω_i, for $i = B+1, \ldots, L$.

In practice, for the case $w = 2$, the parity check equations can be found in a very simple way as follows. We simply put each column of G_{LFSR} into different "buckets", according to the value of the last $L - B$ positions. Each pair of columns in each bucket will provide one parity check equation in Ω^*. And for any two buckets, if only the ith value is different in last $L - B$ positions, then each pair of these different buckets will provide us with one parity check equation in Ω_i, for $i = B+1, \ldots, L$. For $w \geq 3$, we store the columns in the same way as for $w = 2$. To find a parity check equation, we run through all $w - 1$ columns, add them, and look in the bucket corresponding to the values of last $L - B$ bits. Thus all the parity-check equations in Ω^* and Ω_i can be found.

Lemma 3.3. [6] *A tight approximation about the expected number of $|\Omega^*|$ or $|\Omega_i|$ is*

$$m = 2^{B-L} \binom{N}{w}.$$

As an illustration, note that for $N = 40000$, $L = 40$, $w = 2$, and $B = 18, 19, 20, 21, 22$, Lemma 3.3 yields that the expected cardinality m is equal to 190, 380, 761, 1522, 3045.

Lemma 3.3 implies that the expected cardinalities of the parity-check sets specified by Definitions 3.1 and 3.2 do not depend on the LFSR feedback polynomial, and particularly on its weight, since the expected cardinalities of $|\Omega_i|$ and $|\Omega^*|$ are the same. For convenience, we assume that all the sets $|\Omega_i|$ and $|\Omega^*|$ have the identical cardinalities, denoted by m in the sequel.

4. The Random Variables of the Number of Passed-Parity-Check Equations

Assume $(x_1, x_2, \ldots, x_B)$ is the first B information bits of a codeword x in the linear $[N, L]$-code. By (3.3) and (3.4), we have that

$$c'_{i_j} = \sum_{k=1}^{B} a_{j_k} x_k \oplus b_{i_j} \tag{4.1}$$

$$c'_j = \sum_{k=1}^{B} a_{j_k} x_k \oplus b_j \tag{4.2}$$

Let $(u_1, u_2, \ldots, u_N)$ be the target LFSR sequence. We have the following lemmas.

Lemma 4.1. *Let* $p = \frac{1}{2} - \delta = P(z_n \neq u_n)$, $p' = \frac{1}{2} - \varepsilon = P(c'_j \neq 0)$, $p'_i = \frac{1}{2} - \varepsilon_i = P(c'_{i_j} \neq 0)$. *If* $(u_1, u_2, \ldots, u_B) = (x_1, x_2, \ldots, x_B)$, *then* $\varepsilon = 2^{w-1}\delta^w$ *and* $|\varepsilon_i| = \varepsilon$.

Proof. The first part is proved in [5]. We now prove the second part. Since

$$p' = \frac{1}{2} - \varepsilon = P(c'_j \neq 0) = P(\sum_{k=1}^{w} z_{j_k} \neq \sum_{k=1}^{B} a_{j_k} u_k),$$

it implies that p' is the probability of the equations (4.2) with weight w being not a passed-parity-check equation. Similarly, since

$$p'_i = \frac{1}{2} - \varepsilon_i = P(c'_{i_j} \neq 0) = P(\sum_{k=1}^{w} z_{j_k} \neq z_i \oplus \sum_{k=1}^{B} a_{j_k} u_k),$$

and if $z_i = u_i$ then

$$p'_i = P(\sum_{k=1}^{w} z_{j_k} \neq u_i \oplus \sum_{k=1}^{B} a_{j_k} u_k),$$

namely, p'_i is the probability of the equations (4.1) with weight w being not a passed-parity-check equation. Therefore $p'_i = p'$ and $\varepsilon_i = \varepsilon$.

If $z_i \neq u_i$ then

$$p_i' = P\left(\sum_{k=1}^{w} z_{j_k} \neq u_i \oplus 1 \oplus \sum_{k=1}^{B} a_{j_k} u_k\right) = P\left(\sum_{k=1}^{w} z_{j_k} = u_i \oplus \sum_{k=1}^{B} a_{j_k} u_k\right)$$

namely, p_i' is the probability of the equations (4.1) with weight w being a passed-parity-check equation. Thus $p_i' = 1 - p'$ and $\varepsilon_i = -\varepsilon$. □

Lemma 4.2. *Let* $p = \frac{1}{2} - \delta = P(z_n \neq u_n)$, $p' = \frac{1}{2} - \varepsilon = P(c_j' \neq 0)$, $p_i' = \frac{1}{2} - \varepsilon_i = P(c_{i_j}' \neq 0)$. *Suppose that* $a_{j_1}, a_{j_2}, \ldots, a_{j_B}$ *are pairwise independent random variables with* $P(a_{j_k} = 0) = P(a_{j_k} = 1) = \frac{1}{2}$ *for an arbitrary integer* k $(1 \leq k \leq B)$. *If* $(u_1, u_2, \ldots, u_B) \neq (x_1, x_2, \ldots, x_B)$, *then* $\varepsilon_i = \varepsilon = 0$.

Proof. If $(u_1, u_2, \ldots, u_B) \neq (x_1, x_2, \ldots, x_B)$ then there exists at least one integer i such that $u_i \neq x_i$. Let t be an integer such that $u_t \neq x_t$. Without loss of generality, let $x_1 \neq u_1, \ldots, x_t \neq u_t$, and $x_j = u_j$ for $j = t+1, \ldots, B$. Thus

$$p' = \frac{1}{2} - \varepsilon = P(c_j' \neq 0) = P\left(\sum_{k=1}^{w} z_{j_k} = \sum_{k=1}^{B} a_{j_k} u_k) P(a_{j_1} + \cdots + a_{j_t} = 1\right)$$
$$+ P\left(\sum_{k=1}^{w} z_{j_k} \neq \sum_{k=1}^{B} a_{j_k} u_k) P(a_{j_1} + \cdots + a_{j_t} = 0\right)$$

Since random variables $a_{j_1}, a_{j_2}, \ldots, a_{j_B}$ are pairwise independent, and

$$P(a_{j_k} = 0) = P(a_{j_k} = 1) = \frac{1}{2}$$

for k $(1 \leq k \leq B)$, then

$$P(a_{j_1} + \cdots + a_{j_t} = 1) = P(a_{j_1} + \cdots + a_{j_t} = 0) = 1/2$$

and

$$p' = (1/2 + \varepsilon) \times 1/2 + (1/2 - \varepsilon) \times 1/2 = 1/2.$$

Therefore $\varepsilon = 0$. Similarly $\varepsilon_i = 0$. □

Remark 4.3. $a_{j_1}, a_{j_2}, \ldots, a_{j_B}$ are values in the first B coordinates of the sum of some w columns of G_{LFSR}. By experiment results, we can say $a_{j_1}, a_{j_2}, \ldots, a_{j_B}$ are pairwise independent, and $P(a_{j_k} = 0) = P(a_{j_k} = 1) = \frac{1}{2}$.

Let S and S_i be defined by the following equations.

$$S = \sum_{j=1}^{m} (c_j' \oplus 1), \tag{4.3}$$

$$S_i = \sum_{j=1}^{m} (c_{i_j}' \oplus z_i \oplus 1). \tag{4.4}$$

Thus S and S_i are the numbers of passed-parity-check equations in Ω^* and Ω_i respectively.

Theorem 4.4.

(i) *If* $(u_1, u_2, \ldots, u_B) = (x_1, x_2, \ldots, x_B)$, *then* S *has binomial distribution* $B(m, 1/2+\varepsilon)$, *and* S_i *has binomial distribution* $B(m, 1/2+\varepsilon)$ *or* $B(m, 1/2-\varepsilon)$.

(ii) *If* $(u_1, u_2, \ldots, u_B) \neq (x_1, x_2, \ldots, x_B)$, *then both* S *and* S_i *have binomial distribution* $B(m, 1/2)$.

Proof. Let a_{j_k} be the kth value of the vector $(g_{j_1} \oplus g_{j_2} \oplus \cdots \oplus g_{j_w})$ in the jth parity-check equation in sets Ω_i, where $i = B+1, \ldots, L$, $k = 1, 2, \ldots, B$. Since $j_1, j_2, \ldots, j_w$ are selected randomly from the set $\{1, 2, \ldots, N\}$, thus $(g_{j_1} \oplus g_{j_2} \oplus \cdots \oplus g_{j_w})$ can be regarded as a random variable on $GF(2^L)$. Therefore we have $P(a_{j_k} = 0) = P(a_{j_k} = 1) = \frac{1}{2}$ and $a_{j_1}, a_{j_2}, \ldots, a_{j_B}$ are pairwise independent.

Suppose that $(u_1, u_2, \ldots, u_B) = (x_1, x_2, \ldots, x_B)$. By Lemma 4.1, $P(c'_j = 0) = 1/2 + \varepsilon$, and

$$P(c'_{i_j} + z_i = 0) = P(c'_{i_j} = 0)P(z_i = 0) + P(c'_{i_j} = 1)P(z_i = 1).$$

For a given integer i, $B+1 \leq i \leq L$, z_i is a constant. If $z_i = 1$, then $P(c'_{i_j} + z_i = 0) = P(c'_{i_j} = 1) = 1/2 - \varepsilon_i$. If $z_i = 0$, then $P(c'_{i_j} + z_i = 0) = P(c'_{i_j} = 0) = 1/2 + \varepsilon_i$. Thus S has binomial distribution $B(m, 1/2+\varepsilon)$ and S_i has binomial distribution $B(m, 1/2+\varepsilon)$ or $B(m, 1/2-\varepsilon)$.

When $(u_1, u_2, \ldots, u_B) \neq (x_1, x_2, \ldots, x_B)$, by Lemma 4.2, we get $\varepsilon_i = \varepsilon = 0$. Therefore S and S_i have binomial distribution $B(m, 1/2)$. □

Lemma 4.5. (Demoivre-Lapalace central limit theorem [10]) *Suppose* p $(0 < p < 1)$ *is the probability of success on each trial in* n *Bernoulli trials,* ξ_n *is the number of successes, then* $(\xi_n \sim B(n, p))$,*and when* $n \to \infty$,

$$\frac{\xi_n - np}{\sqrt{npq}} \sim N(0, 1)$$

i.e.,

$$\lim_{n\to\infty} P\left(\frac{\xi_n - np}{\sqrt{npq}} < x\right) = \frac{1}{\sqrt{2\pi}} \int_{-\infty}^{x} e^{-\frac{t^2}{2}} dt.$$

Corollary 4.6. *Let* $\xi = \frac{S - \frac{m}{2}}{\frac{\sqrt{m}}{2}}, \xi_i = \frac{S_i - \frac{m}{2}}{\frac{\sqrt{m}}{2}}, \eta = \xi^2 + \sum_{i=B+1}^{L} \xi_i^2$. *Let* m *be sufficiently large. Then:*

(i) *If* $(u_1, u_2, \ldots, u_B) \neq (x_1, x_2, \ldots, x_B)$, *then* η *has chi-square distribution* $\chi^2(L - B + 1)$ *with* $L - B + 1$ *degrees of freedom. The expectation* $E(\eta) = L - B + 1$. *Here we denote* η *as* η_1.

(ii) *If* $(u_1, u_2, \ldots, u_B) = (x_1, x_2, \ldots, x_B)$, *then the expectation* $E(\eta) = (L - B + 1)(1 - 4\varepsilon^2 + 4m\varepsilon^2)$. *We denote this* η *as* η_2.

Proof. When $(u_1, u_2, \ldots, u_B) \neq (x_1, x_2, \ldots, x_B)$, by Theorem 4.4, both S and S_i have binomial distribution $B(m, 1/2)$. Because m is large enough, by Lemma 4.5, both ξ and ξ_i have distribution $N(0, 1)$. Since the parity check equations in Ω_i and Ω^* are constructed independently, we can regard ξ and ξ_i as independent random variables. By the definition of chi-square distribution, we have the conclusion (i).

When $(u_1, u_2, \ldots, u_B) = (x_1, x_2, \ldots, x_B)$, we know

$$E(\eta) = E(\xi^2) + \sum_{i=B+1}^{L} E(\xi_i^2).$$

Clearly, we have

$$\begin{aligned} E(\xi^2) &= E(\frac{S^2 - mS + \frac{m^2}{4}}{\frac{m}{4}}) \\ &= \frac{E(S^2) - mE(S) + \frac{m^2}{4}}{\frac{m}{4}} \\ &= \frac{D(S) + (E(S))^2 - mE(S) + \frac{m^2}{4}}{\frac{m}{4}}. \end{aligned}$$

By Theorem 4.4, S has distribution $B(m, 1/2 + \varepsilon)$, and thus

$$D(S) = m(\frac{1}{2} + \varepsilon)(\frac{1}{2} - \varepsilon)$$

and

$$E(S) = m(\frac{1}{2} + \varepsilon).$$

Therefore we have

$$E(\xi^2) = 1 - 4\varepsilon^2 + 4m\varepsilon^2.$$

Similarly, $E(\xi_i^2) = E(\xi^2) = 1 - 4\varepsilon^2 + 4m\varepsilon^2$. Thus

$$E(\eta) = (L - B + 1)(1 - 4\varepsilon^2 + 4m\varepsilon^2).$$

□

By Corollary 4.6, we know that when ε is a constant and the cardinality m of the parity check equation sets is large enough, random variable η_1 has distribution $\chi^2(L - B + 1)$ which is irrelevant to m. But the expectation of η_2 linearly increases with m. Therefore when m is large, the distinction between η_2 and η_1 is obvious. Thus the distinction between η_2 and η_1 can be used to determine whether the hypothesis $(u_1, u_2, \ldots, u_B) = (x_1, x_2, \ldots, x_B)$ is correct or not. Clearly the larger $E(\eta_2)$ is, the better the performance becomes. However if ε is quite small or m is comparatively small, this distinction is not credible.

In practice, we can find a critical value T_1 as a threshold such that $P(\eta_1 \geq T_1) = \partial$. It means that the probability of correctly judging $(x_1, x_2, \ldots, x_B) = (u_1, u_2, \ldots, u_B)$ is $1 - \partial$. We call $1 - \partial$ the *distinguishable probability* between η_1 and η_2. Hence if there are 2^B possibilities to be exhaustively searched for information bits $(x_1, x_2, \ldots, x_B)$, the number of the remaining possibilities which need to be further judged is $2^B \partial$.

To intuitively understand the distinguishability of η_1 and η_2 on statistic, we give the following experimental data in Table 1 with $N = 40000, L = 40, w = 2$, different noise ratio p, and different B bits for exhaustive search. In the table,

$E(\eta_2)$ p	$B = 18$ $m = 190$	$B = 19$ $m = 380$	$B = 20$ $m = 761$	$B = 21$ $m = 1522$	$B = 22$ $m = 3045$
0.30	134.28	235.45	429.58	798.75	1499.60
0.31	113.64	195.86	353.78	654.30	1224.96
0.32	96.01	162.05	289.07	530.94	990.42
0.33	81.09	133.42	234.28	426.51	291.98
0.34	68.58	109.43	188.35	338.98	625.45
0.35	58.21	89.54	150.01	266.40	487.47
0.36	49.71	73.25	119.10	206.98	374.59
0.37	42.86	60.10	93.93	159.01	283.30
0.38	37.42	49.66	73.95	120.92	210.87
0.39	33.18	41.53	58.39	91.26	154.48
0.40	29.95	35.34	46.53	68.67	111.53
T_1	44.18	42.80	41.40	40.00	38.58

Table 1: The distinguishability between η_2 and η_1.

E(η_2) stands for the expectation of η_2, and T_1 is the threshold satisfying $P(\eta_1 \geq T_1) = 0.005$.

Clearly, when $E(\eta_2) < T_1$, η_2 and η_1 are undistinguishable. We call $P(\eta_2 < T_1)$ the undistinguishable probability of η_2 and η_1. To compute $P(\eta_2 < T_1)$, we need the following lemma.

Lemma 4.7. ([9]) *Let u_r be the rth central moment of binomial distribution $B(n,p)$, i.e., $u_r = E(x-np)^r$, where $r \geq 2$. Then we have the following recursion formula*

$$u_r = npq\sum_{i=0}^{r-2} C_{r-1}^i u_i - p\sum_{i=0}^{r-2} C_{r-1}^i u_{i+1}.$$

Lemma 4.8. (Lindeberg-Levy theorem[10]) *If $X_1, X_2, \ldots, X_n$ is a sequence of independent random variables and $E(X_k) = a, D(X_k) = \sigma^2$ $(\sigma^2 > 0), k = 1,2,\ldots,n$, then*

$$\frac{\sum_{k=1}^n (X_k - na)}{\sigma^2} \sim N(0,1),$$

i.e.,

$$\lim_{n\to\infty} p(\frac{\sum_{k=1}^n (X_k - na)}{\sigma\sqrt{n}}) = \frac{1}{\sqrt{2\pi}}\int_{-\infty}^{x} e^{-\frac{t^2}{2}}dt.$$

Theorem 4.9 *Let m be sufficiently large, S and S_i have distribution $B(m, 1/2+\varepsilon)$ or $B(m,1/2-\varepsilon)$, $X_0 = (\frac{S-\frac{m}{2}}{\frac{\sqrt{m}}{2}})^2$, and $X_i = (\frac{S_i-\frac{m}{2}}{\frac{\sqrt{m}}{2}})^2$. Let $\eta_2 = X_0 + \sum_{B+1}^{L} X_i$, $E(X_i) = a$, $D(X_i) = \sigma^2$, and $L - B + 1$ be sufficiently large. Then*

$$a = 1 - 4\varepsilon^2 + 4m\varepsilon^2, \tag{4.5}$$

$$\begin{aligned}\sigma^2 = {} & \tfrac{1}{256m}(-512 - 512m - 12288m\varepsilon^2 + m^3 + 8192\varepsilon^2 \\ & +40960m\varepsilon^4 - 24576\varepsilon^4 + 4096m^2\varepsilon^2 - 16384m^2\varepsilon^4 \\ & -16m^3\varepsilon^2 - 4000m^3\varepsilon^4 - 256m^3\varepsilon^6 + 256m^3\varepsilon^8)\end{aligned} \tag{4.6}$$

and

$$P(\eta_2 < T_1) = \Phi(\frac{T_1 - (L-B+1)a}{\sigma\sqrt{L-B+1}}), \tag{4.7}$$

where $\Phi(\alpha)$ *is the distribution function of* $N(0,1)$.

Proof. By the proof of Corollary 4.6, we know that X_i have the same expectations. Now we consider the variances of X_i. We have

$$D(X_i) = E(X_i^2) - (E(X_i))^2 = \frac{16}{m^2}u_4 + \frac{32}{m}(2p-1)u_3 + 24(2p-1)^2u_2 + 8m(2p-1)^3u_1 + m^2(2p-1)^4 - (E(X_i))^2$$

where u_r is the rth central moment of S_i. By Corollary 4.6, we have

$$a = E(X_i) = 1 - 4\varepsilon^2 + 4m\varepsilon^2.$$

Note that $q = 1-p$, $u_0 = 1$, $u_1 = 0$, $u_2 = mpq$, $pq = (\frac{1}{4} - \varepsilon^2)$, $(2p-1)^2 = 4\varepsilon^2$, and by Lemma 4.7,

$$u_3 = mpq(2p-1).$$

Thus

$$\frac{16}{m^2}u_4 + \frac{32}{m}(2p-1)u_3 = \frac{16}{m^2}(mpq + 3mpqu_2) - \frac{48p}{m^2}(u_2+u_3) + \frac{32}{m}(2p-1)u_3,$$

and

$$-\frac{48p}{m^2}(u_2+u_3) + \frac{32}{m}(2p-1)u_3 = -(\frac{96}{m}(pq)^2 + 32pq(2p-1)^2).$$

Hence

$$\begin{aligned} D(X_i) &= \tfrac{16}{m^2}(mpq + 3mpqu_2) - \tfrac{48p}{m^2}(u_2+u_3) + \tfrac{32}{m}(2p-1)u_3 + 24(2p-1)^2u_2 \\ &\quad +8m(2p-1)^3u_1 + m^2(2p-1)^4 - (E(X_i))^2 \\ &= \tfrac{16}{m^2}(m(\tfrac{1}{4}-\varepsilon^2) + 3m^2(\tfrac{1}{4}-\varepsilon^2)^2) - (\tfrac{96}{m}(\tfrac{1}{4}-\varepsilon^2)^2 + 32(\tfrac{1}{4}-\varepsilon^2)4\varepsilon^2) \\ &\quad +96\varepsilon^2 m(\tfrac{1}{4}-\varepsilon^2) + m^2(\tfrac{1}{4}-\varepsilon^2)^4 - (1-4\varepsilon^2+4m\varepsilon^2)^2 \\ &= \tfrac{1}{256m}(-512 + 512m - 12288m\varepsilon^2 + 8192\varepsilon^2 + 40960m\varepsilon^4 \\ &\quad -24576\varepsilon^4 + m^3 + 4096m^2\varepsilon^2 - 16384m^2\varepsilon^4 - 16m^3\varepsilon^2 \\ &\quad -4000m^3\varepsilon^4 - 256m^3\varepsilon^6 + 256m^3\varepsilon^8)\ . \end{aligned}$$

By Lemma 4.7, we know

$$\frac{\eta_2 - (L-B+1)a}{\sigma\sqrt{L-B+1}} \sim N(0,1),$$

i.e.,

$$\begin{aligned} P(\eta_2 < T_1) &= P\left(\frac{\eta_2 - (L-B+1)a}{\sigma\sqrt{L-B+1}} < \frac{T_1 - (L-B+1)a}{\sigma\sqrt{L-B+1}}\right) \\ &= \Phi\left(\frac{T_1 - (L-B+1)a}{\sigma\sqrt{L-B+1}}\right). \end{aligned}$$ □

By Theorem 4.9, when $L-B+1$ is sufficiently large, the threshold value T_1 can be used to distinguish η_2 and η_1. The error decoding rate $p_e = P(\eta_2 < T_1)$

can be computed according to formula (4.7). Generally, when $L - B + 1 > 20$, the precision of approximation is very satisfying. There are some experimental results on the approximation in Section 6.

Theorem 4.10. *Let $p = 1/2 - \delta = P(z_n \neq u_n), w$ be the weight of parity-check equation, and N, L, B as defined before. Then the performance of our correlation attack by using $(w+1)$-weight parity check equations is better than the one by only using w-weight equations if and only if*

$$2\delta\sqrt{\frac{N}{w+1}} > 1. \tag{4.8}$$

Proof. Let N_w be the expected value of η_2 when the weight of parity check equation is w. By Corollary 4.6, $N_w = (L - B + 1)(1 - 4\varepsilon^2 + 4 \times 2^{(B-L)} \binom{N}{w} \varepsilon^2)$, and $\varepsilon = 2^{w-1}\delta^w$. We consider the difference between N_w and $N_w + 1$,

$$N_{w+1} - N_w = 2^{B-L+2}(2^{w-1}\delta^w) \binom{N}{w} ((2\delta)^2 \frac{N}{w+1} - 1).$$

Thus $2\delta\sqrt{\frac{N}{w+1}} > 1$ if and only if $N_{w+1} - N_w > 0$. Because η_1 has the chi-square distribution with $L - B + 1$ degrees of freedom, it is irrelevant to w. Therefore, the bigger the η_2 of expectation is, the better the distinguishable property between η_2 and η_1 performances. Thus we conclude that the performance of our correlation attack by using $(w+1)$-weight parity check equations is better than the one by only using w-weight equations if and only if $2\delta\sqrt{\frac{N}{w+1}} > 1$. □

5. Our New Algorithms

The main underlying principles for construction of the novel fast correlation attack include the following:

1. A partial exhaustive search for the first B information bits enhances the performance of the fast correlation attack.
2. Statistical threshold ensures a precision decision for efficiently finding the correct first B information bits.
3. Repeatedly using one-step decoding technology makes our new approach fast with low computational complexity.

According to these principles a new algorithm for the fast correlation attack is proposed. The algorithm is based on the parity-check sets in Section 3. The threshold T_1 is used to determine if a hypothesis of the first B information bits is right. T_1 can be calculated according to the method in Section 4 for a given distinguishable probability $1 - \partial$ between η_1 and η_2. The threshold T for correlation checks can be calculated by a method in [8]. The thresholds T_2^i for the ith information bit can be calculated according to a method in [1].

Algorithm A:
INPUT:
The parameters N, L, B, the thresholds T, T_1 and T_2^i;
The received noisy bits $z_1, z_2, \ldots, z_N$;
The parity-check equations sets Ω_i and Ω^* for $i = B+1, \ldots, L$.

PROCESSING STEPS:
Step 1: Setting the hypothesis
From the set of all possible 2^B binary patterns, select a not previously considered pattern $(x_1, x_2, \ldots, x_B)$, for the first B information bits. If no new pattern is available, go to step 3.

Step 2: Decoding
(a) Calculate S_i and S, the number of passed-parity-check equations in Ω_i and Ω^*. Then calculate η. If $\eta < T_1$, go to step 1. (S_i, S, and η are specified in Section 4)
(b) For every i, if $S_i < T_2^i$, then $x_i = z_i \oplus 1$, else $x_i = z_i$.
(c) Check if the current estimation of the information bits $(x_1, x_2, \ldots, x_L)$ is a true one according to the following:
For $(x_1, x_2, \ldots, x_L)$, generate the corresponding sequence $x_1, x_2, \ldots, x_N$, and calculate $S^* = \sum_{n=1}^{N}(x_n \oplus z_n)$. If $S^* < T$, go to OUTPUT (a), otherwise store $(x_1, x_2, \ldots, x_B)$ into the set Λ.

Step 3: Twice-step decoding
(a) For every $(x_1, x_2, \ldots, x_B)$ stored in the set Λ, transform the linear $[N, L]$-code into linear $[N, L-B]$-code.
(b) Decode the linear $[N, L-B]$-code. If the decoding succeeds, goto OUTPUT(a), otherwise goto OUTPUT (b).

OUTPUT:
(a) Output the result $[x_1, x_2, \ldots, x_L]$ as $[u_1, u_2, \ldots, u_L]$;
(b) The correlation attacks fail, the correct information bits are not found.

Remark 5.1. With knowledge of the first B information symbols, the problem of restoring the remaining $L-B$ bits is much more simple compared to the original problem. Hence we can discard the computational complexity of the *twice-step decoding processing.*

Similarly, we present another new algorithm B by a simple ML-decoding procedure.

Let $F_0 = S$. If $S_i \geq m/2$, then $F_i = S_i$, else $F_i = m - S_i$.

Algorithm B:
INPUT: N, L, B; the received noise sequence $z_1, z_2, \ldots, z_N$; the parity-check sets Ω_i and Ω^*.
DECODING: Exhaustively search 2^B possibilities to find a vector $(x_1, x_2, \ldots, x_B)$ such that the sum $F_0 + \sum_{i=B+1}^{L} F_i$ is maximal.

6. Performance Evaluation

6.1. Complexity

The computational complexity can be divided into two parts, the time for preprocessing and the decoding time. In part of preprocessing, the calculation of all parity-check equations is of order $O(N^{w-1}\log N)$. We also need to store each parity check equation, which is composed of its index positions and a B-bits vector, in Ω^* and Ω_i. Thus the storage requirement is at most $(L-B+1)m(B+wlog_2N)$. If we store $\sum_{k=1}^{w} z_{i_k}$ instead of their index positions, then the storage requirement is at most $(L-B+1)m(B+1)$.

The complexity of the decoding step is given as follows:

Corollary 6.1. *Let W be the weight of the LFSR characteristic polynomial, $1-\partial$ the distinguishable probability. Then the complexity of our algorithm* A *is proportional to $2^B[(L-B+1)m+(N-L)W\partial]$ mod 2 additions.*

Corollary 6.2. *The complexity of the proposed algorithm* B *is proportional to $2^B(L-B+1)m$ mod 2 additions.*

6.2. Simulation

We have made plentiful experiments to evaluate the performance of our new algorithms. The LFSR characteristic polynomial we have chosen is $1+x+x^3+x^5+x^9+x^{11}+x^{12}+x^{17}+x^{19}+x^{21}+x^{25}+x^{27}+x^{29}+x^{32}+x^{33}+x^{38}+x^{40}$. $N=40000, w=2$, the distinguishable probability between η_1 and η_2 is $1-\partial=$ 0.995.

Firstly, we compare the restored proportion $P(\eta_1>T_1)$ with given ∂. Table 2 shows that the actual values of $P(\eta_1>T_1)$ are close to the expected values ∂.

B		∂	m	T_1	$p(\eta_1>T_1)$
19	0.36	0.005	380	42.8	0.00495
20	0.34	0.005	761	41.4	0.00497
21	0.38	0.005	1522	40	0.00498

Table 2. Comparison of the simulation result with the theoretic value

Table 3 compares the error decoding probability of algorithm A with our theoretical estimation from formula (4.7) in Theorem 4.9. Notice that although $L-B+1$ is around 20, the actual values are close to our theoretical estimations, the difference is about 0.05.

Table 4 compares the performance of our algorithm A with the algorithm of [3]. The parameters N, w, ∂ and the characteristic polynomial are the same as above. We made 1000 times random experiments under the condition that $B=$ 18, 19, 20, 21, 22 and p is between 0.30 and 0.40.

In Table 4, NewA means our new algorithm A, [3] means the algorithm in [3]. The data in Table 4 shows that the performance of our algorithm is significantly better than that of the algorithm of [3].

p	The error probability of decoding p_e					
	$B = 18, m = 19$		$B = 19, m = 380$		$B = 20, m = 761$	
	theoretic	Simulation	theoretic	Simulation	theoretic	Simulation
0.30	0.000	0.000	0.000	0.000	0.000	0.000
0.33	0.013	0.005	0.000	0.000	0.000	0.000
0.36	0.327	0.252	0.026	0.007	0.000	0.000
0.39	0.883	0.832	0.548	0.490	0.110	0.076

Table 3. Comparison of the actual decoding error probability with our theoretic estimation

p	The error probability of decoding									
	B=18 m=190		B=19 m=380		B=20 m=761		B=21 m=1522		B=22 m=3045	
	NewA	[3]	NewA	[3]	NewA	[3]	NewA	[3]	NewA	[3]
0.30	0.000	0.254	0.000	0.023	0.000	0.000	0.000	0.000	0.000	0.000
0.31	0.000	0.384	0.000	0.041	0.000	0.002	0.000	0.000	0.000	0.000
0.32	0.000	0.569	0.000	0.098	0.000	0.002	0.000	0.000	0.000	0.000
0.33	0.005	0.696	0.000	0.226	0.000	0.020	0.000	0.000	0.000	0.000
0.34	0.020	0.838	0.000	0.356	0.000	0.053	0.000	0.001	0.000	0.000
0.35	0.086	0.915	0.001	0.542	0.000	0.114	0.000	0.002	0.000	0.000
0.36	0.252	0.955	0.007	0.743	0.000	0.225	0.000	0.019	0.000	0.000
0.37	0.471	0.983	0.075	0.865	0.000	0.450	0.000	0.080	0.000	0.001
0.38	0.695	0.990	0.243	0.932	0.007	0.652	0.000	0.210	0.000	0.023
0.39	0.832	0.997	0.490	0.980	0.076	0.850	0.000	0.445	0.000	0.052
0.40	0.921	1.000	0.729	0.988	0.292	0.935	0.005	0.663	0.000	0.267

Table 4. Comparison of the new algorithm A with the algorithm A of [3]

Table 5 compares the performance of the our new algorithms A and B with the algorithm presented in [1].

p	The error probability of decoding		
	N=45000,m=1941 Algorithm in [1]	N=18000,m=308 Algorithm B	N=18000,m=308, $\partial = 0.005$ Algorithm A
0.33	0.39	0.03	0.00
0.34	0.59	0.18	0.00
0.35	0.75	0.42	0.01
0.36	0.89	0.67	0.09
0.37	0.98	0.82	0.27
0.38	1.00	0.99	0.52

Table 5. The performance comparison between our new algorithm and one in [1]

The performance of our new algorithm is superior to the algorithm presented in [3], because the success of the algorithm presented in [3] depends on the successful decoding of every information bit. So long as there is one information bit decoding error, the whole attack will fail. The authors of [4] have partially overcome this shortcoming. But they needed $D - B$ sets of parity-check equations, where $D > L$. When D increases, the complexities of precomputation and decoding will also linearly increase. Moreover, the decoding processing in [4] needs 2^{B+1} initial states to be correlationally checked. This becomes the main part of computational complexity of decoding. Recently, this approach was improved algorithmically by Chose et al. [2].

However, the successful attack of our new algorithm does not rely on the successful decoding of special information bits. It is decided by all the $L - B + 1$ parity-check sets holistically. The initial states to be correlationally checked in our algorithm are less than $2^B \partial$ where $\partial \leq 0.005$. When the cardinality m of the parity-check sets is large, our algorithm can precisely distinguish η_2 and η_1.

The decoding complexity of the algorithm of [3] is $2^B[(L-B+1)mw+(N-L)w]$. The decoding complexity of our algorithm is $2^B[(L-B+1)m+(N-L)w\partial]$. Since ∂ is very small, for example $\partial = 0.005$, and $w \geq 2$, our algorithm improves the decoding complexity.

Compared with the algorithm in [1], our new algorithm uses $L - B$ times more parity-check equations. This is an important reason why the performance of our new algorithm is superior to the algorithm in [1]. But the complexity of the Algorithm in [1] is $2^B m$, which is lower than the new one.

7. Conclusions

We present a new powerful algorithm for fast correlation attacks. It involves more parity-check sets than other algorithms reported. The performance of the new algorithm is significantly improved with relatively low computational complexity. In particular, we use some random variables and their distributions to determine a statistical threshold which guarantees a precise decision for efficiently finding the correct first B information bits. We also find a new formula to describe the relationship between the tendency of attack performance, the weight w of parity check equations, the noise level δ, and the required length N of the observed sequence. We believe that, with a set of parallel PCs and a few weeks of precomputation, our algorithm can carry out the correlation attack on the LFSRs of length 80–100 and $p = 0.4$ in one PC in several hours by using $B = 35$ and $t \geq 3$.

Acknowledgments

The first author was supported by National Natural Science Foundation of China (10171017, 90204013), Special Funds of Authors of Excellent Doctoral Dissertation in China, and Science and Technology Funds of Shanghai (035115019).

References

[1] V. Chepyzhov, T. Johansson, and B. Smeets, A simple algorithm for fast correlation attacks on stream ciphers, Lecture Notes in Computer Science, vol. 1978, 2001, pp. 181–195.

[2] P. Chose, A. Joux, and M. Mitton, Fast correlation attack: an algorithmic point of view, Lecture Notes in Computer Science, vol. 2332, pp. 209–221, April 2002.

[3] M. Mihaljević, M. Fossorier, and H. Imai, A low-complexity and high-performance algorithm for the fast correlation attack, Lecture Notes in Computer Science, vol. 1978, 2001, pp. 196–212.

[4] M. Mihaljević, M. Fossorier, and H. Imai, Fast correlation attack algorithm with list decoding and an application, Lecture Notes in Computer Science, vol. 2355, 2002, pp. 196–210.

[5] T. Johansson and F. Jonsson, Theoretical analysis of a correlation attack based on convolution codes, IEEE Transactions on Information Theory, vol. 48, no. 8, August 2002, pp. 2173–2181.

[6] T. Johansson and F. Jonsson, Improved fast correlation attacks on stream ciphers via convolutional codes, Lecture Notes in Computer Science, vol. 1592, 1999, pp. 347–362.

[7] W. Meier and O. Staffelbach, Fast correlation attacks on certain stream ciphers, J. Cryptology, vol. 1, 1989, pp. 159–176.

[8] T. Siegenthaler, Decrypting a class of stream ciphers using ciphertext only, IEEE Trans. Comput., vol. C-34, 1985, pp. 81–85.

[9] K. Fang, J. Xu, Statistical Distribution, Science in China Press, Beijing, 1987, in Chinese, pp. 67.

[10] Z. Wei, etc., Introduction to Probability Theory and Statistics, Higher Education Press, Beijing, 1983, in Chinese, pp. 209.

Peizhong Lu and Lianzhen Huang
Department of Computer Sciences and Engineering
and Institute of Mathematics
Fudan University
Fengzhen Road 85, Jiangwan Town
Shanghai, Postcode 200434
China
e-mail: pzlu@fudan.edu.cn

Progress in Computer Science and Applied Logic, Vol. 23, 85–110

LDPC Codes: An Introduction

Amin Shokrollahi

Abstract. LDPC codes are one of the hottest topics in coding theory today. Originally invented in the early 1960's, they have experienced an amazing comeback in the last few years. Unlike many other classes of codes, LDPC codes are already equipped with very fast (probabilistic) encoding and decoding algorithms. The question is that of the design of the codes such that these algorithms can recover the original codeword in the face of large amounts of noise. New analytic and combinatorial tools make it possible to solve the design problem. This makes LDPC codes not only attractive from a theoretical point of view, but also perfect for practical applications. In this note I will give a brief overview of the origins of LDPC codes and the methods used for their analysis and design.

Keywords. LDPC codes, graph based codes.

1. Introduction

This note constitutes an attempt to highlight some of the main aspects of the theory of low-density parity-check (LDPC) codes. It is intended for a mathematically mature audience with some background in coding theory, but without much knowledge about LDPC codes.

The idea of writing a note like this came up during conversations that I had with Dr. Khosrovshahi, head of the Mathematics Section of the Institute for Studies in Theoretical Physics and Mathematics in December 2002. The main motivation behind writing this note was to have a written document for Master's and PhD students. The style is often informal, though I have tried not to compromise exactness.

The note is by no means a complete survey. I have deliberately left out a number of interesting aspects of the theory, such as connections to statistical

Most of the work on this paper was done when the author was visiting the Institute for Theoretical Physics and Mathematics (IPM) in Tehran.

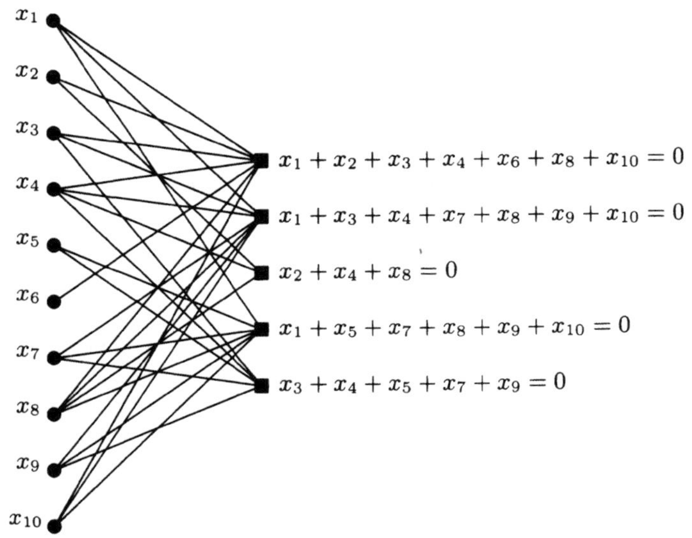

FIGURE 1. An LDPC code

mechanics. The important topics of general Tanner graphs, and factor graphs as well as connections to Turbo codes have also been left untouched.

My emphasis in writing the notes has been on algorithmic and theoretical aspects of LDPC codes, and within these areas on statements that can be proved. I have not discussed any of the existing and very clever methods for the construction of LDPC codes, or issues regarding their implementation.

Nevertheless, I hope that this document proves useful to at least some students or researchers interested in pursuing research in LDPC codes, or more generally codes obtained from graphs.

2. LDPC Codes

LDPC codes were invented by Robert Gallager [13] in his PhD thesis. Soon after their invention, they were largely forgotten, and reinvented several times for the next 30 years. Their comeback is one of the most intriguing aspects of their history, since two different communities reinvented codes similar to Gallager's LDPC codes at roughly the same time, but for entirely different reasons (see [7, 19, 18, 21, 20, 35, 36]).

LDPC codes are linear codes obtained from sparse bipartite graphs. Suppose that $\mathcal{G}$ is a graph with n left nodes (called message nodes) and r right nodes (called check nodes). The graph gives rise to a linear code of block length n and dimension at least $n - r$ in the following way: The n coordinates of the codewords are associated with the n message nodes. The codewords are those vectors $(c_1, \ldots, c_n)$ such that for all check nodes the sum of the neighboring positions among the message nodes is zero. Figure 1 gives an example.

The graph representation is analogous to a matrix representation by looking at the adjacency matrix of the graph: let H be a binary $r \times n$-matrix in which the entry (i, j) is 1 if and only if the ith check node is connected to the jth message node in the graph. Then the LDPC code defined by the graph is the set of vectors $c = (c_1, \ldots, c_n)$ such that $H \cdot c^\top = 0$. The matrix H is called a *parity check matrix* for the code. Conversely, any binary $r \times n$-matrix gives rise to a bipartite graph between n message and r check nodes, and the code defined as the null space of H is precisely the code associated to this graph. Therefore, any linear code has a representation as a code associated to a bipartite graph (note that this graph is not uniquely defined by the code). However, not every binary linear code has a representation by a *sparse* bipartite graph.[1] If it does, then the code is called a low-density parity-check (LDPC) code.

The sparsity of the graph structure is key property that allows for the algorithmic efficiency of LDPC codes. The rest of this note is devoted to elaborating on this relationship.

3. Decoding Algorithms: Belief Propagation

Let me first start by describing a general class of decoding algorithms for LDPC codes. These algorithms are called *message passing algorithms*, and are iterative algorithms. The reason for their name is that at each round of the algorithms messages are passed from message nodes to check nodes, and from check nodes back to message nodes. The messages from message nodes to check nodes are computed based on the observed value of the message node and some of the messages passed from the neighboring check nodes to that message node. An important aspect is that the message that is sent from a message node v to a check node c must not take into account the message sent in the previous round from c to v. The same is true for messages passed from check nodes to message nodes.

One important subclass of message passing algorithms is the *belief propagation* algorithm. This algorithm is present in Gallager's work [13], and it is also used in the Artificial Intelligence community [28]. The messages passed along the edges in this algorithm are probabilities, or beliefs. More precisely, the message passed from a message node v to a check node c is the probability that v has a certain value given the observed value of that message node, and all the values communicated to v in the prior round from check nodes incident to v other than c. On the other hand, the message passed from c to v is the probability that v has a certain value given all the messages passed to c in the previous round from message nodes other than v.

---

[1]To be more precise, sparsity only applies to *sequences* of matrices. A sequence of $m \times n$-matrices is called c-sparse if mn tends to infinity and the number of nonzero elements in these matrices is always less than $c \max(m, n)$.

It is easy to derive formulas for these probabilities under a certain assumption called *independence assumption*, which I will discuss later. It is sometimes advantageous to work with likelihoods, or sometimes even log-likelihoods instead of probabilities. For a binary random variable x let $L(x) = \Pr[x = 0]/\Pr[x = 1]$ be the *likelihood* of x. Given another random variable y, the *conditional likelihood* of x denoted $L(x \mid y)$ is defined as $\Pr[x = 0 \mid y]/\Pr[x = 1 \mid y]$. Similarly, the *log-likelihood* of x is $\ln L(x)$, and the *conditional log-likelihood* of x given y is $\ln L(x \mid y)$.

If x is an equiprobable random variable, then $L(x \mid y) = L(y \mid x)$ by Bayes' rule. Therefore, if $y_1, \ldots, y_d$ are independent random variables, then we have

$$\ln L(x \mid y_1, \ldots, y_d) = \sum_{i=1}^{d} \ln L(x \mid y_i). \tag{3.1}$$

Now suppose that $x_1, \ldots, x_\ell$ are binary random variables and $y_1, \ldots, y_\ell$ are random variables. Denote addition over $\mathbb{F}_2$ by $\oplus$. We would like to calculate $\ln L(x_1 \oplus \cdots \oplus x_\ell \mid y_1, \ldots, y_\ell)$. Note that if $p = 2\Pr[x_1 = 0 \mid y_1] - 1$ and $q = 2\Pr[x_2 = 0 \mid y_2] - 1$, then $2\Pr[x_1 \oplus x_2 = 0 \mid y_1, y_2] - 1 = pq$. (Why?) Therefore, $2\Pr[x_1 \oplus \cdots \oplus x_\ell = 0 \mid y_1, \ldots, y_\ell] - 1 = \prod_{i=1}^{\ell}(2\Pr[x_i = 0 \mid y_i] - 1)$. Since $\Pr[x_i = 0 \mid y_i] = L(x_i \mid y_i)/(1+L(x_i \mid y_i))$, we have that $2\Pr[x_i = 0 \mid y_i] - 1 = (L-1)/(L+1) = \tanh(\ell/2)$, where $L = L(x_i \mid y_i)$ and $\ell = \ln L$. Therefore, we obtain

$$\ln L(x_1 \oplus \cdots \oplus x_\ell \mid y_1, \ldots, y_\ell) = \ln \frac{1 + \left(\prod_{i=1}^{\ell} \tanh(\ell_i/2)\right)}{1 - \left(\prod_{i=1}^{\ell} \tanh(\ell_i/2)\right)}, \tag{3.2}$$

where $\ell_i = \ln L(x_i \mid y_i)$. The belief propagation algorithm for LDPC codes can be derived from these two observations. In round 0, the check nodes send along all the outgoing edges their log-likelihoods conditioned on their observed value. For example, if the channel used is the BSC with error probability p, then the first message sent to all the check nodes adjacent to a message node is $\ln(1-p) - \ln p$ if the node's value is zero, and it is the negative of this value if the node's value is one. In all the subsequent rounds of the algorithm a check node c sends to an adjacent message node v a likelihood according to (3.2). A message node v sends to the check node c its log-likelihood conditioned on its observed value and on the incoming log-likelihoods from adjacent check nodes other than c using the relation (3.1).

Let $\mathtt{m}_{\mathsf{vc}}^{(\ell)}$ be the message passed from message node v to check node c at the ℓth round of the algorithm. Similarly, define $\mathtt{m}_{\mathsf{cv}}^{(\ell)}$. At round 0, $\mathtt{m}_{\mathsf{vc}}^{(0)}$ is the log-likelihood of the message node v conditioned on its observed value, which is independent of c. We denote this value by $\mathtt{m}_{\mathsf{v}}$. Then the update equations for the messages under belief-propagation can be described as

$$\mathtt{m}_{\mathsf{vc}}^{(\ell)} = \begin{cases} \mathtt{m}_{\mathsf{v}}, & \text{if } \ell = 0, \\ \mathtt{m}_{\mathsf{v}} + \sum_{\mathsf{c}' \in C_{\mathsf{v}} \setminus \{\mathsf{c}\}} \mathtt{m}_{\mathsf{c'v}}^{(\ell-1)}, & \text{if } \ell \geq 1, \end{cases} \tag{3.3}$$

$$\mathtt{m}_{\mathtt{cv}}^{(\ell)} = \ln \frac{1 + \prod_{\mathtt{v}' \in V_\mathtt{c} \setminus \{\mathtt{v}\}} \tanh\left(\mathtt{m}_{\mathtt{v}'\mathtt{c}}^{(\ell)}/2\right)}{1 - \prod_{\mathtt{v}' \in V_\mathtt{c} \setminus \{\mathtt{v}\}} \tanh\left(\mathtt{m}_{\mathtt{v}'\mathtt{c}}^{(\ell)}/2\right)}, \tag{3.4}$$

where $C_\mathtt{v}$ is the set of check nodes incident to message node $\mathtt{v}$, and $V_\mathtt{c}$ is the set of message nodes incident to check node $\mathtt{c}$.

The computations at the check nodes can be simplified further by performing them in the log-domain. Since the value of $\tanh(x)$ can be negative, we need to keep track of its sign separately. Let γ be a map from the real numbers $[-\infty, \infty]$ to $\mathbb{F}_2 \times [0, \infty]$ defined by $\gamma(x) := (\mathrm{sgn}(x), -\ln\tanh(|x|/2))$ (we set $\mathrm{sgn}(x) = 1$ if $x \geq 1$ and $\mathrm{sgn}(x) = 0$ otherwise.) It is clear that γ is bijective, so there exists an inverse function γ^{-1}. Moreover, $\gamma(xy) = \gamma(x) + \gamma(y)$, where addition is component-wise in $\mathbb{F}_2$ and in $[0, \infty]$. Then it is very easy to show that (3.4) is equivalent to

$$\mathtt{m}_{\mathtt{cv}}^{(\ell)} = \gamma^{-1}\left(\sum_{\mathtt{v}' \in V_\mathtt{c} \setminus \{\mathtt{v}\}} \gamma\left(\mathtt{m}_{\mathtt{v}'\mathtt{c}}^{(\ell-1)}\right)\right) \tag{3.5}$$

We will use this representation when discussing density evolution later.

In practice, belief propagation may be executed for a maximum number of rounds or until the passed likelihoods are close to certainty, whichever is first. A *certain* likelihood is a likelihood in which $\ln L(x \mid y)$ is either ∞ or $-\infty$. If it is ∞, then $\Pr[x = 0 \mid y] = 1$, and if it is $-\infty$, then $\Pr[x = 1 \mid y] = 1$.

One very important aspect of belief propagation is its running time. Since the algorithm traverses the edges in the graph, and the graph is sparse, the number of edges traversed is small. Moreover, if the algorithm runs for a constant number of times, then each edge is traversed a constant number of times, and the algorithm uses a number of operations that is linear in the number of message nodes!

Another important note about belief propagation is that the algorithm itself is entirely independent of the channel used, though the messages passed during the algorithm are completely dependent on the channel.

One question that might rise is about the relationship of belief propagation and maximum likelihood decoding. The answer is that belief propagation is in general less powerful than maximum likelihood decoding. In fact, it is easy to construct classes of LDPC codes for which maximum likelihood decoding can decode many more errors than belief propagation (one example is given by biregular bipartite graphs in which the common degree of the message nodes is very large but the reader is not required to see this right away).

4. Asymptotic Analysis of Belief Propagation and Density Evolution

The messages passed at each round of the belief propagation algorithm are random variables. If at every round in the algorithm the incoming messages are statistically independent, then the update equation correctly calculates the corresponding log-likelihood based on the observations. (This is what I meant by the independence assumption above.) This assumption is rather questionable, though, especially when the number of iterations is large. In fact, the independence assumption is correct for the ℓ first rounds of the algorithm only if the neighborhood of a message node up to depth ℓ is a tree.

Nevertheless, belief propagation can be analyzed using a combination of tools from combinatorics and probability theory. The first analysis for a special type of belief propagation appeared in [16], and was applied to hard decision decoding of LDPC codes in [18]. The analysis was vastly generalized in [31] to belief propagation over a large class of channels.

The analysis starts by proving that if ℓ is fixed and n and r are large enough, then for random bipartite graphs the neighborhood of depth ℓ of most of the message nodes is a tree. Therefore, for ℓ rounds the belief propagation algorithm on these nodes correctly computes the likelihood of the node. Let us call these nodes the *good* nodes. We will worry about the other nodes later.

Next the expected behavior of belief propagation is calculated by analyzing the algorithm on the tree, and a martingale is used to show that the actual behavior of the algorithm is sharply concentrated around its expectation. This step of the analysis is rather standard, at least in Theoretical Computer Science.

Altogether, the martingale arguments and the tree assumption (which holds for large graphs and a fixed iteration number ℓ) prove that a heuristic analysis of belief propagation on trees correctly mirrors the actual behavior on the full graph for a fixed number of iterations. The probability of error among the good message nodes in the graph can be calculated according to the behavior of belief propagation. For appropriate degree distributions this shows that the error probability of the good message nodes in the graph can be made arbitrarily small. What about the other (non-good) message nodes? Since their fraction is smaller than a constant, they will contribute only a sub-constant term to the error probability and their effect will disappear asymptotically, which means that they are not relevant for an asymptotic analysis. Details can be found in the above mentioned literature.

The analysis of the expected behavior of belief propagation on trees leads to a recursion for the density function of the messages passed along the edges. The general machinery shows that, asymptotically, the actual density of the messages passed is very close to the expected density. Tracking the expected density during the iterations thus gives a very good picture of the actual behavior of the algorithm. This method, called *density evolution* [31, 29, 18], is one of the crown jewels of the asymptotic theory of LDPC codes. In the following, I will briefly discuss this.

As a first remark note that if $X_1, \ldots, X_d$ are i.i.d. random variables over some (additive) group G, and if f is the common density of the X_i, then the density F of $X_1 + \cdots + X_d$ equals the d-fold convolutional power of f. (For any two integrable functions f and g defined over G the convolution of f and g, denoted $f \otimes g$, is defined as $(f \otimes g)(\tau) = \int_G f(\sigma)g(\tau - \sigma)\,\mathrm{d}G$, where $\mathrm{d}G$ is the Haar measure on G.) If G is the group of real numbers with respect to multiplication, then $f \otimes g$ is the well known convolution of real functions.

Let now g_i denote the common density function of the messages $\mathtt{m}_{\mathtt{cv}}^{(i)}$ sent from check nodes to message nodes at round i of the algorithm, and let f denote the density of the messages $\mathtt{m}_v$, i.e., the likelihood of the messages sent at round 0 of the algorithm. Then the update rule for the densities in (3.3) implies that the common density f_{i+1} of the messages sent from message nodes to check nodes at round $i+1$ conditioned on the event that the degree of the node is d equals $f \otimes g_i^{\otimes(d-1)}$.

Next we assume that the graph is random such that each edge is connected to a message node of degree d with probability λ_d, and each edge is connected to a check node of degree d with probability ρ_d. Then the expected density of the messages sent from message nodes to check nodes at round $i+1$ is $f \otimes \lambda(g_i)$, where $\lambda(g_i) = \sum_d \lambda_d g_i^{\otimes(d-1)}$. (All this of course assumes the independence assumption.)

To assess the evolution of the densities at the check nodes, we need to use the operator γ introduced above. For a random variable X on $[-\infty, \infty]$ with density F let $\Gamma(F)$ denote the density of the random variable $\gamma(X)$. $\gamma(X)$ is defined on the group $G := \mathbb{F}_2 \times [0, \infty]$. Therefore, the density of $\gamma(X) + \gamma(Y)$ is the convolution (over G) of $\Gamma(F)$ and $\Gamma(H)$, where H denotes the density of Y. Following (3.5) and assuming independence via the independence assumption, we see that the common density g_i of the messages passed from check to message nodes at round i is $\Gamma^{-1}(\rho(\Gamma(f_i)))$, where $\rho(h) = \sum_d \rho_d h^{\otimes(d-1)}$. All in all, we obtain the following recursion for the densities f_i:

$$f_{i+1} = f \otimes \lambda(\Gamma^{-1}(\rho(\Gamma(f_i)))). \tag{4.1}$$

This recursion is called density evolution. The reason for the naming should be obvious.

I have not made the recursion very explicit. In fact, the operator Γ has not been derived at all. For that I refer the reader to [31] and [29].

Density evolution can be used in conjunction with Fourier Transform techniques to obtain asymptotic thresholds below which belief propagation decodes the code successfully, and above which belief propagation does not decode successfully ([31, 29]).

Density evolution is exact only as long the incoming messages are independent random variables. For a finite graph this can be the case only for a small number of rounds.

5. Decoding on the BEC

Perhaps the most illustrative example of belief propagation is when it is applied to LDPC codes over the BEC with erasure probability p. In fact, almost all the important and interesting features of the belief propagation algorithm are already present on the BEC. A thorough analysis of this special case seems thus to be a prerequisite for the general case.

It is sufficient to assume that the all-zero codeword was sent. The log-likelihood of the messages at round 0, $\mathtt{m}_\mathtt{v}$, is $+\infty$ if the corresponding message bit is not erased, and it is 0 if the message bit is erased. Moreover, consulting the update equations for the messages, we see that if v is not erased, then the message passed from v to any of its incident check nodes is always $+\infty$.

The update equations also imply that $\mathtt{m}_{\mathtt{cv}}$ is $+\infty$ if and only if all the message nodes incident to c except v are not erased. In all other cases $\mathtt{m}_{\mathtt{cv}}$ is zero.

If v is an erased message node, then $\mathtt{m}_v = 0$. The message $\mathtt{m}_{\mathtt{vc}}$ is $+\infty$ if and only if there is some check node incident to v other than c which was sending a message $+\infty$ to v in the previous round.

Because of the binary feature of the messages, belief propagation on the erasure channel can be described much easier in the following:

1. [Initialization]
   Initialize the values of all the check nodes to zero.
2. [Direct recovery]
   For all message nodes v, if the node is received, then add its value to the values of all adjacent check nodes and remove v together with all edges emanating from it from the graph.
3. [Substitution recovery]
   If there is a check node c of degree one, substitute its value into the value of its unique neighbor among the message nodes, add that value into the values of all adjacent check nodes and remove the message nodes and all edges emanating from it from the graph.

This algorithm was first proposed in [17] though connections to belief propagation were not realized then. It is clear that the number of operations that this algorithm performs is proportional to the number of edges in the graph. Hence, for sparse graphs the algorithm runs in time linear in the block length of the code. However, there is *no* guarantee that the algorithm can decode all message nodes. Whether or not this is the case depends on the graph structure.

The decoding algorithm can be analyzed along the same lines as the full belief propagation. First, we need to find the expected density of the messages passed at each round of the algorithm under the independence assumption. In this case, the messages are binary (either $+\infty$ or 0), hence we only need to keep track of one parameter, namely the probability p_i that the messages passed from message nodes to check nodes at round i of the algorithm is 0. Let q_i denote the probability that the message passed from check nodes to message nodes at round i of the algorithm is 0. Then, conditioned on the event that the message node is of degree d, we have $p_{i+1} = p \cdot q_i^{d-1}$. Indeed, a message from a message node v to a check node c is 0 iff v was erased and all the messages coming from the neighboring check nodes other than c are 0, which is q_i^{d-1} under the independence assumption. Conditioned on the event that the check node has degree d we have $q_i = 1-(1-p_i)^{d-1}$: the check node c sends a message $+\infty$ to the message node v iff all the neighboring message nodes except for v send a message $+\infty$ to c in the previous round. Under the independence assumption that probability is $(1-p_i)^{d-1}$, which shows the identity.

These recursions are not in a usable form yet since they are conditioned on the degrees of the message and the check nodes. To obtain a closed form we use again the numbers λ_d and ρ_d defined above. Recall that λ_d is the probability that an edge is connected to a message node of degree d, and ρ_d denotes the probability that an edge is connected to a check node of degree d. Defining the generating functions $\lambda(x) = \sum_d \lambda_d x^{d-1}$ and $\rho(x) = \sum_d \rho_d x^{d-1}$ we obtain the following recursion using the formula for the total probability:

$$p_{i+1} = p \cdot \lambda(1 - \rho(1 - p_i)).$$

Under the independence assumption, and assuming that the underlying graph is random with edge degree distributions given by $\lambda(x)$ and $\rho(x)$, decoding is successful if $p_{i+1} < (1-\varepsilon)p_i$ for all i and some $0 < \varepsilon \leq 1$. This yields the condition

$$p \cdot \lambda(1 - \rho(1 - x)) < x \quad \text{for} \quad x \in (0, p) \tag{5.1}$$

for successful decoding which was first proved in [19] and later reproduced in [17]. It is a useful and interesting exercise for the reader to show that (5.1) is identical to (4.1) in the case of the BEC.

Condition (5.1) was proved in [19] in a completely different way than explained here. A system of differential equations was derived whose solutions tracked the expected fraction of nodes of various degrees as the decoding process evolved. One of the solutions corresponds to the fraction of check nodes of reduced degree one during the algorithm. By keeping this fraction above zero at all times, it is guaranteed that in expectation there are always check nodes of degree one left to continue the decoding process. To show that the actual values of the random variables are sharply concentrated around their computed expectations, a large deviation result was derived which is not unsimilar to Azuma's inequality for martingales.

Condition (5.1) can be used to calculate the maximal fraction of erasures a random LDPC code with given edge degree distributions can correct using the simple decoding algorithm. For example, consider a random biregular graph in which each message node has degree 3 and each check node has degree 6. (Such a graph is called a (3,6)-biregular graph.) In this case $\lambda(x) = x^2$ and $\rho(x) = x^5$. What is the maximum fraction of erasures p? (In fact, this value is a supremum.) You can simulate the decoder on many such random graphs with a large number of message nodes. The simulations will show that on average around 42.9% erasures can be recovered. What is this value? According to (5.1) it is the supremum of all p such that $p(1-(1-x)^5)^2 < x$ on $(0,p)$. The minimum of the function $x/(1-(1-x)^5)^2$ on $(0,1)$ is attained at the unique root of the polynomial $9x^4 - 35x^3 + 50x^2 - 30x + 5$ in the interval $(0,1)$, and this is the supremum value for p. This value can be computed exactly, using formulas for the solution of the quartic [2].

As a side remark, I would like to mention an interesting result. First, it is not hard to see that the ratio r/n between the message and the check nodes equals $\int_0^1 \rho(x)\,dx / \int_0^1 \lambda(x)\,dx$. The rate of the code is at least $1 - r/n$, and since the capacity of the erasure channel with erasure probability p is $1 - p$, (5.1) should imply that $p \leq \int_0^1 \rho(x)\,dx / \int_0^1 \lambda(x)\,dx$ in a purely mechanical way (without using the interpretations above). Can you see how to derive this? (See also [34].)

6. Hard Decision Decoding on the BSC

The belief propagation algorithm is the best algorithm among message passing decoders, and the accompanying density evolution provides a tool for analyzing the algorithm. However, for practical applications on channels other than the BEC the belief propagation algorithm is rather complicated, and often leads to a decrease in the speed of the decoder. Therefore, often times a discretized version of the belief propagation algorithm is used. The lowest level of discretization is achieved when the messages passed are binary. In this case one often speaks of a hard decision decoder, as opposed to a soft decision decoder which uses a larger range of values. In this section I will describe two hard decision decoding algorithms on the BSC, both due to Gallager [13].

In both cases the messages passed between the message nodes and the check nodes consist of 0 and 1. Let me first describe the Gallager A algorithm: in round 0, the message nodes send their received values to all their neighboring check nodes. From that point on at each round a check node c sends to the neighboring message node v the addition (mod 2) of all the incoming messages from incident message nodes other than v. A message node v sends the following message to the check node c: if all the incoming messages from check nodes other than c are the same value b, then v sends the value b to c; otherwise it sends its received value to c.

An exact analysis of this algorithm was first given in [18]. The analysis is similar to the case of the BEC. We first find the expected density of the messages passed at each round. Again, we can assume that the all-zero word was transmitted over the BSC with error probability p. Since the messages are 0 and 1, we only need to track p_i, the probability that the message sent from a message node to a check node at round i is 1. Let q_i denote the probability that the message sent from a check node to a message node at round i is 1. Conditioned on the event that the message node is of degree d, and under the independence assumption, we obtain $p_{i+1} = (1-p)q_i^{d-1} + p \cdot (1 - (1-q_i)^{d-1})$. To see this, observe that the message 1 is passed from message node v to check node c iff one of these two cases occurs: (a) the message node was received in error (with probability p) and at least one of the incoming messages is a 1 (with probability $1-(1-q_i)^{d-1}$), or (b) the message was received correctly (probability $1-p$) and all incoming messages are 1 (probability q_i^{d-1}). To assess the evolution of q_i in terms of p_i, note that a check node c sends a message 1 to message node v at round i iff the addition mod 2 of the incoming messages from message nodes other than v in the previous round is 0. Each such message is 1 with probability p_i, and the messages are independent. Conditioned on the event that the check node is of degree ℓ, there are $\ell - 1$ such messages. The probability that their addition mod 2 is 1 is $q_i = (1 - (1-2p_i)^{\ell-1})/2$ (why?).

These recursions are for the conditional probabilities, conditioned on the degrees of the nodes. Introducing the generating functions $\lambda(x)$ and $\rho(x)$ as above, we obtain the following recursion for the probabilities themselves:

$$p_{i+1} = (1-p)\cdot\lambda\left(\frac{1-\rho(1-2p_i)}{2}\right) + p\cdot\left(1-\lambda\left(\frac{1+\rho(1-2p_i)}{2}\right)\right). \tag{6.1}$$

If $\lambda(x)$, $\rho(x)$, and p are such that p_i is monotonically decreasing, then decoding will be successful asymptotically with high probability, as long as the independence assumption is valid.

For example, consider a (3, 6)-biregular graph. In this case $\lambda(x) = x^2$ and $\rho(x) = x^5$, and the condition becomes

$$(1-p)\cdot\left(\frac{1-(1-2x)^5}{2}\right)^2 + p\cdot\left(1-\left(\frac{1+(1-2x)^5}{2}\right)^2\right) < x$$

for $x \in (0, p)$. A numerical calculation shows that the best value for p is around 0.039.

By Shannon's theorem the maximum error probability that a code of rate 1/2 can correct is the maximum p such that $1 + p\log_2(p) + (1-p)\log_2(1-p) = 0.5$. A numerical approximation shows that p is around 11%, which means that the Gallager A algorithm on the biregular (3, 6)-graph is very far from achieving capacity. Bazzi et al. [2] show that for rate 1/2 the best graph for the Gallager A algorithm is the biregular (4, 8)-graph for which the maximum tolerable error probability is roughly 0.0475 – still very far from capacity. This shows that this algorithm, though simple, is very far from using all the information that can be used.

Gallager's algorithm B is slightly more powerful than algorithm A. In this algorithm, for each degree j and each round i there is a threshold value $b_{i,j}$ (to be determined) such that at round i for each message node v and each adjacent check node c, if at least $b_{i,j}$ neighbors of v excluding c sent the same information in the previous round, then v sends that information to c; otherwise v sends its received value to c. The rest of the algorithm is the same as in algorithm A.

It is clear that algorithm A is a special case of algorithm B, in which $b_{i,j} = j - 1$ independent of the round.

This algorithm can be analyzed in the same manner as algorithm A, and a recursion can be obtained for the probability p_i that a message node is sending the incorrect information to a check node at round i:

$$p_{i+1} = p - \sum_{j\geq 1} \lambda_j \left[p \sum_{t=b_{i,j}}^{j-1} \binom{j-1}{l} \left[\frac{1+\rho(1-2p_i)}{2}\right]^t \left[\frac{1-\rho(1-2p_i)}{2}\right]^{j-1-t} \right.$$
$$\left. + (1-p) \sum_{t=b_{i,j}}^{j-1} \binom{j-1}{t} \left[\frac{1-\rho(1-2p_i)}{2}\right]^t \left[\frac{1+\rho(1-2p_i)}{2}\right]^{j-1-t} \right],$$

where the value of $b_{i,j}$ is the smallest integer that satisfies

$$\frac{1-p}{p} \leq \left[\frac{1+\rho(1-2p_i)}{1-\rho(1-2p_i)}\right]^{2b_{i,j}-j+1}.$$

(See [18] for details of the analysis.)

For another hard decision decoder on the BSC (called "erasure decoder"), see [31].

The above one parameter recursions can be used to design codes that asymptotically perform very well for a given amount of noise. The method of choice in these cases is linear programming. For details I refer the reader to [17, 18].

7. Completing the Analysis: Expander Based Arguments

Density evolution and its instantiations are valid only as long as the incoming messages are independent. The messages are independent for ℓ rounds only if the neighborhoods of depth ℓ around the message nodes are trees. But this immediately puts an upper bound on ℓ (of the order $\log(n)$, where n is the number of message nodes, see Section 9). But this number of rounds is usually not sufficient to prove that the decoding process corrects all errors. A different analysis is needed to complete the decoding.

One property of the graphs that guarantees successful decoding is *expansion*. A bipartite graph with n message nodes is called an (α, β)-expander if for any subset S of the message nodes of size at most αn the number of neighbors of S is at least $\beta \cdot a_S \cdot |S|$, where a_S is the average degree of the nodes in S. In other words, if there are many edges going out of a subset of message nodes, then there should be many neighbors.

Expansion arguments have been used by many researchers in the study of decoding codes obtained from graphs [38, 37, 35, 36]. Later, [17, 18] used expander based arguments to show that the erasure correction algorithm on the BEC and Gallager's hard decision decoding algorithm will decode all the erasures/errors if the fraction of errors is small and the graph has sufficient expansion. Burshtein and Miller [5] generalized these results to general message passing algorithms.

To give the reader an idea of how these methods are used, I will exemplify them in the case of the BEC. Choose a graph with edge degree distributions given by $\lambda(x)$ and $\rho(x)$ at random. The analysis of the belief propagation decoder for the BEC implies that if condition (5.1) is true, then for any $\varepsilon > 0$ there is an n_0

such that for all $n \geq n_0$ the erasure decoder reduces the number of erased message nodes below εn. The algorithm may well decode all the erasures, but the point is that the analysis does not guarantee that.

To complete the analysis of the decoder, we first note the following fact: if the random graph is an $(\varepsilon, 1/2)$-expander, then the erasure decoding algorithm recovers any set of εn or fewer erasures. Suppose that this were not the case and consider a minimal counterexample consisting of a nonempty set S of erasures. Consider the subgraph induced by S, and denote by $\Gamma(S)$ the set of neighbors of S. No node in $\Gamma(S)$ has degree 1, since this neighbor would recover one element in S and would contradict the minimality of S. Hence, the total number of edges emanating from these nodes is at least $2|\Gamma(S)|$. On the other hand, the total number of edges emanating from S is $a_S \cdot |S|$, so $a_S \cdot |S| \geq 2|\Gamma(S)|$ which implies $|\Gamma(S)| \leq a_S \cdot |S|/2$ and contradicts the expansion property of the graph.

In [17] it is shown that for a random bipartite graph without message nodes of degree one or two there is a constant ε depending on the rate of the induced code and on the degrees of the message nodes such that the graph is an $(\varepsilon, 1/2)$-expander with high probability. On random graphs without message nodes of degrees one or two we see that the erasure decoding algorithm succeeds with high probability provided condition (5.1) is satisfied.

8. Achieving Capacity

Recall Shannon's theorem which states the existence of codes that come arbitrarily close to the capacity of the channel when decoded with maximum likelihood decoding. LDPC codes were designed to have decoding algorithms of low complexity, such as belief propagation and its variants. But how close can we get to capacity using these algorithms?

There is no satisfactory answer to this question for arbitrary channels. What I mean by a satisfactory answer is an answer to the question whether subclasses of LDPC codes, for example LDPC codes with an appropriate degree distribution, will provably come *arbitrarily* close to the capacity of the channel. Optimization results for various channels, such as the Additive White Gaussian Noise (AWGN) channel and the BSC have produced specific degree distributions such that the corresponding codes come very close to capacity, see [29, 8].

We call an LDPC code ε-close for a channel $\mathcal{C}$ with respect to some message passing algorithm if the rate of the code is at least $\mathbb{m}(\mathcal{C}) - \varepsilon$ and if the message passing algorithm can correct errors over that channel with high probability. We call a sequence of degree distributions $(\lambda^{(n)}(x), \rho^{(n)}(x))$ *capacity-achieving* over that channel with respect to the given algorithm if for any ε there is some n_0 such that for all $n \geq n_0$ the LDPC code corresponding to the degree distribution $(\lambda^{(n)}(x), \rho^{(n)}(x))$ is ε-close to capacity. Using this notation, the following question is open:

Is there a nontrivial channel other than the BEC and a message passing algorithm for which there exists a capacity-achieving sequence $(\lambda^{(n)}(x), \rho^{(n)}(x))$*?*

I believe that this question is one of the fundamental open questions in the asymptotic theory of LDPC codes.

In [17] the authors describe capacity-achieving sequences for the BEC for any erasure probability p. Let $\varepsilon > 0$ be given, let $D := \lceil 1/\varepsilon \rceil$, and set

$$\lambda(x) = \frac{1}{H(D)} \sum_{i=1}^{D} \frac{x^i}{i}, \quad \rho(x) = \mathrm{e}^{\alpha(x-1)},$$

where $\alpha = H(D)/p$. (Technically, $\rho(x)$ cannot define a degree distribution since it is a power series and not a polynomial. But the series can be truncated to obtain a function that is arbitrarily close to the exponential.) We now apply (5.1):

$$\begin{aligned} p\lambda(1-\rho(1-x)) &< -\frac{p}{H(D)} \ln(\rho(1-x)) \\ &= \frac{\alpha p}{H(D)} x \\ &= x. \end{aligned}$$

This shows that a corresponding code can decode a p-fraction of erasures with high probability. [2]

What about the rate of these codes? Above, we mentioned that the rate of the code given by the degree distributions $\lambda(x)$ and $\rho(x)$ is at least $1 - \int_0^1 \rho(x)\,\mathrm{d}x / \int_0^1 \lambda(x)\,\mathrm{d}x$. In our case, this lower bound equals $1-p(1+1/D)(1-\mathrm{e}^{-\alpha})$ which is larger than $1-p(1+\varepsilon)$.

The degree distribution above is called the *Tornado* degree distribution and the corresponding codes are called *Tornado codes.* These codes have many applications in computer networking which I will not mention here (see, e.g., [6]).

Tornado codes were the first class of codes that could provably achieve the capacity of the BEC using belief propagation. Since then many other distributions have been discovered [34, 27]. The latter paper also discusses general methodologies for constructing such degree distributions, and also discusses optimal convergence speeds to capacity.

9. Graphs of Large Girth

As is clear from the previous discussions, if the smallest cycle in the bipartite graph underlying the LDPC code is of length 2ℓ, then independence assumption is valid for ℓ rounds of belief propagation. In particular, density evolution describes

[2] Actually, as was discussed before, (5.1) only shows that the fraction of erasures can be reduced to any constant fraction of the number of message nodes. To show that the decoding is successful all the way to the end, we need a different type of argument. Expansion arguments do not work for the corresponding graphs, since there are many message nodes of degree 2. For a way to resolve these issues, see [17].

the expected behavior of the density functions of these messages exactly for this number of rounds.

The *girth* of a graph is defined as the length of the smallest cycle in the graph. For bipartite graphs the girth is necessarily even, so the smallest possible girth is 4. It is easy to obtain an upper bound for the girth of a biregular bipartite graph with n message nodes of degree d and r check nodes of degree k: if the girth is 2ℓ, then the neighborhood of depth $\ell - 1$ of any message node is a tree with a root of degree d, and in which all nodes of odd depth have degree $k-1$, while all nodes of even depth have degree $d-1$ (we assume that the root of the tree has depth 0). The number of nodes at even depths in the tree should be at most equal to the message nodes, while the number of nodes at odd depths in the tree should be at least equal to the check nodes. The number of nodes at even depths in the tree equals 1 for depth 0, $d(k-1)$ for depth 2, $d(k-1)D$ for depth 4, $d(k-1)D^2$ for depth 6, etc., where $D = (d-1)(k-1)$. The total number of nodes at even depths is equal to

$$1 + d(k-1)\frac{D^{\lfloor \frac{\ell}{2} \rfloor} - 1}{D - 1}.$$

This number has to be less than or equal to n, the number of message nodes. This yields an upper bound on 2ℓ, the girth of the graph. The bound has order $\log_D(n)$. Similar bounds can be obtained by considering nodes of odd depths in the tree.

A similar bound as above can also be deduced for irregular graphs [1], but I will not discuss it here.

As I said before, graphs of large girth are interesting because of the accuracy of belief propagation. However, this is not interesting for practical purposes, since for obtaining accuracy for many rounds the girth of the graph has to be large which means that the number of nodes in the graph has to be very large.

There are other reasons to study graphs of large girth, however. From the point of view of combinatorics graphs of large girth which satisfy (or come close to) the upper bound on the girth are extremal objects. Therefore, to construct them, methods from extremal graph theory need to be applied. From the point of view of coding theory eliminating small cycles is very similar to eliminating words of small weight in the code. This is because a word of weight d leads to a cycle of length $2d$ or less. (Why?)

How does one construct bipartite graphs of large girth? There are a number of known techniques with origins in algebra and combinatorics. For example, it is very easy to construct optimal bipartite graphs of girth 6. Below I will give such a construction. Let C be a Reed-Solomon code of dimension 2 and length n over the field $\mathbb{F}_q$. By definition, this code has q^2 codewords and the Hamming distance between any two distinct codewords is at least $n-1$. From C we construct a bipartite graph with q^2 message nodes and nq check nodes in the following way: The message nodes correspond to the codewords in C. The check nodes are divided in groups of q nodes each; the nodes in each such group corresponds to the elements of $\mathbb{F}_q$. The connections in the graph are obtained as follows: A message node corresponding to the codewords $(x_1, \ldots, x_n)$ is connected to the check nodes corresponding to x_1 in the first group, to x_2 in the second group, ..., to x_n in the last group. Hence, all message nodes have degree n, and all check nodes have degree q, and the graph has in total nq^2 edges. Suppose that this graph has a cycle of length 4. This means that there are two codewords (corresponding to the two message nodes in the cycle) which coincide at two positions (corresponding to the two check nodes in the cycle). This is impossible by the choice of the code C, which shows that the girth of the graph is at least 6. To show the optimality of these graphs, we compare the number of check nodes to the above bound. Let 2ℓ denote the girth of the graph. If $\ell = 3$, then $1 + n(q-1) \leq q^2$, which shows that $n \leq q + 1$. By choosing $n = q + 1$ we obtain optimal graphs of girth 6.

If $n < q + 1$, the graphs obtained may not be optimal, and their girth may be larger than 6. For $n \neq 2$ it is easy to see that the girth of the graph is indeed 6. For $n = 2$ the girth is 8. (A cycle of length 6 in the graph corresponds to three codewords such that every two coincide in exactly one position. This is possible for $n > 2$, and impossible for $n = 2$.)

There are many constructions of graphs without small cycles using finite geometries, but these constructions are for the most part not optimal (except for cases where the girth is small, e.g., 4, or cases where the message nodes are of degree 2).

The sub-discipline of combinatorics dealing with such questions is called *extremal combinatorics*. One of the questions studied here is that of existence of graphs that do not contain a subgraph of a special type (e.g., a cycle). I will not go deeper into these problems here and refer the reader to appropriate literature (e.g., [3]).

A discussion of graphs of large girth is not complete without at least mentioning Ramanujan graphs which have very large girth in an asymptotic sense. I will not discuss these graphs at all in this note and refer the reader to [22, 15]. For interesting applications of these graphs in coding theory I refer the reader to [33].

10. Encoding Algorithms

An encoding algorithm for a binary linear code of dimension k and block length n is an algorithm that computes a codeword from k original bits $x_1, \ldots, x_k$. To compare algorithms against each other, it is important to introduce the concept of cost, or operations. For the purposes of this note the cost of an algorithm is the number of arithmetic operations over $\mathbb{F}_2$ that the algorithm uses.

If a basis $g_1, \ldots, g_k$ for the linear code is known, then encoding can be done by computing $x_1 g_1 + \cdots + x_k g_k$. If the straightforward algorithm is used to perform the computation (and it is a priori not clear what other types of algorithms one may use), then the number of operations sufficient for performing the computation depends on the Hamming weights of the basis vectors. If the vectors are dense,

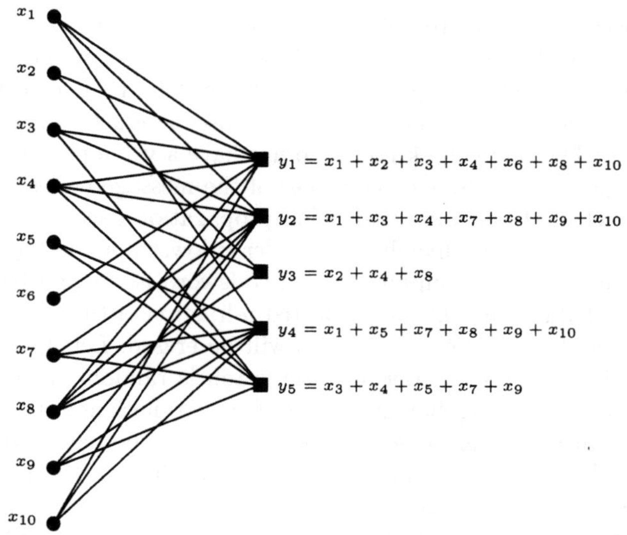

FIGURE 2. Construction with fast encoder

then the cost of the encoding is proportional to nk. For codes of constant rate, this is proportional to n^2, which may be too slow for some applications.

Unfortunately LDPC codes are given as the null space of a sparse matrix, rather than as the space generated by the rows of that matrix. For a given LDPC code it is highly unlikely that there exists a basis consisting of sparse vectors, so that the straightforward encoding algorithm uses a number of operations that is proportional to n^2. However, we would like to design algorithms for which the encoding cost is proportional to n.

At this point there are at least two possible ways to go. One is to consider modifications of LDPC codes which are automatically equipped with fast encoding algorithms. The other is to try to find faster encoding algorithms for LDPC codes. I will discuss both these approaches here, and outline some of the pro's and con's for each approach.

One simple way to obtain codes from sparse graphs with fast encoding is to modify the construction of LDPC codes in such a way that the check nodes have values, and the value of each check node is the addition of the values of its adjacent message nodes. (In such a case, it would be more appropriate to talk about redundant nodes, rather than check nodes, and of information nodes rather than message nodes. But to avoid confusion, I will continue calling the right nodes check nodes and the left nodes message nodes.) Figure 2 gives an example. The number of additions needed in this construction is upper bounded by the number of edges. So, efficient encoding is possible if the graph is sparse. The codewords in this code consist of the values of the message nodes, appended by the values of the check nodes.

This construction leads to a linear time encoder, but it has a major problem with decoding. I will exemplify the problem for the case of the BEC. First, it is not clear that the belief propagation decoder on the BEC can decode all the erasures. This is because the check nodes can also be erased (in contrast to the case of LDPC codes where check nodes do not have a value per-se, but only keep track of the dependencies among the values of the message nodes). This problem is not an artifact of the non-optimal belief propagation decoder. Even the error probability of the maximum likelihood decoder is lower bounded by a constant in this case. Let me elaborate. Suppose that a codeword is transmitted over a BEC with erasure probability p. Then an expected p-fraction of the message nodes and an expected p-fraction of the check nodes will be erased. Let Λ_d be the fraction of message nodes of degree d. Because the graph is random, a message node of degree d will have all its neighbors in the set of erased check nodes with probability p^d. This probability is conditioned on the event that the degree of the message node is d. So, the probability that a message node has all its neighbors within the set of erased check nodes is $\sum_d \Lambda_d p^d$, which is a constant independent of the length of the code. Therefore, no algorithm can recover the value of that message node.

In [36] and [18] the following idea is used the to overcome this difficulty: the redundant nodes will be protected themselves with another graph layer to obtain a second set of redundant nodes; the second set will be protected by a third set, etc. This way a cascade of graphs is obtained rather than a single graph. At each stage the number of message and check nodes of the graphs decreases by a constant fraction. After a logarithmic number of layers the number of check nodes is small enough so the check nodes can be protected using a sophisticated binary code for which we are allowed to use a high-complexity decoder. Details can be found in [17]. If any single graph in the cascade is such that belief propagation can decode a p-fraction of errors, then the entire code will have the same property, with high probability (provided the final code in the cascade has that property, but this can be adjusted). All in all, this construction provides linear time encodable and decodable codes.

The idea of using a cascade, though appealing in theory, is rather cumbersome in practice. For example, in the case of the BEC, the variance of the fraction of erasures per graph-layer will often be too large to allow for decoding. Moreover, maintaining all the graphs is rather complicated and may lead to deficiencies in the decoder. (For some ideas on how to decrease these deficiencies, see [17].)

Another class of codes obtained from sparse graphs and equipped with fast encoders are the *Repeat-Accumulate* (RA) codes of Divsalar et al. [11]. The construction of these codes is somewhat similar to the construction discussed above. However, instead of protecting the check nodes with another layer of a sparse graph, the protection is done via a dense graph, and the check nodes of the first graph are never transmitted. Dense graphs are in general not amenable to fast encoding. However, the dense graph chosen in an RA code is of a special structure which makes its computation easy.

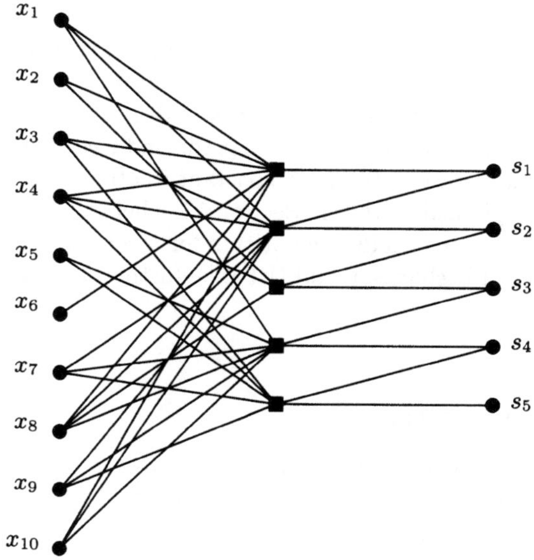

FIGURE 3. An irregular RA code. The left nodes are the information symbols, and the rightmost nodes are the redundant nodes. The squares in between are check nodes. Their values are computed as the addition of the values of their neighbors among the information nodes. The values of the redundant nodes are calculated so as to satisfy the relation that the values of the check nodes is equal to the addition of the values of the neighboring redundant nodes.

More formally, the encoding process for RA codes is as follows. Consider an LDPC code whose graph has n message nodes and r check nodes. The value of the r check nodes is computed using the procedure introduced above, i.e., the value of each check node is the addition of the values of its adjacent message nodes. Let $(y_1, \ldots, y_r)$ denote the values of these check nodes. The redundant values $(s_1, \ldots, s_r)$ are now calculated as follows: $s_1 = y_1, s_2 = s_1 + y_2, \ldots, s_r = s_{r-1} + y_r$. (This explains the phrase "accumulate.") An example of an RA code is given in Figure 3.

The original RA codes used a $(1, k)$-biregular graph for some k as the graph defining the LDPC code. (This explains the phrase "repeat.") RA codes were generalized to encompass *irregular* RA codes for which the underlying graph can be any bipartite graph [14]. The same paper introduces degree distributions for which the corresponding RA codes achieve capacity of the BEC.

We conclude this section by mentioning the work of Richardson and Urbanke [32] which provides an algorithm for encoding LDPC codes. They show that if the degree distribution $(\lambda(x), \rho(x))$ is such that $\rho(1 - \lambda(x)) < x$ for $x \in (0, 1)$, and such that $\lambda_2 \rho'(1) > 1$, then the LDPC code can be encoded in linear time. The condition $\lambda_2 \rho'(1) > 1$ has the following interpretation: consider the graph generated by the message nodes of degree 2. This graph induces a graph on the check nodes, by interpreting the message nodes of degree 2 as edges in that graph (see Section 12). Then $\lambda_2 \rho'(1) > 1$ implies that this induced graph has a connected component whose number of vertices is a constant fraction of the number of check nodes. This will be explained further in Section 12, where this condition is actually used to devise a linear time encoding algorithm for a certain type of graphs. I will not discuss the result of Richardson and Urbanke further, and will refer the reader to [32].

11. Finite-Length Analysis

Density evolution gives a somewhat satisfactory answer to the asymptotic performance of random LDPC codes with a given degree distribution. It is possible to refine the analysis of density evolution to obtain upper bounds on the error probability of the decoder in terms of the degree distributions, and in terms of the number of message and check nodes. However, these bounds are very poor even when the number of message nodes is several tens of thousands large. This is primarily due to two reasons: density evolution is only valid as long as the neighborhood around message nodes is a tree. For small graphs this corresponds to a very small number of iterations, which is usually too small to reduce the fraction of errors in the graph to a reasonable amount. The second source of inaccuracy for the error probability is the set of tools used, since the bounds obtained from the probabilistic analysis are too weak for small lengths.

For these reasons it is important to develop other methods for analyzing the performance of message passing algorithms on small graphs. So far this has only started for the case of the BEC [10]. In this case the analysis is of a combinatorial flavor. Given a bipartite graph, its associated code, and a set of erasures among the check nodes, consider the graph induced by the erased message nodes. A *stopping set* in this graph is a set of message nodes such that the graph induced by these message nodes has the property that no check node has degree one. The number of message nodes in the stopping set is called its *size*. It should be clear that belief propagation for the BEC stops prematurely (i.e., without recovering all the message nodes) if and only if this subgraph has a stopping set. Figure 4 gives some examples of graphs that are themselves stopping sets. Since unions of stopping sets are stopping sets, any finite graph contains a unique maximal stopping set (which may be the empty set). For a random bipartite graph the probability that belief propagation on the BEC has not recovered ℓ message nodes at the point of failure (ℓ can be zero) is the probability that the graph induced by the erased message nodes has a maximal stopping set of size ℓ.

Besides [10] several papers discuss finite-length analysis of LDPC codes on the BEC [25, 26, 30].

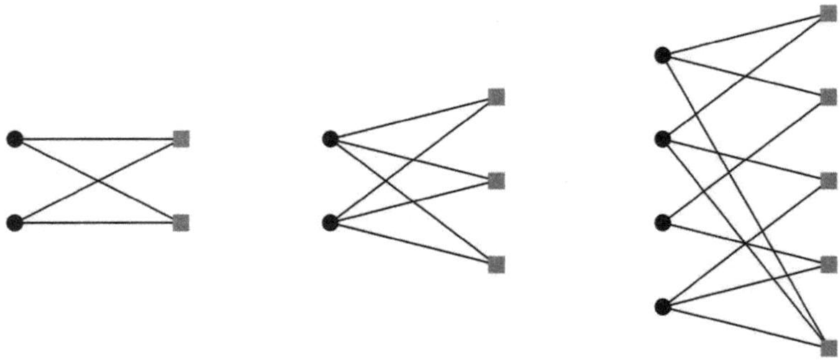

FIGURE 4. Examples of stopping sets

I am not aware of similar analysis tools for channels other than the BEC. Generalizing the concept of stopping sets to other channels would certainly be a worthwhile effort. A recent paper by Feldman et al. [12] gives a different analysis tool using a linear programming relaxation of the decoding problem.

12. An Example

In this section I will exemplify most of the above concepts for a special type of LDPC codes. The codes I will describe in this section certainly do not stand out because of their performance. However, it is rather easy to derive the main concepts for them and this warrants their discussion in the framework of this note. Moreover, it seems that a thorough understanding of their behavior is very important for understanding belief propagation for general LDPC codes. I will try to clarify this more at the end of the section.

For given n and r let $\mathbf{P}(n,r)$ denote the ensemble of bipartite graphs with n message nodes and r check nodes for which each message node has degree 2 and its two neighbors among the check nodes are chosen independently at random. The check node degrees in such a graph are binomially distributed, and if n and r are large, then the distribution is very close to a Poisson distribution with mean $2n/r$. (This is a well-known fact, but the reader may try to prove it for herself.) It turns out that the edge degree distribution of the graph is very close to $\mathrm{e}^{\alpha(x-1)}$ where $\alpha = 2n/r$ is the average degree of the check nodes.

First, let us see how many erasures this code can correct. The maximum fraction of erasures is $1 - R$, where R is the rate, which is at least $1 - r/n$. We should therefore not expect to be able to correct more than an r/n-fraction of erasures, i.e., more than a $2/\alpha$-fraction. We now apply Condition (5.1): p is the maximum fraction of correctable erasures iff $p \cdot (1 - \mathrm{e}^{-\alpha x}) < x$ for $x \in (0, p)$. Replacing x by px, this condition becomes

$$1 - \mathrm{e}^{-\beta x} < x, \quad \beta = p\alpha. \tag{12.1}$$

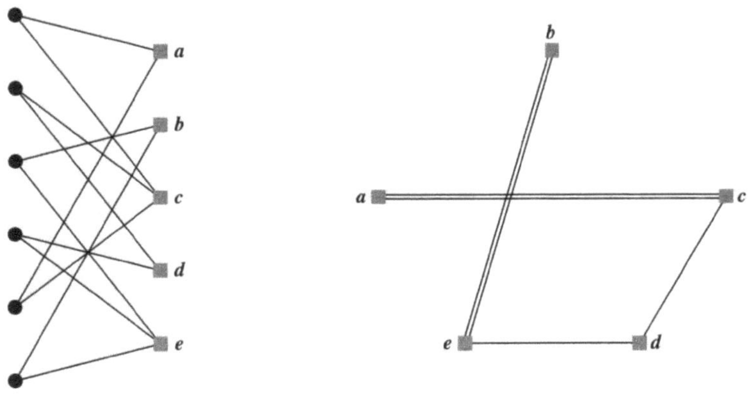

FIGURE 5. A graph with left degree 2 and its induced graph on the check nodes

This latter condition has an interesting interpretation: the graph induced by the p-fraction of erasures is a random graph in the ensemble $\mathbf{P}(e,r)$, where e is the number of erasures, the expected value of which is en. For this graph the edge distribution from the point of view of the check nodes is $e^{-p\alpha(x-1)}$, and thus (5.1) implies (12.1).

Next, I will show that the maximum value of β for which Condition (12.1) holds is $\beta = 1$. For the function $1-e^{-\beta x}-x$ to be less than 0 in $(0,1)$, it is necessary that the derivative of this function be non-positive at 0. The derivative is $\beta e^{-\beta x}-1$, and its value at 0 is $\beta - 1$. Hence, $\beta \leq 1$ is a necessary condition for (12.1) to hold. On the other hand, if $\beta = 1$, then (12.1) is satisfied. Therefore, the maximum fraction of correctable erasures for a code in the ensemble $\mathbf{P}(n,r)$ is $r/(2n)$, i.e., the performance of these codes is at half the capacity. So, the ensemble $\mathbf{P}(n,r)$ is not a very good ensemble in terms of the performance of belief propagation on the BEC.

Before I go further in the discussion of codes in the ensemble $\mathbf{P}(n,r)$, let me give a different view of these codes. A bipartite graph with n message nodes and r check nodes in which each message node has degree 2 defines a (multi-)graph on the set of check nodes by regarding each message node as an edge in the graph in the obvious way. Multi-graphs and bipartite graphs with message degree 2 are in one-to-one correspondence to each other. In the following we will call the graph formed on the check nodes of a bipartite graph G with message degree 2 *induced by G*. Figure 5 gives an example.

For a graph in the ensemble $\mathbf{P}(n,r)$ the corresponding induced graph is a random graph of type $G_{r.n}$, where $G_{m.E}$ denotes the random graph on m vertices in which E edges are chosen randomly and with replacement among all the possible $\binom{m}{2}$ edges in the graph.

For a bipartite graph G with message degree 2 the stopping sets are precisely the edges of a 2-core. Let me define this notion: For any graph and any integer k the k-core of the graph is the unique maximal subgraph of G in which each node has degree k. The k-core may of course be empty.

It is a well-known fact [4] that a giant 2-core exists with high probability in a random graph with E edges in m nodes iff the average degree of a node is larger than 1, i.e., iff $E \geq m$. (A *giant* 2-core in the graph is a 2-core whose size is linear in the number of vertices of the graph.) Condition (12.1) is a new proof for this fact, as it shows that if the average degree of the induced graph is smaller than 1, then with high probability the graph does not contain a 2-core of linear size. It is also a well-known fact that this is precisely the condition for the random graph to contain a *giant component*, i.e., a component with linearly many nodes. Therefore, condition (12.1) can also be viewed as a condition on the graph not having a large component. (This condition is even more precise, as it gives the expected fraction of unrecovered message nodes at the time of failure of the decoder: it is p times the unique root of the equation $1 - x - \mathrm{e}^{-\beta x}$ in the interval $(0, 1)$; incidentally, this is exactly the expected size of the giant component in the graph, as is well known in random graph theory [4].)

More generally, one can study graphs from the ensemble $\mathcal{L}(n, r, \rho(x))$ denoting random graphs with n message and r check nodes with edge degree distribution on the check side given by $\rho(x) = \sum_d \rho_d x^{d-1}$ (i.e., probability that an edge is connected to check node of degree d is ρ_d). The maximum fraction of tolerable erasures in this case is the supremum of all p such that $1 - \rho(1 - px) - x < 0$ for $x \in (0, 1)$. This yields the stability condition $p\rho'(1) < 1$. This condition is also sufficient, since it implies that $p\rho'(1 - px) < 1$ on $(0, 1)$, hence $1 - \rho(1 - px) - x$ is monotonically decreasing, and since this function is 0 at $x = 0$, it is negative for $x \in (0, 1)$.

The condition $p\rho'(1) < 1$ is equivalent to the statement that the graph induced on the check nodes by the bipartite graph has a giant component. This follows from results in [23]. According to that paper, if a graph is chosen randomly on n nodes subject to the condition that for each d the fraction of nodes of degree d is essentially R_d (see the paper for a precise definition), then the graph has almost surely a giant component iff $\sum_d d(d-2)R_d > 0$. Consider the graph obtained from the restriction of the message nodes to a p-fraction, and consider the graph induced by this smaller graph on the check nodes. Then, it is not hard to see that the degree distribution for this graph is $R(px + 1 - p)$, where $R(x) = c \int \rho(x)\,\mathrm{d}$ and c is the average degree of the check nodes in the smaller bipartite graph. Therefore, the condition in [23] for the induced graph to have a giant component equals $pR''(1) < c$, where $R''(x)$ is the second derivative of $R(x)$. This is precisely equal to $p\rho'(1) < 1$, i.e., the stability condition is equivalent to the statement that the induced graph has a giant component. Incidentally, this is also equivalent to the condition that the graph does not have a giant 2-core, since stopping sets are equivalent to 2-cores in this setting. The fraction of nodes in the giant 2-core (if it exists) is equal to the unique solution of the equation $1 - \rho(1 - px) - x = 0$ in

$(0, 1)$. (Compare this also to [24], which obtains formulas for the size of the giant component in a random irregular graph.)

LDPC codes from graphs with left degree 2 play an important role. For example, consider the stability condition proved in [29]. It states that small amounts of noise are correctable by belief propagation for an LDPC code with degree distribution given by $\lambda(x)$ and $\rho(x)$ if and only if $\lambda_2\rho'(1) < \left(\int_{-\infty}^{\infty} f(x)e^{-x/2}\,dx\right)^{-1}$, where $f(x)$ is the density of the log-likelihood of the channel. For example, for the BEC with erasure probability p we obtain $\lambda_2\rho'(1) < 1/p$, and for the BSC with error probability p we obtain $\lambda_2\rho'(1) < 1/\sqrt{p(1-p)}$. The stability condition is actually the condition that belief propagation is successful on the subgraph induced by message nodes of degree 2 (see [9]). This is not surprising, since these message nodes are those that are corrected last in the algorithm. (I do not give a proof of this, but this should sound reasonable, since message nodes of degree 2 receive very few messages in each round of iteration, and hence get corrected only when all the incoming messages are reasonably correct.)

Acknowledgment

Many thanks to Frank Kurth, Mehdi Molkaraie, and Mahmoud Rashidpour for pointing out typos and proof-reading an earlier version of this paper. I would also like to thank the institute for Studies in Theoretical Physics and Mathematics (IPM), and especially Dr. Reza Khosrovshahi for their warm hospitality during my visit in November of 2002.

References

[1] N. Alon, S. Hoory, and N. Linial. The Moore bound for irregular graphs. To appear, 2002.

[2] L. Bazzi, T. Richardson, and R. Urbanke. Exact thresholds and optimal codes for the binary symmetric channel and Gallager's decoding algorithm A. *IEEE Trans. Inform. Theory*, 47, 2001.

[3] B. Bollobas. *Extremal Graph Theory*. Academic Press, 1978.

[4] B. Bollobas. *Random Graphs*. Academic Press, 1985.

[5] D. Burshtein and G. Miller. Expander graph arguments for message-passing algorithms. *IEEE Trans. Inform. Theory*, 47, 2001.

[6] J. Byers, M. Luby, M. Mitzenmacher, and A. Rege. A digital fountain approach to reliable distribution of bulk data. In *proceedings of ACM SIGCOMM '98*, 1998.

[7] J.-F. Cheng, D. MacKay, and R. McEliece. Turbo decoding as an instance of Pearl's belief propagation algorithm. *IEEE J. Sel. Areas Comm.*, 16:140–152, 1998.

[8] S-Y. Chung, D. Forney, T. Richardson, and R. Urbanke. On the design of low-density parity-check codes within 0.0045 dB of the Shannon limit. *IEEE Communication Letters*, 5:58–60, 2001.

[9] L. Decresusefond and G. Zemor. On the error-correcting capabilities of cycle codes of graphs. *Combinatorics, Probability, and Computing*, 6:27–38, 1997.

[10] C. Di, D. Proietti, E. Telatar, T. Richardson, and R. Urbanke. Finite-length analysis of low-density parity-check codes on the binary erasure channel. *IEEE Trans. Inform. Theory*, 48:1570–1579, 2002.

[11] D. Divsalar, H. Jin, and R. McEliece. Coding theorems for 'Turbo-like' codes. In *Proceedings of the 1998 Allerton Conference*, pages 201–210, 1998.

[12] J. Feldman, D. Karger, and M. Wainwright. Using linear programming to decode linear codes. In *Proceedings of the 37th Annual Conference on Information Sciences and Systems (CISS'03)*, 2003.

[13] R. G. Gallager. *Low Density Parity-Check Codes.* MIT Press, Cambridge, MA, 1963.

[14] H. Jin, A. Khandekar, and R. McEliece. Irregular repeat-accumulate codes. In *Proc. 2nd International Symposium on Turbo Codes*, pages 1–8, 2000.

[15] A. Lubotzky, R. Phillips, and P. Sarnak. Ramanujan graphs. *Combinatorica*, 8(3):261–277, 1988.

[16] M. Luby, M. Mitzenmacher, and A. Shokrollahi. Analysis of random processes via and-or tree evaluation. In *Proceedings of the 9th Annual ACM-SIAM Symposium on Discrete Algorithms*, pages 364–373, 1998.

[17] M. Luby, M. Mitzenmacher, A. Shokrollahi, and D. Spielman. Efficient erasure correcting codes. *IEEE Trans. Inform. Theory*, 47:569–584, 2001.

[18] M. Luby, M. Mitzenmacher, A. Shokrollahi, and D. Spielman. Improved low-density parity-check codes using irregular graphs. *IEEE Trans. Inform. Theory*, 47:585–598, 2001.

[19] M. Luby, M. Mitzenmacher, A. Shokrollahi, D. Spielman, and V. Stemann. Practical loss-resilient codes. In *Proceedings of the 29th annual ACM Symposium on Theory of Computing*, pages 150–159, 1997.

[20] D.J.C. MacKay. Good error-correcting codes based on very sparse matrices. *IEEE Trans. Inform. Theory*, 45:399–431, 1999.

[21] D.J.C. MacKay and R.M. Neal. Good codes based on very sparse matrices. In *Cryptography and Coding, 5th IMA Conference*, number 1025 in Lecture Notes in Computer Science, pages 100–111, 1995.

[22] G. A. Margulis. Explicit group-theoretic constructions of combinatorial schemes and their applications in the construction of expanders and concentrators. *Problems of Information Transmission*, 24(1):39–46, 1988.

[23] M. Molloy and B. Reed. A critical point for random graphs with a given degree sequence. *Random Structures and Algorithms*, 6:161–179, 1995. can be downloaded from `http://citeseer.nj.nec.com/molloy95critical.html`.

[24] M. Molloy and B. Reed. The size of the giant component of a random graph with a given degree sequence. *Combin. Probab. Comput.*, 7:295–305, 1998. can be downloaded from `http://citeseer.nj.nec.com/molloy98size.html`.

[25] A. Orlitsky, R. Urbanke, K. Viswanathan, and J. Zhang. Stopping sets and the girth of tanner graphs. In *Proceedings of the International Symposium on Information Theory*, 2002.

[26] A. Orlitsky and J. Zhang. Finite-length analysis of LDPC codes with large left degrees. In *Proceedings of the International Symposium on Information Theory*, 2002.

[27] P. Oswald and A. Shokrollahi. Capacity-achieving sequences for the erasure channel. *IEEE Trans. Inform. Theory*, 48:3017–3028, 2002.

[28] J. Pearl. *Probabilistic Reasoning in Intelligent Systems: Networks of Plausible Inference*. Morgan Kaufmann Publishers, Inc., 1988.

[29] T. Richardson, A. Shokrollahi, and R. Urbanke. Design of capacity-approaching irregular low-density parity-check codes. *IEEE Trans. Inform. Theory*, 47:619–637, 2001.

[30] T. Richardson, A. Shokrollahi, and R. Urbanke. Finite-length analysis of various low-density parity-check ensembles for the binary erasure channel. In *Proceedings of the International Symposium on Information Theory*, 2002.

[31] T. Richardson and R. Urbanke. The capacity of low-density parity-check codes under message-passing decoding. *IEEE Trans. Inform. Theory*, 47:599–618, 2001.

[32] T. Richardson and R. Urbanke. Efficient encoding of low-density parity-check codes. *IEEE Trans. Inform. Theory*, 47:638–656, 2001.

[33] J. Rosenthal and P. Vontobel. Construction of LDPC codes using Ramanujan graphs and ideas from Margulis. In *Proceedings of the 38th Allerton Conference on Communication, Control, and Computing*, pages 248–257, 2000.

[34] A. Shokrollahi. New sequences of linear time erasure codes approaching the channel capacity. In M. Fossorier, H. Imai, S. Lin, and A. Poli, editors, *Proceedings of the 13th International Symposium on Applied Algebra, Algebraic Algorithms, and Error-Correcting Codes*, number 1719 in Lecture Notes in Computer Science, pages 65–76, 1999.

[35] M. Sipser and D. Spielman. Expander codes. *IEEE Trans. Inform. Theory*, 42:1710–1722, 1996.

[36] D. Spielman. Linear-time encodable and decodable error-correcting codes. *IEEE Trans. Inform. Theory*, 42:1723–1731, 1996.

[37] M.R. Tanner. A recursive approach to low complexity codes. *IEEE Trans. Inform. Theory*, 27:533–547, 1981.

[38] V.V. Zyablov and M.S. Pinsker. Estimation of error-correction complexity of Gallager low-density codes. *Probl. Inform. Transm.*, 11:18–28, 1976.

Amin Shokrollahi
Laboratoire d'algorithmique
EPFL
1015 Lausanne
Switzerland

and

Digital Fountain, Inc.
39141 Civic Center Drive
Fremont, CA 94538
USA
e-mail: `amin.shokrollahi@epfl.ch`
e-mail: `amin@digitalfountain.com`

Contributed Papers

Progress in Computer Science and Applied Logic, Vol. 23, 113–127

The New Implementation Schemes of the TTM Cryptosystem Are Not Secure

Jintai Ding and Dieter Schmidt

Abstract. We show that the new TTM implementation schemes have a defect. There exist linearization equations

$$\sum_{i=1,j=1}^{n,m} a_{ij}x_iy_j(x_1,\ldots,x_n)+\sum_{i=1}^{n} b_ix_i+\sum_{j=1}^{m} c_jy_j(x_1,\ldots,x_n)+d=0,$$

which are satisfied by the components $y_j(x_1,\ldots,x_n)$ of the ciphers of the TTM schemes. The inventor of TTM used two versions of the paper [2] to refute a claim in [3]. When we do a linear substitution with the linear equations derived from the linearization equations for a given ciphertext, we can find the plaintext by an iteration of the procedure of first search for linear equations by linear combinations and then linear substitution. The computational complexity of the attack on these two schemes is less than 2^{35} over a finite field of size 2^8.

Keywords. Open-key, multivariable, quadratic polynomials, linearization.

1. Introduction

Recently new methods were invented to construct multivariable cryptosystems, namely cryptosystems based on multivariable functions instead of single variable functions. The security of such systems in general relies on how difficult it is to solve polynomial equations with many variables, a proven NP-hard problem in general.

Matsumoto and Imai suggested one of the first constructions of such cryptosystems [6], which unfortunately has been defeated [8]. Another interesting one is the TTM cryptosystem [7], which was patented in the US in 1998 and is currently marketed by US Data Security Inc. (www.usdsi.com). This system is based on the idea of the composition of invertible polynomial maps, which is closely related to the famous Jacobian Conjecture. Despite the claim of the inventors that the TTM systems are very secure from all standard attacks, the authors of [3] claimed that they completely defeated all possible TTM schemes using the Minrank method and demonstrated it by defeating one of the challenges set by the inventors of

TTM. However the inventors of TTM refuted the claim with [2], where they gave a new implementation scheme to support their claim. In [5], another method was found to defeat the first TTM implementation scheme in [7]. Though this new method can also be applied to other TTM implementation schemes [1], it can not be directly applied to all existing implementation schemes, such as the new ones in two versions of [2]. In this article, we will show that actually all existing implementation schemes for the TTM cryptosystem have a common defect that could make them insecure. For the case of the most recent two TTM implementation schemes in two different versions of the paper [2], we use this defect to defeat the schemes.

The key idea comes from an observation that we can also extend the linearization method by Patarin [8] to attack all TTM implementation schemes.

In all TTM implementation schemes, a cipher F is made of m degree two polynomials of n variables over a finite field K of characteristic 2, namely,

$$F(x_1, \ldots, x_n) = (y_1(x_1, \ldots, x_n), \ldots, y_m(x_1, \ldots, x_n)), \tag{1.1}$$

where $m > n$. These m polynomials y_i are made public. The cipher F is given as a map from K^n to K^m and it is derived from $F = \phi_4 \circ \phi_3 \circ \phi_2 \circ \Phi_1$, where $\circ$ denotes a composition of maps, ϕ_4 and Φ_1 are affine linear maps, ϕ_4 is an invertible map from K^m to K^m, Φ_1 is an injective map from K^n to K^m and ϕ_3 and ϕ_2 are nonlinear maps of the de Jonquières type on K^m. Given an element $X = (z_1, \ldots, z_m)$ in K^m, a de Jonquières map $J(X)$ is defined as a map from K^m to K^m: $J(X) = (z_1+g_1(z_2, \ldots, z_m), z_2+g_2(z_3, \ldots, z_m), \ldots, z_{m-1}+g_{m-1}(z_m), z_m)$, where g_i are polynomial functions.

An affine multiple method uses an equation of the form

$$\sum_{i=1,j=1}^{n,m} a_{ij} x_i y_j(x_1, \ldots, x_n) + \sum_{i=1}^{n} b_i x_i + \sum_{j=1}^{m} c_j y_j(x_1, \ldots, x_n) + d = 0, \tag{1.2}$$

which is satisfied by the set of polynomials y_i of the cipher F and its variables x_i. This equation, which we call 'linearization equation', was used first by Patarin to successfully attack the Matsumoto-Imai cryptosystems.

From the construction of the TTM implementation schemes, we found that all existing TTM implementation schemes have a large number of linearization equations, which are satisfied by the quadratic polynomials y_i of the TTM cipher F. For example, for the most recently proposed implementation scheme [2] (the revised version on IACR e-Archive, the former version has a different implementation scheme), where $m = n + 52$, we found all linearization equations and computed that the dimension of V is actually 347, where V is the linear space of all the linearization equations satisfied by the quadratic polynomials y_i.

This is the source of the common defect among all TTM implementation schemes. The existence of the linearization equations means that for a given ciphertext $(y'_1, \ldots, y'_m)$, we can immediately produce some linear equations satisfied by the plaintext $(x'_1, \ldots, x'_n)$, which is something that a secure open key cryptosystem should not have. For the case of the revised implementation scheme [2], we found that, with the probability $1 - \frac{{}_{17}C_5}{2^{12\times 8}} > 1 - 2^{-82}$, the linearization equations will produce 17 linearly independent linear equations satisfied by x_i.

For this case we can move one step further by performing a substitution of these 17 linear equations into y_i, which makes y_i quadratic polynomials with 17 fewer variables, which we denote by $(x_{v_1}, \ldots, x_{v_{31}})$. Now F becomes a new map $\hat{F}$ from K^{n-17} to K^m, which in the composition form can be equivalently rewritten as: $\hat{F} = \hat{\phi}_4 \circ \phi_2 \circ \hat{\Phi}_1$, where $\hat{\phi}_4$, which is invertible, and $\hat{\Phi}_1$, which is injective, are some affine linear maps. The procedure of the substitution of the 17 linear equations eliminates one of the composition factors of the de Jonquières type. Then solving the equations $\hat{F} = (y'_1, \ldots, y'_m)$ for the given ciphertext becomes straightforward because of the triangular form of the de Jonquières type of maps and it is accomplished by an iteration of the procedure of first search for linear equations by linear combinations and then linear substitution. Finally the plaintext can be derived by substituting the solution of the values of $(x_{v_1}, \ldots, x_{v_{31}})$ into the original 17 linear equations. For the practical example $m = 100$ proposed in [2], we can show that it takes about 2^{32} computations on a finite field of size 2^8 to defeat the scheme. We performed a computation example on a PC (450 MHz) and defeated it in a few hours. Similarly, we can defeat the TTM scheme in the original version of [2].

We arrange the paper in the following way. In Section 2, we will first discuss the basic idea of TTM. Then, we will present the details of our attacks on two different implementation schemes of the TTM: the first one is the one in the revised version (July 2002) of [2], the second one is the one suggested in the first version of (August 2001) [2]. In Section 3, we will present the conclusion.

2. The Common Defect of the TTM Schemes

2.1. Basic Technical Idea of the TTM Schemes

Let $\bar{F}(x_1, \ldots, x_m)$ be a map on the space K^m. It is a composition of several maps G_i on K^m, $i = 1, \ldots, k$, $\bar{F} = G_1 \circ G_2 \circ \cdots \circ G_k$, and has the following properties:

(I) $\bar{F}(x_1, \ldots, x_m)$ is easy and fast to compute if we are given specific values for all x_i.

(II) The factorization of $\bar{F}$ in terms of the composition of G_i is very difficult to compute if we only know the expanded version of $\bar{F}(x_1, \ldots, x_m)$, that is, $\bar{F}^{-1}$ is very difficult to compute without such a decomposition, and G_i are very easy to invert.

With such an $\bar{F}(x_1, \ldots, x_m)$ and if the equation $\bar{F}(x_1, \ldots, x_m) = (a_1, \ldots, a_m)$ is impossible to solve directly, we can use $\bar{F}$ to build an open-key public cryptosystem. The Matsumoto-Imai construction [6] is an attempt of such a type of construction.

For the TTM construction, one uses only the following two types of maps.

1) The Linear Type: Given the space K^m, we can apply all invertible affine linear maps to the m variables: $f(X) = aX + b$, where a is a $m \times m$ invertible matrix, and X and b are in K^m.

2) The de Jonquières Type: These maps give isomorphisms of the corresponding polynomial rings, which are called the tamed transformation in algebraic geometry, and they can be easily inverted. TTM stands for the Tamed Transformation Method.

However due to the consideration of the size of public key and the complexity of public computations, any practical and efficient system requires to have the polynomial components of the cipher to be of degree 2, which seems to be very difficult to accomplish.

In [7], a quadratic construction is obtained by instead using the map

$$F(x_1,\ldots,x_n) = \bar{F}(x_1,\ldots,x_n,0,0,\ldots,0),$$

where $\bar{F}(x_1,\ldots,x_m) = \phi_4 \circ \phi_3 \circ \phi_2 \circ \phi_1(x_1,x_2,\ldots,x_m)$, ϕ_1 and ϕ_4 are of invertible linear type, ϕ_3, ϕ_2 are of the de Jonquières type, ϕ_2 is of degree 2 and ϕ_3 is of a high degree (8). This map F, which can be viewed as a map from K^n to K^m, is an "invertible" map in the sense that it is injective, and given any element in the image of F, we can use $\bar{F}^{-1}$ to recover its preimage easily.

The key component of the construction of the TTM systems is based on a special multivariable polynomial $Q_8(z_1,\ldots,z_l)$ and a special set of quadratic polynomials $q_i(z_1,\ldots,z_k)$, $i = 1,\ldots,l$, such that $Q_8(q_1,\ldots,q_l)$ is still quadratic in z_i. Though the constructions of the TTM schemes are very interesting from a theoretical and a practical point of view, in particular from the point view of algebraic geometry, no principle was given about how Q_8 and q_i are constructed. Our attack starts from an observation of a special property of the polynomials Q_8 and q_i.

2.2. Cryptanalysis of the Revised Version of [2]

2.2.1. The scheme. In this subsection, we will use essentially the notation in the revised version of [2].

First the finite field K is of size 2^8, and $m = n + 52$. The map $\bar{F}$ is made of ϕ_1, ϕ_2, ϕ_3, ϕ_4; $\bar{F} = \phi_4 \circ \phi_3 \circ \phi_2 \circ \phi_1(x_1,x_2,\ldots,x_{n+52})$, which are maps from the $(n+52)$-dimensional space into itself and is defined in [2]. $\phi_1 = (\phi_{1,1},\ldots,\phi_{1,n+52})$, $\phi_4 = (\phi_{4,1},\ldots,\phi_{4,n+52})$ are invertible affine linear maps, and $\phi_{1i} = x_i$, for $i > n$; ϕ_2 and ϕ_3 are nonlinear maps of the de Jonquières type.

The map $F(x_1,\ldots,x_n) = (y_1,\ldots,y_{n+52}) = \phi_4 \circ \phi_3 \circ \phi_2 \circ \phi_1(x_1,\ldots,x_n,0,\ldots,0) = \phi_4 \circ \phi_3 \circ \phi_2 \circ \Phi_1(x_1,\ldots,x_n)$ is the cipher, which is public, but ϕ_1, ϕ_4 are private. $\Phi_1(x_1,\ldots,x_n) = \phi_1(x_1,x_2,\ldots,x_n,0,\ldots,0)$ is an injective map from K^n to K^{n+52}. In the expansion formula, the components y_i of the map F are degree two polynomials of variables $(x_1,\ldots,x_n)$.

To attack this cryptosystem is to solve the set of equations $y_i(x_1,\ldots,x_n) = y_i'$ for $i = 1,\ldots,m$, with the variables x_j, $j = 1,\ldots,n$ and an element in K^m: $(y_1',\ldots,y_{n+52}')$. Here $(y_1',\ldots,y_{n+52}')$ can be viewed as the ciphertext, and the solution $(x_1',\ldots,x_n') \in K^n$ is the plaintext.

In [2] it is claimed that, if $n = 48$, ($m = 100$), no practical methods can work efficiently to attack such a system, in particular, the Minrank method in [3], and the complexity of the attack by Minrank method is far bigger than 2^{84}.

In this scheme, $\phi_2(x_1, \dots, x_n) = (\phi_{2,1}, \dots, \phi_{2,100})$ is given by

$$\begin{aligned}
&\phi_{2,1} = x_1; \\
&\phi_{2,i} = x_i + f_i(x_1, \dots, x_{i-1}), && i = 2, 3, \dots, 41; \\
&\phi_{2,i} = q_{i-41}(x_{38}, \dots, x_{48}), && i = 42, \dots, 48; \\
&\phi_{2,i} = x_i + q_{i-41}(x_{38}, \dots, x_{48}), && i = 49, \dots, 76; \\
&\phi_{2,i} = x_i + q_{i-72}(x_{36}, x_{39}, x_{40}, \dots, x_{45}, x_{37}, x_{47}, x_{48}), && i = 77, \dots, 84; \\
&\phi_{2,i} = x_i + q_{i-80}(x_{34}, x_{39}, x_{40}, \dots, x_{45}, x_{35}, x_{47}, x_{48}), && i = 85, \dots, 92; \\
&\phi_{2,i} = x_i + q_{i-88}(x_{32}, x_{39}, x_{40}, \dots, x_{45}, x_{33}, x_{47}, x_{48}), && i = 93, \dots, 100.
\end{aligned}$$

where a_1 and a_3 can be any nonzero number in the field K,

$$\begin{aligned}
Q_8(q_1, \dots, q_{35}) = {} & (q_5 q_{13} + q_8 q_{14})(q_{19} q_{32} + q_2(q_{18} + q_{24}))^2 (q_{20} q_{19} + q_{23} q_{18}) \\
& +(q_{32} q_3 + (q_{18} + q_{24}) q_{21})^2 \times (q_{22} q_{19} + q_{23} q_{24})(q_9 q_{13} + q_8 q_{15}) \\
& +a_1^8((q_{25} q_{26} + q_{27} q_{28})(q_6 q_{29} + q_7 q_{16}) + (q_{10} q_{30} + q_{11} q_{31})(q_{17} q_1 + q_{18} q_4)) \\
& +a_1^{12}(q_6 q_{33} + q_{34} q_7 + q_5 q_{35} + q_{14} q_{12}),
\end{aligned}$$

and

$$\begin{aligned}
&q_1 = z_4 z_2 + a_1 z_5, && q_2 = z_3 z_4 + a_1 z_6, && q_3 = z_2 z_5 + a_1 z_7, \\
&q_4 = z_4 z_7 + a_1 z_8, && q_5 = z_1 z_5 + a_1 z_9, && q_6 = z_1 z_2 + a_1 z_{10}, \\
&q_7 = z_2 z_9 + a_1 z_{11}, && q_8 = z_3 z_9 + a_1 z_1, && q_9 = z_1 z_3, \\
&q_{10} = z_1 z_7 + a_1 z_9, && q_{11} = z_4 z_9 + a_1 z_1, && q_{12} = z_7 z_9 + a_1 z_1, \\
&q_{13} = z_3 z_{11} + a_1 z_{10}, && q_{14} = z_5 z_{10} + a_1 z_{11}, && q_{15} = z_3 z_{10}, \\
&q_{16} = z_2 z_{10}, && q_{17} = z_7 z_8 + a_1 z_7, && q_{18} = z_5 z_7 + a_1 z_2, \\
&q_{19} = z_2 z_3 + a_1 z_7, && q_{20} = z_5 z_8 + a_1 z_5, && q_{21} = z_4 z_5 + a_1 z_6, \\
&q_{22} = z_3 z_8, && q_{23} = z_3 z_5 + a_1 z_8, && q_{24} = z_3 z_7, \\
&q_{25} = z_6 z_8 + a_3 z_5, && q_{26} = z_2 z_6, && q_{27} = z_5 z_6, \\
&q_{28} = z_6 z_7 + a_3 z_2, && q_{29} = z_2 z_{11}, && q_{30} = z_4 z_{11} + a_1 z_{10}, \\
&q_{31} = z_7 z_{10} + a_1 z_{11}, && q_{32} = z_3 z_6 + z_5 z_6 + a_1 z_4, && q_{33} = z_8 z_{11}, \\
&q_{34} = z_8 z_{10}, && q_{35} = z_7 z_{11} + a_1 z_{10},
\end{aligned}$$

$f_i(x_1, \dots, x_{i-1})$ are randomly chosen quadratic functions.

$\phi_3(x_1, \dots, x_n) = (\phi_{3,1}, \dots, \phi_{3,100})$ is given as: $\phi_{3,i} = x_i$, $i = 5, \dots, 100$; and $\phi_{3,4} = x_4 + R_i(x_1, \dots, x_{100})$, $i = 1, 2, 3, 4$; where $R_i(x_1, \dots, x_{100}) = \sum_1^4 \beta_{ij} P_j$ are linearly independent and the P_i's are given as follows:

$$P_i = Q_8(x_{42}, \dots, x_{45}, x_{101-8i}, \dots, x_{108-8i}, x_{54}, \dots, x_{76}); \quad \text{for } i = 1, 2, 3$$

and $P_4 = Q_8(x_{42}, \dots, x_{76})$.

Remark In the new version of [2], the polynomials Q_8 and q_i actually have three free parameters a_1, a_2 and a_3. We checked the formulas and found out that in order to make the cipher F to be of degree 2, one must make a_1 equal to a_2. We impose this condition on this implementation scheme.

Because of the specific form of ϕ_1, we can write:

$$\phi_1(x_1, x_2, \ldots, x_{48}, 0, \ldots, 0) = \Phi_1(x_1, \ldots, x_{48}) = \pi \circ \hat{\phi}_1(x_1, \ldots, x_{48}),$$

where π is the standard embedding that maps K^{48} into K^{100}:

$$\pi(x_1, \ldots, x_{48}) = (x_1, \ldots, x_{48}, 0, 0, \ldots, 0),$$

and $\hat{\phi}_1(x_1, \ldots, x_{48}) = (\hat{\phi}_{1,1}(x_1, \ldots, x_{48}), \ldots, \hat{\phi}_{1,48}(x_1, \ldots, x_{48}))$ is an invertible affine linear transformation from K^{48} to itself.

Let $\phi_3 \circ \phi_2 \circ \pi = \bar{\phi}_{32}$, then

$$\begin{aligned} F(x_1, \ldots, x_{48}) &= \phi_4 \circ \phi_3 \circ \phi_2 \circ \phi_1(x_1, x_2, \ldots, x_{48}, 0, \ldots, 0) \\ &= \phi_4 \circ \phi_3 \circ \phi_2 \circ \pi \circ \hat{\phi}_1(x_1, \ldots, x_{48}) \\ &= \phi_4 \circ \bar{\phi}_{32} \circ \hat{\phi}_1(x_1, \ldots, x_{48}). \end{aligned}$$

Let $\bar{\phi}_{32}(x_1, \ldots, x_{48}) = (\bar{\phi}_{32,1}, \ldots, \bar{\phi}_{32,100})$, then for the different values of the index i

$$\begin{aligned} \bar{\phi}_{32,i} &= x_i + a_1^{14}\beta_{i4}(x_{38}x_{48} + x_{47}x_{46}) \\ &\quad + a_1^{14}\textstyle\sum_1^3 \beta_{ij}(x_{38-2j}x_{48} + x_{39-2j}x_{47}), && i = 1; \\ \bar{\phi}_{32,i} &= x_i + f_i(x_1, \ldots, x_{i-1}) + a_1^{14}\beta_{i4}(x_{38}x_{48} + x_{37}x_{46}) \\ &\quad + a_1^{14}\textstyle\sum_1^3 \beta_{ij}(x_{38-2j}x_{48} + x_{39-2j}x_{47}), && i = 2, 3, 4; \\ \bar{\phi}_{32,i} &= x_i + f_i(x_1, \ldots, x_{i-1}), && i = 5, 6, \ldots, 41; \\ \bar{\phi}_{32,i} &= q_{i-31}(x_{38}, \ldots, x_{48}), && i = 42, \ldots, 48; \\ \bar{\phi}_{32,i} &= q_{i-31}(x_{38}, \ldots, x_{48}), && i = 49, \ldots, 76; \\ \bar{\phi}_{32,i} &= q_{i-72}(x_{36}, x_{39}, x_{40}, \ldots, x_{45}, x_{37}, x_{47}, x_{48}), && i = 77, \ldots, 84; \\ \bar{\phi}_{32,i} &= q_{i-85}(x_{34}, x_{39}, x_{40}, \ldots, x_{45}, x_{35}, x_{47}, x_{48}), && i = 85, \ldots, 92; \\ \bar{\phi}_{32,i} &= q_{i-93}(x_{32}, x_{39}, x_{40}, \ldots, x_{45}, x_{33}, x_{47}, x_{48}), && i = 93, \ldots, 100. \end{aligned}$$

The formula above is due to the fact that $Q_8(q_1, \ldots, q_{35}) = a_1^{14}(z_9 z_{10} + z_1 z_{11})$, which is the reason why F is of degree 2.

2.2.2. The basic idea of the cryptanalysis. Our attack starts from the observation that all q_i are very simple quadratic polynomials, which have only one quadratic term. In this case, Q_8 has 35 variables and q_i has 11 variables, and we have $q_9 = z_1 z_3, \quad q_{15} = z_3 z_{10}$. This implies that

$$z_{10} q_9 - z_1 q_{15} = 0. \tag{2.1}$$

In this implementation scheme, the map $\bar{\phi}_{32}$ has actually 4 sets of q_i as its components (with intersections). Because F is derived from $\bar{\phi}_{32}$ by composing from both the left side and the right side by an invertible linear map, the equation (2.1) above implies that we must have linearization equations for the y_i, the components of F. This means there is a possibility to actually use such linearization equations to attack this scheme, which is the only method used by Patarin to defeat the Matsumoto-Imai scheme.

Let V denote the linear space of the linearization equations (1.2) satisfied by y_i of F and let D be its dimension.

Let $\bar{V}$ denote the linear space of the linearization equations satisfied by $\bar{\phi}_{32,i}(x_1, \ldots, x_{48})$ of $\bar{\phi}_{32}$:

$$\sum_{i=1,j=1}^{n,m} \bar{a}_{ij} x_i \bar{\phi}_{32,j}(x_1, \ldots, x_{48}) + \sum_{i=1}^{n} \bar{b}_i x_i + \sum_{j=1}^{m} \bar{c}_j \bar{\phi}_{32,j}(x_1, \ldots, x_{48}) + \bar{d} = 0,$$

and let $\bar{D}$ be the dimension of $\bar{V}$.

Let $\hat{\phi}_{32}(x_1, \ldots, x_{48}) = (\hat{\phi}_{32,1}, \ldots, \hat{\phi}_{32,100}) = \bar{\phi}_{32} \circ \hat{\phi}_1(x_1, \ldots, x_{48})$.

Let $\hat{V}$ denote the linear space of the linearization equations satisfied by $\hat{\phi}_{32,i}(x_1, \ldots, x_{48})$ of $\hat{\phi}_{32}$:

$$\sum_{i=1,j=1}^{n,m} \hat{a}_{ij} x_i \hat{\phi}_{32,j}(x_1, \ldots, x_{48}) + \sum_{i=1}^{n} \hat{b}_i x_i + \sum_{j=1}^{m} \hat{c}_j \hat{\phi}_{32,j}(x_1, \ldots, x_{48}) + \hat{d} = 0,$$

and let $\hat{D}$ be the dimension of $\hat{V}$.

Let $\phi_{4,i}$ denote the components of ϕ_4 and $\hat{\phi}_{1,i}$ denote the components of $\hat{\phi}_1$. Let $(\phi)^{-1}_{4,i}$ denote the components of ϕ_4^{-1} and $(\hat{\phi})^{-1}_{1,i}$ denote the components of $\hat{\phi}_1^{-1}$.

Let M be the map from $\hat{V}$ to V given by:

$$\begin{aligned} M : (\Sigma \hat{a}_{ij} x_i \hat{\phi}_{32,j}(x_1, \ldots, x_{48}) + \Sigma \hat{b}_i x_i + \Sigma \hat{c}_j \hat{\phi}_{32,j}(x_1, \ldots, x_{48}) + \hat{d} = 0) \to \\ (\Sigma \hat{a}_{ij} x_i (\phi)^{-1}_{4,j}(y_1(x_1, \ldots, x_{48}), \ldots, y_{100}(x_1, \ldots, x_{48})) + \Sigma \hat{b}_i x_i + \\ \Sigma \hat{c}_j (\phi)^{-1}_{4,j}(y_1(x_1, \ldots, x_{48}), \ldots, y_{100}(x_1, \ldots, x_{48})) + \hat{d} = 0). \end{aligned}$$

Let $\hat{M}$ be the map from $\bar{V}$ to $\hat{V}$ given by:

$$\begin{aligned} \hat{M} : (\Sigma \bar{a}_{ij} x_i \bar{\phi}_{32,j}(x_1, \ldots, x_{48}) + \Sigma \bar{b}_i x_i + \Sigma \bar{c}_j \bar{\phi}_{32,j}(x_1, \ldots, x_{48}) + \bar{d} = 0) \to \\ (\Sigma \bar{a}_{ij} \hat{\phi}_{1,i}(x_1, \ldots, x_{48}) \hat{\phi}_{32,j}(x_1, \ldots, x_{48}) + \Sigma \bar{b}_i \hat{\phi}_{1,i}(x_1, \ldots, x_{48}) \\ + \Sigma \bar{c}_j \hat{\phi}_{32,j}(x_1, \ldots, x_{48}) + \bar{d} = 0). \end{aligned}$$

Theorem 1. *M and $\hat{M}$ are invertible linear maps and $D = \bar{D} = \hat{D}$.*

The proof follows from the fact that both ϕ_4 are $\bar{\phi}_1$ are invertible affine linear maps. Essentially the map $\hat{M}$ is a change of basis of x_i and the map M is an affine linear transformation of the substitution of $\hat{\phi}_{32,i}$ by y_i. This means that we only need to find $\bar{D}$ to find D and we did so by computations.

First we choose the field K to be $K = \mathbf{Z}_2[x]/(x^8 + x^6 + x^5 + x + 1)$. Because a_1 and a_3 can be any nonzero constants, we choose them both to be 1. Then we choose $f_i(x_1, \ldots, x_{i-1})$, $i = 2, \ldots, 41$, randomly as quadratic polynomials over K and β_{ij} randomly in K (but satisfying the condition R_i are linearly independent). We choose 10 different sets of $f_i(x_1, \ldots, x_{48})$ and β_{ij} for testing. For all these 10 choices, our computation showed that the dimension $\bar{D} = 347$ and that all linearization equations are of the form

$$\Sigma_{i>31} \Sigma_{j>41} \bar{a}_{ij} x_i \bar{\phi}_{32,j}(x_1, \ldots, x_{48}) + \Sigma_{i>31} \bar{b}_i x_i + \Sigma_{j>41} c_j \bar{\phi}_{32,j}(x_1, \ldots, x_{48}) = 0,$$

and the polynomials $\phi_{32,j}(x_1,\dots,x_{48})$, for $j>41$ depend only on the 17 variables x_i, with $i>31$.

Though we have such a large number of linearization equations, we are not sure how many linearly independent equations they will produce for a set of given ciphertext y'_i.

Let $(x'_1,\dots,x'_{48})$ be an element in K^{48}. Let $y'_i=y_i(x'_1,\dots,x'_{48})$, $\hat{\phi}'_{32,i}=\hat{\phi}_{32,i}(x'_1,\dots,x'_{48})$. Let U be the space of linear equations derived from substitution of y_i by the values y'_i in V. Let $\hat{U}$ be the linear space of linear equations derived from substitution of $\hat{\phi}_{32,i}$ by the values $\hat{\phi}'_{32,i}$ in $\hat{V}$. Let $\bar{U}$ be the linear space of linear equations derived from substitution of $\bar{\phi}_{32,i}$ by the values $\hat{\phi}'_{32,i}$ and x_i by $(\hat{\phi})^{-1}_{1,i}(x_1,\dots,x_{48})$ in $\bar{V}$.

For a linear equation $\sum_1^{48} a_ix_i+b=0$, we define $\tilde{M}$ to be the linear map:

$$\tilde{M}(\sum_1^{48} a_ix_i+b=0)\rightarrow(\sum_1^{48} a_i(\hat{\phi})_{1,i}(x_1,\dots,x_{48})+b=0).$$

Theorem 2. The dimension of U is equal to the dimension of $\bar{U}$, the dimension of $\hat{U}$ and the dimension of $\tilde{U}$. $\hat{U}=U=\tilde{M}(\bar{U})$.

This is proven easily by using the maps M and $\hat{M}$.

Because all linearization relations in $\bar{V}$ are expressed in the last 59 components $\bar{\phi}_{32,j}(x_1,\dots,x_{48})$, $j>41$ and they are all expressed in terms of the quadratic polynomial q_i; and they involve only the last 17 variables x_i, $i=32,\dots,48$, we did 200 samples of randomly chosen values $\bar{x}'_{32},\bar{x}'_{33},\dots,\bar{x}'_{48}$ for $x_{32},\dots,x_{48}$, computed the corresponding values of $\bar{\phi}_{32,j}, j>41$ for these $\bar{x}'_{32},\bar{x}'_{33},\dots,\bar{x}'_{48}$ and then substituted the values of $\bar{\phi}_{32,j}, j>41$ into the 347 linearization equations. We found out that these 347 linearization equations in $\bar{V}$ actually produce 17 linearly independent equations of x_i, $i>31$, and by solving those equations we have $x_i=\bar{x}'_i$, $i=32,\dots,48$.

Then we notice that if all x_i are set to be zero, which means $\bar{\phi}_{32,i}(0,\dots,0)=0$ for any i, the linearization equations in $\bar{V}$ will not produce 17 linearly independent equations at all. So instead of choosing randomly the values, we chose $(\bar{x}'_{32},\dots,\bar{x}'_{48})$ to be the ones with many zeros, and we found out (with 500 random samples) that as long as at least 5 of $x_{32},\dots,x_{48}$ are not zero, by substituting the corresponding values of $\bar{\phi}_{32,j}, j>41$ into the 347 linearly independent linearization equations in $\bar{V}$, these 347 linearization equations will actually produce 17 linearly independent linear equations of $x_i, i>31$ and by solving those equations we again recover the values of $\bar{x}'_i$ by the solution $x_i=\bar{x}'_i$, $i=32,\dots,48$.

Among all possible values of x_i, $i=32,\dots,48$, the probability that at most 5 of them among x_i, $i=32,\dots,48$ to be non zero is $\frac{{}_{17}C_5 2^{5\times 8}}{2^{17\times 8}}=\frac{{}_{17}C_5}{2^{12\times 8}}<2^{-82}$. Therefore we have a probability $1-\frac{{}_{17}C_5}{2^{12\times 8}}>1-2^{-82}$ that the linearization equations will produce 17 linearly independent equations for a given set of values of $\bar{\phi}_{32,i}$

and solving those equations will recover the values of $\bar{x}'_{32}, \ldots, \bar{x}'_{48}$ if we are given the corresponding values of $\bar{\phi}_{32,j}, j > 41$.

With Theorem 1 and Theorem 2, we conclude that with the probability $1 - \frac{{}_{17}C_5}{2^{12 \times 8}} > 1 - 2^{-82}$, the linearization equations of y_i in V will produce 17 linearly independent equations satisfied by x_i for a given ciphertext $(y'_1, \ldots, y'_{100})$. This is the first step of our attack. Here we would like to emphasize that the statement about the probability to derive 17 linearly independent linear equations from a ciphertext is based on computational experiments not on any theoretical argument and it seems possible to actually prove it.

Let's assume that we now have 17 linearly independent equations in U derived from a ciphertext $(y'_1, \ldots, y'_{100})$ and its substitution in V. Let $(x'_1, \ldots, x'_{48})$ be the corresponding plaintext. This set of linear equations surely is not enough to recover the original plaintext. However, we know that if we have seventeen linearly independent equations, we can use Gaussian elimination method to find two sets: $A = \{u_1, \ldots, u_{17}\}$, $B = \{v_1, \ldots, v_{31}\}$, $A \cap B = \emptyset$ and $A \cup B = \{1, \ldots, 48\}$, such that we can derive 17 linearly independent linear equations in the form $x_{u_j} = h_j(x_{v_1}, \ldots, x_{v_{31}})$.

Then we substitute these 17 equations into the y_i, which will become quadratic polynomials with only 31 variables. We will call this new set of polynomials $\hat{y}_i$. They can be viewed as components of a map from K^{31} to K^{100}, which will be denoted by $\hat{F}$.

Let ϕ_0 be the map from K^{31} to K^{48}, which is given by: $\phi_{0,i}(x_{v_1}, \ldots, x_{v_{31}}) = x_i$ if $i \in B$, otherwise $\phi_{0,i}(x_{v_1}, \ldots, x_{v_{31}}) = h_i(x_{v_1}, \ldots, x_{v_{31}})$, then

$$\hat{F} = \phi_4 \circ \phi_3 \circ \phi_2 \circ \pi \circ \hat{\phi}_1 \circ \phi_0.$$

From the point of view of algebraic geometry, the substitution process is nothing but evaluation of the y_i on the variety defined by the 17 linearly independent linear equations $x_{u_i} = h_i(x_{v_1}, \ldots, v_{31})$ and the existing variables are nothing but the coordinates of this variety.

Because for the case of $\bar{\phi}_{32}$, if the dimension of $\bar{U}$ is 17, the variety is defined by $x_i = \bar{x}'_i$ for $i = 32, \ldots, 48$ and $\bar{x}'_i \in K$, with Theorem 1 and Theorem 2, we know that the variety defined by linear equation in U is the same variety defined by $\hat{\phi}_{1,i}(x_1, \ldots, x_{48}) = \hat{\phi}_{1,i}(x'_1, \ldots, x'_{48})$, for $i > 31$ and we denote this variety by W. The linear equations in U are nothing but linear combinations of this set of linear equations.

Let

$$\hat{\phi}_{32} = \bar{\phi}_{32} \circ \hat{\phi}_1 \circ \phi_0(x_{v_1}, \ldots, x_{v_{31}}) = (\bar{\phi}_{32,1}, \ldots, \bar{\phi}_{32,100})$$

and also define

$$\phi_{10}(x_{v_1}, \ldots, x_{v_{31}}) = \hat{\phi}_1 \circ \phi_0(x_{v_1}, \ldots, x_{v_{31}}) = (\phi_{10,1}, \ldots, \bar{\phi}_{10,100}).$$

Then using the expansion formula of $\bar{\phi}_{32}$, we have:

$$\begin{aligned}
\hat{\phi}_{32,i} &= \phi_{10,i} + a_1^{14}\beta_{i4}(\phi_{10,38}\phi_{10,48} + \phi_{10,47}\phi_{10,46}) \\
&\quad + a_1^{14}\textstyle\sum_1^3 \beta_{ij}(\phi_{10,38-2j}\phi_{10,48} + \phi_{10,39-2j}\phi_{10,47}) = \phi_{10,i} + R_i', \quad \text{for } i=1; \\
\hat{\phi}_{32,i} &= \phi_{10,i} + f_i(\phi_{10,1},\dots,\phi_{10,i-1}) + a_1^{14}\beta_{i4}(\phi_{10,38}\phi_{10,48} + \phi_{10,37}\phi_{10,46}) \\
&\quad + a_1^{14}\textstyle\sum_1^3 \beta_{ij}(\phi_{10,38-2j}\phi_{10,48} + \phi_{10,39-2j}\phi_{10,47}) = \phi_{10,i} + R_i', \quad i=2,3,4; \\
\hat{\phi}_{32,i} &= \phi_{10,i} + f_i(\phi_{10,1},\dots,\phi_{10,i-1}), \quad i=5,\dots,41; \\
\hat{\phi}_{32,i} &= q_{i-31}(\phi_{10,38},\dots,\phi_{10,48}), \quad i=42,\dots,48; \\
\hat{\phi}_{32,i} &= q_{i-31}(\phi_{10,38},\dots,\phi_{10,48}), \quad i=49,\dots,76; \\
\hat{\phi}_{32,i} &= q_{i-72}(\phi_{10,36},\phi_{10,39},\phi_{10,40},\dots,\phi_{10,45},\phi_{10,37},\phi_{10,47},\phi_{10,48}), \quad i=77,\dots,84; \\
\hat{\phi}_{32,i} &= q_{i-85}(\phi_{10,34},\phi_{10,39},\phi_{10,40},\dots,\phi_{10,45},\phi_{10,35},\phi_{10,47},\phi_{10,48}), \quad i=85,\dots,92; \\
\hat{\phi}_{32,i} &= q_{i-93}(\phi_{10,32},\phi_{10,39},\phi_{10,40},\dots,\phi_{10,45},\phi_{10,33},\phi_{10,47},\phi_{10,48}), \quad i=93,.,100
\end{aligned}$$

where $R_i' = \sum_1^4 \beta_{ij}P_i'$, and P_i', for $i = 1,2,3$, is given as

$$P_i' = \hat{\phi}_{1,31+i+1}(x_1',\dots,x_{48}')\hat{\phi}_{1,48}(x_1',\dots,x_{48}') + \hat{\phi}_{1,31+i}(x_1',\dots,x_{48}')\hat{\phi}_{1,47}(x_1',\dots,x_{48}').$$

$$P_4' = \hat{\phi}_{1,42}(x_1',\dots,x_{48}')\hat{\phi}_{1,48}(x_1',\dots,x_{48}') + \hat{\phi}_{1,46}(x_1',\dots,x_{48}')\hat{\phi}_{1,47}(x_1',\dots,x_{48}'),$$

which are constants. Namely $R_i(\bar{\phi}_{32}(x_1,\dots,x_{48}))$ are constants on the variety W. Therefore

$$\hat{F}(x_{v_1},\dots,x_{v_{31}}) = (\hat{y}_1,\dots\hat{y}_{100}) = \phi_4 \circ \hat{\phi}_3 \circ \phi_2 \circ \pi \circ \hat{\phi}_1 \circ \phi_0(x_{v_1},\dots,x_{v_{31}}),$$

where $\bar{\phi}_3 = (\bar{\phi}_{3,1},\dots,\bar{\phi}_{3,100})$ is given by $\hat{\phi}_{3,i} = x_i$, for $i = 5,\dots,100$; and $\hat{\phi}_{3,4} = x_4 + R_i'$, for $i = 1,2,3,4$. Therefore ϕ_3 on the variety W is equivalent to $\hat{\phi}_3$, which is linear and is just a translation.

Then

$$\hat{F}(x_{v_1},\dots,x_{v_{31}}) = (\phi_4 \circ \hat{\phi}_3) \circ \phi_2 \circ (\pi \circ \hat{\phi}_1 \circ \phi_0) = \hat{\phi}_4 \circ \phi_2 \circ \hat{\Phi}_1,$$

where $\hat{\phi}_4 = \phi_4 \circ \hat{\phi}_3$, $\hat{\Phi}_1 = \pi \circ \hat{\phi}_1 \circ \phi_0$ and both $\hat{\phi}_4$, which is invertible, and $\hat{\Phi}_1$, which is injective, are linear maps.

Then $\hat{F}(x_{v_1},\dots,x_{v_{31}}) = (y_1',\dots,y_{100}')$ can be easily solved because of the triangular form of ϕ_2, namely the equation above is equivalent to the equations:

$$\hat{\phi}_4^{-1}(\hat{y}_1,\dots,\hat{y}_{100}) = \phi_2 \circ \hat{\Phi}_1(x_{v_1},\dots,x_{v_{31}}) = \hat{\phi}_4^{-1}(y_1',\dots,y_{100}'),$$

whose first nontrivial equation is always a linear equation.

This shows that the equations can be solved by iteration of the procedure of first searching for linear equations by linear combinations of quadratic equations, and then substituting the linear equations into the quadratic equations. Each time of iteration, we reduce the variable by 1. This eventually will require 31 iterations to find the 31 linearly independent linear equations in the triangular form, whose solution gives the values of the 31 variables x_{v_i}. Then we can substitute the values of x_{v_i}, $i = 1,\dots,31$, back into the first 17 substitution equations $x_{u_i} = h_i(x_{v_1},\dots,v_{31})$, $j = 1,\dots,17$, which recovers the complete set of (x_i'), the plaintext.

Overall, our general method is first to search all linearization equations. Then, for a given ciphertext $(y'_1,\ldots,y'_m)$ corresponding to a plaintext $(x'_1,\ldots,x'_n)$, we use the linearization equations to produce enough (17) linearly independent linear equations satisfied by x_i. Then we do a substitution using these linear equations, which essentially makes ϕ_3 linear on the variety defined by the 17 linear equations. The rest becomes straightforward.

2.2.3. The practical attack procedure and its complexity. We have three steps to derive the plaintext $(x'_1,\ldots,x'_{48})$ from a ciphertext $(y'_1,\ldots,y'_{100})$, and the first step is a common step for any given ciphertext.

Step 1 of the attack

We first look for a basis for the space V, namely the basis of solutions of a_{ij}, b_i, c_j and d for the equations:

$$\sum_{i=1,j=1}^{n,m} a_{ij}x_iy_j(x_1,\ldots,x_n)+\sum_{i=1}^{n} b_ix_i+\sum_{j=1}^{m} c_jy_j(x_1,\ldots,x_n)+d=0.$$

For this set of equations, we have $4949 = 4800 + 48 + 100 + 1$ variables and $19697 = 1 + 48 + (24 \times 47 + 48) + (48 + 24 \times 47 + (8 \times 47 \times 46))$ equations. We know that the dimension of the solutions is 347.

Though we have 19697 equations, we have only 4949 variables, we do not need to use all those equations to find the solutions. We can actually randomly choose 6000 equations, the probability that we will not find the complete solution is essentially zero. To solve these linear equations, is to do row operations on a 6000×4949 matrix. However, because we are working on a finite field with only 2^8 elements, the row operations corresponding the elimination procedure on each column requires at most $2^8 - 1$ multiplication of a given row. To eliminate each variable, on average, it takes $(2^8 - 1) \times 6000/2$ multiplications. Therefore to solve these equations, it requires at most $4600 \times (2^8 - 1) \times 6000/2 \doteq 2^{32}$ computations on the finite field K. This step is also the common step for any attack.

However, because we are working over the fixed field K, we can perform the computation of multiplication on K by finding first a generator g of the multiplicative group of K, and storing the table of elements γ in K as g^k, then computing the multiplication by two searches and one addition. This will improve the speed by at least a factor of 2. Therefore, this step takes at most 2^{31} computations.

Step 2 of the attack

For a given ciphertext $(y'_1,\ldots,y'_{100})$, we substitute the polynomials of y_i by y'_i into the 347 linearly independent solutions of the linearization equations in V and derive 17 linearly independent linear equations of x_i by the Gaussian elimination method in the form of $x_{u_j} = h_j(x_{v_1},\ldots,x_{v_{31}})$, where h_j is a linear function, $A = \{u_1,\ldots,u_{17}\}$, $B = \{v_1,\ldots,v_{31}\}$, $A \cap B = \emptyset$ and $A \cup B = \{1,\ldots,48\}$. We, then, substitute them into y_i to make it into polynomials with only 31 variables $\{v_1,\ldots,v_{31}\}$.

First with a probability 2^{-82}, we might fail to get 17 linearly independent equations, which surely can be neglected.

When we substitute y_i by y'_i, we need to do $347 \times (4800 + 100) \doteq 2^{21}$ computations. Then, to reduce 347 equations to 17 equations for substitution, it takes $(2^8 - 1) \times 48 \times 347/2 \doteq 2^{21}$ computations. Then we perform the substitution of the 17 equations into y_i and it takes $100 \times (2 \times 17^2 + 17 \times 31 + 17) \doteq 2^{17}$ computations.

For the new 100 polynomials with 31 variables, which we denote by $\hat{y}_i$, we will write down first the 100 equations $\hat{y}_i - y'_i = \tilde{y}_{1i}(x_{v_1}, \ldots, x_{v_{31}}) = 0$, and they are linearly dependent and the dimension is only actually 41.

Step 3 of the attack

For the equations $\tilde{y}_{1i}(x_{v_1}, \ldots, x_{v_{31}}) = 0$, $i = 1, \ldots, 100$, we will use Gaussian elimination method, first on the quadratic terms, to derive $\hat{m}$ ($\hat{m} = 41$) linearly independent equations $\hat{y}_{1i}(x_{v_1}, \ldots, x_{v_{31}}) = 0$, $i = 1, \ldots, m$, and the last one is linear. Then we take the linear equation out and substitute it back into the leftover $\hat{m} - 1$ quadratic equations (the linear one is taken out) $\hat{y}_{1i}(x_{v_1}, \ldots, x_{v_{31}}) = 0$, $i = 1, \ldots, m-1$. We denote the new equations $\tilde{y}_{2i}(x_{v_1}, \ldots, x_{v_i}, x_{v_{i+2}}, \ldots, x_{v_{31}}) = 0$. Then we repeat the same process on these new equations, and later again and again for total of 31 times. We then collect all 31 linear equations derived in this process, a set of 31 linearly independent equations in the triangular form. The solution gives us all the values of x_{v_i}, then we plug them back into 17 linear equations $x_{u_j} = h_j(x_{v_1}, \ldots, x_{v_{31}})$ in Step 2, which will give us x_{u_i}. We recover the plaintext.

For the first part of this step, we need at most $100 \times (31 \times 16 + 31) \times 100/2 \doteq 2^{22}$ computations to perform the Gaussian elimination and then the substitution takes at most $42 \times 31^2 \doteq 2^{15}$ computations. Then we need at most to perform 31 times these two procedures and therefore it, at most, takes 2^{25} computations to solve the equations for the 31 variables and then need to do 2^9 computations to find the values for the other 17 variables.

If we add all three steps together, it takes at most 2^{32} computations. We did simulation of an example of this scheme and it took us a few hours to find the plaintext for any given ciphertext. The best way to test our method is definitely to attack the challenges set by T. Moh. However, at this moment the web site (www.usdsi.com), where the challenge is posted, does not allow public access to the challenger's data and we plan to do so once it is available.

2.3. Cryptanalysis for the Scheme in the First Version of [2]

Our work was first done for the scheme in the first version of [2]. When we had finished the work on this scheme in July last year, the new version appeared. In this section, we will present our work on the implementation scheme in the original version of [2]. The construction of the scheme is similar to the revised case above.

We again work on the field K of size 2^8. A map $\bar{F}$ is made of ϕ_1, ϕ_2, ϕ_3, ϕ_4, which are maps from the $(n+68)$-dimensional space to itself. ϕ_1, ϕ_4 are invertible affine maps, and ϕ_2 and ϕ_3 are nonlinear of de Jonquières type but different from that of the section above.

Again a map

$$F(x_1,\ldots,x_n) = \phi_4 \circ \phi_3 \circ \phi_2 \circ \phi_1(x_1,x_2,\ldots,x_n,0,\ldots,0) = (y_1,\ldots,y_{n+68})$$

is the cipher, which is public, but ϕ_1, ϕ_4 are private. To make sure the system is of degree 2, another set of polynomials $Q_8(z_1,\ldots,z_{48})$ and $q_i(z_1,\ldots,z_{14})$ are used. The detail of ϕ_3, ϕ_2 and Q_8 and q_i are given as follows with $m = n+68$.

We will not give the exact detail of ϕ_3 and ϕ_2 (we refer it to the original version of [2]), rather we will give the detail of $\bar{\phi}_{32} = \phi_3 \circ \phi_2 \circ \pi$, where $\pi(x_1,\ldots,x_n) = (x_1,\ldots,x_n,0,\ldots,0)$ is a map from K^n to K^{n+68}:

$$\bar{\phi}_{32,1} = y_1 = x_1 + Q_8(y_{n-7},\ldots,y_{n+34}) + Q_8(y_{n-21},y_{n-14},y_{n+35},\ldots,y_{n+68}) =$$
$$x_1 + x_{n-5}x_{n-6} + x_{n-8}x_n + x_{n-19}x_{n-20} + x_{n-22}x_{n-14},$$
$$\bar{\phi}_{32,2} = y_2 = x_2 + Q_8(y_3,\ldots,y_m) = x_2 + x_1^2 + Q_8(y_{n-21},y_{n-14},y_{n+35},\ldots,y_{n+68}) =$$
$$x_1 + x_2^2 + x_{n-19}x_{n-20} + x_{n-22}x_{n-14},$$

$$\begin{array}{ll}
\bar{\phi}_{32,i} = y_i = x_i + f_i(x_1,\ldots,x_{i-1}), & i = 3,\ldots,n-22; \\
\bar{\phi}_{32,i} = y_i = q_{i-(n-24)}(x_{n-27},\ldots,x_{n-14}), & i = n-23,\ldots,n-14; \\
\bar{\phi}_{32,i} = y_i = x_i + f_i(x_1,\ldots,x_{i-1}), & i = n-13,\ldots,n-8; \\
\bar{\phi}_{32,i} = y_i = q_{i-(n-8)}(x_{n-13},\ldots,x_n), & i = n-7,\ldots,n+34; \\
\bar{\phi}_{32,i} = y_i = q_{i-(n+26)}(x_{n-27},\ldots,x_{n-14}), & i = n+35,\ldots,n+68;
\end{array}$$

where

$$Q_8(q_1,\ldots,q_{42}) = [q_{14}q_{23} + q_{17}q_{24}][q_{10}q_9 + q_6q_4]^2[q_{11}q_{30} + q_1q_{31}] +$$
$$[q_{33}q_{34} + q_{35}q_{36}][q_{15}q_{37} + q_{16}q_{26}] + [q_{19}q_8 + q_{20}q_{38}][q_{13}q_7 + q_{12}q_5] +$$
$$[q_{21}q_{39} + q_{40}q_2 + q_{22}q_{41} + q_{42}q_3];$$

q_i are functions of 14 variables $z_i, z_2, \ldots, z_{14}$:

$$\begin{array}{lll}
q_1 = z_7 + z_2z_5, & q_2 = z_8 + z_6z_7, & q_3 = z_9 + z_6z_5, \\
q_4 = z_2z_4 + z_{10}, & q_5 = z_3z_5 + z_{11}, & q_6 = z_1z_3 + z_{12}, \\
q_7 = z_3z_7 + z_{13}, & q_8 = z_{14} + z_8z_3, & q_9 = z_3 + z_2z_{12}, \\
q_{10} = z_4 + z_{10}z_1, & q_{11} = z_{13} + z_{11}z_2, & q_{12} = z_4z_{13} + z_7, \\
q_{13} = z_4z_{11} + z_5, & q_{14} = z_6 + z_9z_2, & q_{15} = z_{14} + z_9z_{10}, \\
q_{16} = z_8 + z_{10}z_6, & q_{17} = z_9 + z_1z_6, & q_{18} = z_9z_1, \\
q_{19} = z_9z_4 + z_6, & q_{20} = z_6z_3 + z_9, & q_{21} = z_7z_9 + z_{14}, \\
q_{22} = z_9z_{13} + z_6, & q_{23} = z_1z_8 + z_{14}, & q_{24} = z_2z_{14} + z_8, \\
q_{25} = z_{14}z_1, & q_{26} = z_{10}z_{14}, & q_{27} = z_2z_{10}, \\
q_{28} = z_2z_3, & q_{29} = z_1z_{11}, & q_{30} = z_1z_7 + z_5, \\
q_{31} = z_1z_{13} + z_{11}, & q_{32} = z_1z_5, & q_{33} = z_{12}z_{11} + z_{13}, \\
q_{34} = z_{12}z_7, & q_{35} = z_{12}z_{13}, & q_{36} = z_{12}z_5 + z_7, \\
q_{37} = z_{10}z_8, & q_{38} = z_4z_{14} + z_8, & q_{39} = z_8z_{11}, \\
q_{40} = z_{14}z_{11}, & q_{41} = z_8z_5 + z_{14}, & q_{42} = z_{14}z_{13} + z_8.
\end{array}$$

and

$$Q_8(q_1,\ldots,q_{42}) = z_8z_9 + z_6z_{14}.$$

Through computations and similar argument as in the section above, we have:

1) The dimension of V of the space of linearization equations for the components y_i of F is 286;
2) For a given ciphertext $(y'_1, \ldots, y'_{68+n})$, with a probability of $1 - \frac{{}_{14}C_4}{2^{10\times 8}} \doteq 1 - 2^{-70}$, the linearization will produce 28 linearly independent linear equations of x_i.
3) For the case of 28 linearly independent equations, we can again do a substitution using these 28 linear equations into y_i to derive a new operator $\hat{F}$ which is a map from K^{n-28} to K^{n+68} and $\hat{F} = \hat{\phi}_4 \circ \phi_2 \circ \hat{\Phi}_1$, for some linear maps $\hat{\phi}_4$ invertible, and $\hat{\Phi}_1$ injective.

This allows us to use exactly the same attack steps as in the previous section to defeat this scheme.

Here if we choose $n = 52$, $m = 120$, we estimate that it takes about 2^{35} computations on K to defeat the scheme. We performed a computation example on such a scheme on a PC and it took a few days to find the plaintext from a given ciphertext.

By now, there are several implementation schemes that have been suggested by the inventor. We notice that for all cases, due to the fact that the q_i components are all very simple and they never have more than 2 quadratic monomials, it is easy to see that for all schemes, the dimension of linear space of all the linearization equations for the components y_i of F is not small. This is a common defect for the implementation schemes, which is not in any way desirable for a secure open key cryptosystem. However even with those linearization equations, it does not necessarily mean that finding the plaintext from a given ciphertext is easy. One example is the first implementation [7] to demonstrate the situation. But this schemes was defeated by another method [5].

3. Conclusion

A key component of the TTM schemes is a set of quadratic polynomials q_i. These polynomials are very simple and often a q_i consists of just one degree two monomial. Due to this fact we show through computations that in all TTM implementation schemes, the polynomial components y_i of the public cipher F satisfy linearization equations and for a given cipher text y'_i, we can obtain linear equations satisfied by the plaintext x'_i. This is something that a secure open key cryptosystem should not have. This defect does not necessarily allow us to defeat all implementation schemes easily, but for the cases of the two most recent implementation schemes suggested in two different versions of [2], we show that, with a very small probability of failure, this defect allows us to defeat the schemes easily. For the suggested practical example in the revised paper [2], we show that it takes 2^{32} computations on the finite field K to defeat the scheme and we confirmed this by a computation example. For the case of the scheme in the original version of [2], it takes 2^{35} computations to defeat it, and it is confirmed by a computer experiment as well.

Also for the existing TTM implementation schemes, we can even find higher order type of linearization equations, which a secure scheme should avoid as well.

Considering all the attacks on the TTM schemes, [3, 5] and this paper, we conclude that all existing TTM schemes are insecure. But we do not, in any way, suggest that our results imply that there do not exist good TTM schemes. We do conclude that to avoid the defect we found in this paper, more sophisticated construction of Q_8 and q_i is needed. We think this is a very interesting direction to pursue, but it needs some deep insight from algebraic geometry.

References

[1] Chou, G., Guan, J., Chen, J., *A systematic construction of a Q_{2^k}-model in TTM*, Comm. in Algebra, 30(2), 551–562, (2002).

[2] Chen, J., Moh, T., *On the Goubin-Courtois attack on TTM*, Cryptology ePrint Archive (2001/72).

[3] Goubin, L., Courtois, N., *Cryptanalysis of the TTM cryptosystem*, Asiacrypt2000, LNCS 1976, 44–57.

[4] Dickerson, M., *The inverse of an automorphism in polynomial time,* J. Symbolic Comput. 13 (1992), no. 2, 209–220.

[5] Ding, J., Hodges, T., *Cryptanalysis of an implementation scheme of TTM,* Department of Mathematical Sciences, University of Cincinnati, Preprint 2002.

[6] Matsumoto, T., Imai, H., *Public quadratic polynomial-tuples for efficient signature-verification and message-encryption*, Advances in Cryptology – EUROCRYPT '88 (Davos, 1988), 419–453, Lecture Notes in Comput. Sci., 330, Springer, Berlin, 1988.

[7] Moh, T. T., *A fast public key system with signature and master key functions*, Communications in Algebra, 27(5), pp. 2207–2222 (1999) & Lecture Notes at EE Department of Stanford University. (May 1999) & http://www.usdsi.com/ttm.html

[8] Patarin, J., *Cryptanalysis of the Matsumoto and Imai public key scheme of Eurocrypt'88.*, Des. Codes Cryptogr. 20 (2000), no. 2, 175–209.

[9] Patarin, J., *Hidden field equations (HFE) and isomorphism of polynomials (IP): Two new families of asymmetric algorithms*, EuroCrypt'96, Lecture Notes in Comput. Sci., (1996) Ueli Maurer ed., 33–48.

Jintai Ding
Department of Mathematical Sciences
University of Cincinnati
Cincinnati, OH, 45221-0025, USA
e-mail: `ding@math.uc.edu`

Dieter Schmidt
Department of Electrical and Computer Engineering and Computer Science,
University of Cincinnati
Cincinnati, OH, 45221-0030, USA
e-mail: `dieter.schmidt@uc.edu`

Progress in Computer Science and Applied Logic, Vol. 23, 129–137

The Capacity Region of Broadcast Networks with Two Receivers

Elona Erez and Meir Feder

Abstract. According to the max-flow min-cut theorem a source s can transmit information to a sink t in a graph (V, E) at a rate that does not exceed the capacity of the minimal cut that separates the source and the sink. Recently, it has been shown that if the intermediate nodes in the network are allowed to code the information that they receive, then the source s can multicast common information to several sinks at a rate that does not exceed the min-cut between the source and any of the individual sinks. In this paper we find the achievable rate region when there are two receiver nodes t_1 and t_2, but we allow both common information at rate R_0 and private information rates to t_1 and t_2 at rates R_1, R_2, respectively.

Keywords. network codes, broadcast channel, multicast, min-cut max-flow theorem.

1. Introduction

According to the max-flow min-cut theorem a source s can transmit information to a sink t in a directed graph $G = (V, E)$ at a rate that does not exceed the capacity of the minimal cut that separates the source and the sink. Any edge (i, j) in the network is assumed to be free of transmission errors and have capacity of C_{ij} bits per channel use. It has been shown by Ahlswede *et al.* [1] that if the intermediate nodes in the network are allowed to code the information that they receive, then the source s can multicast common information to several sinks at a rate that does not exceed the minimal of the min-cuts between the source and any of the individual sinks. In this work we find the broadcast capacity region when there is a single transmitter s and two receiver nodes t_1 and t_2, but we allow both common information at rate R_0 and private information rates to t_1 and t_2 at rates R_1, R_2, respectively. The special case of $R_2 = 0$ was solved using an algebraic formulation in [2]. We also show how to construct codes that achieve any triplet (R_0, R_1, R_2) in the capacity region. Specifically, we show that this situation is equivalent to the common multicast case in an extended network that we define. The code for the original network is derived from the common multicast code of the extended network. We note that the fact that the entire capacity region for the broadcast network can be easily found is rather surprising since the parallel problem of two-user broadcast channel in multiuser information theory is in fact not solved yet [3], [4], [5]. As it turns out, the resulting capacity region is what one would intuitively expect by postulating achievability of min-cut bounds. In the sequel we term by unicast network the case where there is no common information. We term by multicast network the case of no private information. We term by broadcast network the case of both private and common information.

2. Unicast Network

As mentioned in the introduction, in the multicast case, when all the sinks have to receive the same information, the maximal rate can be achieved through coding at the intermediate nodes.

At the other extreme, which is termed the unicast case, where each receiver is required to receive different information, the capacity region, as noted in [2], can be achieved without coding. Suppose that there are L receivers, $t_1, \ldots, t_L$ required to receive information at rates $R_1, \ldots, R_L$. We want to verify whether the rate requirements are feasible. The original network can be extended to another network, as can be seen in Figure 1, where each sink t_i is connected to the

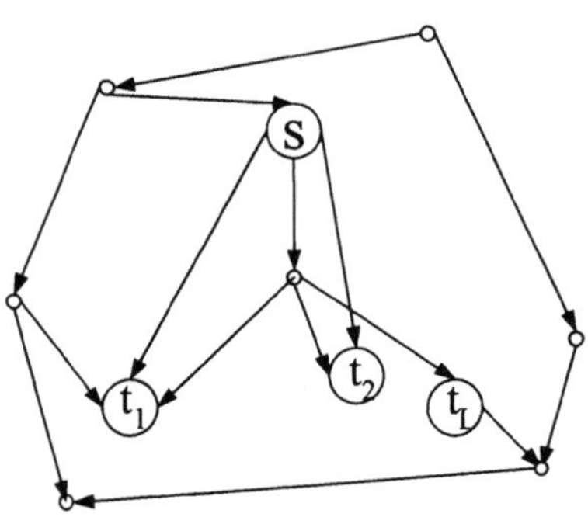

(a) Original Network

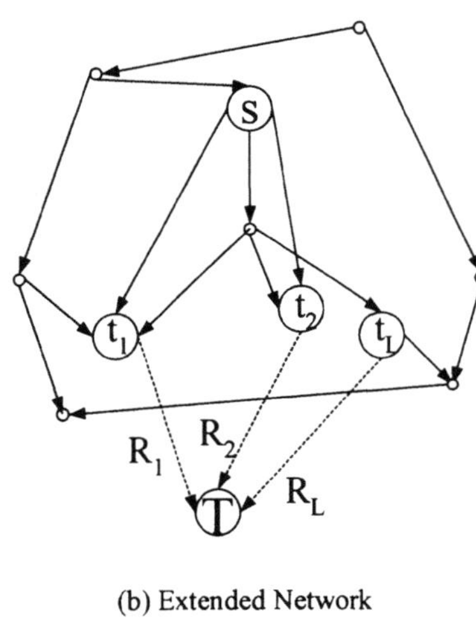

(b) Extended Network

FIGURE 1. Unicast Network

supersink T. The capacity of the link that connects t_i with T is R_i. Clearly, there is one-to-one correspondence between communication from s to T at rate $R_1 + R_2 + \cdots + R_L$ in the extended network and communication from s to $t_1, t_2, \ldots, t_L$ at rates $R_1, R_2, \ldots, R_L$ in the original network. Since the maximal rate in the extended rate is between a single source to a single sink, no coding is required. The min-cut bound can be achieved using Ford-Fulkerson algorithm for maximal flow. Inspecting the various cuts in the extended network, it follows that the capacity region for the original network is:

$$\sum_{t_i \in \Theta} R_i \leq \text{mincut}(s;\Theta), \forall \Theta \subseteq \{t_1, t_2, \cdot, t_L\} \tag{1}$$

where $\text{mincut}(s;\Theta)$ denotes the minimal cut between s and the subset of sinks Θ.

3. Main Result

In the case of two receivers we show how to solve the intermediate case, when there is both common and private information to deliver. Suppose it is required to design a broadcast code with rates (R_0, R_1, R_2). It is first needed to verify that these rates are within the capacity region, which can be found as follows. The original network G can be transformed into an extended network G'. In G' there are two supersinks T_1 and T_2 as in shown in Figure 2. The sink t_1 is connected to

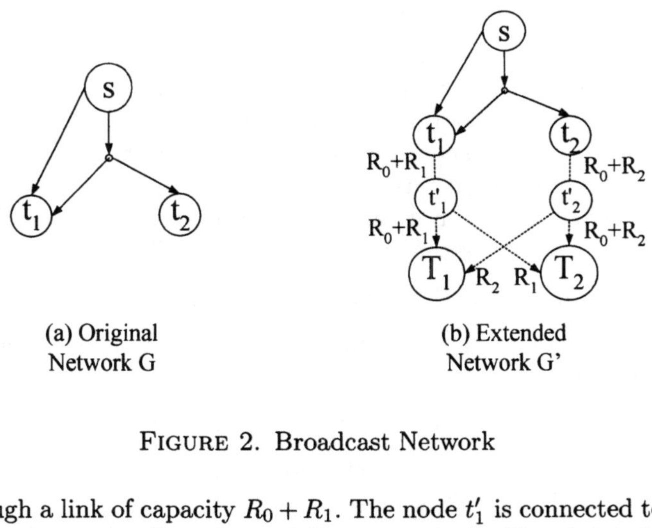

(a) Original Network G

(b) Extended Network G'

FIGURE 2. Broadcast Network

node t_1' through a link of capacity $R_0 + R_1$. The node t_1' is connected to T_1 and T_2 through links of capacity $R_0 + R_1$ and R_1, respectively. The sink t_2 is connected to node t_2' through a link of capacity $R_0 + R_2$. The node t_2' is connected to T_1 and T_2 through links of capacity R_2 and $R_0 + R_2$, respectively.

As was shown in [6],[2],[7] for the common multicast case linear codes achieve the maximal rate. We use notations similar to [7]. We assume that each link has capacity of 1 bit per channel use and that there are C_{ij} parallel links from node i to node j. Denote by h bits per channel use the rate of the code. Any link e has a vector $\mathbf{b}(e)$ of dimension h associated with it. We use the notation $\Gamma_I(v)$ and $\Gamma_O(v)$ for the set of edges reaching and leaving node v, respectively. The source node s gets h binary input symbols[1] denoted $X_1, \ldots, X_h$. The vector $\mathbf{b(e)} \in 2^h$ associated with edge e is given by:

$$\mathbf{b(e)} = \sum_{e' \in \Gamma_I(v)} m_e(e')\mathbf{b(e')} \tag{2}$$

where $\mathbf{b(e')}$ is the vector associated with the incoming edge e' into e and $m_e(e')$ is the coding coefficient.

Assume $y(e)$ is the symbol transmitted on a link e which is given by

$$y(e) = \sum_{e' \in \Gamma_I(v)} m_e(e')y(e') = \mathbf{b(e)}^T\mathbf{x} \tag{3}$$

where $\mathbf{x} = (X_1, \ldots, X_h)^T$ denotes the input vector. Each node v has a subspace associated with it $U(v)$ given by:

$$U(v) = \text{span}\{\mathbf{b(e)} : e \in \Gamma_I(v)\} \tag{4}$$

If the dimension of $U(t_i)$ is h, then sink t_i can reconstruct the transmitted message [6],[2].

Lemma 3.1. *The rate (R_0, R_1, R_2) is achievable in the original network G if and only if $R_0 + R_1 + R_2$ is an achievable multicast rate in the extended graph G'.*

Proof: If (R_0, R_1, R_2) is achievable in G, then t_1 can transmit $R_0 + R_1$ bits to T_1 and t_2 can transmit its private R_2 bits to T_1. Thus T_1 can reconstruct $R_0+R_1+R_2$ bits. Likewise, T_2 can reconstruct $R_0 + R_1 + R_2$ bits. Note that no processing is required in this case by the supersinks, except detecting the incoming information. For the opposite direction, if rate $R_0 + R_1 + R_2$ is an achievable multicast rate in the extended graph G' then it can be achieved by a linear code. For this code, since the entire message is reconstructible by T_1 and T_2, it follows that:

$$\dim\{U(T_1)\} = \dim\{U(T_2)\} = R_0 + R_1 + R_2 \tag{5}$$

where $U(\cdot)$ is defined in (4) and is a subspace of a $R_0+R_1+R_2$-dimensional vector space. It also follows that

$$\dim\{U(t_1') + U(t_2')\} = R_0 + R_1 + R_2 \tag{6}$$

where in the LHS '+' denotes direct sum subspace. Since the capacity in the incoming link to t_1' is $R_0 + R_1$ we have:

$$\dim\{U(t_1')\} \leq R_0 + R_1. \tag{7}$$

From the following relation

$$\dim\{U(T_1)\} \leq \dim\{U(t_1')\} + R_2 \tag{8}$$

and (5) it follows that:

$$\dim\{U(t_1')\} \geq R_0 + R_1 \tag{9}$$

[1]As shown in [7] in the case of two receivers a binary field suffices for the code.

Thus,

$$\dim\{U(t_1')\} = R_0 + R_1. \tag{10}$$

Likewise

$$\dim\{U(t_2')\} = R_0 + R_2. \tag{11}$$

Recall the following general relation:

$$\dim\{U(t_1') \cap U(t_2')\} = \dim\{U(t_1')\} + \dim\{U(t_2')\} - \dim\{U(t_1') + U(t_2')\} \tag{12}$$

From (6), (10), (11) and (12) it follows that:

$$\dim\{U(t_1') \cap U(t_2')\} = R_0 \tag{13}$$

We show now by construction that there exists a (R_0, R_1, R_2) code for G. Denote by U_c the intersection subspace:

$$U_c = U(t_1') \cap U(t_2') = \mathrm{span}\{\mathbf{c_1}, \ldots, \mathbf{c_{R_0}}\} \tag{14}$$

where $\{\mathbf{c_1}, \ldots, \mathbf{c_{R_0}}\}$ is the basis of U_c. Similarly denote by U_a and U_b the following subspaces:

$$\begin{aligned} U_a &= U(t_1')\backslash U_c = \mathrm{span}\{\mathbf{a_1}, \ldots, \mathbf{a_{R_1}}\} \\ U_b &= U(t_2')\backslash U_c = \mathrm{span}\{\mathbf{b_1}, \ldots, \mathbf{b_{R_2}}\} \end{aligned} \tag{15}$$

Clearly, U_a, U_b and U_c have an empty intersection. We have

$$\begin{aligned} U(t_1') &= \mathrm{span}\{\mathbf{a_1}, \ldots, \mathbf{a_{R_1}}, \mathbf{c_1}, \ldots, \mathbf{c_{R_0}}\} \\ U(t_2') &= \mathrm{span}\{\mathbf{b_1}, \ldots, \mathbf{b_{R_2}}, \mathbf{c_1}, \ldots, \mathbf{c_{R_0}}\} \end{aligned} \tag{16}$$

Define the matrix M and the vector $\mathbf{d}$ by the following matrix relation:

$$M\mathbf{x} = \begin{pmatrix} \longleftarrow & \mathbf{a_1}^T & \longrightarrow \\ & \vdots & \\ \longleftarrow & \mathbf{a_{R_1}}^T & \longrightarrow \\ \longleftarrow & \mathbf{b_1}^T & \longrightarrow \\ & \vdots & \\ \longleftarrow & \mathbf{b_{R_2}}^T & \longrightarrow \\ \longleftarrow & \mathbf{c_1}^T & \longrightarrow \\ & \vdots & \\ \longleftarrow & \mathbf{c_{R_0}}^T & \longrightarrow \end{pmatrix} \begin{pmatrix} X_1 \\ \vdots \\ X_{R_0+R_1+R_2} \end{pmatrix} = \begin{pmatrix} d_1 \\ \vdots \\ d_{R_1} \\ \hline d_{R_1+1} \\ \vdots \\ d_{R_1+R_2} \\ \hline d_{R_1+R_2+1} \\ \vdots \\ d_{R_0+R_1+R_2} \end{pmatrix} = \mathbf{d} \tag{17}$$

From (6) it follows that M is full rank. Thus, if $\{X_1, \ldots, X_{R_0+R_1+R_2}\}$ are statistically independent and uniformly distributed, so are $\{d_1, \ldots, d_{R_0+R_1+R_2}\}$ and vice versa. The first R_1 symbols of $\mathbf{d}$ can be reconstructed at t_1' only and are therefore the private data to t_1', and hence also to t_1. The next R_2 symbols of $\mathbf{d}$ are the private data to t_2', and hence also to t_2. The last R_0 symbols of $\mathbf{d}$ are the common data. Therefore, (R_0, R_1, R_2) is achievable in G. □

It follows that in order to find the achievable region for G, we have to find the achievable multicast rate for G', according to the min-cut. The following theorem can be proved immediately using the lemma.

Theorem 3.1. *The achievable rate region* (R_0, R_1, R_2) *for the broadcast network* G *is given by:*

$$R_0 + R_1 \leq \text{mincut}\,(s; t_1) \quad (18)$$
$$R_0 + R_2 \leq \text{mincut}\,(s; t_2) \quad (19)$$
$$R_0 + R_1 + R_2 \leq \text{mincut}\,(s; t_1, t_2) \quad (20)$$

Proof. The bounds (18) and (19) are trivial. The potential minimal cuts between s and T_1, as shown in Figure 3, can be divided into three characteristic types. Cuts of type a yield the bound $R_0 + R_1 + R_2 \leq R_0 + R_1 + \text{cut}(s, t_1; t_2)$ or $R_2 \leq \text{cut}(s, t_1; t_2)$. Cuts of type b yield the bound $R_0 + R_1 + R_2 \leq R_2 + \text{cut}(s, t_2; t_1)$ or $R_0 + R_1 \leq \text{cut}(s, t_2; t_1)$. Cuts of type c yield the bound $R_0 + R_1 + R_2 \leq \text{cut}(s; t_2, t_1)$. Similar bounds hold for T_2. It follows that the restrictive cuts can be only of type b and c. The bounds of the theorem follow. □

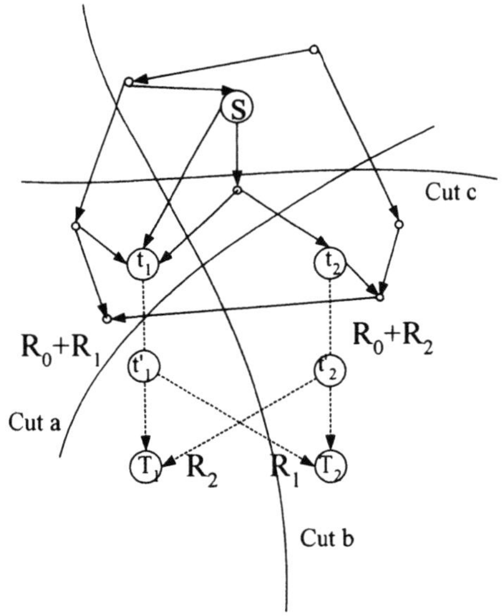

FIGURE 3. Characteristic Cuts

Note that for the multicast case, i.e., $R_1 = R_2 = 0$, the capacity region becomes $R_0 \leq \min\{\text{mincut}(s, t_1), \text{mincut}(s; t_2)\}$, as expected. For the unicast case, i.e., $R_0 = 0$, the supersinks T_1 and T_2 are equivalent and therefore a single supersink suffices and no coding is required, as expected.

There is a one-to-one correspondence between a multicast code in G' and a broadcast code G. In order to design a broadcast code for G, we begin by designing a multicast code in G'. The multicast code can be designed using the polynomial time algorithm developed in [7]. Given a multicast code for G', a broadcast code for G is derived using simple processing, as explained in the proof of Lemma 3.1. The procedure is illustrated in the following example.

Example. Figure 4 shows the graph, already with its extension (dashed lines). Unless otherwise specified, the capacities of each edge is 1 bit. The capacity region

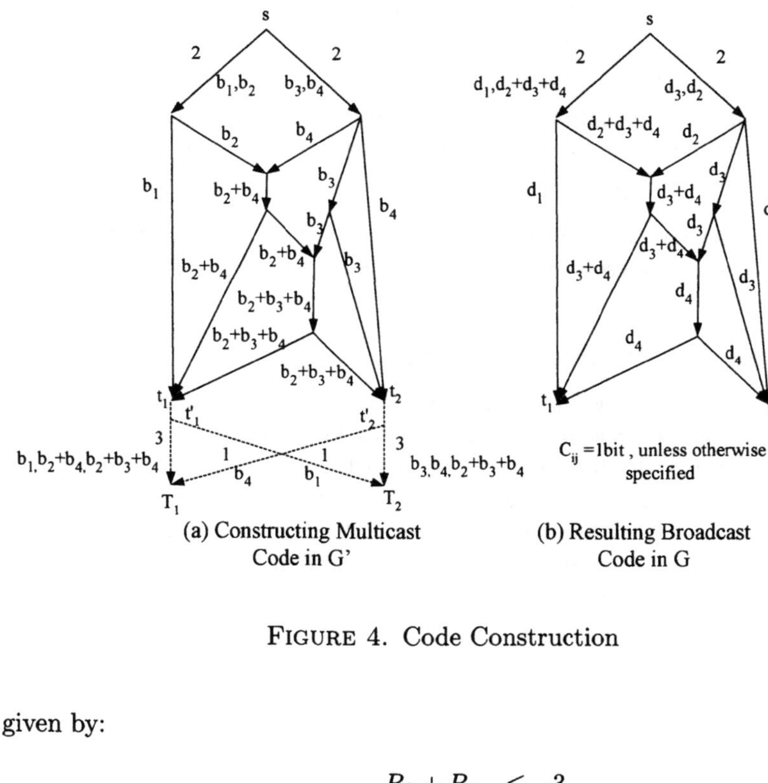

FIGURE 4. Code Construction

is given by:

$$\begin{aligned} R_0 + R_1 &\leq 3 \\ R_0 + R_2 &\leq 3 \\ R_0 + R_1 + R_2 &\leq 4 \end{aligned} \tag{21}$$

Therefore, the rate $(R_0, R_1, R_2) = (2, 1, 1)$ is in the capacity region. We design the multicast code for the extended graph, as shown in Figure 4. The information received by t_1' for this code is $b_1, b_2 + b_4, b_2 + b_3 + b_4$ and the information received by t_2' is $b_3, b_4, b_2 + b_3 + b_4$. Thus we have:

$$\begin{aligned} U(t_1') &= \text{span}\{(1,0,0,0)^T, (0,0,1,0)^T, (0,1,1,1)^T\} \\ U(t_2') &= \text{span}\{(0,0,0,1)^T, (0,0,1,0)^T, (0,1,1,1)^T\} \end{aligned} \tag{22}$$

Thus,

$$\begin{aligned} U_a &= \text{span}\{(1,0,0,0)^T\} \\ U_b &= \text{span}\{(0,0,0,1)^T\} \\ U_c &= \text{span}\{(0,0,1,0)^T, (0,1,1,1)^T\} \end{aligned} \tag{23}$$

and

$$M\mathbf{x} = \begin{pmatrix} 1 & 0 & 0 & 0 \\ 0 & 0 & 0 & 1 \\ 0 & 0 & 1 & 0 \\ 0 & 1 & 1 & 1 \end{pmatrix} \begin{pmatrix} b_1 \\ b_2 \\ b_3 \\ b_4 \end{pmatrix} = \begin{pmatrix} d_1 \\ d_2 \\ d_3 \\ d_4 \end{pmatrix} \tag{24}$$

It follows that d_1, d_3, d_4 can be reconstructed at t_1 and d_2, d_3, d_4 at t_2. Thus d_3 and d_4 are the common information, whereas d_1 and d_2 are the private information. Since M is guaranteed to be full rank, matrix inversion in M of (24) yields:

$$b_1 = d_1, b_2 = d_2 + d_3 + d_4, b_3 = d_3, b_4 = d_2 \tag{25}$$

The final code constructed is also given in Figure 4. Note that unlike the code for the extended graph, the code for G requires (pre)coding at the source.

4. Conclusion and Further Research

In this paper we have shown how an extension of a graph enabled us to find the capacity region and to construct network codes for more general scenarios than multicast. Unfortunately, it is not known whether it is possible to extend this technique to more than two receivers, and what is the capacity region for that case. For example, in [6] the scenario of Figure 5 is given. The random process X_0 is the common information with rate R_0 and the random process X_2 is the private information of t_2 with rate R_2. By inspection, it can be seen that the bounds on

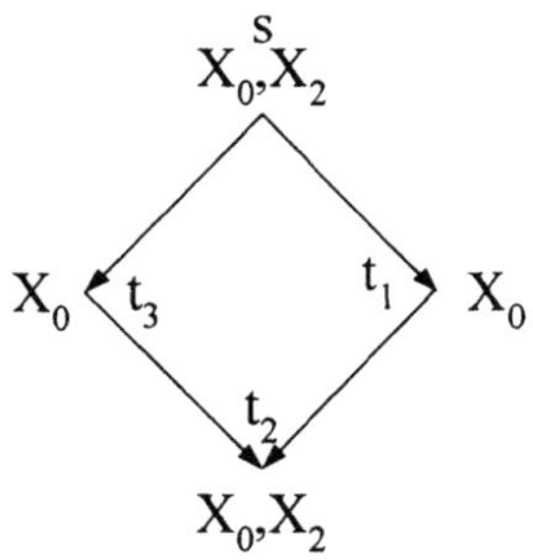

FIGURE 5. Broadcast Network With Three Receivers [Yeung 2002]

the rates are given by $2R_0 + R_2 \leq 2$. However, we have not found an extension of the graph that will enable to solve it. The problem seems to be that whereas for the two receivers case, the only requirement is that a certain amount of information will be private, and a certain amount common, here we have stricter requirements on exactly which information is common and which is private.

Interference is a different scenario in a network, when a source s_1 has to transmit information to a sink t_1 with rate R_1 and a different source s_2 has to transmit information to t_2 with rate R_2. A possible code, which is not necessarily optimal, can be achieved by joining s_1 and s_2 to a super source S with links of capacities R_1 and R_2, respectively. The super source S is then required to multicast the same information to t_1 and t_2. Better rates can be shown to be achieved using our method, where only a certain part of the information is common, and the rest is private.

References

[1] R. Ahlswede, N. Cai, S.-Y.R. Li, and R.W. Yeung, "Network information flow," *IEEE Transactions on Information Theory*, vol. 46, no. 4, pp. 1204–1216, July 2000.

[2] R. Koetter and M. Medard, "An algebraic approach to network coding," *Proceedings of INFOCOM*, 2002.

[3] T.M. Cover, "Comments on broadcast channels," *IEEE Transactions on Information Theory*, vol. 44, pp. 2524–2530, 1998.

[4] K. Marton, "A coding theorem for the discrete memoryless broadcast channel," *IEEE Transactions on Information Theory*, vol. 25, pp. 306–311, 1979.

[5] T. M. Cover, "An achievable rate region for the broadcast channel," *IEEE Transactions on Information Theory*, vol. 21, pp. 399–404, 1975.

[6] R. Yeung, *A First Course in Information Theory*, Kluwer Academic/Plenum Publishers, March 2002.

[7] P. Sanders, S. Egner, and L. Tolhuizen, "Polynomial time algorithms for network information flow," 2002.

Elona Erez and Meir Feder
Dept. of Electrical Engineering-Systems
Tel Aviv University
Tel Aviv, 69978, Israel
e-mail: `elona@eng.tau.ac.il`
e-mail: `meir@eng.tau.ac.il`

Progress in Computer Science and Applied Logic, Vol. 23, 139–152

Constructions of Nonbinary Codes Correcting t-Symmetric Errors and Detecting All Unidirectional Errors: Magnitude Error Criterion

Fang-Wei Fu, San Ling and Chaoping Xing

Abstract. In this paper, based on residue rings of polynomials, we present a general construction for nonbinary codes capable of correcting t or fewer symmetric errors and detecting all unidirectional errors with the magnitude error criterion. Some new lower bounds for such codes are obtained from this general construction.

Mathematics Subject Classification (2000). Primary 94B15; Secondary 94B60.

Keywords. Coding theory, nonbinary codes, code construction, magnitude error criterion, unidirectional errors, residue rings of polynomials.

1. Introduction

Let $V = \{0, 1, \cdots, m-1\}$ be a finite set where $m \geq 2$ is a positive integer. Let V^n be the set of n-tuples over V, i.e.,

$$V^n = \{(x_1, x_2, \cdots, x_n) \mid x_i \in V,\ i = 1, 2, \cdots, n\}.$$

For $\mathbf{x} = (x_1, x_2, \cdots, x_n) \in V^n$ and $\mathbf{y} = (y_1, y_2, \cdots, y_n) \in V^n$, the Hamming distance $d_H(\mathbf{x}, \mathbf{y})$ between $\mathbf{x}$ and $\mathbf{y}$ is the number of coordinates in which they differ, i.e.,

$$d_H(\mathbf{x}, \mathbf{y}) = \mid \{i \mid x_i \neq y_i\} \mid .$$

The L^1-distance $d_1(\mathbf{x}, \mathbf{y})$ between $\mathbf{x}$ and $\mathbf{y}$ is defined as

$$d_1(\mathbf{x}, \mathbf{y}) = \sum\nolimits_{i=1}^{n} \mid x_i - y_i \mid .$$

This research work is supported in part by the DSTA project (POD 0103223), the National Natural Science Foundation of China under the Grant 60172060, the Trans-Century Training Program Foundation for the Talents by the Education Ministry of China, and the Foundation for University Key Teacher by the Education Ministry of China, the MOE-ARF grant R-146-000-029-112 and the 100-Talents program of the Chinese Academy of Science.

Obviously, for $m = 2$, $d_1(\mathbf{x}, \mathbf{y}) = d_H(\mathbf{x}, \mathbf{y})$. Denote

$$N(\mathbf{x}, \mathbf{y}) = \sum_{i=1}^{n} \max\{y_i - x_i, 0\}.$$

The asymmetric distance $d_a(\mathbf{x}, \mathbf{y})$ between $\mathbf{x}$ and $\mathbf{y}$ is defined as

$$d_a(\mathbf{x}, \mathbf{y}) = \max\{N(\mathbf{x}, \mathbf{y}), N(\mathbf{y}, \mathbf{x})\}.$$

Clearly,

$$d_1(\mathbf{x}, \mathbf{y}) = N(\mathbf{x}, \mathbf{y}) + N(\mathbf{y}, \mathbf{x}).$$

For $\mathbf{x} = (x_1, \cdots, x_n) \in V^n$ and $\mathbf{y} = (y_1, \cdots, y_n) \in V^n$, we say $\mathbf{x} \le \mathbf{y}$ if $x_i \le y_i$ for all i. Note that if $\mathbf{x} \le \mathbf{y}$, then $N(\mathbf{y}, \mathbf{x}) = 0$ and $d_1(\mathbf{x}, \mathbf{y}) = d_a(\mathbf{x}, \mathbf{y}) = N(\mathbf{x}, \mathbf{y})$.

A nonempty subset C of V^n is called an m-ary code of length n. Furthermore, if $|C| = M$, then C is called an $(n, M)_m$ code. Any word $\mathbf{c}$ in C is called a codeword of C. The code C is used to transmit information in digital communication systems. In classical coding theory, when a codeword $\mathbf{c} \in C$ is sent and a vector $\mathbf{y} \in V^n$ is received, the number of errors occurred is defined as the number of coordinates in which they differ, i.e., $d_H(\mathbf{c}, \mathbf{y})$. We call this error criterion the Hamming error criterion. Note that the magnitude of the difference at each of these coordinates is not important in this definition. If one wishes to take into account the magnitude of each symbol error, a suitable and widely used definition for the number of errors occurred is $\sum_{i=1}^{n} |y_i - c_i|$, i.e., $d_1(\mathbf{c}, \mathbf{y})$. We call this error criterion the magnitude error criterion. In this paper, we study the constructions of codes with the magnitude error criterion. This topic has been dealt with in [10], [12], [15], [29] and [32].

Three types of errors, asymmetric errors, unidirectional errors and symmetric errors, are defined as follows (see [2]). Suppose a codeword $\mathbf{c} \in C$ is sent and a vector $\mathbf{y} \in V^n$ is received. The number of errors occurred is $d_1(\mathbf{c}, \mathbf{y})$.

(i) We say that $\mathbf{c}$ has suffered asymmetric errors if $\mathbf{y} \le \mathbf{c}$.
(ii) We say that $\mathbf{c}$ has suffered unidirectional errors if either $\mathbf{y} \le \mathbf{c}$ or $\mathbf{c} \le \mathbf{y}$.
(iii) In general we say that $\mathbf{c}$ has suffered symmetric errors.

Note that for the symmetric errors, we do not impose any specific relation between $\mathbf{c}$ and $\mathbf{y}$ (such as $\mathbf{y} \le \mathbf{c}$ or $\mathbf{c} \le \mathbf{y}$). The following theorem (see [10], [15], [32]) gives necessary and sufficient conditions on block codes correcting/detecting certain types of errors.

Theorem 1.1. *With the magnitude error criterion,*

(i) *a code C is capable of correcting t or fewer symmetric errors if and only if $d_1(\mathbf{x}, \mathbf{y}) \ge 2t + 1$ for all $\mathbf{x}$, $\mathbf{y} \in C$ and $\mathbf{x} \ne \mathbf{y}$;*
(ii) *a code C is capable of correcting t or fewer asymmetric errors if and only if $d_a(\mathbf{x}, \mathbf{y}) \ge t + 1$ for all $\mathbf{x}$, $\mathbf{y} \in C$ and $\mathbf{x} \ne \mathbf{y}$;*
(iii) *a code C is capable of correcting t or fewer symmetric errors and detecting all unidirectional errors if and only if $N(\mathbf{x}, \mathbf{y}) \ge t + 1$ and $N(\mathbf{y}, \mathbf{x}) \ge t + 1$ for all $\mathbf{x}$, $\mathbf{y} \in C$ and $\mathbf{x} \ne \mathbf{y}$.*

Note that for $m = 2$, the magnitude error criterion is just the Hamming error criterion. Hence, Theorem 1.1 generalizes the corresponding theorems for binary codes to m-ary codes with the magnitude error criterion. Actually, one can prove Theorem 1.1 by extending and modifying the arguments used for proving the corresponding theorems for binary codes (see [2] and [32]). Clearly, one can see from the definitions of errors of the three types that if a code C is capable of correcting t or fewer symmetric errors, then it is capable of correcting t or fewer unidirectional errors; if a code C is capable of correcting t or fewer unidirectional errors, then it is capable of correcting t or fewer asymmetric errors. Note that nonbinary codes for correcting asymmetric errors have been studied in [10], [12], [15], [16], [27], [29] and [32].

Remark 1.2. Weber et al. [32] gave necessary and sufficient conditions for a block code to be capable of correcting up to t_1 symmetric errors, up to t_2 unidirectional errors, and up to t_3 asymmetric errors, as well as detecting from $t_1 + 1$ to d_1 symmetric errors that are not of the unidirectional type, from t_2+1 to d_2 unidirectional errors that are not of the asymmetric type, and from t_3+1 to d_3 asymmetric errors. As special cases, Theorem 1.1 follows directly from these general necessary and sufficient conditions. In this paper, we only need to use Theorem 1.1 to establish our results.

Let $\Gamma_m(n,t)$ denote the maximum size of an $(n, M)_m$ code which is capable of correcting t or fewer symmetric errors and detecting all unidirectional errors with the magnitude error criterion. By Theorem 1.1, we know that $\Gamma_m(n,t)$ is the maximum size of an $(n, M)_m$ code C satisfying $N(\mathbf{x},\mathbf{y}) \geq t+1$ and $N(\mathbf{y},\mathbf{x}) \geq t+1$ for all $\mathbf{x}$, $\mathbf{y} \in C$ and $\mathbf{x} \neq \mathbf{y}$. Note that binary codes capable of correcting t or fewer symmetric errors and detecting all unidirectional errors have been studied in [1]–[7] and [10]–[36].

In this paper, based on residue rings of polynomials, we present a general construction for nonbinary codes capable of correcting t or fewer symmetric errors and detecting all unidirectional errors with the magnitude error criterion. Some new lower bounds for such codes are obtained from this general construction. This paper is organized as follows. In Section 2, we review and derive some basic properties of generalized binomial coefficients introduced and studied in [8] and [9]. In Section 3, we present a general construction for nonbinary codes for correcting t or fewer symmetric errors and detecting all unidirectional errors. In Section 4, some new lower bounds for $\Gamma_m(n,t)$ are given.

2. Generalized Binomial Coefficients

In this section, we review and derive some basic properties of generalized binomial coefficients introduced and studied in [8] and [9].

Given three integers $m, n \geq 1$ and $r \geq 0$, the generalized binomial coefficient $\binom{n}{r}_m$ is defined as follows:

$$\binom{1}{r}_m = \begin{cases} 1, & \text{if } 0 \leq r \leq m-1 \\ 0, & \text{otherwise} \end{cases}$$

and

$$\binom{n}{r}_m = \sum_{i=0}^{m-1} \binom{n-1}{r-i}_m \quad \text{for } n \geq 2.$$

The following basic properties of generalized binomial coefficients are listed in [8, pp. 215-216].

Properties:

(i) $\binom{n}{r}_m$ is the number of integer solutions to the equation

$$x_1 + x_2 + \cdots + x_n = r$$

with $0 \leq x_i \leq m-1$ for each $i = 1, 2, \cdots, n$;

(ii) $\binom{n}{0}_m = 1$;

(iii) $\binom{n}{1}_m = n$, where $m \geq 2$;

(iv) $\binom{n}{r}_m = \sum_{i=0}^{n}(-1)^i \binom{n}{i}\binom{n-1+r-mi}{n-1}$;

(v) $\binom{n}{r}_m = \binom{n}{s}_m$, where $r + s = n(m-1)$.

By Property (i), we have

$$\binom{n}{r}_m = 0 \quad \text{for } r < 0 \text{ or } r > n(m-1).$$

Hence, we only need to consider $\binom{n}{r}_m$ for the case $0 \leq r \leq n(m-1)$. Obviously, by Property (i), $\binom{n}{r}_m \geq 1$ for $0 \leq r \leq n(m-1)$. Below we derive the following unimodal property of $\binom{n}{r}_m$.

Lemma 2.1. *For $n \geq 2$,*

(i) *if $n(m-1)$ is even, then*

$$\binom{n}{0}_m < \binom{n}{1}_m < \cdots < \binom{n}{n(m-1)/2}_m > \cdots$$
$$\cdots > \binom{n}{n(m-1)-1}_m > \binom{n}{n(m-1)}_m;$$

(ii) *if $n(m-1)$ is odd, then*

$$\binom{n}{0}_m < \binom{n}{1}_m < \cdots < \binom{n}{[n(m-1)-1]/2}_m = \binom{n}{[n(m-1)+1]/2}_m > \cdots$$
$$\cdots > \binom{n}{n(m-1)-1}_m > \binom{n}{n(m-1)}_m.$$

Proof. By Property (v), we only need to prove that for $n \geq 2$

$$\binom{n}{r-1}_m < \binom{n}{r}_m \quad \text{for } 1 \leq r \leq \frac{n(m-1)}{2}. \tag{2.1}$$

Below we prove (2.1) by mathematical induction. From the definition of $\binom{n}{r}_m$, we have

$$\begin{aligned}\binom{n}{r}_m - \binom{n}{r-1}_m &= \sum_{i=0}^{m-1}\binom{n-1}{r-i}_m - \sum_{i=0}^{m-1}\binom{n-1}{r-1-i}_m \\ &= \binom{n-1}{r}_m - \binom{n-1}{r-m}_m. \end{aligned}\tag{2.2}$$

For $n = 2$, it follows from (2.2) and the definition of $\binom{1}{r}_m$ that for $1 \leq r \leq m-1$,

$$\binom{2}{r}_m - \binom{2}{r-1}_m = \binom{1}{r}_m - \binom{1}{r-m}_m = 1 - 0 = 1 > 0.$$

Hence, (2.1) is true for $n = 2$. Assume that (2.1) is true for $n = k-1$. Now we prove that (2.1) is true for $n = k$. By (2.2),

$$\binom{k}{r}_m - \binom{k}{r-1}_m = \binom{k-1}{r}_m - \binom{k-1}{r-m}_m, \quad 1 \leq r \leq \frac{k(m-1)}{2}. \tag{2.3}$$

Next we consider three cases:

(A) If $1 \leq r \leq m-1$, then $\binom{k-1}{r-m}_m = 0$. Hence, by (2.3), $\binom{k}{r}_m > \binom{k}{r-1}_m$.

(B) If $m \leq r \leq \frac{(k-1)(m-1)}{2}$, then $r - m < r \leq \frac{(k-1)(m-1)}{2}$. Hence, by induction assumption, we have

$$\binom{k-1}{r}_m > \binom{k-1}{r-1}_m > \cdots > \binom{k-1}{r-m}_m.$$

By (2.3), this implies that

$$\binom{k}{r}_m > \binom{k}{r-1}_m.$$

(C) If $\frac{(k-1)(m-1)}{2} < r \leq \frac{k(m-1)}{2}$, then

$$\frac{(k-2)(m-1)}{2} \leq (k-1)(m-1) - r < \frac{(k-1)(m-1)}{2}$$

and

$$r - m \leq \frac{k(m-1)}{2} - m < \frac{(k-2)(m-1)}{2}.$$

Hence,

$$r - m < (k-1)(m-1) - r < \frac{(k-1)(m-1)}{2}.$$

By (2.3), Property (v) and the induction assumption,

$$\binom{k}{r}_m - \binom{k}{r-1}_m = \binom{k-1}{(k-1)(m-1)-r}_m - \binom{k-1}{r-m}_m > 0.$$

From the discussion in (A), (B) and (C), we see that (2.1) is true for $n = k$. Hence, by induction, (2.1) is true. □

The following result follows from Lemma 2.1 immediately.

Proposition 2.2.

$$\binom{n}{\lfloor n(m-1)/2 \rfloor}_m = \max\nolimits_{0 \le r \le n(m-1)} \binom{n}{r}_m$$

where $\lfloor x \rfloor$ *is the greatest integer less than or equal to* x.

3. A General Construction

Xing [33] gave a construction of binary constant weight codes. By modifying his method, Fu, Ling and Xing [11] presented a general construction for binary asymmetric error-correcting codes. Bose and Rao [6] gave a construction of binary codes capable of correcting t or fewer symmetric errors and detecting all unidirectional errors by using binary constant weight codes. By modifying and generalizing the methods in [6], [11] and [33], we present a general construction for nonbinary codes capable of correcting t or fewer symmetric errors and detecting all unidirectional errors with the magnitude error criterion.

Let $\mathbf{F}_q$ be a finite field of q elements, where q is a prime power. Let $\mathbf{F}_q^*$ be the set of nonzero elements of $\mathbf{F}_q$. For a monic polynomial $f(x) \in \mathbf{F}_q[x]$, consider the residue class ring

$$R = \mathbf{F}_q[x]/(f(x)).$$

For simplicity, in this paper, we can also make the following identification:

$$R = \{g(x) \in \mathbf{F}_q[x] : \deg(g(x)) < \deg(f(x))\}.$$

The addition and multiplication operations in R are the polynomial addition and multiplication modulo $f(x)$.

Let $f(x)$ have the factorization

$$f(x) = \prod\nolimits_{i=1}^{k} p_i^{e_i}(x),$$

where $p_1(x), \cdots, p_k(x)$ are distinct monic irreducible polynomials in $\mathbf{F}_q[x]$ and $e_1, \cdots, e_k$ are positive integers. It is known that all invertible polynomials of the ring R form a multiplicative group, denoted by G. It is a finite abelian group and consists of all polynomials in R which are co-prime to $f(x)$, that is

$$G = \{g(x) \in \mathbf{F}_q[x] : \deg(g(x)) < \deg(f(x)) \text{ and } (g(x), f(x)) = 1\}. \tag{3.1}$$

The multiplication operation $\bigodot$ over G is the polynomial multiplication modulo $f(x)$. This group contains exactly

$$\Phi(f(x)) \triangleq \prod\nolimits_{i=1}^{k} (q^{d_i} - 1) q^{d_i(e_i - 1)} \tag{3.2}$$

elements, where d_i is the degree of $p_i(x)$. Below we use the group G to construct nonbinary codes capable of correcting t or fewer symmetric errors and detecting all unidirectional errors with the magnitude error criterion.

For $0 \le r \le n(m-1)$, denote

$$V^n(r) = \{\mathbf{y} = (y_1, y_2, \cdots, y_n) \in V^n : \sum_{i=1}^{n} y_i = r\}.$$

By Proposition 2.2 and Property (i) of $\binom{n}{r}_m$, we have

$$|V^n(r)| = \binom{n}{r}_m$$

and

$$\begin{aligned} |V^n(\lfloor n(m-1)/2 \rfloor)| &= \binom{n}{\lfloor n(m-1)/2 \rfloor}_m \\ &= \max_{0 \le r \le n(m-1)} \binom{n}{r}_m \\ &= \max_{0 \le r \le n(m-1)} |V^n(r)|. \end{aligned}$$

Construction. Let m, n and t be three positive integers satisfying $m \ge 2$, $n \le q$ and $1 \le t < n(m-1)$. Let $f(x) \in \mathbf{F}_q[x]$ be a monic polynomial of degree t such that there exist n distinct elements $\alpha_1, \alpha_2, \cdots, \alpha_n \in \mathbf{F}_q$ with $f(\alpha_i) \ne 0$ for all $i = 1, 2, \cdots, n$.

Since $f(\alpha_i) \ne 0$ for $i = 1, 2, \cdots, n$, then $(x - \alpha_i)$ is co-prime to $f(x)$ for $i = 1, 2, \cdots, n$. Hence

$$(x - \alpha_i) \in G, \quad i = 1, 2, \cdots, n.$$

Consider the map

$$\Omega : V^n(\lfloor n(m-1)/2 \rfloor) \to G, \quad (c_1, c_2, \cdots, c_n) \mapsto \prod_{i=1}^{n} \bigodot (x - \alpha_i)^{c_i} \in G.$$

For every $g(x) \in G$, denote

$$C_g = \Omega^{-1}(g(x)).$$

For every $g \in G$, if $C_g \ne \emptyset$, then C_g is an m-ary code C of length n capable of correcting t or fewer symmetric errors and detecting all unidirectional errors with the magnitude error criterion.

Proof of the construction. By Theorem 1.1, we want to show that

$$N(\mathbf{u}, \mathbf{v}) \ge t + 1 \quad \text{and} \quad N(\mathbf{v}, \mathbf{u}) \ge t + 1$$

for all $\mathbf{u}$, $\mathbf{v} \in C_g$ and $\mathbf{u} \ne \mathbf{v}$.

Let $\mathbf{u} = (u_1, u_2, \cdots, u_n)$ and $\mathbf{v} = (v_1, v_2, \cdots, v_n)$. Since

$$\mathbf{u}, \mathbf{v} \in C_g \subseteq V^n(\lfloor n(m-1)/2 \rfloor),$$

then

$$\sum_{i=1}^{n} u_i = \sum_{i=1}^{n} v_i = \lfloor n(m-1)/2 \rfloor \tag{3.3}$$

and

$$\Omega(\mathbf{u}) = \Omega(\mathbf{v}) = g(x) \in G. \tag{3.4}$$

It follows from (3.3) that

$$N(\mathbf{u}, \mathbf{v}) = N(\mathbf{v}, \mathbf{u}). \tag{3.5}$$

By (3.4), the element $\Omega(\mathbf{u})/\Omega(\mathbf{v})$ is the identity of G. This implies that in the group G

$$\frac{\Omega(\mathbf{u})}{\Omega(\mathbf{v})} = \frac{\prod_{i=1}^{n} \bigodot(x-\alpha_i)^{u_i}}{\prod_{i=1}^{n} \bigodot(x-\alpha_i)^{v_i}} = 1. \tag{3.6}$$

Denote

$$S = \{i : \; v_i > u_i\}$$

and

$$T = \{i : \; u_i > v_i\}.$$

Then $S \bigcap T = \emptyset$, and either $S \neq \emptyset$ or $T \neq \emptyset$ since $\mathbf{u} \neq \mathbf{v}$. Furthermore,

$$N(\mathbf{u}, \mathbf{v}) = \sum\nolimits_{i \in S} (v_i - u_i), \tag{3.7}$$

$$N(\mathbf{v}, \mathbf{u}) = \sum\nolimits_{j \in T} (u_j - v_j). \tag{3.8}$$

It is easy to see from (3.6) that

$$\frac{\Omega(\mathbf{u})}{\Omega(\mathbf{v})} = \frac{\prod_{j\in T} \bigodot(x-\alpha_j)^{u_j-v_j}}{\prod_{i\in S} \bigodot(x-\alpha_i)^{v_i-u_i}} = 1$$

in the group G. This is equivalent to the fact that $f(x)$ divides the polynomial

$$A(x) \triangleq \prod\nolimits_{j\in T} (x-\alpha_j)^{u_j-v_j} - \prod\nolimits_{i\in S} (x-\alpha_i)^{v_i-u_i} \in \mathbf{F}_q[x].$$

The roots of the polynomial $\prod_{j\in T}(x-\alpha_j)^{u_j-v_j}$ are $\alpha_j, j \in T$, and the roots of the polynomial $\prod_{i\in S}(x-\alpha_i)^{v_i-u_i}$ are $\alpha_i, i \in S$. Since

$$\{\alpha_i : i \in S\} \bigcap \{\alpha_i : i \in T\} = \emptyset$$

and either $S \neq \emptyset$ or $T \neq \emptyset$, we have

$$\prod\nolimits_{j\in T} (x-\alpha_j)^{u_j-v_j} \neq \prod\nolimits_{i\in S} (x-\alpha_i)^{v_i-u_i}.$$

Hence, $A(x) \neq 0$. By (3.7), (3.8) and the fact that $N(\mathbf{u}, \mathbf{v}) = N(\mathbf{v}, \mathbf{u})$, we know that the degree of $A(x)$ is at most $N(\mathbf{u}, \mathbf{v}) - 1$. Since $A(x) \neq 0$, we have

$$N(\mathbf{u}, \mathbf{v}) - 1 \geq \deg(A(x)) \geq \deg(f(x)) = t.$$

Hence,

$$N(\mathbf{u}, \mathbf{v}) = N(\mathbf{v}, \mathbf{u}) \geq t + 1.$$

This completes the proof. □

From the construction, we know that $C_g,\ g \in G$ form a partition of $V^n(\lfloor n(m-1)/2 \rfloor)$. Since $|\, G \,| = \Phi(f(x))$, we can find one element $\pi(x) \in G$ such that

$$|\, C_\pi \,| \geq \frac{\binom{n}{\lfloor n(m-1)/2 \rfloor}_m}{\Phi(f(x))}. \tag{3.9}$$

Hence, we obtain the following result.

Theorem 3.1. *Let $\mathbf{F}_q$ be a finite field of q elements, where q is a prime power. Let m, n and t be three positive integers satisfying $m \geq 2$, $n \leq q$ and $1 \leq t < n(m-1)$. Let $f(x) \in \mathbf{F}_q[x]$ be a monic polynomial of degree t such that there exist n distinct elements $\alpha_1, \alpha_2, \cdots, \alpha_n \in \mathbf{F}_q$ with $f(\alpha_i) \neq 0$ for all $i = 1, 2, \cdots, n$. Then there exists an m-ary code C of length n and size*

$$|\, C \,| \geq \frac{\binom{n}{\lfloor n(m-1)/2 \rfloor}_m}{\Phi(f(x))}, \tag{3.10}$$

which can correct t or fewer symmetric errors and detect all unidirectional errors with the magnitude error criterion.

From the construction, it is easy to see that

Corollary 3.2. *With notations as in the construction, we have*

$$\Gamma_m(n,t) \geq \max\nolimits_{g \in G} |\, C_g \,|. \tag{3.11}$$

Bound (3.11) is in general stronger than Bound (3.10), but it is less explicit and requires more computation to determine.

Remark 3.3. In the proof of the construction, if we define the map $\Omega : V^n(r) \to G$ in the same way, we obtain a code with at least $\binom{n}{r}_m / \Phi(f(x))$ codewords. By Proposition 2.2, we take $r = \lfloor n(m-1)/2 \rfloor$ in order to make the code size big.

4. New Lower Bounds for $\Gamma_m(n,t)$

In this section, we show that some new lower bounds for $\Gamma_m(n,t)$ can be obtained from Theorem 3.1. Note that the lower bounds for $\Gamma_m(n,t)$ obtained by Theorem 3.1 depend on the selection of $f(x)$. It seems that the following selections of $f(x)$ are optimal for the corresponding cases.

Theorem 4.1.

(i) *If n is a prime power, $n \geq m$ and $2 \leq t < n(m-1)$, then*

$$\Gamma_m(n,t) \geq \frac{\binom{n}{\lfloor n(m-1)/2 \rfloor}_m}{(n^2-1)^r (n^3-1)^s}. \tag{4.1}$$

where r and s are the two unique non-negative integers satisfying $t = 2r + 3s$ and $s \in \{0, 1\}$.

(ii) *If n is not a prime power and $n \geq m$, denote k as the least positive integer such that $q = n + k$ is a prime power. If $2 \leq t \leq k$, then*

$$\Gamma_m(n,t) \geq \frac{\binom{n}{\lfloor n(m-1)/2 \rfloor}_m}{(q-1)^t}. \tag{4.2}$$

If $k < t < n(m-1)$, then

$$\Gamma_m(n,t) \geq \frac{\binom{n}{\lfloor n(m-1)/2 \rfloor}_m}{(q-1)^k q^{s'} (q^2-1)^{r'}}, \tag{4.3}$$

where r' and s' are the two unique non-negative integers satisfying $t - k = 2r' + s'$ and $s' \in \{0,1\}$.

Proof. (i) Let $q = n$ in Theorem 3.1 since n is a prime power. Let

$$\mathbf{F}_q = \{\alpha_1, \alpha_2, \cdots, \alpha_q\}.$$

Note that the number of monic quadratic irreducible polynomials over $\mathbf{F}_q$ is $q(q-1)/2$. Since $n \geq m$, we have

$$r \leq \frac{t}{2} < \frac{n(m-1)}{2} \leq \frac{n(n-1)}{2} = \frac{q(q-1)}{2}.$$

Hence, we can choose r distinct monic quadratic irreducible polynomials

$$p_1(x), p_2(x), \cdots, p_r(x)$$

in $\mathbf{F}_q[x]$ and a monic cubic irreducible polynomial $p(x)$ in $\mathbf{F}_q[x]$. Let

$$f(x) = p^s(x) \prod_{i=1}^{r} p_i(x).$$

Then $\deg(f(x)) = t$ and

$$\Phi(f(x)) = (q^2-1)^r (q^3-1)^s = (n^2-1)^r (n^3-1)^s.$$

It is easy to see that $f(\alpha_i) \neq 0$ for all $i = 1, 2, \cdots, n$. Hence, (4.1) follows from Theorem 3.1.

(ii) In Theorem 3.1, let

$$\mathbf{F}_q = \{\beta_1, \beta_2, \cdots, \beta_k, \alpha_1, \alpha_2, \cdots, \alpha_n\}.$$

If $2 \leq t \leq k$, let

$$f(x) = (x-\beta_1)(x-\beta_2)\cdots(x-\beta_t).$$

Then

$$\Phi(f(x)) = (q-1)^t.$$

If $k < t < n(m-1)$, by the fact that $t - k = 2r' + s'$, we have

$$r' \leq \frac{t}{2} < \frac{n(m-1)}{2} \leq \frac{n(n-1)}{2} \leq \frac{q(q-1)}{2}.$$

Hence, we can choose r' distinct monic quadratic irreducible polynomials

$$p_1(x), p_2(x), \cdots, p_{r'}(x)$$

in $\mathbf{F}_q[x]$. Let

$$f(x) = (x-\beta_1)^{1+s'}(x-\beta_2)\cdots(x-\beta_k)\prod_{i=1}^{r'} p_i(x).$$

Then $\deg(f(x)) = t$ and

$$\Phi(f(x)) = (q-1)^k q^{s'}(q^2-1)^{r'}.$$

It is easy to see that $f(\alpha_i) \neq 0$ for all $i = 1, 2, \cdots, n$. Hence, (4.2) and (4.3) follow from Theorem 3.1. □

Letting $k = 1, 2$ in Theorem 4.1(ii), we obtain

Corollary 4.2.

(i) *If $n+1$ is a prime power and $n \geq m$, then for $2 \leq t < n(m-1)$*

$$\Gamma_m(n,t) \geq \frac{\binom{n}{\lfloor n(m-1)/2 \rfloor}_m}{n(n+1)^s[(n+1)^2-1]^r} \tag{4.4}$$

where r and s are the two unique non-negative integers satisfying $t-1 = 2r+s$ and $s \in \{0,1\}$.

(ii) *If $n+2$ is a prime power and $n \geq m$, then for $3 \leq t < n(m-1)$*

$$\Gamma_m(n,t) \geq \frac{\binom{n}{\lfloor n(m-1)/2 \rfloor}_m}{(n+1)^2(n+2)^s[(n+2)^2-1]^r} \tag{4.5}$$

where r and s are the two unique non-negative integers satisfying $t-2 = 2r+s$ and $s \in \{0,1\}$.

The lower bound (4.1) in Theorem 4.1 can be rewritten as the following form. If n is a prime power, $n \geq m$ and $2 \leq t < n(m-1)$, then

$$\Gamma_m(n,t) \geq \frac{\binom{n}{\lfloor n(m-1)/2 \rfloor}_m}{(n^2-1)^{\frac{t}{2}}}, \quad t \text{ even,} \tag{4.6}$$

$$\Gamma_m(n,t) \geq \frac{\binom{n}{\lfloor n(m-1)/2 \rfloor}_m}{(n^2-1)^{\frac{(t-3)}{2}}(n^3-1)}, \quad t \text{ odd.} \tag{4.7}$$

The lower bound (4.4) in Corollary 4.2 can be rewritten as the following form. If $n+1$ is a prime power and $n \geq m$, then for $2 \leq t < n(m-1)$

$$\Gamma_m(n,t) \geq \frac{\binom{n}{\lfloor n(m-1)/2 \rfloor}_m}{n(n+1)[(n+1)^2-1]^{\frac{(t-2)}{2}}}, \quad t \text{ even,} \tag{4.8}$$

$$\Gamma_m(n,t) \geq \frac{\binom{n}{\lfloor n(m-1)/2 \rfloor}_m}{n[(n+1)^2-1]^{\frac{(t-1)}{2}}}, \quad t \text{ odd.} \tag{4.9}$$

The lower bound (4.5) in Corollary 4.2 can be rewritten as the following form. If $n+2$ is a prime power and $n \geq m$, then for $3 \leq t < n(m-1)$

$$\Gamma_m(n,t) \geq \frac{\binom{n}{\lfloor n(m-1)/2 \rfloor}_m}{(n+1)^2[(n+2)^2-1]^{\frac{(t-2)}{2}}}, \quad t \text{ even}, \tag{4.10}$$

$$\Gamma_m(n,t) \geq \frac{\binom{n}{\lfloor n(m-1)/2 \rfloor}_m}{(n+1)^2(n+2)[(n+2)^2-1]^{\frac{(t-3)}{2}}}, \quad t \text{ odd}. \tag{4.11}$$

Acknowledgment

The authors would like to thank the anonymous reviewer and Professor Harald Niederreiter for their valuable suggestions and comments that helped to improve the paper.

References

[1] D.K. Bhattacharyya and S.J. Nandi, *Theory and Design of SEC-DED-AUED Codes.* IEE Proceedings-Computers and Digital Techniques **145** (1998), 121–126.

[2] M. Blaum, *Codes for Detecting and Correcting Unidirectional Errors.* IEEE Computer Society Press, Los Alamitos, California, 1993.

[3] M. Blaum and H. van Tilborg, *On t-Error Correcting/All Unidirectional Error Detecting Codes.* IEEE Trans. Computers **38** (1989), 1493–1501.

[4] F.J.H. Boinck and H. van Tilborg, *Constructions and Bounds for Systematic tEC/AUED Codes.* IEEE Trans. Inform. Theory **36** (1990), 1381–1390.

[5] B. Bose, *On Unordered Codes.* IEEE Trans. Computers **40** (1991), 125–131.

[6] B. Bose and T.R.N. Rao, *Theory of Unidirectional Error Correcting/Detecting Codes.* IEEE Trans. Computers **31** (1982), 521–530.

[7] J. Bruck and M. Blaum, *New Techniques for Constructing EC/AUED Codes.* IEEE Trans. Computers **41** (1992), 1318–1324.

[8] C.C. Chen and K.M. Koh, *Principles and Techniques in Combinatorics.* World Scientific, Singapore, 1992, pp. 215–216.

[9] C. Cooper and R.E. Kennedy, *A Dice-Tossing Problem.* Crux Mathematicorum **10** (1984), 134–138.

[10] P. Delsarte and P. Piret, *Spectral Enumerators for Certain Additive-Error-Correcting Codes over Integer Alphabets.* Inform. Contr. **48** (1981), 193–210.

[11] F.-W. Fu, S. Ling and C.P. Xing, *New Lower Bounds and Constructions for Binary Codes Correcting Asymmetric Errors.* IEEE Trans. Inform. Theory **48**(12) (2003), to appear.

[12] T. Helleseth and T. Kløve, *On Group-Theoretic Codes for Asymmetric Channels.* Inform. Contr. **49** (1981), 1–9.

[13] R.S. Katti, *A Note on SEC/AUED Codes.* IEEE Trans. Computers **45** (1996), 244–246.

[14] R.S. Katti and M. Blaum, *An Improvement on Constructions of t-EC/AUED Codes.* IEEE Trans. Computers **45** (1996), 607–608.

[15] T. Kløve, *Error Correcting Codes for the Asymmetric Channel.* Rep. 18-09-07-81, Dept. Mathematics, University of Bergen, July 1981 (revised in 1983 and updated in 1995).

[16] T. Kløve, *On Robinson's Coding Problem.* IEEE Trans. Inform. Theory **29** (1983), 450–454.

[17] S. Kundu and S.M. Reddy, *On Symmetric Error Correcting and All Unidirectional Error Detecting Codes.* IEEE Trans. Computers **39** (1990), 752–761.

[18] C.-S. Laih and C.-N. Yang, *On the Analysis and Design of Group Theoretical t-SYEC/AUED Codes.* IEEE Trans. Computers **45** (1996), 103–108.

[19] D.J. Lin and B. Bose, *Theory and Design of t-Error Correcting and $d(d > t)$-Unidirectional Error Detecting (t-EC d-UED) Codes.* IEEE Trans. Computers **37** (1988), 433–439.

[20] D.J. Lin and B. Bose, *On the Maximality of the Group Theoretic Single Error Correcting and All Unidirectional Error Detecting (SEC-AUED) Codes.* Sequences: Combinatorics, Compression, Security, and Transmission, Springer-Verlag (Editor: R. Capocelli), pp. 506–529, 1990.

[21] M.-C. Lin, *Constant Weight Codes for Correcting Symmetric Errors and Detecting Unidirectional Errors.* IEEE Trans. Computers **42** (1993), 1294–1302.

[22] B.L. Montgomery and B.V. Kumar, *Systematic Random Error Correcting and All Unidirectional Error Detecting Codes.* IEEE Trans. Computers **39** (1990), 836–840.

[23] S.J. Nandi and P.P. Chaudhuri, *New Class of t-Error Correcting and All Unidirectional Error Detecting (t-EC/AUED) Codes.* IEE Proceedings-Computers and Digital Techniques **142** (1995), 32–40.

[24] D. Nikolos, *Theory and Design of t-Error Correcting/d-Error Detecting ($d > t$) and All Unidirectional Error Detecting Codes.* IEEE Trans. Computers **40** (1991), 132–142.

[25] D. Nikolos and A. Krokos, *Theory and Design of t-Error Correcting, k-Error Detecting and d-Unidirectional Error Detecting Codes with $d > k > t$.* IEEE Trans. Computers **41** (1992), 411–419.

[26] T.R.N. Rao and E. Fujiwara, *Error-Control Coding for Computer Systems.* Englewood Cliffs, NJ: Prentice-Hall Inc., 1989.

[27] J.P. Robinson, *An Asymmetric Error-Correcting Ternary Code.* IEEE Trans. Inform. Theory **24** (1978), 258–261.

[28] D.L. Tao, C.R.P. Harmann, and P.K. Lala, *A Note on t-EC/d-UED Codes.* IEEE Trans. Computers **40** (1991), 660–663.

[29] R.R. Varshamov, *A General Method of Constructing Asymmetric Coding Systems, related to the Solution of a Combinatorial Problem Proposed by Dixon.* Doklady Akad. Nauk. USSR **194** (1970), 284–287 (trans: Soviet Physics-Doklady **15** (1970), 811–813).

[30] J.H. Weber, *Asymptotic Results on Codes for Symmetric, Unidirectional, and Asymmetric Error Control.* IEEE Trans. Inform. Theory **40** (1994), 2073–2075.

[31] J.H. Weber, C. de Vroedt and D.E. Boekee, *Bounds and Constructions for Codes Correcting Unidirectional Errors.* IEEE Trans. Inform. Theory **35** (1989), 797–810.

[32] J.H. Weber, C. de Vroedt and D.E. Boekee, *Necessary and Sufficient Conditions on Block Codes Correcting/Detecting Errors of Various Types.* IEEE Trans. Computers **41** (1992), 1189–1193.

[33] C.P. Xing, *Constructions of Codes from Residue Rings of Polynomials.* IEEE Trans. Inform. Theory **48** (2002), 2995–2997.

[34] C.-N. Yang and C.-S. Laih, *DC_m Codes for Constructing t-EC/AUED Codes.* IEEE Trans. Computers **47** (1998), 492–495.

[35] Z. Zhang and X.-G. Xia, *LYM-Type Inequalities for tEC/AUED Codes.* IEEE Trans. Inform. Theory **39** (1993), 232–238.

[36] Z. Zhang and X.-G. Xia, *On the Construction of Systematic tEC/AUED Codes.* IEEE Trans. Inform. Theory **39** (1993), 1662–1669.

Fang-Wei Fu
Temasek Laboratories, National University of Singapore
Engineering Drive 3, 10 Kent Ridge Crescent
Singapore 119260, Republic of Singapore

on leave from

the Department of Mathematics, Nankai University
Tianjin 300071, P. R. China
e-mail: `tslfufw@nus.edu.sg`

San Ling
Department of Mathematics
National University of Singapore
2 Science Drive 2
Singapore 117543, Republic of Singapore
e-mail: `matlings@nus.edu.sg`

Chaoping Xing
Department of Mathematics
National University of Singapore
2 Science Drive 2
Singapore 117543, Republic of Singapore

and

Department of Mathematics
University of Science and Technology of China
Hefei, Anhui 230026, P. R. China
e-mail: `matxcp@nus.edu.sg`

Progress in Computer Science and Applied Logic, Vol. 23, 153–168

On the Propagation Criterion of Boolean Functions

Aline Gouget

Abstract. The propagation criterion is one of the main cryptographic criteria on Boolean functions used in block ciphers. Quadratic Boolean functions satisfying the propagation criterion of high degree were given by Preneel *et al.*, but their algebraic degree is too small for a cryptograhic use. Then designing Boolean functions of high algebraic degree and high degree of propagation has been the goal of several papers. In this paper, we investigate the work of Kurosawa and Satoh in order to optimize the algebraic degree and the degree of propagation, and the work of Honda, Satoh, Iwata, and Kurosawa, by giving in particular a construction of Boolean functions satisfying $PC(3)$ and having a very large algebraic degree. We also show that among symmetric functions, only the quadratic ones satisfy the propagation criterion of degree greater than 1. A particular case of this result is that symmetric bent functions must be quadratic – a result that needed a whole paper to be proved before.

Keywords. Boolean functions, Block-Cipher, Propagation criterion, Symmetric functions.

1. Introduction

The security of block ciphers, (*e.g.*, DES, AES) is often discussed by viewing their encrypting functions (more precisely, their S-boxes) as a set of Boolean functions. Part of the security of the system relies on the choice of these Boolean functions which must fulfil several cryptographic criteria. Such a Boolean function f must be *balanced*, *i.e.*, take the value 1 and the value 0 with the same probability 1/2. The function f must also have high *algebraic degree* (the degree of its polynomial representation on n binary variables, called the algebraic normal form), and must satisfy the *propagation criterion*.

In this paper, we focus on the construction of Boolean functions which satisfy the propagation criterion of degree l, have a high algebraic degree, and, in some cases, are balanced. In Section 2, we introduce the definitions and notation that are needed in the paper. We briefly recall the most important results on the propagation criterion.

In Section 3, we recall Kurosawa-Satoh's construction [12] which is the first construction of nonquadratic Boolean functions satisfying $PC(l)$ of order k (for some values l and k). This construction is a particular case of the famous Maiorana-MacFarland's construction of those functions which can be written in the form $f(x,y) = x \cdot \phi(y) \oplus g(y)$, where ϕ is a mapping from $\mathbb{F}_2^t$ into $\mathbb{F}_2^s$ and g is any s-variable Boolean function, where s and t are two positive integers. Kurosawa and Satoh set the function ϕ to be linear. By using the properties of linear error-correcting codes, they obtained some n-variable Boolean functions satisfying the propagation criterion with high degree of propagation ($l \approx n/4$) and small order (k is a constant near 3) or *vice versa*. The functions they obtained have algebraic degree at most $n/2$. We show that a slight modification of Kurosawa-Satoh's construction leads to a construction of Boolean functions which have a degree of propagation at least as good as, and have a higher algebraic degree than the previous constructions; the order of propagation is no longer ensured. Carlet [4] generalized Kurosawa-Satoh's construction by using a not necessarily linear mapping ϕ; he constructed Boolean functions satisfying $PC(l)$ of order k by using nonlinear codes. We give a table of values for the different parameters (number of variables, algebraic degree, order and degree of propagation) which can be achieved by using the linear and nonlinear codes proposed by Kurosawa, Satoh, and Carlet. We propose to use another linear code, the parity check code, which allows us to obtain a higher degree of propagation ($l \approx n/2$) than the codes previously proposed. Furthermore, we give the values obtained by using several nonlinear codes (see [2, 15]) constructed by means of the Hensel lifting to $\mathbb{Z}_4$ of quadratic residue codes and the application of the Gray map. We also give a new construction which provides balanced Boolean functions having an odd number of variables and almost the same values of parameters than these obtained with the parity check code. Next, Honda, Satoh, Iwata, and Kurosawa [11] obtained a construction of Boolean functions satisfying the propagation criterion of degree 2 and having high algebraic degree ($d \approx n - \log_2 n$ where n is the number of variables) by using the generator matrix of the simplex code. We generalize this construction by using a not necessarily linear mapping ϕ. Furthermore, we adapt this construction to obtain Boolean functions satisfying $PC(3)$ and having the same algebraic degree. Finally, we propose a general construction for Boolean functions satisfying *Odd-PC*, a criterion introduced by Bernasconi [1].

In Section 4, we focus on the propagation criterion of symmetric functions, *i.e.*, the functions which are invariant under any permutation of input coordinates. We show that the construction of $PC(1)$ symmetric functions is equivalent to the construction of balanced functions. We recall a trivial construction of balanced symmetric functions for every odd number of variables n. By exhaustive search, we check that, for almost every odd integer n lower than 26, this trivial construction generates all symmetric balanced functions. Von zur Gathen and Roche [10] proposed several constructions of Boolean functions having numerical degree (defined in Section 2) equal to $n-1$. We use these constructions to provide $PC(1)$ Boolean functions. Savicky [19] proved that the only symmetric bent functions are the quadratic symmetric functions. His proof needed a whole paper. We prove more generally, and in a few lines, that the symmetric functions satisfying $PC(l)$ where $l \geq 2$, are the four quadratic functions $\bigoplus_{1\leq i<j\leq n} x_i x_j \oplus a_0$, and $\bigoplus_{1\leq i<j\leq n} x_i x_j \bigoplus_{1\leq i\leq n} x_i \oplus a_0$, where a_0 is in $\mathbb{F}_2$.

2. Preliminaries

Let n be any positive integer. We denote by $\bigoplus$ the usual addition in $\mathbb{F}_2$ and in $\mathbb{F}_2^n$. The *Hamming weight* $w_H(u)$ of a word u in $\mathbb{F}_2^n$ is the number of its components equal to 1. We denote by $\preceq$ the partial order on the words of $\mathbb{F}_2^n$, *i.e.*, $(u_1,\ldots,u_n) \preceq (v_1,\ldots,v_n)$ if and only if $(u_i = 1) \Rightarrow (v_i = 1)$ for every $i = 1,\ldots,n$. The *Hamming weight* $w_H(f)$ of an n-variable Boolean function f is the size of its support, *i.e.*, the size of the set $\{x \in \mathbb{F}_2^n | f(x) = 1\}$.

2.1. Representation of Boolean Functions

Any Boolean function f in n variables, $f : \mathbb{F}_2^n \mapsto \mathbb{F}_2$, admits a unique algebraic normal form (ANF), $f(x_1,\ldots,x_n) = \bigoplus_{u\in F_2^n} a_u(\prod_{i=1}^n x_i^{u_i}) = \bigoplus_{u\in\mathbb{F}_2^n} a_u x^u$. The function $g : u \mapsto a_u$ is called the binary Möbius transform of f. For any word u, the coefficient a_u belongs to $\mathbb{F}_2$, and can be computed thanks to the formula $a_u = \bigoplus_{v\in\mathbb{F}_2^n, v\preceq u} f(v)$. The *algebraic degree* of a Boolean function f is the degree of its algebraic normal form. Every Boolean function f can also be uniquely represented by its *Numerical Normal Form* (NNF) [6], *i.e.*, its polynomial representation over $\mathbb{Z}$, $f(x_1,\ldots,x_n) = \sum_{u\in F_2^n} \lambda_u (\prod_{i=1}^n x_i^{u_i}) = \sum_{u\in\mathbb{F}_2^n} \lambda_u x^u$. For any Boolean function f and for every word u in $\mathbb{F}_2^n$, the coefficient λ_u can be computed thanks to the formula $\lambda_u = (-1)^{w_H(u)} \sum_{v\in\mathbb{F}_2^n|v\preceq u} (-1)^{w_H(v)} f(v)$. The *numerical degree* of a Boolean function f is the degree of its NNF representation.

2.2. Criteria on Boolean Functions

An n-variable Boolean function f is *balanced* if its Hamming weight equals 2^{n-1}. The function $x \mapsto f(x) \oplus f(x \oplus u)$, denoted by $D_u f$, is called the derivative of f over u. The *Strict Avalanche Criterion (SAC)* was introduced by Webster and Tavares [21] in 1985; it has been later generalized by Forré [9] who defined an order over it and by Preneel, Van Leekwijck and Van Linden [16] who defined the propagation criterion of degree l and order k.

Definition 1. *Let f be an n-variable Boolean function and l a positive integer. The function f satisfies the* Propagation Criterion of degree l, *denoted $PC(l)$, if for all words $u \in \mathbb{F}_2^n$ such that $0 < w_H(u) \leq l$, the function $D_u f$ is balanced.*

Definition 2. *Let l and k be two positive integers and f an n-variable Boolean function. The function f satisfies the* propagation criterion of degree l and order k ($PC(l)$ of order k) *if any function obtained from f by keeping constant k of its input coordinates satisfies $PC(l)$.*

The notion of order on the propagation criterion is related to the correlation-immunity, another cryptographic criterion, introduced by Siegenthaler [20]. An n-variable Boolean function f whose output distribution probability does not change when at most k input coordinates are kept constant is called *correlation-immune of order* k. Furthermore, if the function f is balanced, then it is called k-*resilient*. The nonlinearity of f is its minimum distance to the set of all affine functions (the functions having algebraic degree at most equal to 1). A Boolean function f is called *bent* if its nonlinearity equals $2^{n-1} - 2^{\frac{n}{2}-1}$, *i.e.*, the maximum possible value (n must be even).

2.3. Properties and Constructions of Boolean Functions

We are interested in the construction of Boolean functions having high algebraic degrees and satisfying the propagation criterion; these two criteria are partially opponent. Siegenthaler [20] showed that the algebraic degree of any function f satisfying correlation immunity of order k ($0 \leq k < n$) is upper bounded by $n-k$. Furthermore, if f is balanced and $0 \leq k < n-1$, then d is at most $n-k-1$ and $d = 1$ if $k = n-1$. The following upper bound on the algebraic degree of Boolean functions satisfying $PC(1)$ of order k is a direct consequence of this bound.

Proposition 1. [16] *Let f be an n-variable Boolean function. If f satisfies $PC(1)$ of order k, where $0 \leq k \leq n-2$, then f has algebraic degree at most $n-k-1$.*

Rothaus [18] proved that a Boolean function f satisfies $PC(n)$ if and only if it is a bent function; consequently, bent functions are also called *perfectly nonlinear* [13]. Rothaus also proved that the algebraic degree of any bent function is upper bounded by $n/2$. Zheng and Zhang established an explicit lower bound on the nonlinearity N_f of a function f satisfying $PC(l)$ which shows that, the higher the degree of propagation, the higher the minimum nonlinearity.

Proposition 2. [22] *If f is an n-variable Boolean function satisfying $PC(l)$, then the nonlinearity N_f of f satisfies $N_f \geq 2^{n-1} - 2^{n-1-\frac{1}{2}l}$.*

We say that a Boolean function f *linearly depends on* x_i if the function f actually depends on the variable x_i and x_i occurs in the ANF of f only in the monomial of degree 1. If a Boolean function f linearly depends on at least one variable, then f is trivially balanced. This property is often used in order to prove the propagation criterion of Boolean functions.

One of the most important classes of Boolean functions is obtained by Maiorana-MacFarland's construction. This construction was introduced in the 70's by Dillon [8] in order to design perfectly nonlinear functions and it was later extended [3] to design resilient functions. Maiorana-MacFarland's functions are defined by $f(x,y) = x \cdot \phi(y) \oplus g(y)$, where $x \in \mathbb{F}_2^s$, $y \in \mathbb{F}_2^{n-s}$, g is any $(n-s)$-variable Boolean function and ϕ is any mapping from $\mathbb{F}_2^{n-s}$ into $\mathbb{F}_2^s$.

3. Design of Boolean Functions Satisfying $PC(l)$

3.1. Maiorana-MacFarland's Construction

Kurosawa and Satoh studied in [12] a class of functions within the general class of the so-called Maiorana-MacFarland's functions, $f(x,y) = x\cdot\phi(x,y)\oplus g(y)$, where ϕ is a mapping from $\mathbb{F}_2^t$ into $\mathbb{F}_2^s$ and g is any s-variable Boolean function, where s and t are two positive integers. They set the function ϕ to be linear and gave a sufficient condition for such Maiorana-MacFarland's Boolean functions to satisfy $PC(l)$ of order k. This construction allowed them to construct Boolean functions having high degree of propagation and low order of propagation, or *vice versa*, according to their choices of the linear code used for the construction. Indeed, order and degree of propagation depend on the minimal distance and the dual distance of the linear code. Furthermore, this construction is the first one to provide functions having algebraic degree greater than 2. Carlet [4] generalized Kurosawa-Satoh's construction for not necessarily linear mappings ϕ and defined sufficient conditions on it to construct Boolean functions satisfying $PC(l)$ of order k.

Proposition 3. [4] *Let f be a Maiorana-MacFarland function $f(x,y) = x\cdot\phi(y)\oplus g(y)$ where $x\in\mathbb{F}_2^s$, $y\in\mathbb{F}_2^t$ and let g be any t-variable function. If the mapping ϕ from $\mathbb{F}_2^t$ into $\mathbb{F}_2^s$ satisfies the following conditions:*

1. *the sum of at least one and at most l coordinates of ϕ is k-resilient,*
2. *if $b\in\mathbb{F}_2^t$ is such that $0 < w_H(b)\le l$, then for every $y\in\mathbb{F}_2^t$, at least $k+1$ coordinates of the words $\phi(y\oplus b)$ and $\phi(y)$ differ,*

then f satisfies the propagation criterion of degree l and order k.

We recall that Kurosawa and Satoh set the function ϕ to be linear. In order to construct a linear mapping ϕ, they proposed to use generator matrices of linear codes. A binary linear $[n,k,d]$ code C is a k-dimensional vector subspace of $\mathbb{F}_2^n$. Its minimal distance d equals the smallest positive Hamming weight of the words of C. The dual code $C^\perp$ is a linear code $[n, n-k, d^\perp]$ defined by $C^\perp = \{u\in\mathbb{F}_2^n \mid u\cdot v = 0, \quad \forall v\in C\}$. The *dual distance* of C is the minimal distance of $C^\perp$. Usually, a linear code C is represented by a $k\times n$ generator matrix G, whose rows form a basis of the vector space. Then, for all $x\in\mathbb{F}_2^k$, xG (usual matrix product) is a codeword. Furthermore, G is a parity check matrix of the code $C^\perp$; that means, y is a codeword of $C^\perp$ if and only if $Gy^T = 0$. Furthermore, if H is an $(n-k)\times n$ generator matrix of $C^\perp$, then for all $y\in\mathbb{F}_2^{n-k}$, yH is a codeword of $C^\perp$ and x is a codeword of C if and only if $Hx^T = 0$.

Proposition 4. [12] *Let G_1 (resp. G_2) be the generator matrix of a linear $[n_1,k_1,d_1]$ (resp. $[n_2,k_2,d_2]$) code C_1 (resp. C_2) of dual distance $d_1^\perp$ (resp. $d_2^\perp$). Then the function $f(x,y) = x\cdot\phi(y)\oplus g(y)$, where $x\in\mathbb{F}_2^n$, $y\in\mathbb{F}_2^n$, g is any n-variable Boolean function and ϕ is a mapping from $\mathbb{F}_2^n$ into $\mathbb{F}_2^n$ such that $\phi(y) = G_2^T G_1 y^T$, satisfies the propagation criterion of degree $\min(d_1^\perp, d_2^\perp)-1$ and order $\min(d_1,d_2)-1$.*

Kurosawa-Satoh's construction can be used to construct $(2n)$-variable Boolean functions of algebraic degrees at most n and satisfying $PC(d^\perp-1)$ of order $d-1$. The maximum algebraic degree can be achieved by choosing a function g having maximum algebraic degree. Kurosawa and Satoh [12] gave a necessary condition on the function f defined in Proposition 4 to be balanced. Indeed, f is balanced if:

$$\#\{y \mid g(y) = 0, G_2^T G_1 y = 0\} = \#\{y \mid g(y) = 1, G_2^T G_1 y = 0\}.$$

In order to maximize the algebraic degree and the degree of propagation (with respect to the number of variables), we adapt Kurosawa-Satoh's construction.

Proposition 5. *Let G be the generator matrix of a linear $[n,k]$ code C of dual distance $d^\perp$ and ϕ the mapping from $\mathbb{F}_2^n$ into $\mathbb{F}_2^k$ defined by $\phi : y \mapsto Gy^T$. Then, the function $f(x,y) = x \cdot \phi(y) \oplus g(y)$, where $x \in \mathbb{F}_2^k$ and $y \in \mathbb{F}_2^n$, has algebraic degree at most n and satisfies the propagation criterion of degree $d^\perp - 1$.*

Proof. We have to check Conditions 1 and 2 of Proposition 3 (with $k = 0$ and $l = d^\perp - 1$). In the present case, Condition 1 is equivalent to saying that for every a in F_2^n such that $0 < w_H(a) \leq d^\perp - 1$, we must have $w_H(aG) \geq 1$. The rows of G form a basis of C, then Condition 1 is fulfilled. Condition 2 is equivalent to saying that for every b in $\mathbb{F}_2^n$ such that $0 < w_H(b) \leq d^\perp - 1$, we must have $w_H(Gb^T) > 0$. This condition is clearly fulfilled because b is not a codeword of $C^\perp$ and G is a parity check matrix of $C^\perp$. □

Kurosawa and Satoh used several linear codes: the Hamming Code $\mathcal{H} = [2^m - 1, 2^m - m - 1, 3]$ and its dual code the Simplex code $\mathcal{H}^\perp = [2^m - 1, m, 2^{m-1}]$, the first order Reed-Muller Code $R(1,m) = [2^m, m+1, 2^{m-1}]$ and its dual the extended Hamming code $R^\perp(1,m) = [2^m, 2^m - m - 1, 4]$. We propose to consider another linear code which is the parity check code $[m, m-1, 2]$ whose dual code has parameters $[m, 1, m]$. We give in Figure 1 the values of the different parameters (number of variables, degree of propagation and algebraic degree) for the constructions of Proposition 4 (by taking $G_1 = G_2$) and Proposition 5, according to the choice of the linear code (the degree of propagation and the algebraic degree do not change between the two constructions, only the number of variables). Figure 1 shows that the functions obtained by Proposition 5 have very high algebraic degrees. Furthermore, the use of the parity check code leads to the construction of $(2m-1)$-variable Boolean functions (for $m \geq 2$) having algebraic degree m and

Linear Codes	Nb. of Variables (Proposition 4)	Nb. of Variables (Proposition 5)	Algebraic Degree	Degree of Propagation
$\mathcal{H}_m$	$2^{m+1} - 2$	$2^{m+1} - m - 2$	$2^m - 1$	$2^{m-1} - 1$
$\mathcal{H}_m^\perp$	$2^{m+1} - 2$	$2^m + m - 1$	$2^m - 1$	2
$R(1,m)$	2^{m+1}	$2^m + m + 1$	2^m	3
$R(1,m)^\perp$	2^{m+1}	$2^{m+1} - m - 1$	2^m	$2^{m-1} - 1$
Parity check	$2m$	$2m - 1$	m	$m - 1$

FIGURE 1. Applications of Proposition 4 and 5 for linear codes

satisfying $PC(m-1)$; the degree l of propagation is near $n/2$ instead of at most $n/4$ for the linear codes previously used.

Carlet [4] generalized Kurosawa-Satoh's construction by using two systematic nonlinear codes C_1 and C_2 (the notion of dual distance being still valid for nonlinear codes), and proposed to use the $(2^m, 2^{2^m-2m}, 6)$ Preparata code $\mathcal{P}_m$ whose dual distance is $2^{m-1} - 2^{\frac{m}{2}-1}$ and the $(2^m, 2^{2m}, 2^{m-1} - 2^{\frac{m}{2}-1})$ Kerdock code $\mathcal{K}_m$ whose dual distance is 6 (m even ≥ 4; we give here the length, the cardinality and the minimum distance). We complete the table (see Figure 2) by considering four

Nonlinear Codes	Nb. of Variables	Degree of Propagation	Order of Propagation
$C_1 = \mathcal{K}_m,\ C_2 = \mathcal{P}_m$	2^{m+1}	$2^{m-1} - 2^{\frac{m}{2}-1} - 1$	5
$C_1 = \mathcal{P}_m,\ C_2 = \mathcal{K}_m$	2^{m+1}	5	$2^{m-1} - 2^{\frac{m}{2}-1} - 1$
$(36, 2^{18}, 8)$ [2]	72	7	7
$(48, 2^{24}, 12)$ [2]	96	11	11
$(64, 2^{32}, 14)$[15]	128	13	13
$(96, 2^{48}, 18)$ [15]	192	17	17

FIGURE 2. Values of parameters for Carlet's construction

nonlinear codes obtained by Bonnecaze *et al.* [2] and Pless and Qian [15] from Hensel lifting to $\mathbb{Z}_4$ of quadratic residue codes and applying the Gray map. The best value for the degree of propagation obtained using those codes is near $n/4$.

3.2. A New Construction of Balanced Boolean Functions Satisfying $PC(l)$

We now propose another construction which allows us to obtain almost the same values of parameters (algebraic degree and the degree of propagation) than these obtained with the parity check code, for an odd number of variables. Furthermore, we give a necessary and sufficient condition on the function to be balanced.

Proposition 6. *Let n be any positive integer and f a $(2n+1)$-variable Boolean function such that $f(x,y,z) = z(g(x) \oplus y_1 \oplus \cdots \oplus y_n) \oplus x \cdot y$, where $x \in \mathbb{F}_2^n$, $y \in \mathbb{F}_2^n$, $z \in \mathbb{F}_2$ and g is any n-variable Boolean function. Then f has algebraic degree at most n and satisfies $PC(n)$. Furthermore, the function f is balanced if and only if $g(\vec{1}) = 1$.*

Proof. For all (a,b,c) such that $0 < w_H(a) + w_H(b) + w_H(c) \leq n$ and $a \in \mathbb{F}_2^n$, $b \in \mathbb{F}_2^n$, $c \in \mathbb{F}_2$, we have $D_{a,b,c}f(x,y,z) = z(g(x) \oplus g(x \oplus a)) \oplus c(g(x \oplus a) \oplus y_1 \oplus \cdots \oplus y_n \oplus b_1 \oplus \cdots \oplus b_n) \oplus z(b_1 \oplus \cdots \oplus b_n) \oplus a \cdot y \oplus b \cdot x \oplus a \cdot b$. If $0 < w_H(a) < n$, then $a \cdot y \oplus c(y_1 \oplus \cdots \oplus y_n)$ is not equal to the null function and $D_{a,b,c}f$ linearly depends on at least one variable y_i. If $w_H(a) = n$ then $w_H(b) = w_H(c) = 0$ and the derivated function is balanced. If $w_H(a) = 0$ and $w_H(c) \neq 0$, then the derivated function linearly depends on $y_1, y_2, \ldots, y_n$ and it is balanced. At last, $D_{0,b,0}f(x,y,z) = z(b_1 \oplus \cdots \oplus b_n) \oplus b \cdot x$ is balanced.

The function f can be decomposed as follows: $f(x,y,z) = (1\oplus z)f_1(x,y)\oplus z f_2(x,y)$, where $f_1(x,y) = x \cdot y$ and $f_2(x,y) = g(x) \oplus (x \oplus \vec{1}) \cdot y$. Then, we have $w_H(f) = w_H(f_1) + w_H(f_2)$. The Hamming weight of f_1 is equal to $2^{2n-1} - 2^{n-1}$ (see [14]). Thus, the function f is balanced if and only if the Hamming weight of f_2 equals $2^{2n-1}+2^{n-1}$. We have $w_H(f_2) = 2^n w_H(g) + (2^{2n-1} - 2^{n-1}) - 2\#\{(x,y) \in \mathbb{F}_2^n \times \mathbb{F}_2^n \mid g(x) = 1, (x \oplus \vec{1}) \cdot y = 1\}$. For every $x \in \mathbb{F}_2^n$ such that $g(x) = 1$ and $x \neq \vec{1}$, we have $\#\{y \in \mathbb{F}_2^n \mid (x \oplus \vec{1}) \cdot y = 1\} = \#\{y \in \mathbb{F}_2^n \mid (x \oplus \vec{1}) \cdot y = 0\} = 2^{n-1}$. So, $w_H(f_2) = 2^n w_H(g) + (2^{2n-1} - 2^{n-1}) - 2^n\#\{x \in \mathbb{F}_2^n \mid g(x) = 1, x \neq \vec{1}\}$. Thus, $w_H(f_2)$ equals $2^{2n-1} + 2^{n-1}$ if and only if $g(x) = 1$ where $x = \vec{1}$. □

Since there is no strong condition on the function g (only for $x = \vec{1}$, $g(x) = 1$), this construction provides balanced Boolean functions in $(2n+1)$ variables, having algebraic degree equal to n and satisfying $PC(n)$.

3.3. Honda *et al*'s Construction and Improvements

Honda *et al.* [11] studied a class of functions also related to linear codes but which are not Maiorana-MacFarland's functions. They set the linear code to be the binary Simplex code and then got a construction of n-variable Boolean functions satisfying $PC(2)$ and having algebraic degree near $n - \log_2 n$.

Proposition 7. [11] *Let m be a positive integer and G the generator matrix of the $[2^m - 1, m, 2^{m-1}]$ simplex code. We assume that the ith column of G is the binary representation of the integer i. Let f be a $(2^m + m - 1)$-variable Boolean function such that $f(x,y,z) = f_1(x) \oplus f_2(y) \oplus f_3(z) \oplus xG[y_1, \dots, y_{2^m-2}, z]^T$, where $x \in \mathbb{F}_2^m$, $y \in \mathbb{F}_2^{2^m-2}$, $z \in \mathbb{F}_2$, and f_1, f_2 and f_3 are any Boolean functions. Then f has algebraic degree at most $2^m - 2$ and satisfies $PC(2)$.*

This proposition shows the existence of n-variable Boolean functions having algebraic degrees near $n - \log_2 n$ and satisfying $PC(2)$. We can generalize this construction by replacing the mapping $(y,z) \mapsto G[y_1, \dots, y_{2^m-2}, z]^T$ by the mapping $(y,z) \mapsto \phi(y,z)$; note that this mapping is not necessarily linear.

Proposition 8. *Let f be an $(s+t)$-variable Boolean function defined as follows:*

$$f(x,y,z) = f_1(x) \oplus f_2(y) \oplus f_3(z) \oplus x \cdot \phi(y,z),$$

where $x \in \mathbb{F}_2^s$, $y \in \mathbb{F}_2^{t-1}$, $z \in \mathbb{F}_2$, f_1, f_2 and f_3 are Boolean functions and ϕ is a mapping from $\mathbb{F}_2^t$ into $\mathbb{F}_2^s$. If the mapping ϕ satisfies the following conditions, then f satisfies $PC(2)$.

1. *Every component of ϕ linearly depends on z,* i.e., *$\phi_i(y,z) = h_i(y) \oplus z$ where h_i is a $(t-1)$-variable Boolean function,*
2. *$\phi_i(y,z)$ and $\phi_i(y,z) \oplus \phi_j(y,z)$ are balanced for $i \neq j$ where i and j are in $\{1, \dots, s\}$,*
3. *$\phi(y,z) \neq \phi(y \oplus b, z \oplus c)$ for every b in $\mathbb{F}_2^{t-1}$ such that $0 < w_H(b) + w_H(c) \leq 2$.*

Proof. The function f satisfies $PC(2)$ if the function $x,y,z \mapsto D_a f_1(x) \oplus D_b f_2(y) \oplus D_c f_3(z) \oplus x \cdot (\phi(y,z) \oplus \phi(y \oplus b, z \oplus c)) \oplus a \cdot \phi(y \oplus b, z \oplus c)$ is balanced for all (a,b,c)

such that $0 < w_H(a) + w_H(b) + w_H(c) \leq 2$. If $w_H(a) = 0$, then the function is balanced thanks to Condition 3. Otherwise, if $w_H(b) = 0$, then $D_{a,b,c}f(x,y,z) = D_a f_1(x) \oplus D_c f_3(z) \oplus x \cdot (\phi(y,z) \oplus \phi(y, z \oplus c)) \oplus a \cdot \phi(y, z \oplus c)$. Thanks to Condition 1, $\phi(y,z) \oplus \phi(y, z \oplus c)$ is constant, and the function is balanced thanks to Condition 2 (and $D_c f_3(z)$ is constant). Finally, if $w_H(a) = w_H(b) = 1$, then $w_H(c) = 0$ and $D_{a,b,c}f(x,y,z) = D_a f_1(x) \oplus D_b f_2(y) \oplus x \cdot (\phi(y,z) \oplus \phi(y \oplus b, z)) \oplus a \cdot \phi(y \oplus b, z)$. Since $x \cdot (\phi(y,z) \oplus \phi(y \oplus b, z))$ does not depend on z and $a \cdot \phi(y \oplus b, z)$ linearly depends on z (thanks to Condition 1), the function is balanced. □

The functions constructed by the previous proposition are not necessarily $PC(3)$. We propose an adapted construction for Boolean functions satisfying $PC(3)$.

Proposition 9. *Let f be a Boolean function defined as follows:*

$$f(x,y,z) = f_1(x) \oplus f_2(y) \oplus x \cdot \phi(y) \oplus z_1(x_1 \oplus \cdots \oplus x_s) \oplus z_2(y_1 \oplus \cdots \oplus y_t),$$

where $x \in \mathbb{F}_2^s$, $y \in \mathbb{F}_2^t$, $z \in \mathbb{F}_2^2$, f_1, f_2 are Boolean functions and ϕ is a mapping from $\mathbb{F}_2^t$ into $\mathbb{F}_2^s$. If the mapping ϕ satisfies the following conditions, the function f satisfies $PC(3)$:

1. *for every i and j such that $1 \leq i < j \leq s$, the functions $y \mapsto \phi_i(y) \oplus \phi_j(y)$ and $y \mapsto \phi_i(y) \oplus \phi_j(y) \oplus y_1 \oplus \cdots \oplus y_t$ are balanced.*
2. *If $b \in \mathbb{F}_2^t$ is such that $w_H(b) = 2$, then for every $y \in \mathbb{F}_2^t$, at least one and at most $t-1$ coordinates of the words $\phi(y \oplus b)$ and $\phi(y)$ differ.*

Proof. If $w_H(a)$ or $w_H(b)$ is odd then $D_{a,b,c}f$ linearly depends on z_1 and/or z_2 and it is balanced. In the following, assume that $w_H(a)$ and $w_H(b)$ are even. If $w_H(a) = w_H(b) = 0$, then $D_{0,0,c}f(x,y,z) = c_1(x_1 \oplus \cdots \oplus x_s) \oplus c_2(y_1 \oplus \cdots \oplus y_t)$ is an affine nonconstant function since $w_H(c)$ is positive. If $w_H(a) = 2$ then $w_H(b) = 0$ and $D_{a,0,c}f(x,y,z) = D_a f_1(x) \oplus a \cdot \phi(y) \oplus c_1(x_1 \oplus \cdots \oplus x_s) \oplus c_2(y_1 \oplus \cdots \oplus y_t)$, and the function is balanced thanks to Condition 1. Indeed, either $c_2 = 0$ and then we know that $\phi_i(y) \oplus \phi_j(y)$ is balanced for $i \neq j$, or $c_2 = 1$ and we know that $\phi_i(y) \oplus \phi_j(y) \oplus y_1 \oplus \cdots \oplus y_t$ is balanced. If $w_H(b) = 2$ then $w_H(a) = 0$ and $D_{0,b,c}f(x,y,z) = D_b f_2(y) \oplus x \cdot (\phi(y) \oplus \phi(y \oplus b)) \oplus c_1(x_1 \oplus \cdots \oplus x_s) \oplus c_2(y_1 \oplus \cdots \oplus y_t)$, and either $c_1 = 0$ or $c_1 = 1$. Condition 2 ensures that $w_H(\phi(y) \oplus \phi(y \oplus b)) \notin \{0, t\}$, and so the function linearly depends on at least one variable x_i. □

3.4. Odd-Propagation Criterion

Zheng and Zhang [22] proposed several constructions of Boolean functions satisfying the propagation criterion on almost all vectors of $\mathbb{F}_2^n$, *i.e.*, $f(x) \oplus f(x \oplus u)$ is balanced for all word u in $\mathbb{F}_2^n \backslash A$ where A is a subset of $\mathbb{F}_2^n$. Next, Bernasconi introduced the notion of *odd-PC*, that is, the property for a Boolean function to satisfy the propagation criterion for every word u of odd Hamming weight. The main motivation for introducing the class *odd-PC* is that bent functions both achieve the highest nonlinearity $2^{n-1} - 2^{\frac{n}{2}-1}$ and satisfy the propagation criterion with respect to *all* nonzero vectors (these two conditions are equivalent to each other). But bent functions are not balanced and have algebraic degree at most $n/2$. So they can not be used for cryptographic applications. The class of *odd-PC* functions includes bent functions and balanced functions with high algebraic degree.

Definition 3. *A Boolean function f belongs to the class* odd-PC *if and only if it satisfies the propagation criterion with respect to any word $u \in \mathbb{F}_2^n$ of odd Hamming weight,* i.e., *for every word u such that $w_H(u) \equiv 1 \mod 2$, the function $x \mapsto f(x) \oplus f(x \oplus u)$ is balanced.*

Zheng and Zhang constructed n-variable Boolean functions satisfying PC for every word u in $\mathbb{F}_2^n$ except for u in $\{(0\ldots0),(10\ldots0)\}$: the functions $f = x_1 \oplus g(x_2,\ldots,x_n)$ where g is a bent function. Using this construction, Bernasconi proposed a way to obtain *odd-PC* functions having the best algebraic degree available, *i.e.*, $n-1$. Indeed, since an odd-PC function is $PC(1)$, the bound on the algebraic degree of $PC(1)$ functions is also valid for *odd-PC* functions.

Proposition 10. [1] *For any $n \geq 3$, there exists an explicit balanced and* odd-PC *Boolean function whose algebraic degree d is equal to $n-1$.*

In order to prove this proposition, Bernasconi constructed one such function by induction from a 3-variable function (see [1] for more details). We give a general direct construction of *odd-PC* functions having algebraic degree at most $n-2$.

Proposition 11. *Let $x \in \mathbb{F}_2^{n-2}$ and $z \in \mathbb{F}_2^2$. If f is an n-variable function such that $f(x,z) = g(x) \oplus z_1(x_1 \oplus \cdots \oplus x_{n-2} \oplus z_2)$, where g is any $(n-2)$-variable Boolean function, then f satisfies* odd-PC. *Furthermore, f is balanced if and only if g is balanced.*

Proof. We have to show that the function $D_{a,c}f$ is balanced for all (a,c) such that $a \in \mathbb{F}_2^{n-2}$, $c \in \mathbb{F}_2^2$ and $w_H(a) + w_H(c) \equiv 1 \mod 2$. If $c_1 = 0$, then $w_H(a) + c_2$ is odd and the function $D_{a,c}f$ linearly depends on z_1 and is balanced. If $c_1 = 1$ then the derivated function linearly depends on z_2.
For every x in $\mathbb{F}_2^{n-2}$ such that $g(x) = 1$, we have $\#\{z \in \mathbb{F}_2^2 \mid z_1(x_1 \oplus \cdots \oplus x_{n-2} \oplus z_2) = 0\} = 3$, and for every x in $\mathbb{F}_2^{n-2}$ such that $g(x) = 0$, we have $\#\{z \in \mathbb{F}_2^2 \mid z_1(x_1 \oplus \cdots \oplus x_{n-2} \oplus z_2) = 0\} = 1$. So $w_H(f) = 3w_H(g) + (2^{n-2} - w_H(g)) = 2^{n-2} + 2w_H(g)$ and f is balanced if and only if g is balanced. □

4. Propagation Criterion for Symmetric Boolean Functions

To make the computation of the ciphertext from the plaintext more efficient, Daemen *et al.* [7] proposed to use symmetric functions. This obviously presents the risk of allowing attacks using the specificities of these functions. For instance, Savicky proved in [19] that all symmetric bent functions are quadratic, *i.e.*, have algebraic degree 2 (so they are not proper for cryptographic use). He needed a whole paper to give such proof. We prove in a shorter way that, more generally, nonquadratic symmetric functions cannot satisfy $PC(2)$. Furthermore, we give constructions of nonquadratic symmetric Boolean functions satisfying $PC(1)$.

Definition 4. *An n-variable Boolean function f is called a* symmetric *function if it is invariant under permutation of the variables,* i.e., $\forall \pi \in S_n$, $f(x_{\pi(1)}, \ldots, x_{\pi(n)}) = f(x_1, \ldots, x_n)$.

The algebraic normal form of a symmetric function is of the form:

$$f(x) = \bigoplus_{i=0}^{n} a_i \Big(\bigoplus_{\substack{u \in \mathbb{F}_2^n \\ w_H(u) = i}} x^u \Big),$$

where a_i is in F_2. Also, there exists a function $f^{\#} : \{0, \ldots, n\} \mapsto \mathbb{F}_2$ such that $f(x) = f^{\#}(w_H(x))$ for every $x \in \mathbb{F}_2^n$.

4.1. Construction of Symmetric $PC(1)$ Boolean Functions

In this paper, we are interested in the propagation criterion for symmetric functions and more precisely in the construction of such functions. We first notice a link between the constructions of balanced symmetric functions and of $PC(1)$ symmetric functions. Every symmetric n-variable Boolean function f can be written $f = x_1 f_1 \oplus f_2$ where f_1 and f_2 are two symmetric $(n-1)$-variable Boolean functions. Since f is a symmetric function, it satisfies the propagation criterion of degree 1 if and only if $D_{e_1} f$ is balanced, *i.e.*, if f_1 is balanced. Thus, the construction of $PC(1)$ symmetric functions in n variables is equivalent to the construction of balanced symmetric $(n-1)$-variable functions. Indeed, the knowledge of f_1 uniquely determines f_2, up to constants. More precisely, if $(a_0, \ldots, a_{n-1})$ are the coefficients of the elementary symmetric functions $\bigoplus_{u | w_H(u) = i} x^u$ in the ANF of f_1, then f_2 can be computed as follows.

$$f_2(x) = \bigoplus_{i=0}^{n-2} a_i \Big(\bigoplus_{\substack{u \in \mathbb{F}_2^{n-1} \\ w_H(u) = i+1}} x^u \Big) + cst.$$

Furthermore, the construction of balanced functions is equivalent to the construction of Boolean functions having numerical degree at most $n-1$. Indeed, Carlet and Guillot [6] showed that a function $f(x)$ is balanced if and only if $f(x) \oplus x_1 \oplus \cdots \oplus x_n$ has numerical degree at most $n-1$. Von zur Gathen and Roche [10] proposed several constructions of symmetric Boolean functions having numerical degree at most $n-1$ while they were working on the degree of polynomials in $\mathbb{R}[x]$ that take only two values on the domain $\{0, \ldots, n\}$.

For every positive integer n, the symmetric affine functions are balanced. For an odd number of variables, we first recall the *trivial construction* of balanced symmetric functions.

Proposition 12. *Let n be an odd positive integer, and f an n-variable symmetric Boolean function. If $f(u) = f(u \oplus \vec{1}) \oplus 1$ for every word u in $\mathbb{F}_2^n$, then the function f is balanced.*

Proof. First recall that $\binom{n}{i} = \binom{n}{n-i}$ for every $i = 0, \dots, n$. Then, the support of such symmetric function f contains for all i in $\{0, \dots, n\}$ either the words of Hamming weight i or the words of Hamming weight $n-i$ (but not both). We need $i \neq n-i$ for every i, that is, n odd. The function is then balanced. □

It can be checked that, for any odd integer n lower than or equal to 25 and different from 13, the symmetric functions constructed by Proposition 12 are the only symmetric ones. For $n = 13$, von zur Gathen and Roche obtained a nontrivial construction of n-variable symmetric balanced functions having algebraic degree equal to $n-1$.

Proposition 13. [10] *Let n be a positive odd integer, k an integer such that $2 \leq k \leq (n-3)/2$. Let f be an n-variable Boolean function whose support is such that: $supp(f) = \left\{u \in \mathbb{F}_2^n | w_H(u) \in \{k-2, k-1, n-k-1, n-k\}\right\}$. The function f has numerical degree less than n if and only if $n = 4t^2 - 3$ and $k = 2t^2 - t - 1$ when $t \geq 2$.*

A Boolean function has numerical degree less than n if and only if its support contains the same number of words of odd and even Hamming weights. The proof consists in solving the equation $\binom{n}{k-2} + \binom{n}{n-k-1} = \binom{n}{k-1} + \binom{n}{n-k}$ to get the possible values of k and n when n is odd. From the previous proposition and Proposition 12, we can deduce a construction of balanced symmetric functions. Indeed, instead of taking a balanced Boolean function f whose support is such that, for every i in $\{0, \dots, n\}$, the words whose Hamming weights are either i or $n-i$ (but not both) are in the support of f, we search the functions f whose support contains the words of Hamming weight either $k-2, n-k+2, k+1, n-k-1$ or $k-1, n-k+1, k, n-k$, and either i or $n-i$ for the other values.

Corollary 1. *Let n be a positive odd integer, k and t two integers and b an element of $\mathbb{F}_2$. The following function f is balanced if and only if $n = 4t^2 - 3$ and $k = 2t^2 - t - 1$.*

$$f(x) = \begin{cases} b & \text{if } w_H(x) \in \{k-2, k+1, n-k-1, n-k+2\} \\ b \oplus 1 & \text{if } w_H(x) \in \{k-1, k, n-k, n-k+1\} \\ f(x \oplus \vec{1}) \oplus 1 & \text{otherwise} \end{cases}$$

Proof. Let f' be a symmetric Boolean function such that $f'(x) = f(x) \oplus 1$ for all x of Hamming weight in $\{v_1, v_2, v_3, v_4\}$ where $v_1 = k-2$ or $v_1 = n-k+2$, and $v_2 = k+1$ or $v_2 = n-k-1$, $v_3 = k-1$ or $v_3 = n-k+1$, and $v_4 = k$ or $v_4 = n-k$, and $f(x) = f'(x)$ for other values. We choose the values v_1, v_2, v_3 and v_4 in order to get the property $f'(x) \oplus 1 = f'(x \oplus \vec{1})$ for all $x \in \mathbb{F}_2^n$. Then, we have $w_H(f') = 2^{n-1}$ and $w_H(f) = w_H(f') \pm \left[\binom{n}{k} + \binom{n}{k-1} - \binom{n}{k+1} - \binom{n}{k-2}\right]$. The function f is balanced if and only if $\binom{n}{k} + \binom{n}{k-1} = \binom{n}{k+1} + \binom{n}{k-2}$. The solutions of this equation are given by Proposition 13. □

All the 13-variable balanced symmetric functions can be constructed from Proposition 12 and Corollary 1. Furthermore, Corollary 1 gives a way to construct balanced symmetric Boolean functions for a number of variables n equal to $13, 33, 61, 97, \ldots$, and then a way to construct $PC(1)$ symmetric functions for a number of variables $n = 14, 34, 62, 98, \ldots$. For an even number of variables n, von zur Gathen and Roche's constructions (presented in [10]) of Boolean functions having numerical degree at most $n-1$ provide balanced symmetric functions thanks to a result of Carlet and Guillot previously recalled. By using exhaustive search of balanced symmetric functions for a low number of variables, we observe that all the functions are described by the different constructions which can be found in [10]. The first balanced symmetric functions which are not characterized by von zur Gathen and Roche exist for $n = 24$. We give the truth-tables of all balanced symmetric functions for $n = 24$ (except affine functions) in Figure 3 where b_i is in $\mathbb{F}_2$ and $\underline{b_i} = b_i \oplus 1$. The *Walsh spectrum* of an n-variable symmetric function f is

$w_H(x)$	0	1	2	3	4	5	6	7	8	9	10	11	12
$f_1(x)$	0	1	0	b_1	1	1	0	b_2	0	b_3	1	0	1
$f_2(x)$	0	1	1	1	0	b_1	1	b_2	0	b_3	1	b_4	0

$w_H(x)$	13	14	15	16	17	18	19	20	21	22	23	24
$f_1(x)$	0	1	$\underline{b_3}$	0	$\underline{b_2}$	0	1	1	$\underline{b_1}$	0	1	0
$f_2(x)$	$\underline{b_4}$	1	$\underline{b_3}$	0	$\underline{b_2}$	1	$\underline{b_1}$	0	1	1	1	0

FIGURE 3. Balanced symmetric function for $n = 24$ (ANF 1 and ANF 2)

defined by the list of the following $n+1$ values:

$$\left(\sum_{x \in \mathbb{F}_2^n} (-1)^{f(x)}, \sum_{x \in \mathbb{F}_2^n} (-1)^{f(x) \oplus x_i}, \sum_{x \in \mathbb{F}_2^n} (-1)^{f(x) \oplus x_i \oplus x_j}, \ldots \right.$$
$$\left. \ldots, \sum_{x \in \mathbb{F}_2^n} (-1)^{f(x) \oplus x_1 \oplus \cdots \oplus x_n} \right).$$

The Walsh spectrum is often studied because several cryptographic criteria (as, *e.g.*, resiliency) can be characterized by it. The Walsh spectrum of the functions having ANF 1 and ANF 2 has the particularity that it contains only one zero. We give the Walsh spectrum of the function f_1 with $b_1 = b_2 = 1$ and $b_3 = 0$:
$(0, 362296, 123568, -62552, -12448, 8152, -3184, 72, 3264, -1672, -1936, 2024, 1056, -2024, -688, 1672, 384, -72, 3632, -8152, -44832, 62552, 538384, -362296, -9814464)$.

4.2. Propagation Criterion of Degree Greater Than 1

We are finally interested in the construction of symmetric Boolean functions satisfying $PC(l)$ where $l \geq 2$. Savicky [19] proved that the symmetric bent functions (*i.e.*, satisfying $PC(n)$) are quadratic. Preneel *et al.* [16] proved that quadratic symmetric functions satisfying $PC(l)$ of order k exist if $k+l \leq n-1$ or if $k+l=n$ and k even. Furthermore, they proved that quadratic functions are the only (not necessarily symmetric) functions satisfying $PC(2)$ of order $n-2$. Carlet [4] showed that the Boolean functions f which satisfy $PC(l)$ of order $n-l$ are the four symmetric quadratic Boolean functions. We prove here that the symmetric quadratic functions are the only symmetric functions satisfying $PC(l)$ when $l \geq 2$.

Theorem 1. *The only symmetric Boolean functions which satisfy the propagation criterion of degree l where $2 \leq l < n$ are the quadratic symmetric functions. Furthermore, if n is even, the quadratic functions also satisfy $PC(n)$.*

Proof. Any symmetric function f can be written $f(x) = x_1x_2f_1 \oplus x_1f_2 \oplus x_2f_3 \oplus f_4$ where f_1, f_2 f_3 and f_4 are $(n-2)$-variable symmetric functions. As f is symmetric, only $D_{e_1}f$, $D_{e_2}f$ and $D_{e_1+e_2}f$ have to be considered. Then, $D_{e_1+e_2}f = (1 \oplus x_1 \oplus x_2)f_1 \oplus f_2 \oplus f_3$. Since the function f is symmetric, we have $f_2 = f_3$ and $D_{e_1+e_2}f(x) = (1 \oplus x_1 \oplus x_2)f_1$. Thus, $D_{e_1+e_2}f$ is balanced if and only if f_1 identically equals 1. We can deduce that every symmetric Boolean function f satisfying $PC(2)$ has algebraic degree 2. Conversely, the derivative D_uf of a symmetric quadratic function f is a non-constant affine function except if n is odd and $u = \vec{1}$. □

Remark 1. *Conversely, the same proof can be used to show that if f is a quadratic Boolean function satisfying $PC(2)$, then f is a symmetric Boolean function.*

Corollary 2. *Let f be an n-variable Boolean function satisfying $PC(l)$. If f can be decomposed in one of the following two forms :*

1. $f(x) = f_1(x_1, \ldots, x_p) \oplus f_2(x_{p+1}, \ldots, x_n)$, *where f_1 is a p-variable symmetric function and f_2 an $(n-p)$-variable Boolean function; or*
2. $f(x_1, \ldots, x_n) = f_1(x_1, \ldots, x_n) \oplus f_2(x_1, \ldots, \hat{x_i}, \ldots, \hat{x_j}, \ldots, x_n)$, *where f_1 is an n-variable symmetric function and f_2 is an $(n-2)$-variable Boolean function;*

then either f_1 is quadratic or l is at most 1.

Proof. Suppose $l \geq 2$. Since f is $PC(l)$, the function $D_{e_i+e_j}f$ is balanced. For any (i,j) such that $1 \leq i < j \leq p$, we have $D_{e_i+e_j}f = D_{e_i+e_j}f_1$. Thus, the symmetric function f_1 satisfies $PC(2)$. From Theorem 1, either the function f_1 is quadratic or the hypothesis $l \geq 2$ is false. □

References

[1] A. Bernasconi, On Boolean functions satisfying odd order propagation criteria, 3rd International Workshop on Boolean Problems, IWSBP'98, (1998), 117–124.

[2] A. Bonnecaze, P. Solé, and A.R. Calderbank, Quaternary quadratic residue codes and unimodular lattices, IEEE *Transactions on Information Theory,* **41** (1995), 366–377.

[3] P. Camion, C. Carlet, P. Charpin, and N. Sendrier, On correlation-immune functions, in *Advances in Cryptology, Proc. of* Crypto '91, LNCS **576** (1991), 86–100.

[4] C. Carlet, On the propagation criterion of degree l and order k, in *Advances in Cryptology, Proc. of* EUROCRYPT'98, LNCS **1403** (1998), 462–474.

[5] C. Carlet, On cryptographic propagation criteria for Boolean functions, Special Issue on Cryptology of *Information and Computation* **150** (1999), 32–56.

[6] C. Carlet and P. Guillot, A new representation of Boolean functions, AAECC, (1999), 94–103.

[7] J. Daemen, R. Govaerts, and J. Vandewalle, A practical approach to the design of high speed self-synchronizing stream ciphers, Singapore ICCS/ISITA '92 Conference Proceedings, IEEE, (1992), pp. 279–283.

[8] J. F. Dillon. Elementary Hadamard Difference sets, Ph. D. Thesis, Univ. of Maryland, 1974.

[9] R. Forré, The strict avalanche criterion: spectral properties of Boolean functions and an extended definition, in *Advances in Cryptology, Proc. of* CRYPTO'88, LNCS **403** (1989), 450–468.

[10] J. von zur Gathen and J. Roche, Polynomials with two values, *Combinatorica,* **17 (3)** (1997), 345–362.

[11] T. Honda, T. Satoh, T. Iwata, and K. Kurosawa, Balanced Boolean functions satisfying $PC(2)$ and very large degree, in *Proceedings of* SAC'97, (1997), 64–72.

[12] K. Kurosawa and T. Satoh, Design of $SAC/PC(l)$ of order k Boolean functions and three other cryptographic criteria, in *Advances in Cryptology, Proc. of* EUROCRYPT '97, LNCS **1223** (1997), 434–449.

[13] W. Meier and O. Staffelbach, Nonlinearity criteria for cryptographic functions, in *Advances in Cryptology, Proc. of* EUROCRYPT'89, LNCS **434** (1990), 549–562.

[14] V.S. Pless and W.C. Huffman (Eds.), The Handbook of Coding Theory, North-Holland, New York, 1998.

[15] V.S. Pless and Z. Qian, Cyclic codes and quadratic residue codes over $\mathbb{Z}_4$, *IEEE Transactions on Information Theory,* **42(5)** (1996), 1594–1600.

[16] B. Preneel, W. Van Leekwijck, L. Van Linden, R. Govaerts, and J. Vandewalle, Propagation characteristics of Boolean functions, in *Advances in Cryptology, Proc. of* EUROCRYPT'90, LNCS **473** (1991), 161–173.

[17] B. Preneel, R. Govaerts, and J. Vandewalle, Boolean functions satisfying higher-order propagation criterion, in *Advances in Cryptology, Proc. of* Eurocrypt'91, LNCS **547** (1991), 141–152.

[18] O.S. Rothaus, On bent functions, *Journal of Combinatorial Theory (A),* **20**, (1976), 300–305.

[19] P. Savicky, On the bent functions that are symmetric, *European J. of Combinatorics,* **15** (1994), 407–410.

[20] T. Siegenthaler, Correlation-immunity of nonlinear combining functions for cryptographic applications, *IEEE Transactions on Information Theory,* **30(5)** (1984), 776–780.

[21] A.F. Webster and S.E. Tavares, On the design of S-box, in *Advances in Cryptology, Proc. of* CRYPTO'85, LNCS **218** (1986), 523–534.

[22] Y. Zheng and X. M. Zhang, On relationships among avalanche, nonlinearity, and correlation-immunity, in *Advances in Cryptology, Proc. of* ASIACRYPT'00, LNCS **1976** (2000), 470–482.

Aline Gouget
GREYC, Université de Caen
F-14032 Caen Cedex, France
e-mail: gouget@info.unicaen.fr

Progress in Computer Science and Applied Logic, Vol. 23, 169–176

On Certain Equations over Finite Fields and Cross-Correlations of m-Sequences

Tor Helleseth, Jyrki Lahtonen and Petri Rosendahl

Abstract. We study the number of solutions to certain equations over finite fields and show how this gives a family of four-valued cross-correlation functions of binary m-sequences. This new family includes both of the four-valued cross-correlations found by Niho.

Mathematics Subject Classification (2000). Primary 11T55; Secondary 94A55.

Keywords. Finite fields, Cross-correlation, m-sequences.

1. Introduction

For the theory of finite fields, their equations and characters we refer to [7] and [6].

The finite field with $q = p^k$ elements is denoted by $GF(q)$. Later, when studying cross-correlation functions of binary m-sequences, we will restrict ourselves to the case $p = 2$.

Let $y \in GF(q^2) \setminus \{0\}$, and denote $y^q = \overline{y}$. We will find the possible number of solutions to

$$\begin{cases} x^{p^s+1} + yx^{p^s} - \overline{y}x - 1 & = & 0 \\ x^{q+1} & = & 1. \end{cases} \tag{1.1}$$

The motivation to study this kind of equations comes from a cross-correlation problem for m-sequences. However, this equation is interesting in itself. We will see that it, in some sense, behaves like an affine equation over the subfield. In fact, our treatment is based on this idea.

In the binary case, the possible number of solutions to the above equation gives the possible values taken by the cross-correlation function of two binary m-sequences of period $2^n - 1$ which differ by the decimation

$$d = (2^{2k} + 2^{s+1} - 2^{k+1} - 1)/(2^s - 1), \tag{1.2}$$

where we have assumed that $n = 2k$ and that $2s$ divides k. It turns out that the cross-correlation function is four-valued.

Finding the distribution of the values taken by the cross-correlation corresponding to the decimation above involves solving another equation, namely

$$(x+1)^d + x^d + 1 = 0. \tag{1.3}$$

For a given d, this is usually more or less a routine task. We give the number of solutions for a family of decimations d.

The cross-correlation function between two cyclically distinct m-sequences takes at least three values, see [2], and all known three-valued cases are covered by theoretical results. Previously, only three families of four-valued cross-correlation functions have been found. These correspond to the decimations

A. $d = 2^{n/2+1} - 1$, with $n \equiv 0 \pmod 4$,
B. $d = (2^{n/2}+1)(2^{n/4}-1)+2$, with $n \equiv 0 \pmod 4$, and
C. $d = \sum_{i=0}^{n/2} 2^{im}$, with $n \equiv 0 \pmod 4$, $0 < m < n$, $\gcd(n,m) = 1$.

The cases A. and B. are due to Niho [9] and case C. is due to Dobbertin [1]. The decimations in C. include the decimations in A.

Our family of decimations includes the decimations both in A. and B., and in addition case C. leads to the same pair of equations. Thus all *known* infinite families of four-valued cross-correlations arise from the same equation!

2. The Equation

Suppose n is even, say $n = 2k$. We denote $q = p^k$. In analogy with the usual complex conjugation we will denote

$$\overline{y} = y^q$$

for $y \in GF(q^2)$. The usual properties of conjugation carry over to the finite case. For instance, we have $\overline{u+v} = \overline{u} + \overline{v}$ and $u + \overline{u} \in GF(q)$ for all $u, v \in GF(q^2)$. A less trivial property is presented in the following lemma.

We define the unit circle of $GF(q^2)$ to be the set

$$S = \left\{x \in GF(q^2) : x\overline{x} = 1\right\}.$$

Lemma 2.1.

(i) *Let $z \in GF(q^2) \setminus GF(q)$ be fixed. Then*

$$S \setminus \{1\} = \left\{\frac{z+u}{\overline{z}+u} : u \in GF(q)\right\}.$$

(ii) *Let $\beta \in S \setminus \{\pm 1\}$ be fixed. Then*

$$S \setminus \{\beta\} = \left\{\frac{\alpha\beta+1}{\alpha+\beta} : \alpha \in GF(q)\right\}.$$

Proof. Since $u = \overline{u}$ for $u \in GF(q)$ we have $\overline{x} = x^{-1}$ for $x = (z+u)/(\overline{z}+u)$. Furthermore, $z \in GF(q^2) \setminus GF(q)$ implies that the elements of this form are distinct and different from 1. This proves (i) and (ii) is equally simple. □

We note here that both of these parameterizations have a geometric interpretation, and this is the way they were found.

Lemma 2.2. *Let α be a nonzero element in some extension of the field $GF(p)$. If the equation*

$$x^{p^s-1}=\alpha$$

has a solution in $GF(p^k)$, then it has exactly $p^{\gcd(k,s)}-1$ solutions in the field $GF(p^k)$.

Proof. Assume $x_0 \in GF(p^k)$ satisfies the above equation. Then any ux_0, with $u \in GF(p^s)$, is a solution and every solution is obtained in this way. The claim follows from the fact $GF(p^k)\bigcap GF(p^s)=GF(p^r)$, where $r=\gcd(k,s)$. □

Theorem 2.3. *Let $n=2k$ and $y \in GF(q^2)\setminus\{0\}$. The equation*

$$x^{p^s+1}+yx^{p^s}-\overline{y}x-1=0 \tag{2.1}$$

has either 0, 1, 2 or $p^{\gcd(s,k)}+1$ solutions $x \in S$.

Proof. The proof is divided into two cases.

Case 1. Assume first that $y \in GF(q)$, i.e., $y=\overline{y}$. In this case $x=1\in S$ is a solution to (2.1). We apply the parameterization (i) of Lemma 2.1 to the equation (2.1), and then multiply it by $(\overline{z}+u)^{p^s+1}$ (note that the coefficient of u^{p^s+1} disappears) to get

$$(z-\overline{z}+y\overline{z}-yz)u^{p^s}+(z^{p^s}-\overline{z}^{p^s}+yz^{p^s}-y\overline{z}^{p^s})u=-(z^{p^s+1}-\overline{z}^{p^s+1}+yz^{p^s}\overline{z}-yz\overline{z}^{p^s}).$$

Every solution $x \in S\setminus\{1\}$ to (2.1) corresponds to a solution $u \in GF(2^k)$ of the previous equation.

If $z-\overline{z}+y\overline{z}-yz=0$, there is nothing to prove. Otherwise we have an affine equation of the form

$$u^{p^s}+\alpha_1 u=\alpha_2, \tag{2.2}$$

where $\alpha_1,\alpha_2 \in GF(q)$. Lemma 2.2 implies that the corresponding linear equation

$$u^{p^s}+\alpha_1 u=0 \tag{2.3}$$

has either exactly one root or exactly $p^{\gcd(k,s)}$ roots in $GF(q)$. From linear algebra (or the theory of linearized polynomials, see [7]) we know that the affine equation (2.2) either has no solutions or it has the same number of solutions as (2.3). Hence, in the case $y \in GF(q)$, the equation (2.1) has either 1, 2 or $p^{\gcd(k,s)}+1$ solutions in S.

Case 2. For the rest of the proof, we assume that $y \notin GF(q)$. If (2.1) has no solution in S, we are through. Suppose now that there is such a solution. We apply the parameterization (ii) of Lemma 2.1 to the equation (2.1). Since $y \notin GF(q)$, the fixed element β can be chosen to be one of the solutions. Multiplied by $(\alpha+\beta)^{p^s+1}$ the equation (2.1) transforms to

$$\begin{aligned}(\beta^{p^s+1}+y\beta^{p^s}-\overline{y}\beta-1)\alpha^{p^s+1} &+ (\beta^{p^s}+y\beta^{p^s+1}-\overline{y}-\beta)\alpha^{p^s}\\ &+ (\beta+y-\overline{y}\beta^{p^s+1}-\beta^{p^s})\alpha\\ &+ (1+y\beta-\overline{y}\beta^{p^s}-\beta^{p^s+1})=0.\end{aligned}$$

Here the leading coefficient is zero. We should now find the solutions in $GF(q)$.

If the coefficient of α^{p^s} above is zero, then we are through. Otherwise we have again an affine equation of the form

$$u^{p^s}+\alpha_1 u=\alpha_2, \tag{2.4}$$

where $\alpha_1,\alpha_2\in GF(q^2)$. To complete the proof, we may now proceed similarly as in the case $y\in GF(p^k)$. □

The binary case of this theorem is proved in [5], and in that paper only parameterization (i) is used. Actually, either one of the parameterizations would be enough in the binary case. In this more general case some difficulties occur if we try to use either (i) or (ii) only.

An easy computation shows that $\overline{\alpha_i}=\alpha_i$ in the equation (2.4). Hence we have in fact $\alpha_1,\alpha_2\in GF(q)$ although this is not needed.

It may seem difficult to find the number of times each possibility happens, e.g., how many times (2.1) has exactly one solution in S. However, in the binary case the equation is related to certain cross-correlation functions, and the question above can be answered by solving an equation of the type (1.3). We will do this for a more general class of decimations d after giving some background.

3. An Application

For basic properties of m-sequences we refer to [8] and [4].

From now on we assume that $p=2$. Recall that the cross-correlation function between two binary sequences $u(t)$ and $v(t)$ of the same period ϵ is by definition

$$C_{u,v}(\tau)=\sum_{t=0}^{\epsilon-1}(-1)^{u(t)+v(t+\tau)}.$$

An important problem in sequence analysis is to determine the values and the number of their occurrences taken by the cross-correlation function.

Assume now that $u(t)$ and $v(t)$ are m-sequences of period 2^n-1. We may assume that $u(t)$ is given by

$$u(t)=tr_1^n(\gamma^t),$$

where tr_1^n denotes the trace from $GF(2^n)$ onto $GF(2)$ and γ is a primitive element of $GF(2^n)$. Furthermore, $v(t)$ can be shifted cyclically in such a way that $v(t)=$

$u(dt)$ for some d satisfying $\gcd(d, 2^n - 1) = 1$. As usual, we denote the cross-correlation function of these sequences by $C_d(\tau)$, i.e.,

$$C_d(\tau) = \sum_{t=0}^{2^n-2} (-1)^{tr_1^n(\gamma^t + \gamma^{d(t+\tau)})}.$$

It is well known that the values (and the number of their occurrences) of $C_d(\tau)$ depend only on d, and not on the choice of the primitive element.

The main technique used in [9] is given by the following theorem. Again we assume that $n = 2k$.

Theorem 3.1. *Assume that the integer d satisfies*

(i) $\gcd(d, 2^n - 1) = 1$,
(ii) $d \equiv 1 \pmod{2^k - 1}$, *and*
(iii) $ed \equiv f \pmod{2^k + 1}$,

for some f and some e for which $\gcd(e, 2^k + 1) = 1$. Then $C_d(\tau)$ assumes exactly the values

$$-1 + 2^k(N(y) - 1), \tag{3.1}$$

where $N(y)$ is the number of solutions to the pair of equations

$$\begin{cases} x^{2f} + yx^{f+e} + \overline{y}x^{f-e} + 1 &= 0 \\ x^{2^k+1} &= 1, \end{cases} \tag{3.2}$$

and y runs through the nonzero elements of $GF(2^n)$.

The proof of Theorem 3.1 is based on the transitivity of the trace and the observation that every $x \in GF(2^n) \setminus \{0\}$ can be represented uniquely as $x = \alpha\beta^f$, where $\alpha \in GF(2^k) \setminus \{0\}$ and $\beta \in S$.

The assumption (i) is needed only to guarantee that the decimated sequence is indeed an m-sequence. Without this condition, the theorem would still be useful in determining cross-correlation functions or weight distributions of cyclic codes.

Let

$$d = \frac{2^{k-1}}{2^s - 1}(2^{2k} + 2^{s+1} - 2^{k+1} - 1), \tag{3.3}$$

where it is assumed $n = 2k$ and $2s$ divides k. It is straightforward to see, that d satisfies the conditions of Theorem 3.1 for $e = 2^s - 1$ and $f = 2^k - 2^s$. Now the corresponding equation is exactly the binary special case of (2.1).

In view of (3.1), Theorem 2.3 now implies that for the d in question, $C_d(\tau)$ is indeed four-valued, and that the cross-correlation values are $-1 - 2^k$, -1, $-1 + 2^k$, and $-1 + 2^{k+s}$. In order to find the distribution of the values (or the number of occurrences of each possibility in Theorem 2.1), we will use the following lemma.

Lemma 3.2. *We have*

(i) $\sum_{\tau=0}^{2^n-2}(C_d(\tau) + 1) = 2^n$
(ii) $\sum_{\tau=0}^{2^n-2}(C_d(\tau) + 1)^2 = 2^{2n}$
(iii) $\sum_{\tau=0}^{2^n-2}(C_d(\tau) + 1)^3 = 2^{2n}b$,

where b is the number $x \in GF(2^n)$ such that

$$(x+1)^d + x^d = 1.$$

The equations (i) and (ii) are well known and proofs can be found, e.g., in [9]. The equation (iii) is proved in [2].

We will now find the number b for a family of decimations.

Lemma 3.3. *Let $\beta, \gamma \in S$. Then $\beta + \gamma \in GF(2^k)$ if and only if $\beta = \gamma$ or $\beta = \gamma^{-1}$.*

We omit the simple proof.

Theorem 3.4. *Assume that $d \equiv 1 \pmod{2^k - 1}$. If $\gcd(d-1, 2^k+1) = \gcd(d+1, 2^k+1) = 1$, then the equation*

$$(x+1)^d = x^d + 1 \tag{3.4}$$

has exactly 2^k solutions in $GF(2^n)$.

Proof. Every $x \in GF(2^k)$ is a solution to (3.4) since $d \equiv 1 \pmod{2^k - 1}$. We now assume that $x \neq 0$ satisfies (3.4).

The equation (3.4) implies $(\overline{x}+1)^d = \overline{x}^d + 1$, and hence

$$(x+1)^d (\overline{x}+1)^d = \left(x^d + 1\right)\left(\overline{x}^d + 1\right),$$

that is

$$(x\overline{x} + x + \overline{x} + 1)^d = (x\overline{x})^d + x^d + \overline{x}^d + 1.$$

For $a \in GF(2^k)$ we have $a^d = a$, and thus

$$x^d + x = \overline{x}^d + \overline{x}.$$

This is equivalent to

$$x^d + x \in GF(2^k). \tag{3.5}$$

Representing $x = \alpha\beta$, where $\alpha \in GF(2^k)$ and $\beta \in S$, gives that $\beta \in S$ satisfies $\beta^d + \beta \in GF(2^k)$. Lemma 3.3 implies $\beta^d = \beta$ or $\beta^d = \beta^{-1}$, i.e., $\beta^{d\pm 1} = 1$. By assumption, this is possible if and only if $\beta = 1$, and thus $x \in GF(2^k)$. □

Lemma 3.5. *We have $\gcd(d \pm 1, 2^k + 1) = 1$ for d in (3.3).*

Proof. Since now $\gcd(2^s - 1, 2^k + 1) = 1$, we have $\gcd(d \pm 1, 2^k + 1) = \gcd((2^s - 1)(d \pm 1), 2^k + 1)$. The lemma follows easily from the congruence $(2^s - 1)d \equiv 2^k - 2^s \pmod{2^k + 1}$. □

Finally, $C_d(\tau)$ is as follows.

Theorem 3.6. *Let $n = 2k$, where $2s$ divides k, and let $d = (2^{2k} + 2^{s+1} - 2^{k+1} - 1)/(2^s - 1)$. Then the cross-correlation function $C_d(\tau)$ between two m-sequences takes the following values:*

$$\begin{array}{llll}
-1 - 2^k & \textit{occurs} & \frac{2^{2k+s-1} - 2^{k+s-1}}{2^s+1} & \textit{times} \\
-1 & \textit{occurs} & \frac{2^{2k} - 2^k - 2^s}{2^s} & \textit{times} \\
-1 + 2^k & \textit{occurs} & \frac{2^{2k+s-1} - 2^{2k} + 2^{k+s-1}}{2^s - 1} & \textit{times} \\
-1 + 2^{k+s} & \textit{occurs} & \frac{2^{2k} - 2^k}{2^{3s} - 2^s} & \textit{times.}
\end{array}$$

Proof. Theorem 2.3 shows that $C_d(\tau)$ is four-valued and gives the values. Furthermore, Theorem 3.4 gives the number b of Lemma 3.2. Denote by N_i the number of times (2.1) has exactly i solutions in S. We have a system of linear equations

$$\begin{aligned} N_0 + N_1 + N_2 + N_{2^s+1} &= 2^{2k} - 1 \\ -2^k N_0 + 2^k N_2 + 2^{k+s} N_{2^s+1} &= 2^{2k} \\ 2^{2k} N_0 + 2^{2k} N_2 + 2^{2k+2s} N_{2^s+1} &= 2^{4k} \\ -2^{3k} N_0 + 2^{3k} N_2 + 2^{3k+3s} N_{2^s+1} &= 2^{5k}. \end{aligned}$$

The first equation comes from the number of equations of the form (2.1), and the other ones are simple consequences of Lemma 3.2. Straightforward calculations give the claimed distribution. □

Remark 3.7. It is a routine matter to verify that $s = 1$ (resp. $s = k/2$) corresponds to the case A. (resp. B.) given in the introduction. We note here that Niho's proof of B. is somewhat complicated. In fact, it is incomplete in the sense that it essentially depends on a result due to Welch, and this result does not seem to be published. An earlier simple proof of B. can be found in [3].

The case C. by Dobbertin [1] leads to the same equation but with the restriction $\gcd(s, k) = 1$. The proof presented by Dobbertin is based on Niho's technique but is different otherwise. Thus we have an alternative proof also in this case. It should be noted, that according to the computed results, there are four-valued cross-correlations which are not related to the equation studied in Section 2.

Niho [9] gave tables of binary cross-correlation functions up to $n = 16$, and now all at most four-valued cross-correlation functions of binary m-sequences within this table belong to a known infinite family.

Lastly we mention the well-known fact that the problem of determining the cross-correlation function of m-sequences is equivalent to determining the weight distribution of certain cyclic codes. This connection is explained in detail in [6].

References

[1] H. Dobbertin: *One-to-one highly nonlinear power functions on $GF(2^n)$*, AAECC Applicable Algebra in Engineering, Communication and Computing 9 (1998) 139–152.

[2] T. Helleseth: *Some results about the cross-correlation function between two maximal linear sequences*, Discrete Mathematics 16 (1976), 209–232.

[3] T. Helleseth: *A note on the cross-correlation function between two binary maximal length linear sequences*, Discrete Mathematics 23 (1978), 301–307.

[4] T. Helleseth, P.V. Kumar: *Sequences with low correlation*, in Handbook of Coding Theory (ed. V.S. Pless, W.C. Huffman), Elsevier Science (1998), 1765–1853.

[5] T. Helleseth, P. Rosendahl: *New pairs of m-sequences with 4-level cross-correlation*, submitted to Finite Fields and Their Applications.

[6] I. Honkala, A. Tietäväinen: *Codes and number theory*, in Handbook of Coding Theory (ed. V.S. Pless, W.C. Huffman), Elsevier Science (1998), 1141–1194.

[7] R. Lidl, H. Niederreiter: *Finite Fields*, Encyclopedia of Mathematics and Its Applications vol. 20, Addison-Wesley, Reading (1983).

[8] R.J. McEliece: *Finite Fields for Computer Scientists and Engineers*, Kluwer Academic Publishers, Boston (1987).

[9] Y. Niho: *Multivalued cross-correlation functions between two maximal linear recursive sequences*, PhD Thesis, University of Southern California (1972).

[10] H.M. Trachtenberg: *On the cross-correlation functions of maximal linear sequences*, PhD Thesis, University of Southern California (1970).

Tor Helleseth
Department of Informatics
University of Bergen
N-5020 Bergen, Norway
e-mail: torh@ii.uib.no

Jyrki Lahtonen
Department of Mathematics
University of Turku
SF-20014 Turku, Finland
e-mail: lahtonen@utu.fi

Petri Rosendahl
Department of Mathematics
University of Turku
SF-20014 Turku, Finland
e-mail: perosen@utu.fi

Progress in Computer Science and Applied Logic, Vol. 23, 177–192

A Polly Cracker System Based on Satisfiability

Françoise Levy-dit-Vehel and Ludovic Perret

Abstract. This paper presents a public-key cryptosystem based on a subclass of the well-known *satisfiability* problem from propositional logic, namely the *doubly-balanced* 3-SAT problem. We describe the construction of an instance of our system – which is a modified Polly Cracker scheme – starting from such a 3-SAT formula. Then we discuss security issues: this is achieved on the one hand by exploring best methods to date for solving this particular problem, and on the other hand by studying (systems of multivariate) polynomial equation solving algorithms in this particular setting. The main feature of our system is the resistance to intelligent linear algebra attacks.

Keywords. Combinatorial-algebraic cryptosystems, systems of polynomial equations, 3-SAT, hard instances generation.

1. Introduction

Since the failure of knapsack-based cryptosystems [Od, Sh], a widely accepted opinion was that NP-complete problems were not suited for the construction of secure trapdoor one-way functions. In 1993, M. Fellows and N. Koblitz [FK] proposed to further investigate the use of those problems for designing public-key cryptosystems, and proposed a general framework, called CA-systems[1], the main illustration of which was the Polly Cracker cryptosystem. In this system, the public-key is a set $S = \{p_1, \ldots, p_\ell\}$ of multivariate polynomials over a finite field $\mathbb{F}_q$, and the secret-key is a zero α of S. To encrypt a message $M \in \mathbb{F}_q$, Bob chooses an element $e_S = \sum_{i=1}^{\ell} h_i p_i$ of the ideal generated by the polynomials of S, and sends $c = e_S + M$ to Alice. Knowledge of α then allows Alice to decrypt the ciphertext just by evaluating it on α.
The (public-key, secret-key) pair is derived from an instance of an NP-complete combinatorial[2] problem, in such a way that knowing the public-key is equivalent to knowing the considered instance, and that finding a secret-key from the public-key is equivalent to finding a solution for this particular instance. M. Fellows and

[1] For "combinatorial-algebraic" cryptosystems.
[2] In a broad sense, i.e., this includes graph theory, boolean logic, ...

N. Koblitz suggest several NP-complete problems for use in this context, mainly based on graph theory (e.g., 3-colorability, perfect codes in graphs, ...) but do not really investigate the way of generating "hard" instances of these problems with a fixed solution. As pointed out by R. Steinwandt et al. [GS], a naive technique to generate such instances yields very weak public-keys.
Here, we follow the CA-systems line of research by proposing a public-key cryptosystem based on the well-known SATISFIABILITY problem from propositional logic. More precisely, we use the 3-SAT problem. One advantage of using this underlying hard problem is that it has been extensively studied, mainly due to the fact that it is of interest in other research areas, such as planning or scheduling, see, e.g., [CMi, CMo].
Although proven to be NP-complete, this problem admits many "easy" instances, where deterministic algorithms (such as the recursive DPLL[DLL]) perform quite well in practice. Indeed, let n (resp. m) denote the number of variables (resp. clauses) of the problem, and set $m = cn$ with $c \in \mathbb{R}^{*+}$. Then, as c increases, it has been shown experimentally that the probability of an instance of 3-SAT being satisfiable shifts from almost one to almost zero. The range of c over which this transition occurs is[3] $3.003 < c < 4.598$. This is known as the *threshold conjecture*. In this range, there is a value of c corresponding to a complexity peak at which on average half of the instances are satisfiable. The exact value of c yielding this peak can be numerically determined for each instance distribution.
Non-deterministic methods have also been devised, that often give better results on satisfiable instances (e.g., Walksat, [SKC]), especially near the threshold region. They are known as *local search methods*.
The hardness of this problem is tightly located in the critical range for c, and for (very) large values of n. Having this in mind, and also that the parameter sizes and generation times of our system have to be polynomial[4], we chose to restrict ourselves to a particular class of the 3-SAT instances, namely the class of so-called *doubly-balanced* 3-SAT [DB], a.k.a. *literal-regular* 3-SAT [BS]. Formulae in this class have the particularity that every variable appears (almost) equally often, and (almost) as often negated as unnegated. Instances from this class are much more difficult to solve in general than random 3-SAT instances, as they are designed to have structural regularities, thus confusing variable selection heuristics that are used by most solvers (for example, DPLL-like algorithms treat the variables with a small number of occurrences first).
Note that for random 3-SAT the complexity peak occurs for $c \approx 4.25$, while for doubly-balanced 3-SAT, it has been shown to be $c \approx 3.5$ (both values experimentally determined).
The paper is organized as follows: in the next section, we begin by providing the necessary background to understand the basics of the 3-SAT problem, as well

[3]For 3-SAT; For k-SAT with higher values of k, this range is shifted. Also, the higher n is, the sharper the range becomes.
[4]In the size of the input of 3-SAT, namely $n\lg(n)$, denoting by lg() the base-two logarithm.

as methods for generating random instances, and doubly-balanced ones. Then we show how to translate this problem into a system of polynomial equations, in order to use it in our cryptographic setting. We exhibit the correspondence existing between the models of 3-SAT and the solutions of the system, and we link particular 3-SAT formulae with reduced Gröbner bases. In Section 3, we describe the cryptographic scheme we propose and present an original method to encrypt messages, the security of which is addressed in Section 4. We address carefully the single break attacks found by H.W. Lenstra Jr. [Ko], and show that they cannot be conducted in our context. We also consider the differential attack proposed in [SG], which is a very powerful tool to attack generic Polly Cracker systems. In addition, we suggest an extension of this attack. On the other hand, we investigate total break methods on the system. They are of two types: the first type is the use of 3-SAT solvers to break the considered instances, from which we protected ourselves by carefully choosing the instances. The second type is to run algorithms computing (an element of) the variety of the set of polynomials involved. One of the best algorithm known to us – namely F_4 [Fa] – does in fact more: it computes a Gröbner basis of the set of polynomials. For the considered sizes, it appears that such an algorithm is of no help.

We end the paper by a section concerning implementation aspects. We would like to mention that, when investigating Polly Cracker-type systems, our intention was not to design a scheme that was likely to compete with the public-key systems in use. What we were interested in was mainly to design a new Polly Cracker system offering resistance to linear algebra attacks. Moreover, our approach of the SATISFIABILITY problem in this cryptographic setting appears quite interesting, as the public keys arising from this problem can be chosen strong.

2. CNF Formulae and Systems of Polynomial Equations

2.1. 3-SAT and Instance Generation Methods

We begin by recalling what the 3-SAT problem is. Let $X = \{x_1, \dots, x_n\}$ be a set of variables and let $\wedge$, $\vee$, $\bar{}$ denote logical *and, or, not* respectively. A *truth assignment* for X is a function $t : X \mapsto \{True, False\}$. For all j,$1 \leq j \leq n$, a *literal* u_j is either x_j or $\bar{x_j}$. For a variable $x_j \in X$, a literal x_j (resp. $\bar{x_j}$) is true if $t(x_j) = True$ (resp. $t(\bar{x_j}) = False$). A *clause* over X is the disjunction of a set of literals over X. It is satisfied by a truth assignment if, and only if, at least one of its literals is true under that assignment. A clause containing only three literals will be called a 3-*clause*. For instance, $C = x_{j_1} \vee \bar{x_{j_2}} \vee x_{j_3}$, $1 \leq j_1, j_2, j_3 \leq n$, is a 3-clause, and is satisfied unless $t(x_{j_1}) = False$, $t(x_{j_2}) = True$, $t(x_{j_3}) = False$. A *CNF-formula*[5] $\mathcal{C}$ is the conjunction of arbitrarily many clauses $C_1, \dots, C_m$, $m \in \mathbb{N}^*$. It is satisfiable if, and only if there exists some truth assignment for X that simultaneously satisfies all the clauses in $\mathcal{C}$. Such a truth assignment is called a *satisfying truth assignment*, or a *model* for the formula $\mathcal{C}$. If $\mathcal{C}$ contains only 3-clauses, then we say that $\mathcal{C}$ is a

[5] Conjunctive Normal Form.

3-CNF formula. For instance, $\mathcal{C} = \wedge_{j=1}^{m} C_j$ where $C_j = u_{j_1} \vee u_{j_2} \vee u_{j_3}$, $m \in \mathbb{N}^*$, is such a formula.
In the sequel, we shall denote a CNF-formula either as a conjunction of clauses as above, or equivalently as a collection of clauses, the conjunction then being implicit.
The 3-satisfiability problem can then be stated as follows:

INSTANCE: a collection $\mathcal{C} = \{C_1, \ldots, C_m\}$ of 3-clauses on X.

QUESTION: is there a satisfying truth assignment for $\mathcal{C}$?

The random 3-SAT problem which we referred to in the introduction is the 3-SAT problem in which instances are generated according to the following procedure[6]: *The number of variables n and the number of clauses m being fixed, randomly select three distinct variables out of n, then negate each variable with probability 1/2. Combine these literals in a 3-clause. Repeat this process until the desired number m of clauses is reached. Conjoin them to form a CNF-formula.*
The restriction of 3-SAT to balanced formulae is the one in which a formula $\mathcal{C}$ is such that, for all i, $1 \leq i \leq n$, each variable x_i appears equally often[7], i.e., in $\lfloor 3m/n \rfloor$ clauses (there are $3m$ positions to fill, corresponding to the m 3-clauses). But then, it can be that some variables appear more often negated than unnegated (or the converse). The doubly-balanced 3-SAT subclass is precisely the class of formulae that do not present this type of irregularity; namely, a formula in this class is such that each literal appears (almost) $3m/(2n)$ times (there are $2n$ possible literals). Such instances can be generated with the following algorithm:
The number of variables n and the number of clauses m is being fixed. Place $\lfloor 3m/(2n) \rfloor$ occurrences of each of the $2n$ literals in a bag. To reach exactly $3m$ literals in the bag, add randomly some literals, not twice the same. To construct each clause, remove three literals on distinct variables from the bag. At some point, if the literals remaining in the bag concern only one or two distinct variables, then randomly add distinct variables in the bag, negating each of them with probability 1/2. Keep on the construction of the clauses until the desired number is reached.
Note that to generate a (doubly-balanced) formula admitting a particular model y, one simply modifies the above procedure by throwing away the 3-clauses that are not satisfied by y.

2.2. Constructing a System of Polynomial Equations from 3-SAT

We shall now explain how to translate an instance of the 3-SAT problem into a system of polynomial equations. A similar description already appeared in [Ba].
We shall denote by $K[X]$, the polynomial ring $K[x_1, \ldots, x_n]$ over the field K. We choose two field values $T, F \in K$, representing $True$ and $False$ respectively. To a 3-clause c involving the three literals u_j, u_k, u_ℓ, $1 \leq j, k, \ell \leq n$, one can associate a total degree 3 polynomial in $K[X]$ as follows: if $u_j = x_j$, then we replace u_j by $(x_j - T)$; if $u_j = \bar{x}_j$, then we replace it by $(x_j - F)$. Replace $\vee$ by

[6] Fixed Clause Length generation.
[7] Almost: occurrences of some variables must be added if $3m/n$ is not an integer.

multiplication. For instance, the polynomial[8] $p_c(X) \in K[X]$ corresponding to the clause $c = x_j \vee \bar{x_k} \vee x_\ell$ is $p_c(X) = (x_j - T)(x_k - F)(x_\ell - T)$. It is then clear that a satisfying truth assignment of X for c corresponds to a zero of the polynomial $p_c(X)$. With this construction, we have:

Theorem 2.1. *A 3-CNF formula $\mathcal{C} = \wedge_{i=1}^m C_i$ admits a model if, and only if, the corresponding system of polynomial equations $\{p_1(X) = 0, \ldots, p_m(X) = 0\}$ has a solution over the algebraic closure of K.*

Let $k \in \{1, \ldots, m\}, \{i_1, \ldots, i_k\} \subset \{1, \ldots, m\}$ and $\{C_{i_j}\}_{1 \le j \le k}$ be a set of clauses. We shall say that $\{Var(C_{i_j})\}_{1 \le j \le k}$ is a *disjoint set* if for all $a, b \in \{i_1, \ldots, i_k\}$, $a \neq b$, $Var(C_a) \cap Var(C_b) = \emptyset$. We will give now a simple connection between a set of clauses and a Gröbner basis. For a detailed description of Gröbner bases, we refer to [BW].

In order to prove the next proposition, we introduce a few notations, that will be useful throughout the paper.

We shall denote by $Term = \{x_1^{\nu_1} \ldots x_n^{\nu_n}, (\nu_1, \ldots, \nu_n) \in \mathbb{N}^n\}$ the set of *terms* in $\{x_1, \ldots, x_n\}$. We define the *total degree* of a term $x_1^{\nu_1} \ldots x_n^{\nu_n} \in Term$ as the sum $\sum_{i=1}^n \nu_i$, $Term(f)$ as the set of terms of the polynomial $f \in K[X]$ and $HT(f)$ as the head term of f (with respect to some fixed order on the terms). A *monomial* at is simply a term t multiplied by a constant $a \in \mathbb{F}_q$.

Proposition 2.2. *Let $\mathcal{C} = \wedge_{i=1}^m C_i$ be a 3-CNF formula, $\{p_1, \ldots, p_m\}$ be polynomials of $K[X]$ constructed from this formula as explained above, with $T, F \in K$.*
If $\{Var(C_{i_j})\}_{1 \le j \le k}$ is a disjoint set, then $\{p_{i_j}\}_{1 \le j \le k}$ is a reduced Gröbner basis of $\langle p_{i_j} \rangle_{1 \le j \le k}$ for the degree lexicographical (deglex) order.

Proof. The fact that $\{Var(C_{i_j})\}_{1 \le j \le k}$ is a disjoint set implies that any two $p, p' \in \{p_{i_j}\}_{1 \le j \le k}$ have disjoint head terms. It follows, by the Buchberger's first criterion, that $\{p_{i_j}\}_{1 \le j \le k}$ is a Gröbner basis of $\langle p_{i_j} \rangle_{1 \le j \le k}$ for the *deglex* order. Moreover, by construction, all these polynomials are monic. Finally, suppose that there exists two different indices $a, b \in \{i_1, \ldots, i_k\}$ for which $t_b = t * HT(p_a)$ with $t \in Term$ and $t_b \in Term(p_b)$, i.e., $t = \frac{t_b}{HT(p_a)}$.
It is then necessary that $Var(C_a) \cap Var(C_b) \neq \emptyset$, contradicting the assumption. □

We shall here use for K a finite field $\mathbb{F}_q$. We ask that T and F be two non-zero field elements, so we set $q \ge 3$.

3. The System

Selecting the Public-key/Secret-key pair

Alice chooses a finite field $\mathbb{F}_q$ with $q \ge 3$, and positive integers m and n. She also takes a vector y of $\{T, F\}^n$ at random. This is her secret-key.

She then generates an instance $\mathcal{C} = \wedge_{i=1}^m C_i$ of doubly-balanced 3-SAT admitting y as model. For this, she uses a generation method due to E. Hirsch [Hi] and

[8] Letting X stand for $x_1, \ldots, x_n$.

called *hgen2*. This method follows the one described in Section 2.1, but with some other constraints, that aim to generate formulae with as independent clauses as possible. For instance, if a clause involves literals u_j, u_k and u_ℓ, then his algorithm is designed such that no other clause of $\mathcal{C}$ involves any two of them.
Having done this, Alice publishes the formula $\mathcal{C}$, together with m, n and q (values T and F are also publicly known). In Section 5, we shall explain how we represent $\mathcal{C}$. Indeed, as shown in Section 2.2, it would have been equivalent – from an information theoretic viewpoint – to publish the m polynomials corresponding to these m clauses, but the "clause-representation" allows for a more compact form.

Encryption
The encryption phase follows the idea of a regular Polly Cracker scheme. But the practical realization is quite different from [FK]. We shall denote by $\{p_1, \dots, p_m\}$, the polynomials constructed from the clauses $\{C_1, \dots, C_m\}$, and by I the ideal generated by these polynomials. We shall now explain how to use Proposition 2.2 to construct, in a very simple way, an element of I. In the next sections, we shall motivate this construction. The algorithm is the following:

Algorithm 1
Input: $f \in \mathbb{F}_q[X]$, $l \geq 2$, $\{\lambda_1, \dots, \lambda_l\}, \lambda_i \in \mathbb{F}_q$ with $\sum_{i=1}^{l} \lambda_i \equiv 0[q]$ and $\mathfrak{D} = \{\mathfrak{d}_1, \dots, \mathfrak{d}_l\}$ a set of subset indexes such that $\forall 1 \leq i \leq l$, $\{Var(C_{i_j})\}_{j \in \mathfrak{d}_i}$ is a disjoint set.
Output: An element of the ideal I.
For i from 1 to l **do**
  Compute $N_i(f)$, the normal form of f modulo $\{p_j\}_{j \in \mathfrak{d}_i}$.
End For
Return $e_I = \sum_{i=1}^{l} \lambda_i N_i(f)$.

Theorem 3.1 (Correctness). *With the inputs given in the preceding algorithm, e_I is an element of I.*

Proof. Note that, according to Proposition 2.2, at each step i, $1 \leq i \leq l$, of the algorithm, $\langle p_j \rangle_{j \in \mathfrak{d}_i}$ is a Gröbner basis. Hence, $N_i(f)$ being the normal form of f modulo $\{p_j\}_{j \in \mathfrak{d}_i}$, we have that $f_i = N_i(f) - f$ reduces to 0 modulo $\{p_j\}_{j \in \mathfrak{d}_i}$. Thus:

$$\forall 1 \leq i \leq l, f_i \in \langle p_j \rangle_{j \in \mathfrak{d}_i} \subset \langle p_1, \dots, p_m \rangle.$$

We conclude the proof by noticing that, due to the choice of $\{\lambda_1, \dots, \lambda_l\}$:

$$e_I = \sum_{i=1}^{l} \lambda_i N_i(f) = \sum_{i=1}^{l} \lambda_i (N_i(f) - f) = \sum_{i=1}^{l} \lambda_i f_i \in \langle p_1, \dots, p_m \rangle. \qquad \square$$

For $e_I(X) = \sum_{\alpha \in \mathbb{N}^n} a_\alpha x^\alpha$, with almost all the αs being zero, we define *supp*(e_I) as the set $\{\alpha \in \mathbb{N}^n : a_\alpha \neq 0\}$.

To encrypt M, Bob chooses $\beta = (\beta_1, \ldots, \beta_n) \in supp(e_I)$, $\beta \neq (0, \ldots, 0)$, computes the ciphertext defined by:

$$c(X) = e_I(X) + Mx_1^{\beta_1} \ldots x_n^{\beta_n} = e_I(X) + Mx^{\beta} \in \mathbb{F}_q[X]$$

and sends to Alice $(c(X), \beta)$.

Decryption

Upon receiving $(c(X), \beta)$, Alice evaluates:

$$\frac{c(y)}{y^{\beta}} = \frac{e_I(y) + My^{\beta}}{y^{\beta}} = M$$

and recovers[9] the plaintext.

4. Security Issues

4.1. Total Break

It is clear that the crucial point in using the 3-SAT problem in a Polly Cracker system lies in the method chosen for generating hard satisfiable instances. While it remains an open problem to generate hard solved instances [ILL], the doubly-balanced 3-SAT formulae are among the hardest 3-SAT instances to solve by currently known methods: this is due to the fact that they are not completely random, as instances from the random 3-SAT problem can be, nor completely "structured" (this terminology refers to 3-SAT instances arising from the modelling of real-life phenomena occurring in, e.g., planning or scheduling). Thus, efficient algorithms on random formulae such as UnitWalk or OKsolver [Sa] will be defeated by the regularity of those formulae, whereas algorithms that perform well on structured instances – like Zchaff or Sato [Sa] – will then behave poorly, those formulae being too "random" to handle. The ones chosen by us for the construction of our public-keys come from the *hgen2* generator of E. Hirsch [Hi]. The formulae of this family have been confronted, in the SAT'2002 and in the SAT'2003 competitions, to all the best solvers (see again [Sa]). The result is that formulae generated by this algorithm have proven to be the ones that best resist to known solvers. Besides, instances from this generation method have won the smallest (in terms of n) satisfiable unsolved instance challenge of this competition: the smallest such instance had parameters $n = 500$ and $m = 1750$. These formulae, available in a benchmark [Hi], still remains unsolved. For this system, we recommend $700 \leq n \leq 900$ and $m = 3.5n$, which makes instances of these sizes far beyond reach of the current best solvers.

On the other hand, the security of our scheme relies on the difficulty of finding a solution of a system of m polynomial equations of maximal degree 3 in n variables over a finite field $\mathbb{F}_q$. In other words, if I denotes the ideal generated by these polynomials, the problem is to find an element of the variety $V_{\overline{\mathbb{F}_q}}(I)$. This problem can be solved by means of computing a Gröbner basis of I. In this case, this gives

[9] T and F being two non-zero field elements, it follows that $y^{\beta} \neq 0$ for any choice of β.

in fact all the elements of $V_{\overline{\mathbb{F}_q}}(I)$. The complexity of computing a Gröbner basis of a system of polynomials – although theoretically doubly exponential in the number of variables – depends in practice very much on the nature of the system, and of the algorithm used. We have run the F_4 algorithm [Fa] on instances of our scheme via the web interface [10] of Fgb. Practically, yet for polynomial systems corresponding to $n = 100$ variables and $m = 350$ clauses, such an algorithm fails computing a Gröbner basis: indeed, we have noticed that, after some iterations, the algorithm cannot terminate, due to the handling of huge matrices (typically square matrices of a hundred thousand entries). Thus, it appears that the sizes we consider are far out of reach of this type of algorithms.

Chosen ciphertext attack
Recently, a chosen-ciphertext attack on Polly Cracker system was designed [SG], whereby it is possible to retrieve the secret key by n queries to a decryption oracle. It is well known that homomorphic cryptosystems are vulnerable to this type of attacks. For cryptosystems over the integers (e.g., RSA), padding schemes like REACT [OP] address the problem. For polynomial-based cryptosystems, it remains an open problem to adapt those paddings, especially how to represent plaintexts in order to perform operations like hashing or "xoring" on them.

4.2. Single Break

The second approach to cryptanalysis looks for weaknesses in Bob's construction of the ciphertext rather than in Alice's construction of the public-key. We recall that this attack, as opposed to the total break one, consists of recovering the cleartext from a particular ciphertext, but does not recover a secret key, thus in principle not compromising other uses of the system.

Linear algebra attack
The method is as follows. Since $e_I \in I$, there exists $\{h_i\}_{1 \leq i \leq m}$ in $\mathbb{F}_q[X]$, such that:

$$e_I = \sum_{i=1}^{m} h_i p_i.$$

We call these polynomials the decomposition of the polynomial e_I under I. Moreover:

$$c = e_I + Mx^{\beta},$$

we then have the following equation:

$$c = \sum_{i=1}^{m} h_i p_i, \text{ except for the term } x^{\beta}.$$

We can solve this equation by regarding the coefficients of h_is as unknowns and get linear equations by identifying the coefficients of the terms of c with the coefficients of the terms of $\sum_{i=1}^{m} h_i p_i$ (except for the term x^{β}). Due to the huge number of unknowns in the linear system, this attack is in general intractable [Ko].

[10] http://calfor.lip6.fr/~jcf/Software/Fgb/index.html

4.2.1. Intelligent Linear Algebra Attack. In order to decrease the number of unknowns, H.W. Lenstra Jr. [Ko] proposed a improvement of this method[11]. Let:

$$H(c) = \{t \in Term : \exists t_p \in \cup_{i=1}^m Term(p_i), \exists t_c \in Term(c) \text{ such that } t_c = tt_p\}.$$

Roughly speaking, $H(c)$ denotes the set of terms that Bob can potentially use to construct the given ciphertext c. If:

$$\cup_{i=1}^m Term(h_i) \subseteq H(c) \quad (C1),$$

i.e., for all i, every term of h_i divides at least one term of the ciphertext c, then, it is possible to recover e_I by solving a system of linear equations (constructed with the method described above) involving only $\#H(c)$ unknowns.

In order to avoid this attack, Koblitz in [Ko]($Ch.$ 5) proposed a clever construction of the ciphertext for which the condition $C1$ is not achieved, i.e., there exists at least one term $t \in \cup_{i=1}^m Term(h_i)$ which does not divide any of the terms of the ciphertext.

With carefully chosen parameters of the system, we now show that our construction is resistant to this attack (corollary 4.3). For this, we need a couple of intermediate results:

Theorem 4.1. *Let $\{p_1, \ldots, p_m\}$ be the polynomials of the public-key. We shall denote by:*
$I \subset \{1, \ldots m\}$, $\{p_j\}_{j\in I}$ a subset of the public-key polynomials corresponding to a disjoint set of clauses,
$f = ax^\alpha$ with $(a, \alpha) \in \mathbb{F}_q^* \times \mathbb{N}^n$,
$N(f)$ the normal form of f modulo $\langle p_j\rangle_{j\in I}$ w.r.t. the degree lexicographic order,
D the set of terms of the decomposition of $N(f) - f$ under $\langle p_j\rangle_{j\in I}$.
If $k = |I| > 3$ and if x^α is a multiple of $\prod_{i=1}^n x_i$ then there exists at least one term $t \in D$ of total degree strictly larger than any term of $N(f)$.

Proof. We shall give a constructive proof of this theorem. First, we outline the different steps realized during the reduction process, for a more detailed description of this process, we refer to [BW].

$$\begin{aligned}
N^{(1)}(f) &= f - a_{(1)}t_{(1)}p_{(1)}, \\
N^{(2)}(f) &= N^{(1)}(f) - a_{(2)}t_{(2)}p_{(2)} &&= f - \textstyle\sum_{p=1}^2 a_{(p)}t_{(p)}p_{(p)}, \\
&\vdots && \vdots \\
N^{(l)}(f) &= N^{(l-1)}(f) - a_{(l)}t_{(l)}p_{(l)} &&= f - \textstyle\sum_{p=1}^l a_{(p)}t_{(p)}p_{(p)},
\end{aligned}$$

where $N^{(l)}(f)$ is the l-th reduction of f modulo $\{p_j\}_{j\in I}$, $p_{(l)}$ a polynomial of $\{p_j\}_{j\in I}$ used at the l-th step of the reduction process. The term $t_{(l)}$ and the constant $a_{(l)}$ are chosen in such a way as to eliminate from $N^{(l-1)}(f)$ a term t multiple of the head term of $p_{(l)}$, and more precisely, we have $t = t_{(l)}HT(p_{(l)})$ and $a_{(l)} = \frac{Coeff(t, N^{(l-1)}(f))}{HC(p_{(l)})}$ with $Coeff(t, N^{(l-1)})$ the coefficient of t in $N^{(l-1)}(f)$. We

[11] In fact, we present here an adaptation of this attack to our scheme.

shall say that $t_{(l)}$ is the l-th term of the decomposition, chosen by the reduction algorithm, of $N(f) - f$ under $\langle p_j \rangle_{j \in I}$. We define $\widetilde{l}$ as the minimum index for which there does not exist a term in $N^{(\widetilde{l}+1)}(f)$ divisible by one of the head terms of $\{p_j\}_{j \in I}$, meaning that the reduction process ends after step $\widetilde{l}$ is performed. Hence:

$$N(f) = N^{(\widetilde{l})}(f) = f - \sum_{p=1}^{\widetilde{l}} a_{(p)} t_{(p)} p_{(p)}.$$

In the sequel, we suppose that the reduction process is performed with respect to the *deglex* order.

We prove that at least $t_{(1)}$, the first term of the decomposition of $N(f) - f$ (under $\langle p_j \rangle_{j \in I}$) cannot be recovered, with the intelligent linear algebra attack described previously, from the terms of $N(f)$. By showing that all the terms of $N(f)$ are of total degree strictly smaller than the total degree of $t_{(1)}$. For this, we will show that all the terms generated during the reduction process of f, of total degree equal or larger than the total degree of $t_{(1)}$, are cancelled. Remark that due to the regular shape of the polynomials of the public-key and the particular form of f, we can give the total degree of the terms occurring at each step of this process.

First step

At the first step, $t_{(1)}$ is chosen in order to remove multiples of $HT(p_{(1)})$ from the term x^α. If we denote by d the total degree of x^α, one sees at once that the total degree of $t_{(1)}$ is equal to $d - 3$. Since x^α is divisible by the product of n distinct variables, $t_{(1)}$ is divisible by the product of at least $n - 3$ distinct variables. Let x_i, x_j and x_k be variables such that $x_i x_j x_k t_{(1)} = x^\alpha$. We have:

$$Term(N^{(1)}(f)) = \{t_{(1)} x_i x_j, t_{(1)} x_i x_k, t_{(1)} x_j x_k, t_{(1)} x_i, t_{(1)} x_j, t_{(1)} x_k, t_{(1)}\}.$$

The terms of $N^{(1)}(f)$ of total degree $d - 1$ (resp. $d - 2$ and $d - 3$) are divisible by the product of at least $n - 1$ (resp. $n - 2$ and $n - 3$) distinct variables. Moreover $k > 3$ and the polynomials $\{p_j\}_{j \in I}$ are constructed from a disjoint set of clauses, therefore all the terms of $Term(N^{(1)}(f))$ are divisible by at least one of the head terms of $\{p_j\}_{j \in I}$. Since the reduction process is confluent, we can suppose without loss of generality, that the algorithm first eliminates all the terms of total degree $d - 1$ then those of total degree $d - 2$ and finally the terms of total degree $d - 3$.

Total degree $d - 1$

In order to cancel the terms of $N^{(1)}(f)$ of total degree $d - 1$, the algorithm chooses terms of total degree $d - 4$. Since all the terms of total degree $d - 1$ are divisible by the product of at least $n - 1$ distinct variables, the terms chosen to cancel them are divisible by the product of at least $n - 4$ distinct variables. The terms generated during this step are of total degree $d - 2$ (resp. $d - 3$ and $d - 4$) and are divisible by the product of at least $n - 2$ (resp. $n - 3$ and $n - 4$) distinct variables.

Total degree $d - 2$

This step is slightly different from the two steps above since the terms of total degree $d - 2$ come from the elimination of the terms of total degree d and $d - 1$.

But all the terms of total degree $d-2$ computed during the previous steps are divisible by the product of at least $n-2$ distinct variables. Hence, all these terms are eliminated and lead to the generation of terms of total degree $d-3$ (resp. $d-4$ and $d-5$) which are divisible by the product of at least $n-3$ (resp. $n-4$ and $n-5$) distinct variables.

Total degree $d-3$

The terms of total degree $d-3$ come from the elimination of the terms of total degree d, $d-1$ and $d-2$. We have shown that up to this point, all the terms of total degree $d-3$ generated during the previous steps are divisible by the product of at least $n-3$ distinct variables. Hence, all these terms are also eliminated. Remark that after this step, no term of total degree $d-3$ is generated by the algorithm. Hence, in this setting, the terms of the polynomial $N(f)$ are of total degree strictly smaller than $d-3$. □

More generally, we have:

Corollary 4.2. *Let $\{p_1, \ldots, p_m\}$ be the polynomials of the public-key, $I \subset \{1, \ldots m\}$, $\{p_j\}_{j \in I}$ be a subset of the public-key polynomials corresponding to a disjoint set of clauses and x^α be a term of total degree d.*
We define $f'(X) = ax^\alpha + g(X) \in \mathbb{F}_q[X]$ with $(a, \alpha) \in \mathbb{F}_q^* \times \mathbb{N}^n$ such that all the terms of the polynomial $g \in \mathbb{F}_q[X]$ are of total degree strictly smaller than $d-3$. If $k = |I| > 3$ and if x^α is a multiple of $\prod_{i=1}^n x_i$ then there exists at least one term t in the set of terms of the decomposition of $N(f') - f'$ under $\langle p_j \rangle_{j \in I}$ of total degree strictly larger than any term of $N(f')$.

Proof. The proof is similar to the one given above. Since the reduction process is confluent, we can suppose without loss of generality, that it begins by cancelling x^α and due to the particular choice of the terms of g, one sees at once that the total degree of the first term of the decomposition of $N(f') - f'$ (under $\langle p_j \rangle_{j \in I}$) is equal to $d-3$. Moreover, all the terms generated during the reduction process of f', of total degree equal or larger than $d-3$, are cancelled. Hence, the terms of $N(f')$ are of total degree strictly less than $d-3$. □

This result is very interesting since in our context, the ciphertext is a linear combination of normal forms. Finally, we have the following security result:

Corollary 4.3. *Let $\{p_1, \ldots, p_m\}$ be the polynomials of the public-key and x^α be a term of total degree d. We set $d \leq q$, we define $f'(X) = ax^\alpha + g(X) \in \mathbb{F}_q[X]$ with $(a, \alpha) \in \mathbb{F}_q^* \times \mathbb{N}^n$ and such that all the terms of the polynomial $g \in \mathbb{F}_q[X]$ are of total degree strictly smaller than $d-3$. We also set:*

- $l \geq 2$,
- $\{\lambda_1, \ldots, \lambda_l\}, \lambda_i \in \mathbb{F}_q$ *such that* $\sum_{i=1}^l \lambda_i \equiv 0[q]$ *and,*
- $\mathfrak{D} = \{\mathfrak{d}_1, \ldots, \mathfrak{d}_l\}$ *a set of subset indexes such that:*
  $\forall 1 \leq i \leq l$, $\{p_{i_j}\}_{j \in \mathfrak{d}_i}$ *is constructed from a disjoint set of clauses,*
  $\forall 1 \leq i \leq l$, $\mathfrak{d}_i$ *is a set of indexes of cardinality*[12] $3 < |\mathfrak{d}_i| \leq \lfloor \frac{n}{3} \rfloor$.

[12] $\lfloor \frac{n}{3} \rfloor$ is the maximum cardinality of a disjoint set

If e_I is an element of $\langle p_1, \dots, p_m\rangle$ computed by Algorithm 1 with these parameters, then we have:

$$D \not\subseteq H(e_I),$$

D being the set of terms of the decomposition of e_I under $\langle p_i\rangle_{1\le i\le m}$.

Proof. Recall that e_I is a linear combination of normal forms. If we denote by $\{h_j^{(i)}\}_{j\in\mathfrak{d}_i}$ the decomposition of $N_i(f) - f$ under $\langle p_j\rangle_{j\in\mathfrak{d}_i}$, we have:

$$e_I = \sum_{i=1}^{l} \lambda_i N_i(f) = -\sum_{i=1}^{l} \lambda_i \sum_{j\in\mathfrak{d}_i} h_j^{(i)} p_j.$$

Moreover, with the parameters given in corollary 4.3 and according to corollary 4.2, for all $i, 1 \le i \le l$ all the terms of the normal forms $N_i(f')$ modulo $\{p_j\}_{j\in\mathfrak{d}_i}$ computed by Algorithm 1 are of total degree strictly smaller than $d-3$. Hence, all the terms of e_I are of total degree strictly smaller than $d-3$. Therefore, the terms t of total degree $d-3$ of the polynomials $\{h_j^{(i)}\}_{j\in\mathfrak{d}_i}^{1\le i\le l}$ cannot have a decomposition of the form:

$$t_{e_I} = t't, \text{ with } t_{e_I} \in Term(e_I) \text{ and } t' \in Term,$$

since the terms of e_I are of total degree strictly smaller than $d-3$.
Hence, we get that:

$$D \not\subseteq H(e_I).$$

Finally, all ciphertexts generated with such an element e_I are resistant to the intelligent linear algebra attack. □

4.2.2. Differential Attack. Hofheinz and Steinwandt propose in [HS], a method to enhance the feasibility of the intelligent linear algebra attack previously described. In particular, their attack permits to recover "hidden monomials" in the Koblitz's graph perfect code instance of Polly-Cracker [Ko](*Ch.* 5). We detail here the ideas of this attack.
For $p = \sum_{\alpha\in\mathbb{N}^n} a_\alpha x^\alpha$, we denote by $|c|$ the number of monomials of c and we also set:

$$\Delta(p) = \{\frac{a_\mu}{a_\nu} x^{\mu-\nu} : x^\mu \succ x^\nu, a_\mu \cdot a_\nu \neq 0\},$$

$\succ$ denoting here the lexicographic order on the terms.
Suppose that for some $i, 1 \le i \le m$, there exists a "characteristic difference" δ_i, i.e., $\delta_i = \frac{a_{\mu_i}}{a_{\nu_i}} x^{\mu_i-\nu_i}$, with $a_{\mu_i} x^{\mu_i}, a_{\nu_i} x^{\nu_i}$ monomials in p_i and such that:

$$\delta_i \in \Delta(p_i) \setminus \Big(\cup_{j\neq i} \Delta(p_j)\Big).$$

Suppose in addition that there exists a monomial $a_{\eta_i} x^{\eta_i}$ in h_i such that $x^{\eta_i} x^{\mu_i}$ and $x^{\eta_i} x^{\nu_i}$ do not occur among the monomials of $c - a_{\eta_i} x^{\eta_i} q_j$. If for this "characteristic difference", an adversary can find monomials m_1, m_2 in the ciphertext with $x^{\mu_i} | m_1$ and m_1/m_2 being equal to δ_i, then we can identify a potential monomial m_h of h_i as:

$$m_h = \frac{m_1}{a_{\mu_i} x^{\mu_i}} = \frac{m_2}{a_{\nu_i} x^{\nu_i}}.$$

The adversary can not be sure about the correctness of his guess (i.e., if m_h is really a monomial of h_i). But, he can check it by computing the number of monomials in the simplified ciphertext $c' = c - m_h p_i$. Indeed, if the number of monomials in c' is smaller than in c, it is then very likely that t_h is a monomial of h_i. Notice that c and c' encrypt the same plaintext.
An adversary repeats this simplification process of the ciphertext for each "characteristic difference" in the set $\Delta(p_i) \setminus \Big(\cup_{j \neq i} \Delta(p_j) \Big)$ and for all $i, 1 \leq i \leq m$. If at some point of this simplification process c' is a monomial of the form $a_\beta x^\beta$, then the encrypted plaintext has been recovered successfully. Otherwise, he can try to perform an intelligent linear algebra attack on the simplified ciphertext.
Subtracting a polynomial of the ciphertext can reveal "hidden monomials". Indeed, the fact that a monomial m_{h_j} in h_j is hidden in the ciphertext c implies that for all $i, 1 \leq i \leq m$ there exist two monomials m_{h_i} in h_i and m_{p_i} in p_i such that:

$$m_{h_j} p_j + \sum_{i=1}^{m} m_{h_i} m_{p_i} = 0.$$

Therefore, if one can find a monomial $m_{h_i} \in \{m_{h_1}, \dots, m_{h_m}\}$, then we know that the simplified polynomial $c' = c - m_{h_i} p_i$ contains a monomial of the form $m_{h_j} m_{p_j}$, m_{p_j} being a monomial of p_j. Therefore, the monomial which was hidden in the ciphertext c is no longer hidden in the simplified ciphertext c'.
We also would like to emphasize that it is not clear that the sets $\{\Delta(p_i) \setminus \Big(\cup_{j \neq i} \Delta(p_j) \Big)\}_{1 \leq i \leq m}$ always contain enough characteristic differences to recover all the "hidden monomials".
Following these remarks, we propose an improvement of the differential attack. In particular, we no longer consider characteristic differences. Given a ciphertext c, we first compute – for a monomial m_i occurring in a decomposition of the form $m_c = m_i m_{p_i}$, with m_c being a monomial of c and for some monomial m_{p_i} of p_i – the polynomial $c' = c - m_i p_i$. This polynomial can validate the choice of the guess (we don't know if m_i is really a monomial of h_i). Indeed, if $|c'| = |c| - |m_i p_i|$, then this can be taken as evidence that m_i is a monomial of h_i. If this equality on the number of monomials is not true, the polynomial c' can also be useful to reveal hidden monomials: if there exists a monomial m'_j in c' which is not a monomial of c, and which occurs in a decomposition of the form $m_{c'} = m'_j m_{p_j}$, for some monomial m_{p_j} of p_j (indeed, we then have $m_{c'} = m'_j m_{p_j} = m_i m_{p_i}$) then, in addition to the fact that m_i is probably a monomial of h_i, it is also very likely that m'_j was a monomial of h_j that was hidden in the ciphertext c. In all other cases, m_i is not a monomial of h_i, and we then set $c' = c$.
At the second step, we select a monomial $m_k \neq m_i$ having a decomposition of the form $m_{c'} = m_k m_{p_k}$, with $m_{c'}$ a monomial of c' and for some monomial m_{p_k} of p_k. We compute $c'' = c' - m_k p_k$ and we verify as previously whether m_k is a correct guess. We iterate this process while the simplified ciphertext is not a monomial of the form $a_\beta x^\beta$ (when it equals $a_\beta x^\beta$, then a_β is the plaintext corresponding to

c, according to our encryption procedure). Notice that even if there are hidden monomials in the ciphertext, it is very likely that these monomials can be guessed by considering simplified ciphertext.
As presented here, the attack of [HS] and the improvement we have described above appear to be quite generic, and thus apply to our system too.

5. Practical Considerations

The generation of the set C of clauses has been performed using the algorithm *hgen2*. Apart from that, the complete implementation of instances of our system has been done using the MAGMA symbolic language – which we found best suited for the manipulation of multivariate polynomials (multiplication, evaluation on a vector of $\mathbb{F}_q^n$, computation of the number of terms ...) – with interfaces in C.
The public-key consists of m 3-clauses in the variables x_i, $1 \leq i \leq n$. It can thus be stored using $3m\lg(n)$ bits, that is $O(n\lg(n))$ bits with $m = cn$.
The secret-key is n bits long, as we can identify T with 1 and F with 0 for its storage.
In practice, we choose a large d (e.g., $d \simeq 200$), and q roughly of the same size as d, with $q \geq d$.
Our construction of the ciphertext presents some practical advantages. First it allows to construct a relatively short ciphertext (in comparison with a regular Polly Cracker scheme) in a quite efficient way. We can control the size of the ciphertext with the parameters $\{\lambda_1, \ldots, \lambda_l\}$ of Algorithm 1 (by setting λ_i to zero when having reached a certain size). Moreover, increasing the size of the public-key does not increase the size of the ciphertext, and hence does not degrade the performance of the system.
To have a more precise idea of the characteristics of this system, we give an example of real-time implementations. For $n = 700$ and $m = 2450$, we obtain: 4.3s to generate a public-key of size 6.9KBytes, an encryption time of 3.22s, 1527 terms in the ciphertext and a decryption time of 0.13s.

6. Concluding Remarks

We have presented a cryptographic scheme of Polly-cracker type, the underlying problem of which is based on a subclass of the family of SATISFIABILITY problems. We have proposed a specific method to construct the ciphertext. We have examined its security on the one hand by considering single break attacks, and on the other hand by exploring the best known methods to date to attack the hard problem. Concerning single break attacks, the results obtained are quite interesting because resistance to intelligent linear algebra attacks has always been a concern for Polly Cracker type schemes. On the other hand, differential attack and our extension of it seem hard to defeat, as they are a generic tool to handle all Polly Cracker-like ciphertext constructions. Finally, we believe that our approach – namely the investigation of sharp methods from propositional logic and the setting of results in a cryptographic context – is quite new.

Acknowledgments

The authors would like to thank Laurent Simon, Olivier Bayeux and Yacine Boufkhad for very helpful discussions.

References

[BS] R.J. Bayardo Jr., R. Schrag. *Using CSP look-back techniques to solve exceptionnally hard SAT instances.* Proceedings of 2nd Int. conference on Principles and Practice of constraint Programming, 1996, pp. 46–60.

[Ba] D. Bayer. *The division algorithm and the Hilbert scheme.* PhD. Thesis, Harvard University, Cambridge, Massachussets, 1982.

[BW] T. Becker and V. Weispfenning. *Gröbner Bases, A Computational Approach to Commutative Algebra.* In cooperation with Heinz Kredel. Graduate Texts in Mathematics, 141. Springer-Verlag, New York, 1993.

[CMo] S. Cocco, R. Monasson. *Statistical physics analysis of the computational complexity of solving random satisfiability problems using backtrack algorithms.* The European Physical Journal B 22, 2001, pp. 505–531.

[CMi] S.A. Cook. D.G. Mitchell. *Finding hard instances of the satisfiability problem: a survey.* DIMACS Series in discrete mathematics and theoretical computer science, 1997.

[DLL] M. Davis, G. Logemann, D. Loveland. *A machine program for theorem proving.* Communications of the ACM, 5, 1962, pp. 394–397.

[DB] O. Dubois, Y. Boufkhad. *From very hard doubly balanced SAT formulae to easy unbalanced SAT formulae, variations of the satisfiability threshold.* Proceedings of the DIMACS workshop on the satisfiability problem: theory and applications, March 1996.

[Fa] J.-C. Faugère. *A new efficient algorithm for computing Gröbner basis: F_4.* Journal of pure and applied algebra, vol. 139, 1999, pp. 61–68.

[FK] M. Fellows, N. Koblitz. *Combinatorial cryptosystems galore !* Proceedings of the second international conference on "Finite Fields: theory, applications and algorithms", Las Vegas 1993, Contemporary Mathematics, vol. 168, 1994, pp. 51–61.

[GS] W. Geiselmann, R. Steinwandt. *Some cracks in Polly Cracker.* Europäisches Institut für Systemsicherheit, Universität Karlsruhe, Tech. Report 01/01, 2001.

[Hi] E. Hirsch. http://logic.pdmi.ras.ru/~ hirsch/

[HS] D. Hofheinz and R. Steinwandt. *A "Differential" Attack on Polly Cracker.* Proceedings of 2002 IEEE International Symposium on Information Theory ISIT 2002, extended abstract, p. 211, 2002.

[ILL] R. Impagliazzo, L. Levin, M. Luby. *Pseudo-random number generation from one-way functions.* Proceedings of 21st STOC, 1989, pp. 12–24.

[Ko] N. Koblitz. *Algebraic aspects of cryptograhy.* Algorithms and Computation in Mathematics, 3. Springer-Verlag 1998.

[Le] L. Van Ly. *Polly Two – a public-key cryptosystem based on Polly Cracker.* Thèse de l'université de Bochum, Faculté de Mathématiques, Décembre 2002.

[Od] A. Odlyzko. *The rise and fall of knapsack cryptosystems.* Cryptology and computational number theory, Proceedings of Symposium on Applied Mathematics 42, AMS 1990, pp. 75–88.

[OP] T. Okamoto, D. Pointcheval. *REACT: Rapid Enhanced-Security Asymmetric Cryptosystem Transform.* CT-RSA 2001: 159–175

[Sa] http://www.satlive.org/SATCompetition

[SKC] B. Selman, H. Kautz, B. Cohen. *Noise strategies for improving local search.* Proceedings of AAAI-94, 1994, pp. 337–343.

[Sh] A. Shamir. *A polynomial-time algorithm for breaking the basic Merkle-Hellman cryptosystem.* IEEE Transactions on Information Theory IT-30, 1984, pp. 699–704.

[SG] R. Steinwandt and W. Geiselmann. *Cryptanalysis of Polly Cracker.* IEEE Transactions on Information Theory 48(11): 2990–2991, 2002.

Françoise Levy-dit-Vehel and Ludovic Perret
ENSTA
32 boulevard Victor
F-75739 Paris cedex 15, France
e-mail: `levy@ensta.fr`
e-mail: `lperret@ensta.fr`

Progress in Computer Science and Applied Logic, Vol. 23, 193–208

Combinatorially Designed LDPC Codes Using Zech Logarithms and Congruential Sequences

Jing Li

Abstract. We investigate a systematic construction of regular low density parity check (LDPC) codes based on $(\gamma\rho^{\gamma-1}, \rho^{\gamma}, \rho, \gamma, \{0,1\})$ combinatorial designs. The proposed (γ, ρ)-regular LDPC ensemble has rate $(1-\frac{1}{\rho})^{\gamma}$, girth $\geq 2^{\gamma+1}$, and exists for all $\gamma \geq 2$. The codes are a subset of Gallager's random ensemble, but contains a good combination of structure and (pseudo)randomness. In particular, the simple case of $\gamma = 2$ results in a class of codes that are high-rate, systematic, quasi-cyclic, linear-time encodable and decodable, and free of length-4 and length-6 cycles. Analysis on distance spectrum shows that they are better than the Gallager ensemble of the same parameters. Simulation of the proposed codes on intersymbol interference channels show that they perform comparably to random LDPC codes. Unlike random codes, the proposed structured LDPC codes can lend themselves to a low-complexity implementation for high-speed applications.

Keywords. Codes on graphs, Combinatorial design, Low density parity check (LDPC) codes, Inter-symbol interference (ISI).

1. Introduction

Considerable work has been done recently about the design and analysis of low density parity check (LDPC) codes. The original LDPC codes proposed by Gallager use random matrices [1]. Although research work indicates that randomness is important for capacity-approaching performance, codes with structure and regularity are preferred for ease in implementation. In addition to the random construction of LDPC codes like bit filling and/or optimization of degree profiles using density evolution, systematic constructions are also being proposed, which include the approaches from combinatorial designs [4, 7, 5], finite geometries [6], congruential

This project was supported in part by a grant from Seagate Technology and a grant from the Commonwealth of Pennsylvania, Department of Community and Economic Development, through the Pennsylvania Infrastructure Technology Alliance (PITA).

sequences [8], Ramanujan graph [9] and lattice designs [7]. It has been shown that in some cases (especially for short lengths and/or high rates, like those used for digital recording systems), structured LDPC codes are a better choice than random LDPC codes with comparable performance, less memory requirement, and more implementable structure [4, 6].

In general, an LDPC code is either represented using a parity check matrix H or its corresponding Tanner graph. Major parameters for an LDPC code includes the column weight γ and the row weight ρ in H matrix, and the *girth* (the length of the shortest cycle) in the Tanner graph. An LDPC code is said to be (γ, ρ)-*regular* if all columns in H have weight γ and all rows have weight ρ. The girth is important to LDPC codes because the existing decoder is an iterative message-passing decoder whose efficiency is sensitive to short cycles.

In this work, we investigate a class of structured LDPC codes from $(\gamma\rho^{\gamma-1}, \rho^{\gamma}, \rho, \gamma, \{0,1\})$ combinatorial design $(\gamma \geq 2)$ [5]. In addition to their regular and thus easily-implementable structure, a key merit of considering combinatorial designs is that length-4 cycles can be systematically avoided [2, 4, 7, 5]. In fact, for the specific design proposed here, the girth is at least $2^{\gamma+1}$, i.e., length-6 cycles are also systematically avoided. The proposed LDPC codes are a subset of Gallager's random ensemble, but contains a good combination of structure and pseudo-randomness. Zech logarithm in Galois field (GF) and congruential sequences are used to facilitate the implementation. In particular, the simple case of $\gamma = 2$ results in a class of systematic, quasi-cyclic, high-rate $(2, \rho)$-regular codes, which are linear time encodable and linear time decodable. Computation of distance spectrum reveals that they are (slightly) better than the Gallager ensemble of the same parameters.

In [11] it is shown that the thresholds of regular Gallager codes over the dicode channel approaches the i.i.d capacity[1] of the channel at high rates. This means that high-rate regular LDPC codes are asymptotically optimal for dicode channels. We expect it to be true for a general intersymbol interference (ISI) channel also. The later part of the paper investigates the application of the proposed LDPC codes on partial response maximum likelihood (PRML) models that are used in digital recording systems. We show that the proposed structured LDPC codes perform as well as random LDPC codes, yet with more implementable structure.

2. Preliminaries

Definition 2.1. (Combinatorial Design)

[1] A *combinatorial design* is an arrangement of a set of m *points* into n subsets, called *blocks*, which satisfy certain regularity constraints.

[2] The *incidence* matrix of a combinatorial design gives the (0,1)-matrix (of dimensionality $n \times m$) which has a row for each point v and a column for each block B, and $(v, B) = 1$ iff point v is incident with block B.

---

[1] We use capacity to loosely denote the information rate.

[3] The *covalency* λ_{v_1,v_2} of two points v_1 and v_2 is the number of blocks that contain both of them.

[4] A design is said to be *regular* if the number of points contained in each block (denoted as γ) is the same for every block and the number of blocks each point is incident with (denoted as ρ) is the same for every point.

[5] A design is said to be *balanced* if the covalency λ_{v_1,v_2} of the point pair (v_1, v_2) are the same for all pairs. A regular and balanced design can be denoted as a $(m, n, \rho, \gamma, \lambda)$-design, where $m\rho = n\gamma$.

It follows from the above definitions that a combinatorial design with favorable constraints can define a binary LDPC code. The transpose of the incidence matrix can be used as the parity check matrix H, where *points* and *blocks* in the combinatorial design correspond to rows and columns in the H matrix. The H matrix has m rows, n columns (codeword length), with row weight ρ and column weight γ. The code rate is given by $R = 1 - \frac{\text{rank}(H)}{n} \leq 1 - \frac{m}{n}$ (all rows in H matrix may not be independent). Further, covalency $\lambda < 2$ guarantees that the corresponding Tanner graph is free of length-4 cycles.

An example is given in Fig. 1. $m = 8$ points are grouped in $n = 16$ blocks with each point incident with 4 blocks and each block containing 2 points, where $B_1 = (v_1, v_2)$, $B_2 = (v_1, v_4)$, $\ldots$, $B_{16} = (v_7, v_8)$. Fig. 1(a) shows the combinatorial design (where a line connecting 2 dots is used to denote a block containing two points) and (b) the corresponding H matrix. The design has covalency $\lambda = \{0, 1\}$ for all pairs of points and, hence, is free of length-4 cycles. In fact, the code shown here is also free of length-6 cycles.

Some popular classes of combinatorial designs that have already been studied for generating LDPC codes are *Steiner systems* or $(m, n, \rho, \gamma, 1)$-designs [2], *Kirkman triple systems* (KTS) or $(m, n, \rho, 3, \{0, 1\})$-designs (which are resolvable Steiner triple systems (STS)) [4, 7], and balanced incomplete block designs (BIBD) [12, 13]. Others designs from lattice [7] and Ramanujan graphs [9] are also proposed. These systematically-designed LDPC codes share the same desirable properties like simplicity in construction and regularity in code structure. Some of these codes were shown to perform within 1 dB from the capacity on AWGN channels, and other have been evaluated for use in magnetic recording channels. Below we present a new design which results in a class of (γ, ρ)-regular LDPC codes of rate $(1 - \frac{1}{\rho})^\gamma$.

3. $(\gamma\rho^{\gamma-1}, \rho^\gamma, \rho, \gamma, \{0, 1\})$-Designed LDPC Ensemble

3.1. $(\gamma\rho^{\gamma-1}, \rho^\gamma, \rho, \gamma, \{0, 1\})$-Design

Consider $\gamma\rho^{\gamma-1}$ points and ρ^γ blocks. For ease of proposition, we label blocks with γ-tuple subscripts, i.e., $B_{(x_1, x_2, \ldots, x_\gamma)}$, where $x_1, x_2, \ldots, x_\gamma \in \{0, 1, \ldots, \rho - 1\}$. Two blocks are said to be in the same plane if at least $(\gamma - 2)$ coefficients in the γ-tuple subscript are the same. For each direction along the axis of x_i (call it "principal" direction), $\rho^{\gamma-2}$ parallel planes (call them "principal" planes) can be

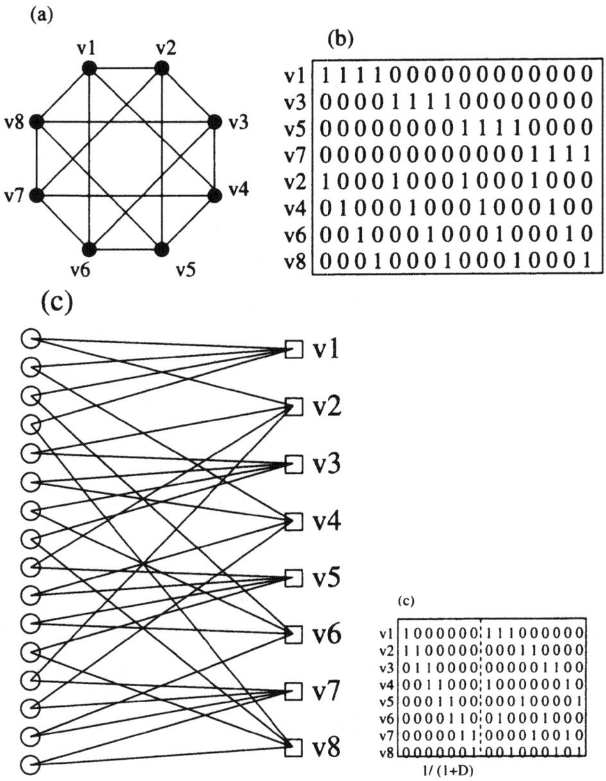

FIGURE 1. (a) $(m, n, \rho, 2, \{0, 1\})$ Combinatorial design, where $\rho = 4$, $m = 2\rho = 8$, and $n = \rho^2 = 16$; (b) Resulting H matrix for an LDPC code with code length $n = 16$; (c) A form of linear time encodable LDPC codes; (d) A form of turbo product codes.

selected each containing ρ^2 blocks (and collectively covering all blocks). In each principal plane, the ρ^2 blocks can be evenly divided into ρ discrete "bundles" according to a predefined "bundle-rule" (which will be discussed later). Hence, there are altogether $\gamma\rho^{\gamma-2}$ principal planes and $\gamma\rho^{\gamma-1}$ bundles where no two bundles contain a same pair of blocks. Each bundle then uniquely determines a point, in other words, a point is incident only with blocks in the same bundle. We thus have $\gamma\rho^{\gamma-1}$ well-defined points and ρ^γ well-defined blocks, where each point is incident with γ blocks and each block ρ points. Further, the overlap of any two blocks is at the most one point. This results in a $(\gamma\rho^{\gamma-1}, \rho^\gamma, \rho, \gamma, \{0, 1\})$-design. Fig. 1 gives an example of a $(8, 16, 4, 2, \{0, 1\}0$-design.

Lemma 3.1. *The above combinatorial construction results in* $(\gamma\rho^{\gamma-1}, \rho^\gamma, \rho, \gamma, \{0,1\})$*-designs that have the following properties:*

[1] *The resulting (γ, ρ)-regular LDPC code has code length $n = \rho^\gamma$ and rate $R = (1 - 1/\rho)^\gamma$.*

[2] *The girth of the corresponding Tanner graph* $\geq 2^{\gamma+1}$.

The *proof* of Lemma 3.1 is straightforward and, hence, is omitted. A comment is that, since the code rate is $(1-1/\rho)^\gamma$, to get a reasonable rate (i.e., not too small), either γ is small or ρ is large. Hence for practical applications, $\rho! \geq \gamma\rho^{\gamma-1}$ is almost always satisfied which makes good construction possible and likely.

Clearly, the performance of the above design under message-passing decoder is much affected by how blocks are bundled in a plane (i.e., how incidence matrix is defined). The desired bundle-rule contains enough structure to ease the construction and implementation, and enough randomness to avoid recurrence of a bad pattern (like short cycles) in the design. Below we discuss two effective ways of using Zech logarithm and congruential sequences to construct bundle-rules. Other approaches like cyclic patterns from finite geometry are also possible.

3.2. Zech Logarithm

When $\rho = p^t$ where p is a prime number and t an integer, the bundle-rule can be described using a pseudo-random permutation table generated systematically from Zech logarithm arithmetics in Galois field GF(p^t). This is how it works. Let α be a (predetermined) primitive element in GF(p^t), the elements in GF(p^t) can be represented as $0, 1, \alpha, \alpha^2, \ldots, \alpha^{\rho-2}$, or equivalently $-\infty, 0, 1, \ldots, \rho-2$ where $\log \alpha^k = k$ for $0 \leq k \leq \rho-2$ and $\log 0 = -\infty$. A permutation vector with seed i_0, denoted as π_{i_0}, is constructed using Zech logarithm as follows

$$\pi_{i_0}(j) = \begin{cases} \log(\alpha^{i_0} + \alpha^j), & \text{for j} = 0, 1, 2, \ldots, \rho-2, \\ \log(\alpha^{i_0}), & \text{for j} = \rho-1, \end{cases} \tag{3.1}$$

where $i_0 \in \{-\infty, 0, 1, \ldots, \rho-2\}$. Each permutation vector π_{i_0} uniquely specifies a bundle in a plane, such that ρ blocks with subscripts $(j, \pi_{i_0}(j))$ where $j = 0, 1, \ldots, \rho-1$ (for ease of proposition, we omit the irrelevant $\gamma-2$ indexes in the subscript) belong to the same bundle. Since different seed i_0 results in a different permutation vector, there are altogether ρ permutation vectors which can be used to bundle the ρ^2 blocks in a plane. We illustrate this through the following example.

Example. (Permutation Table from Zech Logarithm in GF(2^3))
Consider $\rho = 2^3$. We take the root of the minimal polynomial $P(x) = x^3 + x + 1$ in GF(2^3) as the primitive element α. Tab. 1 summarizes the Zech logarithm arithmetic and the resulting permutation table ($-\infty$ is denoted and interpreted as position $\rho-1 (= 7)$ in the table). Fig. 2 shows how bundles are defined by permutation vector π_0, π_2 and π_7. We use boxes to denote blocks, and those connected to the same line are considered in one bundle. Hence, a permutation table uniquely defines ρ bundles (corresponding to ρ points) in a plane.

Further, note that any two rows in the a permutation table can be exchanged which results in a different permutation table (and thus different bundle-rule).

TABLE 1. Permutation Table Constructed Using Zech Logarithm in GF(2^3)

β	Log $\log_\alpha \beta$	Zech Log $\log_\alpha(\alpha^{i_0}+\beta)$ $i_0=0$	1	2	3	4	5	6	7
1 001	0	7	3	6	1	5	4	2	0
α 010	1	3	7	4	0	2	6	5	1
α^2 100	2	6	4	7	5	1	3	0	2
α^3 100	3	1	0	5	7	6	2	4	3
α^4 100	4	5	2	1	6	7	0	3	4
α^5 100	5	4	6	3	2	0	7	1	5
α^6 100	6	2	5	0	4	3	1	7	6
0 100	7	0	1	2	3	4	5	6	7
		π_0	π_1	π_2	π_3	π_4	π_5	π_6	π_7

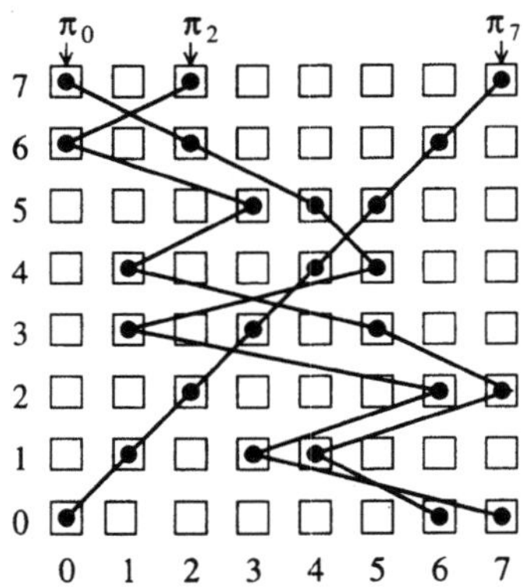

FIGURE 2. Illustration of how permutation table defines bundle-rule

When $\rho! \geq \gamma\rho^{\gamma-1}$, i.e., the number of permutation tables is larger than the number of principal planes, we have sufficient choices such that no two principal planes use the same bundle rule. This brings a good amount of pseudo-randomness into the construction to prevent short cycles. In the meantime, the basic permutation table (i.e., Zech logarithm arithmetic) can be implemented using simple hardware with small memory.

3.3. Congruential Sequences

Another efficient way to design pseudo-random permutation tables of dimension $\rho \times \rho$ is to use maximal length congruential sequences (or M-sequences) of length $M = \rho^2$.

If $M = 2^k$ for some integer k, a k-tap linear feedback shift register (LFSR) can be used to generate an M-sequence. The *characteristic polynomial* of the LFSR

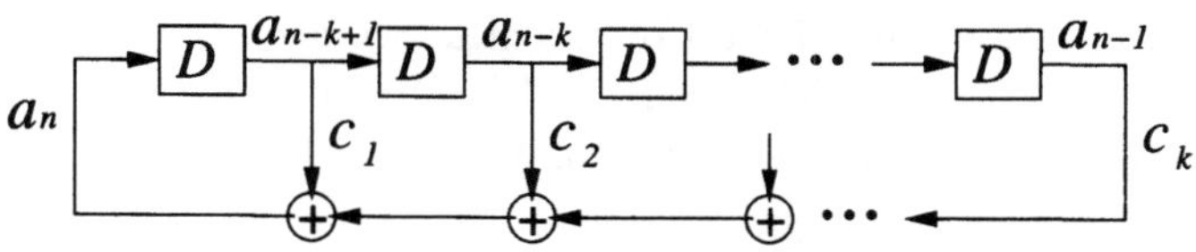

FIGURE 3. System diagram of a k-tap linear feedback shift register

depicted in Fig. 3 is given by:

$$f(D) = 1 - \sum_{i=1}^{k} c_i D^i, \tag{3.2}$$

where c_i's are connection variables, and D is a delay operator. When $f(D)$ is a primitive polynomial (which exists for all $k \geq 1$), and when the initial values, $a_0, a_1, \ldots, a_{k-1}$, are not all zeros, the corresponding binary LFSR sequence generated through $a_n = \bigoplus_{i=1}^{k} c_i a_{n-i}$ (where $\bigoplus$ stands for binary summation), has period $2^k - 1$ (see for example [17]). This is what is used in code division multiple access (CDMA) systems to generate binary pseudo-random sequences (also known as *PN codes*) of period $2^k - 1$. It can be conveniently shown that an integer pseudo-random sequence of period $2^k - 1$, $\{A_n\}$, can be generated by combining k consecutive terms in the binary LFSR sequence, namely,

$$A_n = [a_n, a_{n+1}, \ldots, a_{n+k-1}]_{binary} = \sum_{i=0}^{k-1} a_{n+i} 2^{k-1-i}. \tag{3.3}$$

Any $2^k - 1$ consecutive integers in the sequence lead to a length $2^k - 1$ (integer) M-sequence. Notice that this sequence covers numbers from 1 to $2^k - 1$ (without number 0). Hence, inserting a 0 to any position of this sequence leads to a pseudo-random M-sequence of length $M = 2^k$.

A different approach that is just as simple but more general (i.e., applicable to any non-zero N) is to use the algebraic formula [18]:

$$A_n = aA_{n-1} + b \bmod M. \tag{3.4}$$

To ensure that the resulting sequence is of maximal length, the parameters a and b need to satisfy:

- $a < M$, $b < M$, b be relatively prime to M;
- $(a$-$1)$ be a multiple of p, for every prime p dividing M;
- $(a$-$1)$ be a multiple of 4 if M is a multiple of 4.
- (optional) a be relatively prime to M.

The mapping of a length $N = \rho^2$ maximal congruential sequence to a $\rho \times \rho$ permutation table can be defined arbitrarily in principle, but it is desirable for the mapping rule to contain both structure (for easy description and implementation) and randomness (for good performance). For example, a simple way is to fill the M-sequence in the permutation table and sort and order the elements in each row.

Example. (Permutation Table from Congruential Sequences)
For the case of $\rho = 6$, let us pick $a = 13$ and $b = 5$. The length $M = 36$ sequence generated using (3.4) is as follows (starting with $A_0 = 0$):

00, 05, 34, 15, 20, 13, 30, 35, 28, 09, 14, 07, 24, 29, 22, 03, 08, 01,
18, 23, 16, 33, 02, 31, 12, 17, 10, 27, 32, 25, 06, 11, 04, 21, 26, 19.

There are many ways to fill the M-sequence in the table, like row-wise, column-wise, diagonal-wise or any other cyclic pattern. Tab. 2(A) illustrates the zig-zag filling pattern that starts from the top-left corner and proceeds diagonally from top-right to bottom-left. Then, sorting and ordering each row, we obtain a permutation table as shown in Tab. 2(B). By cyclically shifting any one row or several rows, a new permutation table (and therefore a new bundle rule) will result.

TABLE 2. Permutation Table Constructed Using Congruential Sequences

00	05	15	30	14	03
34	20	35	07	08	33
13	28	24	01	02	10
09	29	18	31	27	06
22	23	12	32	11	21
16	17	25	04	26	19

0	2	4	5	3	1
4	2	5	0	1	3
3	5	4	0	1	2
1	5	3	2	4	0
3	4	1	5	0	2
1	2	4	0	5	3

3.4. The Simplest Case of $\gamma = 2$

In this subsection, we discuss a simple case, $\gamma = 2$, of the above design (since it is easily analyzable) and compare it to the Gallager ensemble. First, we note that this case (Fig. 1) is somewhat special in that the resulting LDPC ensemble contains only one code for each given ρ (if the relevant order of the bits in the codeword is ignored). Second, instead of using the aforementioned procedure and labelling blocks with 2-tuples, all points and blocks can be labelled using a scaler and their relations can be conveniently specified as follows: for a set of point containing even number of points, denoted as $V = \{v_1, v_2, \dots, v_{2\rho-1}, v_{2\rho}\}$, a block B is composed of two points from V, v_i and v_j, where

$$i = j + k \bmod 2\rho, \quad \forall \mathrm{k} = 1, 3, 5, \dots 2\lceil \frac{\rho}{2} \rceil - 1. \tag{3.5}$$

The example of $\gamma = 2$, $\rho = 4$ is shown in Fig. 1. We have the following lemma for this class of $(2, \rho)$-regular LDPC codes:

Lemma 3.2. *The $(2, \rho)$-regular LDPC codes from the above design have the following properties:*

[1] *They are a class of high-rate, systematic codes with code length $n = \rho^2$, rate $R = (1 - 1/\rho)^2$ and girth 8.*
[2] *They are quasi-cyclic LDPC codes.*
[3] *They are linear time encodable and linear time decodable.*

The above properties can be conveniently verified. Here are a few comments. First, we note that shifting a valid codeword leftward or rightward by ρ bits produces another valid codeword (quasi-cyclicity). However, the codewords are not M-sequences since the codeword length ρ^2 is a multiple of the period ρ. Second, the encoder can be implemented with a linear shift register with feedback connections based on its generator polynomial which eliminates the necessity of storing the generator matrix. Third, the linear time encodability (a property that is not readily attainable for random LDPC codes [3]) can be either inferred from quasi-cyclicity or from the following lemma:

Lemma 3.3. *For an LDPC code specified by an $m \times n$ parity check matrix H, if there are at least $(m-1)$ weight-2 columns which do not complete a cycle among them, then encoding can be performed with linear time in n.*

Proof: As shown in Fig. 1(c), we can rearrange these weight-2 columns to make the corresponding matrix diagonal or sub-diagonal. Clearly, this realization can be encoded linear time using back substitution [16]. Furthermore, the parity check matrix in Fig. 1(c) also presents a form of irregular repeat accumulate (IRA) codes [14], where the left sub-diagonal part of the H matrix plays the role of an accumulator $1/(1 \oplus D)$, and the right part functions to repeat data bits and form checks among them. It is well known that IRA codes are linear time encodable.

It is worth mentioning that the above $(2, \rho)$-regular LDPC codes can also be viewed as a special type of 2-dimensional turbo product codes (TPC) constructed from arrays of single-parity check (SPC) codes. Fig. 1(d) presents the same $(2, 4)$-regular LDPC code in an equivalent TPC/SPC format. We note, however, that the general case (γ, ρ)-regular codes ($\gamma \geq 3$) from the proposed design are not TPC/SPC codes. The major differences include that 1) a TPC/SPC code is deterministic and rigid in structure, where the proposed LDPC ensemble contains a variety of realizations and pseudo-randomness for $\gamma \geq 3$; and 2) a γ-dimensional TPC/SPC code of length n has girth $2^{\gamma+1}$ and contains approximately $\frac{n^2}{2^\gamma}$ cycles of length $2^{\gamma+1}$ (for large $n \gg 2^\gamma$), whereas the proposed LDPC ensemble has girth $\geq 2^{\gamma+1}$ (worse case construction has girth $2^{\gamma+1}$), and the number of length $2^{\gamma+1}$ cycles is small (due to pseudo-randomness in bundle rule, we expect this number to decrease with the increase of n).

3.5. Distance Spectrum Analysis

In his original construction [1], Gallager specified an ensemble of (γ, ρ)-regular LDPC codes whose $m \times n$ parity check matrix H can be horizontally split into γ sub-matrices of dimensionality $\frac{m}{\gamma} \times n$ each, where each sub-matrix has uniform column weight 1 and row weight ρ (denote such a sub-matrix as $H_{(1,\rho)}$). We refer to this ensemble as the *Gallager ensemble*, since Gallager has used it to derive many useful results concerning the properties of LDPC codes and the iterative decoding. It can be seen from the construction procedure (as well as Fig. 1(b)) that the proposed (γ, ρ)-regular LDPC ensemble is a subset of the Gallager ensemble.

Below we evaluate and compare the distance spectrum of the proposed subset with the whole set.

Distance spectrum is useful in evaluating the ensemble average performance (assuming an optimal decoder), but is generally hard to compute for an LDPC ensemble. For the Gallager ensemble with random constructions, the expectation (i.e., average) of the output weight enumerator function (OWEF) can be derived fairly easily. For the structured (γ, ρ)-regular ensemble proposed above, a closed-form expression for OWEF involves tedious mathematics. Hence, we consider only the simple case of $\gamma = 2$.

Example. (Gallager Ensemble)
Considering Gallager ensemble of (γ, ρ)-regular codes with code length n, the parity check matrix, $H_{(\gamma,\rho)}$, constitutes of γ sub-matrices, $H_{(1,\rho)}$, each of which has output weight enumerator function [1]

$$A_{(1,\rho)}(w) = \underbrace{B(w) * B(w) * \cdots * B(w)}_{\rho}, \tag{3.6}$$

where $*$ denotes convolution operation and

$$B(w) = \begin{cases} \binom{\rho}{w}, & w \text{ even}, \\ 0, & w \text{ odd}. \end{cases} \tag{3.7}$$

The average OWEF of (γ, ρ)-regular Gallager ensemble is thus given by

$$A^{gall}_{(\gamma,\rho)}(w) = A_{(1,\rho)} \cdot \left(\frac{A_{(1,\rho)}}{\binom{n}{w}} \right)^{\gamma-1}. \tag{3.8}$$

Example. ($(2\rho, \rho^2, \rho, 2, \{0, 1\})$-Designed LDPC Ensemble)
The codes resulted from this design are an alternative form of 2-dimensional TPC/SPC codes. Hence, the exact OWEF (rather than the ensemble average) can be computed using [19]

$$A^{prop}_{(2,\rho)}(w) = \frac{1}{2^\rho} \sum_{\alpha=0}^{\rho} \binom{\rho}{\alpha} \left(\sum_{\beta \text{ even}, \beta=0}^{\rho} P(\beta, \alpha, \rho) w^\beta \right)^\rho, \tag{3.9}$$

where

$$P(\beta, \alpha, \rho) = \sum_{k=0}^{\beta} (-1)^k \binom{\alpha}{k} \binom{\rho - \alpha}{m - k}. \tag{3.10}$$

In Tab. 3, we compare the output weight enumerators (in logarithm scale) of the Gallager ensemble $(2, \rho)$-regular (random) LDPC codes and the proposed $(2, \rho)$-regular (structured) LDPC codes. We observe that the proposed structured codes are better than the ensemble average of random codes, with fewer codewords at the low weight end of the distance spectrum. In other words, the proposed codes are above average in the maximum likelihood sense.

It is worth noting that the number of weight-2 columns in an LDPC code usually needs to be limited in order for the code to be asymptotically "good" on

TABLE 3. Comparing the Output Weight Enumerator of Gallager Ensemble (Random) LDPC Codes and the Proposed (Structured) Combinatorial Designed LDPC Codes ((2,16)-regular, n=256, Logarithm Scale)

Output weight w	Gallager $\log_{10}(A_w)$	Proposed $\log_{10}(A_w)$
2	2.0529	-
4	4.2471	4.1584
6	6.4509	6.2745
8	8.6383	8.5254
10	10.7988	10.7074
12	12.9265	12.8570
14	15.0175	14.9651
16	17.0689	17.0300
18	19.0781	19.0500
20	21.0431	21.0232
22	22.9620	22.9483
24	24.8333	24.8242
26	26.6558	26.6499
28	28.4286	28.4249
30	30.1510	30.1488

AWGN channels ("good" in the sense as MacKay defined in [2]). For irregular codes, the upper limit of the weight-2 columns is determined by the stability condition. For regular codes, the column weight needs to be at least 3, since it was shown by Gallager that only with $\gamma \geq 3$ will the average minimum distance of a regular LDPC ensemble increase linearly with the code length [1]. Hence, $(2, \rho)$-regular LDPC codes are not "good" codes on AWGN channels. However, when $(2, \rho)$-regular codes are used with a modulation or channel that has memory, the modulation/channel will provide another level of parity check (either binary or nonbinary) to the coded bits from LDPC codes. In other words, $(2, \rho)$-regular LDPC codes can be "good" codes in such cases. Further, recent work has shown that regular LDPC codes are asymptotically optimal (i.e., reaching to the i.i.d. capacity of the channel) on a dicode channel [11]. We conjecture the result to be valid on general ISI channels too.

4. Application on PRML Channels

One possible application for the proposed structured LDPC codes, especially the $(2, \rho)$-regular codes that are both simple and high-rate, is the digital recording systems. This section evaluates their performance on ideal PR magnetic recording channels.

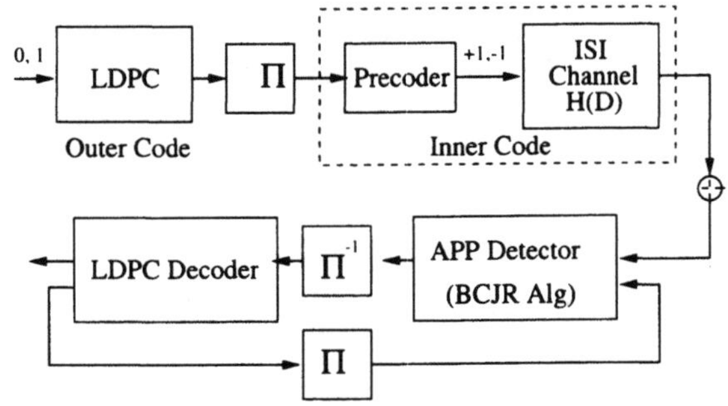

FIGURE 4. System model for LDPC-coded PRML channels.

Typical PR channel models used in magnetic recording systems include PR-IV channel family whose channel response takes the form of $H(D) = (1+D)(1-D)^q$, where $q=1$ is PR4, $q=2$ is EPR4 channel, and $q=3$ is E^2PR4 channel. A block diagram of the LDPC-coded PRML channel and a matching decoder is shown in Fig. 4. We consider a soft-in soft-out iterative decoding and equalization (IDE, also known as *turbo equalization*) receiver which composes of an inner BCJR decoder matched to the PR channel and an outer message-passing decoder matched to the LDPC code. Further, a random interleaver is inserted between the LDPC code and the PR channel to break up the correlation among code bits and to bring up possible interleaving gain (the interleaver size is an integer multiple of the outer LDPC code length).

The PR channel in a magnetic recording system is usually binary precoded whose traditional role is to limit error propagation in the threshold detectors for ISI channels, but has recently acquired another important role of improving the distance spectrum in an iterative process. To facilitate the choice of a good precoder, the i.i.d. capacity is computed using density evolution with Gaussian approximation. The i.i.d. capacity of the system is computed as the maximum mutual information between input and output LLRs of the system

$$I = \frac{1}{2}\sum_{d=\pm 1}\int_{-\infty}^{\infty} f_d^{(code)}(l)\log\frac{2f_d^{(code)}(l)}{f_{+1}^{(ch)}(l)+f_{-1}^{(ch)}(l)}dl, \tag{4.1}$$

where $d=\pm 1$, $f_d^{(ch)}(l)$ and $f_d^{(code)}(l)$ are the pdf's of the input LLRs (from the channel) to the LDPC-coded PR system and the output LLRs from the system after joint decoding/detection, respectively.

The i.i.d. capacity of the proposed $(2,\rho)$-LDPC codes on EPR4 systems is plotted in Fig. 5 where several binary precoders are evaluated. We see from the plot that $1/(1\oplus D\oplus D^2)$ and $1/(1\oplus D^2\oplus D^3)$ are apparently worse precoders than the other two. Whereas $1/(1\oplus D)$ and $1/(1\oplus D^2)$ present almost identical i.i.d. capacities, simulation results with finite lengths shows that $1/(1\oplus D^2)$ seems to yield slightly better performance and, hence, will be used throughout the simulations.

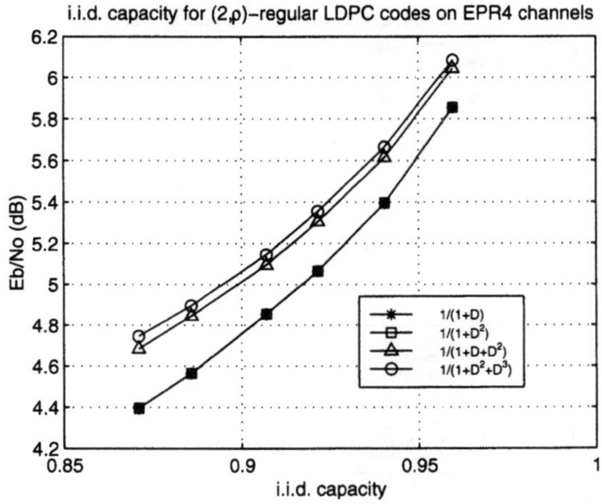

FIGURE 5. I.i.d. capacity of $(2\rho, \rho^2, \rho, 2, \{0,1\})$-designed LDPC codes on PRML channels with different precoding.

The performance of PRML magnetic recording channels employing the proposed regular LDPC codes from $(2\rho, \rho^2, \rho, 2, \{0,1\})$-design is evaluated via computer simulations. We consider 2 basic code rates of $R = 0.88$ and 0.94, 3 interleaver sizes of 1024, 2048 and 4096 bits, and 3 typical channel models of PR4, EPR4 and E^2PR4, respectively. Fig. 6 plots the bit error rate (BER) curves of a rate 0.88 code from $(32, 256, 16, 2, \{0,1\})$-design on PR4, EPR4, and E^2PR4 channels with a precoder $1/(1 \oplus D^2)$. Although not shown, for uncoded PRML system to reach BER of 10^{-5}, 10.25, 10.5 and 10.8 dB are required for PR4, EPR4, and E^2PR4 channels, respectively. Hence, we see that $4-5$ dB gains are achievable by the proposed codes. Interleaving gain phenomenon is also observed from the plot, where the increase of the interleaver length from 1K to 4K brings an additional 0.5 dB gain.

Fig. 7 compares the BER performance of structured LDPC codes (from $(32, 256, 16, 2, \{0,1\})$-design and $(64, 1024, 32, 2, \{0,1\})$-design) with that of random (3,ρ)-regular LDPC codes. Computation of I.i.d. capacity reveals that the structured LDPC codes used here perform best with a precoder $1/(1 \oplus D^2)$ and that the random LDPC codes perform best without a precoder. The BER curves of the best cases for both codes are plotted, which shows that they are comparable in performance; however, the proposed structured LDPC codes are simpler in structure.

5. Conclusion

We propose and discuss in this work a systematic construction of regular LDPC codes from $(\gamma\rho^{\gamma-1}, \rho^\gamma, \rho, \gamma, \{0,1\})$ combinatorial design. The resulting LDPC codes contain a good combination of pseudo-randomness and structure. Investigation on

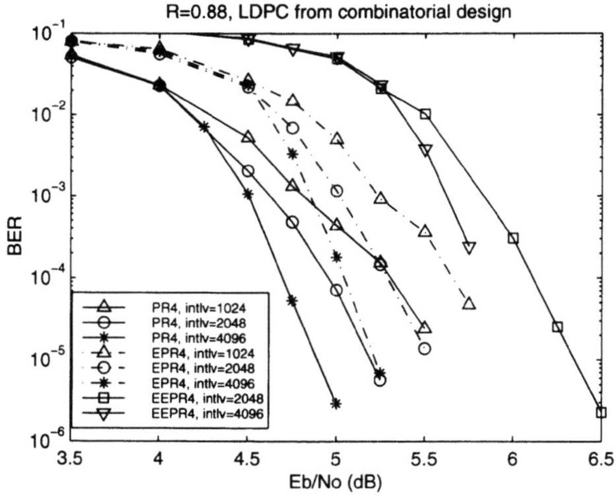

FIGURE 6. BER performance of rate 0.88 LDPC codes from combinatorial designs

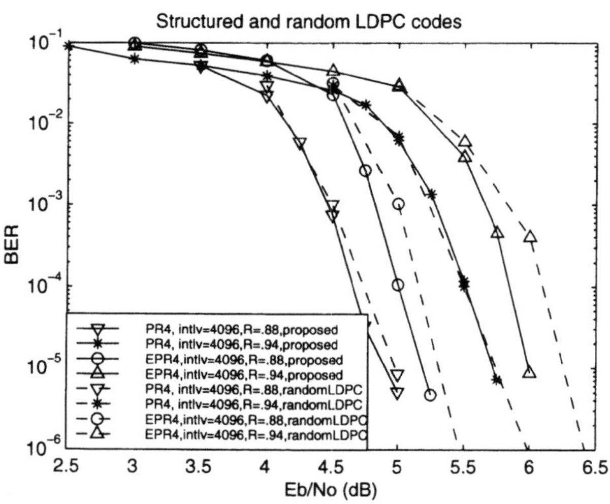

FIGURE 7. Comparison of the proposed structured LDPC codes and random LDPC codes

their performance on PR magnetic recording channels shows that they perform as well as or slightly better than the average random LDPC codes, yet their well-defined structure allows them to be implemented at a much lower cost than random codes. This is also in agreement with the result from [11] that regular LDPC codes are asymptotically optimal on ISI channels.

References

[1] R.G. Gallager, *Low-density parity-check codes* MIT press, Cambridge, MA, 1963.

[2] D.J. MacKay and M.C. Davey, *Evaluation of Gallager codes for short block length and high rate applications* Proc. of the IMA Workshop on Codes, System and Graphical Models, (1999).

[3] T. Richardson, and R. Urbanke, *Efficient encoding of low-density parity-check codes* IEEE Trans. Inform. Theory, Feb. 2001.

[4] S.J. Johnson, and S.R. Weller, *Construction of low-density parity-check codes from Kirkman Triple Systems* Proc GLOBECOM, San Antonio, Nov. 2001, 770–974.

[5] J. Li and E. Kurtas, *A class of* $(\gamma\rho^{\gamma-1}, \rho^{\gamma}, \rho, \gamma, \{0,1\})$ *combinatorially designed LDPC codes with applications to ISI channels* , Proc. IEEE Intl. Symp. Inform. Theory, Yokohama, Japan, June 2003, 29–29.

[6] Y. Kou, S. Lin, and M.P.C. Fossorier, *Low-density parity-check codes based on finite geometries: a rediscovery and new results*, IEEE Trans. Inform. Theory, Vol 47, Nov. 2001. 2711–2736.

[7] E. Kurtas, B. Vasic, and A.V. Kuznetsov, *Design and analysis of low density parity check codes for applications to perpendicular recording channels* The Wiley Encyclopedia of Telecom., (invited chapter), 2002.

[8] A. Prabhakar, and K.R. Narayanan, *Pseudo-random construction of low density parity check codes using linear congruential sequences* IEEE Trans. Commun., vol. 50, Sept. 2002, 1389–1396.

[9] I.J. Rosenthal, and P. Vontobel, *Construction of LDPC codes using Ramanujan graphs and ideas from Margulis* Proc Intl. Symp. on Inform. Theory, 2001.

[10] R.M. Tanner, D. Srkdhara, and T. Fuja, *A class of group-structured LDPC codes* Proc. Intl. Conf. on Inform. Tech. and Applications, Ambleside, England.

[11] A. Kavcic, B. Marcus, M. Mitzenmacher, and B. Wilson, *Deriving performance bounds for ISI channels using Gallager codes* Proc. Intl. Symp. Inform. Theory, June 2001, 345–345.

[12] R.C. Bose, *On the construction of balanced incomplete block designs* Ann. Eugenics 9, 1939, 353–399.

[13] B. Ammar, B. Honary, Y. Kou, and S. Lin, *Construction of low density parity check codes* Intl. Symp. Inform. Theory, Switzerland, June 2002, 311–311.

[14] H. Jin, A. Khandekar and R. McEliece, *Irregular repeat-accumulate codes* 2nd Intl. Symp. on Turbo Codes and Related Topics, Brest, France, Sept 2000.

[15] J. Li, K.R. Narayanan, E. Kurtas, and C.N. Georghiades, *On the performance of high-rate TPC/SPC codes and LDPC codes over partial response channels* IEEE Trans. Commun., May 2002, vol. 50, 723–734.

[16] L. Ping, W.K. Leung, and N. Phamdo, *Low density parity check codes with semi-random parity check matrix* Electronics Letters, vol. 35, no. 1, Jan. 1999, 38–39.

[17] Andrew Viterbi, *CDMA: Principles of Spread Spectrum Communication*, Prentice Hall, 1995.

[18] G.C. Clark, Jr. and J.B. Cain, *Error-correction coding for digital communications*, Plenum Press, NY, 1981.

[19] G. Caire, and C. Taricco, *Weight distribution and performance of the iterated product of single-parity-check codes*, Proc. GLOBECOM Conf., 1994, 206–211.

Jing Li
Department of Electrical and Computer Engineering
Lehigh University
19 Memorial Dr. W.
Bethlehem, PA 18015, USA
e-mail: JingLi@ece.lehigh.edu

Progress in Computer Science and Applied Logic, Vol. 23, 209–222

New Constructions of Constant-Weight Codes

Lei Li and Shoulun Long

Abstract. By generalizing a propagation rule for binary constant-weight codes, we present three constructions of binary constant-weight codes. It turns out that our constructions produce binary constant-weight codes with good parameters.

Mathematics Subject Classification (2000). Primary 94B60; Secondary 94B65.
Keywords. Constant-weight codes, sets, linear spaces, free modules.

1. Introduction

Constant-weight codes have a very long history because of both practical applications and theoretical interests. Various methods from algebra, finite geometry, combinatorics, etc., have been employed to construct good codes. The reader may refer to [6] and [2] for a good survey on this topic.

In this paper, we first give a simple propagation rule by identifying a binary constant-weight code with a family of subsets. This idea is further generalized to linear spaces and free modules to construct binary constant-weight codes with reasonable parameters.

2. Preliminaries

In this section, we introduce some concepts and definitions that will be used in the next sections.

2.1. Constant-Weight Codes

A binary *constant-weight code* $\mathcal{C} \subseteq \mathbb{F}_2^n$ is a set of codewords that have the same (Hamming) weight. $\mathcal{C}$ is called an $(n, M, d; w)$ constant-weight code if $\mathcal{C}$ is a set of cardinality M, such that each codeword has the same weight w, and the distance between any two codewords is at least d. Given n, d and w, to determine the maximum possible size $A(n, d, w)$ of an $(n, M, d; w)$ binary constant-weight code is an important problem in coding theory.

In calculating the distance between two codewords we have a useful formula:

Proposition 2.1 ([8], **Lemma 4.3.4**). *For any two codewords* $\mathbf{x} = (x_1, x_2, \ldots, x_n)$ *and* $\mathbf{y} = (y_1, y_2, \ldots, y_n)$ *in* $\mathbb{F}_2^n$, *put* $\mathbf{x} * \mathbf{y} = (x_1y_1, x_2y_2, \ldots, x_ny_n)$, *then*

$$d(\mathbf{x},\mathbf{y}) = wt(\mathbf{x}+\mathbf{y}) = wt(\mathbf{x}) + wt(\mathbf{y}) - 2wt(\mathbf{x}*\mathbf{y}). \tag{2.1}$$

2.2. Gaussian Coefficients

Given a prime power q and two positive integers k, r with $k \leq r$, the number

$$\begin{bmatrix} r \\ k \end{bmatrix}_q \triangleq \frac{\prod_{i=r-k+1}^{r}(q^i - 1)}{\prod_{i=1}^{k}(q^i - 1)}$$

is called a *Gaussian coefficient.* For the convenience of later usage, we define $\left[\begin{smallmatrix} r \\ k \end{smallmatrix}\right]_q = 0$ if $k > r$. The significance of Gaussian coefficients is described in the following proposition.

Proposition 2.2 ([8], **Theorem 5.1.12**). *Let* $\mathbb{F}_q$ *be a finite field and* V *a linear space of dimension* r *over* $\mathbb{F}_q$. *Then the number of dimension* $k(\leq r)$ *subspaces of* V *is*

$$\begin{bmatrix} r \\ k \end{bmatrix}_q = \frac{(q^r - 1)(q^r - q)\cdots(q^r - q^{k-1})}{(q^k - 1)(q^k - q)\cdots(q^k - q^{k-1})}.$$

2.3. Linearized Polynomials and Rank Distance Codes

We first review rank distance codes studied by Gabidulin in [3]. Let $\Lambda = \{A_i\}$ be a set of $t \times m$ matrices over a finite field $\mathbb{F}_q$. The distance $d(A, B)$ between two matrices A and B in Λ is defined by $d(A, B) = \operatorname{rank}(A - B)$ and the minimum distance of Λ, denoted by $d(\Lambda)$, is defined as $d(\Lambda) = \min\{d(A, B) : A \neq B \in \Lambda\}$. Let $d = d(\Lambda)$ and $M = |\Lambda|$. We call Λ a $(t \times m, M, d)$ *rank distance code.* For a $(t \times m, M, d)$ rank distance code Λ, the Singleton bound is valid, i.e.,

$$d(\Lambda) \leq t - l + 1, \tag{2.2}$$

where $l = \log_{q^m} M$. Codes for which equality holds in (2.2) are referred to as *MRD-codes*(Maximum-Rank-Distance codes).

In [5], Johansson presents a method for constructing MRD-codes from linearized polynomials. Let $1 \leq l \leq t \leq m$ be positive integers. A polynomial of the form

$$F(x) = \sum_{i=0}^{t} f_i x^{q^i},$$

where $f_i \in \mathbb{F}_{q^m}$ is called a *linearized polynomial.* Denote all linearized polynomials of degree not higher than q^{l-1} as

$$P_{l,t,m} = \{F(x) = \sum_{i=0}^{t} f_i x^{q^i} : f_i \in \mathbb{F}_{q^m},\ \deg(F(x)) \leq q^{l-1}\}.$$

Assume $g_1, g_2, \ldots, g_t$ are specified elements in the field $\mathbb{F}_{q^m}$ which are linearly independent over $\mathbb{F}_q$, and for each $F(x) \in P_{l,t,m}$, put

$$A_F = (F(g_1), F(g_2), \ldots, F(g_t))^T.$$

Fix a basis of $\mathbb{F}_{q^m}$ over $\mathbb{F}_q$, write each $F(g_i) = (a_{i1}, a_{i2}, \ldots, a_{im})$ expressed in this fixed basis as a row vector, where each entry $a_{ij} \in \mathbb{F}_q$. Therefore, each A_F can be viewed as a $t \times m$ matrix (a_{ij}) over $\mathbb{F}_q$, and $\Lambda = \{A_F : F(x) \in P_{l,t,m}\}$ can be viewed as a rank distance code. Moreover Johansson proved that Λ is an MRD-code.

Theorem 2.3 ([5], **Lemma 3**). *$\Lambda = \{A_F : F(x) \in P_{l,t,m}\}$ is an MRD-code. That is, Λ is a $(t \times m, q^{ml}, t-l+1)$ rank distance code.*

2.4. Free Modules

Let R be a ring and M an R-module (cf. [4], Ch. IV). A subset S of M is said to be *R-linearly independent* provided that for distinct $x_1, x_2, \ldots, x_n \in S$ and $r_i \in R$, $r_1x_1 + r_2x_2 + \cdots + r_nx_n = 0$ deduces $r_1 = r_2 = \cdots = r_n = 0$. An R-linearly independent subset of M that spans M is called a *basis* of M, and M is called a *free R-module* if M has a basis.

If R is a commutative ring with identity and M is a free R-module, then each basis of M has the same cardinality. In this case, the number of elements in a basis of M is called the *rank* of M, denoted by rank$M$.

Proposition 2.4 ([4], **Ch. IV, Theorem 2.1**). *Let R be a commutative ring with identity and M a free R-module. If rankM=r, then $M \cong \underbrace{R \times R \times \cdots \times R}_{r}$.*

In the next sections, we always consider free modules over the congruence class ring $\mathbb{Z}_m$, which is of course a commutative ring with identity.

3. New Constructions of Constant-Weight Codes

We present our main work in this section. In the first part, we identify each binary constant-weight code with a family of subsets, then give a propagation rule for constant-weight codes from this relationship. In the next two parts, we generalize the rule to linear spaces and free modules, respectively, which will lead to new constructions of binary constant-weight codes.

3.1. Sets

Suppose $A = \{a_1, a_2, \ldots, a_n\}$ is a set of cardinality n. Denote $2^{|A|} = \{S : S \subset A\}$ to be the set of all subsets of A, then we can define a map $\psi : \mathbb{F}_2^n \longrightarrow 2^{|A|}$ as follows:

$$\psi((x_1, x_2, \ldots, x_n)) = \{a_i : x_i = 1\}.$$

It is easy to verify that ψ is a bijection. For each $(n, M, d; w)$ binary constant-weight code $\mathcal{C} \subset \mathbb{F}_2^n$, $\psi(\mathcal{C}) = \{A_1, A_2, \ldots, A_M\} \subset 2^{|A|}$ satisfies

$$|A_i| = w \leq n, \quad \text{for } i = 1, 2, \ldots, M; \tag{3.1}$$

and

$$|A_i| + |A_j| - 2|A_i \cap A_j| = 2w - 2|A_i \cap A_j| \geq d, \quad \text{for all } 1 \leq i \neq j \leq M. \tag{3.2}$$

Therefore, given an $(n, M, d; w)$ binary constant-weight code $\mathcal{C}$, we can find a family of subsets of A satisfying (3.1) and (3.2).

On the other hand, if there is a family of subsets of A, denoted by $\{A_1, A_2, \ldots, A_M\}$, satisfying the above conditions (3.1) and (3.2), we can also construct a family of binary constant-weight codes.

For each fixed s with $1 \leq s \leq w$, suppose $B_1, B_2, \ldots, B_{\binom{n}{s}}$ are all subsets of cardinality s of A. Construct a binary constant-weight code $\mathcal{C}' = \{\mathbf{c}_1, \mathbf{c}_2, \ldots, \mathbf{c}_M\} \subset \mathbb{Z}_2^{n'}$ as follows:

$$\mathbf{c}_i = (c_{i_1}, c_{i_2}, \ldots, c_{i_{n'}}), \quad \text{where } n' = \binom{n}{s}, \; c_{i_j} = \begin{cases} 1 & , \quad B_j \subset A_i \\ 0 & , \quad B_j \not\subset A_i \end{cases} . \tag{3.3}$$

It's obvious that $\mathrm{wt}(\mathbf{c}_i) = \binom{w}{s}$ for all i, and for any $1 \leq i \neq j \leq M$, we have

$$\begin{aligned} d(\mathbf{c}_i, \mathbf{c}_j) &= \mathrm{wt}(\mathbf{c}_i) + \mathrm{wt}(\mathbf{c}_j) - 2\mathrm{wt}(\mathbf{c}_i * \mathbf{c}_j) \\ &= 2\binom{w}{s} - 2\binom{|A_i \cap A_j|}{s} \\ &\geq 2\binom{w}{s} - 2\binom{\frac{2w-d}{2}}{s} \triangleq d' \qquad \text{(by (3.2))}. \end{aligned}$$

Thus, $\mathcal{C}'$ is an $(n', M, d'; w')$ constant-weight code, where $n' = \binom{n}{s}$, $d' = 2\binom{w}{s} - 2\binom{\frac{2w-d}{2}}{s}$, and $w' = \binom{w}{s}$.

Therefore, we can obtain an $(n', M, d'; w')$ binary constant-weight code $\mathcal{C}'$ with the above parameters from an $(n, M, d; w)$ binary constant-weight code $\mathcal{C}$. Given a lower bound N on some $A(n, d, w)$, there must exist at least one $(n, N, d; w)$ constant-weight code, then we can construct new constant-weight codes from this $(n, N, d; w)$ code. In conclusion, we have the following theorem:

Theorem 3.1. *Given a lower bound N on an $A(n, d, w)$, then for all $1 \leq s \leq w$ there exists an $(n', N, d'; w')$ constant-weight code, where $n' = \binom{n}{s}$, $d' = 2\binom{w}{s} - 2\binom{\frac{2w-d}{2}}{s}$, and $w' = \binom{w}{s}$. Thus, $A(n', d', w') \geq N$.*

3.2. Linear Spaces

In this part, we introduce a construction of binary constant-weight codes from linear spaces similar to the one from sets in the previous part.

Let V be a linear space of dimension r over a finite field $\mathbb{F}_q$ and s, t, r positive integers satisfying $1 \leq s \leq t \leq r$. Put $V_1, V_2, \ldots, V_{\left[{r \atop s}\right]_q}$ to be all dimension s subspaces of V. Then given some dimension t subspaces $W_1, W_2, \ldots, W_M$ $(M \leq \left[{r \atop t}\right]_q)$ of V, we can construct a binary constant-weight code $\mathcal{C} = \{\mathbf{c}_1, \mathbf{c}_2, \ldots, \mathbf{c}_M\}$ in a similar way:

$$\mathbf{c}_i = (c_{i_1}, c_{i_2}, \ldots, c_{i_n}), \quad \text{where } n = \begin{bmatrix} r \\ s \end{bmatrix}_q, \; c_{i_j} = \begin{cases} 1 & , \quad V_j \subset W_i \\ 0 & , \quad V_j \not\subset W_i \end{cases} . \tag{3.4}$$

The parameters of $\mathcal{C}$ can be easily determined. Obviously, $\text{wt}(\mathbf{c}_i) = \begin{bmatrix} t \\ s \end{bmatrix}_q$ for all i. Denote $\theta \triangleq \max\{\dim(W_i \cap W_j) : 1 \le i \ne j \le M\} \le t-1$, then for any $1 \le i \ne j \le M$,

$$\begin{aligned} d(\mathbf{c}_i, \mathbf{c}_j) &= \text{wt}(\mathbf{c}_i) + \text{wt}(\mathbf{c}_j) - 2\text{wt}(\mathbf{c}_i * \mathbf{c}_j) \\ &= 2\begin{bmatrix} t \\ s \end{bmatrix}_q - 2\begin{bmatrix} \dim(W_i \cap W_j) \\ s \end{bmatrix}_q \\ &\ge 2\begin{bmatrix} t \\ s \end{bmatrix}_q - 2\begin{bmatrix} \theta \\ s \end{bmatrix}_q . \end{aligned} \tag{3.5}$$

So $\mathcal{C}$ is an $\left(\begin{bmatrix} r \\ s \end{bmatrix}_q, M, 2\begin{bmatrix} t \\ s \end{bmatrix}_q - 2\begin{bmatrix} \theta \\ s \end{bmatrix}_q; \begin{bmatrix} t \\ s \end{bmatrix}_q\right)$ constant-weight code. Hitherto, we have already proven the following theorem:

Theorem 3.2. *Let V be a linear space of dimension r over a finite field $\mathbb{F}_q$. If there exists a set of subspaces of V, denoted by $\Omega(r, M, t, \theta)$, satisfying*

(i) $|\Omega| = M$;

(ii) $\dim(W) = t, \forall W \in \Omega$;

(iii) $\dim(W \cap W') \le \theta, \forall W \ne W' \in \Omega$,

then there exists an $\left(\begin{bmatrix} r \\ s \end{bmatrix}_q, M, 2\begin{bmatrix} t \\ s \end{bmatrix}_q - 2\begin{bmatrix} \theta \\ s \end{bmatrix}_q; \begin{bmatrix} t \\ s \end{bmatrix}_q\right)$ binary constant-weight code for all $1 \le s \le t$.

Now our problem turns to be how to find $\Omega(r, M, t, \theta)$ with M as large as possible. We partly solve this problem by obtaining $\Omega(r, M, t, \theta)$ from rank distance codes. The relationship between $\Omega(r, M, t, \theta)$ and rank distance codes has been established by Theorem 1 and 3 in [9].

Theorem 3.3 ([9]). *If there exists a $(t \times m, M, t - \theta)$ rank distance code over $\mathbb{F}_q$, then there exists an $\Omega(m + t, M, t, \theta)$ over $\mathbb{F}_q$.*

We adopt Johansson's way ([5]) to construct rank distance codes with good parameters. As we've restated in Section 2.3, the code $\Lambda = \{A_F : F(x) \in P_{l,t,m}\}$ in Theorem 2.3 is a $(t \times m, q^{ml}, t - l + 1)$ rank distance code where $l \le t \le m$. By Theorem 3.3 there must exist an $\Omega(t + m, q^{ml}, t, l - 1)$ and hence a family of $\left(\begin{bmatrix} m+t \\ s \end{bmatrix}_q, q^{ml}, 2\begin{bmatrix} t \\ s \end{bmatrix}_q - 2\begin{bmatrix} l-1 \\ s \end{bmatrix}_q; \begin{bmatrix} t \\ s \end{bmatrix}_q\right)$ constant-weight codes for all $1 \le s \le t$. So we get the following proposition:

Proposition 3.4. *Let $\mathbb{F}_q$ be a finite field and $1 \le l \le t \le m$ positive integers, then for all $1 \le s \le t$,*

$$A(n, d, w) \ge q^{ml}, \tag{3.6}$$

where $n = \begin{bmatrix} m+t \\ s \end{bmatrix}_q, d = 2\begin{bmatrix} t \\ s \end{bmatrix}_q - 2\begin{bmatrix} l-1 \\ s \end{bmatrix}_q$, and $w = \begin{bmatrix} t \\ s \end{bmatrix}_q$.

We claim this family of $\Omega(t+m, q^{ml}, t, l-1)$ are "optimal", in sense that they will reach their maximum possible sizes gradually as q goes to infinity. To show this, we give an upper bound on the size of $\Omega(r, M, t, \theta)$ at first, then compare the parameters of $\Omega(t + m, q^{ml}, t, l - 1)$ with the upper bound.

Theorem 3.5 (Upper bound on $|\Omega(r,M,t,\theta)|$). *Let $M(r,t,\theta)$ be the maximum possible size of $\Omega(r,M,t,\theta)$, then*

$$M(r,t,\theta) \leq \frac{\begin{bmatrix} r \\ \theta+1 \end{bmatrix}_q}{\begin{bmatrix} t \\ \theta+1 \end{bmatrix}_q}. \tag{3.7}$$

Proof. Suppose V is a dimension r linear space over $\mathbb{F}_q$ and $\Omega(r,M,t,\theta)=\{W_1,W_2,\ldots,W_M\}$ a set of subspaces of V with $M = M(r,t,\theta)$. Let U be a dimension $\theta+1$ subspace of V. If U was contained in a W_i, we assert that U cannot be contained in any other W_j with $j \neq i$. Otherwise, $\dim(W_i \cap W_j) \geq \dim U = \theta+1$, which contradicts with the definition of $\Omega(r,M,t,\theta)$.

The number of dimension $\theta+1$ subspaces of V is $\begin{bmatrix} r \\ \theta+1 \end{bmatrix}_q$, and there are $\begin{bmatrix} t \\ \theta+1 \end{bmatrix}_q$ dimension $\theta+1$ subspaces contained in each W_i, therefore

$$M(r,t,\theta) \leq \frac{\begin{bmatrix} r \\ \theta+1 \end{bmatrix}_q}{\begin{bmatrix} t \\ \theta+1 \end{bmatrix}_q}.$$

□

For any fixed r and k, as $q \to +\infty$ we have

$$\begin{aligned} \begin{bmatrix} r \\ k \end{bmatrix}_q &= \frac{(q^r-1)(q^{r-1}-1)\cdots(q^{r-k+1}-1)}{(q^k-1)(q^{k-1}-1)\cdots(q-1)} \\ &\approx q^{(r-k)k}. \end{aligned} \tag{3.8}$$

It follows that

$$\begin{aligned} M(r,t,\theta) &\leq \frac{\begin{bmatrix} r \\ \theta+1 \end{bmatrix}_q}{\begin{bmatrix} t \\ \theta+1 \end{bmatrix}_q} \\ &\approx \frac{q^{(r-\theta-1)(\theta+1)}}{q^{(t-\theta-1)(\theta+1)}} \\ &\approx q^{(r-t)(\theta+1)}, \end{aligned} \tag{3.9}$$

when $q \to +\infty$. Clearly, the set $\Omega(t+m, q^{ml}, t, l-1)$ will reach the upper bound on $M(r,t,\theta)$ gradually as q goes to infinity.

Since $\Omega(t+m, q^{ml}, t, l-1)$ is "optimal", the corresponding constant-weight codes obtained from our construction should have good parameters, too. To show this, we compare the lower bound on $A(n,d,w)$ in Proposition 3.4 with an upper bound introduced in [1].

Proposition 3.6 ([1], **Theorem 12).** *Let $u = w - d/2 + 1$. Then*

$$A(n,d,w) \leq \frac{\binom{n}{u}}{\binom{w}{u}}. \tag{3.10}$$

Still using notations in Proposition 3.4 and 3.6, we get that when $q \to +\infty$,

$$n = \begin{bmatrix} m+t \\ s \end{bmatrix}_q \approx q^{(m+t-s)s}; \tag{3.11}$$

$$w = \begin{bmatrix} t \\ s \end{bmatrix}_q \approx q^{(t-s)s}; \tag{3.12}$$

$$u = w - d/2 + 1 = \begin{bmatrix} l-1 \\ s \end{bmatrix}_q + 1 \approx q^{(l-1-s)s}. \tag{3.13}$$

By Stirling's Formula $n! \approx \sqrt{2\pi n}(\frac{n}{e})^n, (n \to +\infty)$, we have

$$\begin{aligned} A(n,d,w) &\leq \frac{\binom{n}{u}}{\binom{w}{u}} \\ &= \frac{n!(w-u)!}{w!(n-u)!} \\ &\approx \left(\frac{n}{w}\right)^{n+\frac{1}{2}} \left(\frac{w}{w-u}\right)^{n-w} \left(\frac{w-u}{n-u}\right)^{n-u+\frac{1}{2}} \\ &= q^{ms(n+\frac{1}{2})} \left(1 + \frac{1}{q^{(t-l+1)s}-1}\right)^{n-w} \\ &\quad \cdot \left[q^{ms}\left(1 + \frac{1-q^{-ms}}{q^{(t-l+1)s}-1}\right)\right]^{-(n-u+\frac{1}{2})} \\ &\approx q^{msu}, \end{aligned} \tag{3.14}$$

when q goes to infinity.

When q is large, if we choose s that is close to l (especially we can take $s = l$), then q^{ml} can be very close to the upper bound q^{msu}. Therefore, our construction can produce binary constant-weight codes with good parameters in such cases. In the next section, we will give some examples to illustrate the significance of these codes.

3.3. Free $\mathbb{Z}_m$-Module

In this part, we give a construction of binary constant-weight codes from free $\mathbb{Z}_m$-modules. The reason why we choose $\mathbb{Z}_m$ as the coefficient ring is that the number of free submodules of a free $\mathbb{Z}_m$-module of finite rank is finite and easy to calculate, thus the parameters of binary constant-weight codes constructed from free $\mathbb{Z}_m$-modules can be easily determined.

Let $m > 1$ be a positive integer, and $\mathbb{Z}_m$ be the congruence class ring mod m. Then $\mathbb{Z}_m^r \triangleq \underbrace{\mathbb{Z}_m \times \mathbb{Z}_m \times \cdots \mathbb{Z}_m}_{r}$ is a free module over $\mathbb{Z}_m$ of rank r. Denote the number of rank $k (\leq r)$ free submodules of $\mathbb{Z}_m^r$ by $F_{m,r}(k)$. Suppose $1 \leq s \leq t \leq r$ and $\{A_1, A_2, \ldots, A_n\}$ is the set of all rank s free submodules of $\mathbb{Z}_m^r$, where $n = F_{m,r}(s)$. Similarly, given some rank t free submodules of $\mathbb{Z}_m^r$: $B_1, B_2, \ldots, B_M (M \leq$

$F_{m,r}(t)$) such that $\text{rank}(B_i \cap B_j) \leq \theta \leq t-1$ for any $1 \leq i \neq j \leq M$, we can construct a binary constant-weight code $\mathcal{C} = \{\mathbf{c}_1, \mathbf{c}_2, \ldots, \mathbf{c}_M\}$ as follows:

$$\mathbf{c}_i = (c_{i_1}, c_{i_2}, \ldots, c_{i_n}), \qquad \text{where} \quad c_{i_j} = \begin{cases} 1 & , \quad A_j \subset B_i \\ 0 & , \quad A_j \not\subset B_i \end{cases}. \tag{3.15}$$

It is easy to verify that $\mathcal{C}$ is an $(n, M, 2F_{m,t}(s) - 2F_{m,\theta}(s); F_{m,t}(s))$ binary constant-weight code. To determine the parameters of $\mathcal{C}$, we just need to calculate $F_{m,r}(k)$.

In the following, we are going to get a formula to calculate $F_{m,r}(k)$ in several steps. Firstly, we have a useful lemma.

Lemma 3.7. *Write m in the form $m = p_1^{e_1} p_2^{e_2} \cdots p_t^{e_t}$, where p_i's are different primes and $e_i > 0$ for all i. Then*

$$F_{m,r}(k) = F_{p_1^{e_1},r}(k) F_{p_2^{e_2},r}(k) \cdots F_{p_t^{e_t},r}(k). \tag{3.16}$$

Proof. By the Chinese Remainder Theorem, $\mathbb{Z}_m \cong \mathbb{Z}_{p_1^{e_1}} \oplus \mathbb{Z}_{p_2^{e_2}} \oplus \cdots \oplus \mathbb{Z}_{p_t^{e_t}}$. Put $R \triangleq \mathbb{Z}_{p_1^{e_1}} \oplus \mathbb{Z}_{p_2^{e_2}} \oplus \cdots \oplus \mathbb{Z}_{p_t^{e_t}}$, then $R^r \triangleq \underbrace{R \times R \times \cdots R}_{r} \cong \mathbb{Z}_m^r$ and $F_{m,r}(k)$ equals to the number of rank k free R-submodules of R^r.

For any $\mathbf{x} = (x_1, x_2, \ldots, x_r)^T \in R^r$, we write $\mathbf{x}$ in the form

$$\mathbf{x} = \begin{pmatrix} x_{11} & x_{12} & \cdots & x_{1t} \\ x_{21} & x_{22} & \cdots & x_{2t} \\ \cdots & \cdots & \cdots & \cdots \\ x_{r1} & x_{r2} & \cdots & x_{rt} \end{pmatrix},$$

where $x_j = (x_{j1}, x_{j2}, \ldots, x_{jt}) \in R$ is the jth row of $\mathbf{x}$ for all $j = 1, 2, \ldots, r$. Let $\text{col}(\mathbf{x})_i = (x_{1i}, x_{2i}, \ldots, x_{ri})^T$ denote the ith column of $\mathbf{x}$ for all $i = 1, 2, \ldots, t$. Then $\text{col}(\mathbf{x})_i \in \mathbb{Z}^r_{p_i^{e_i}}$, and for any $\lambda = (\lambda_1, \lambda_2, \ldots, \lambda_t) \in R$,

$$\lambda \mathbf{x} = (\lambda_1 \text{col}(\mathbf{x})_1, \lambda_2 \text{col}(\mathbf{x})_2, \ldots, \lambda_t \text{col}(\mathbf{x})_t).$$

Suppose A is a rank $k \geq 1$ free R-submodule of R^r. Put $A_i \subset \mathbb{Z}^r_{p_i^{e_i}}$ to be the set of all the ith columns of elements in A, i.e.,

$$A_i \triangleq \{\text{col}(\mathbf{x})_i \in \mathbb{Z}^r_{p_i^{e_i}} : \mathbf{x} = (x_1, x_2, \ldots, x_r) \in A \subset R^r\}, i = 1, 2, \ldots, t. \tag{3.17}$$

We assert that A_i is a rank k free $\mathbb{Z}_{p_i^{e_i}}$-submodule of $\mathbb{Z}^r_{p_i^{e_i}}$. In fact, take an R-basis $\{\mathbf{x}^{(1)}, \mathbf{x}^{(2)}, \ldots, \mathbf{x}^{(k)}\}$ of A, it's easy to check that $\{\text{col}(\mathbf{x}^{(1)})_i, \text{col}(\mathbf{x}^{(2)})_i, \ldots, \text{col}(\mathbf{x}^{(k)})_i\}$ is a $\mathbb{Z}_{p_i^{e_i}}$-basis of A_i.

Put $M = \{A \subset R^r : \text{rank}A = k\}$ and $N = \{(B_1, B_2, \ldots, B_t) : B_i \subset \mathbb{Z}^r_{p_i^{e_i}}, \text{rank}B_i = k\}$, then we can define a map

$$\varphi : M \longrightarrow N, \quad A \mapsto (A_1, A_2, \ldots, A_t).$$

It is sufficient to prove that φ is bijective.

We show that φ is injective at first. Suppose there exist $A, B \in M$ satisfying $\varphi(A) = \varphi(B) = (A_1, A_2, \ldots, A_t)$. Take an R-basis $\{\mathbf{x}^{(1)}, \mathbf{x}^{(2)}, \ldots, \mathbf{x}^{(k)}\}$ of A and

$\{\mathbf{y}^{(1)}, \mathbf{y}^{(2)}, \ldots, \mathbf{y}^{(k)}\}$ of B. Then $\{\mathrm{col}(\mathbf{x}^{(n)})_i : n = 1, 2, \ldots, k\}$ and $\{\mathrm{col}(\mathbf{y}^{(n)})_i : n = 1, 2, \ldots, k\}$ are two $\mathbb{Z}_{p_i^{e_i}}$-bases of A_i. So there exist some $\lambda_{i1}, \lambda_{i2}, \ldots, \lambda_{ik} \in \mathbb{Z}_{p_i^{e_i}}$ for each $i = 1, 2, \ldots, t$ such that

$$\mathrm{col}(\mathbf{y}^{(1)})_i = \sum_{n=1}^{k} \lambda_{in} \mathrm{col}(\mathbf{x}^{(n)})_i.$$

Thus

$$\begin{aligned} \mathbf{y}^{(1)} &= (\mathrm{col}(\mathbf{y}^{(1)})_1, \mathrm{col}(\mathbf{y}^{(1)})_2, \ldots, \mathrm{col}(\mathbf{y}^{(1)})_t) \\ &= \sum_{n=1}^{k} \lambda_n \mathbf{x}^{(n)} \in A, \end{aligned}$$

where $\lambda_n = (\lambda_{1n}, \lambda_{2n}, \ldots, \lambda_{tn}) \in R$ $(n = 1, 2, \ldots, k)$. Similarly, we can prove $\mathbf{y}^{(2)}, \ldots, \mathbf{y}^{(k)} \in A$, which deduces $B \subset A$. In the same way, we can get $A \subset B$. So $A = B$, i.e., φ is injective.

Conversely, for each $(A_1, A_2, \ldots, A_t) \in N$, we are going to find an $A \in M$ such that $\varphi(A) = (A_1, A_2, \ldots, A_t)$. Assume $\{\alpha_{i1}, \alpha_{i2}, \ldots, \alpha_{ik}\}$ is a $\mathbb{Z}_{p_i^{e_i}}$-basis of A_i. For each $n = 1, 2, \ldots, k$, take $\alpha_{in} \in \mathbb{Z}^r_{p_i^{e_i}}$ as the ith coordinate of $\mathbf{x}^{(n)}$ to form $\mathbf{x}^{(n)} = (\alpha_{1n}, \alpha_{2n}, \ldots, \alpha_{tn}) \in R^r$. We assert that $\{\mathbf{x}^{(1)}, \mathbf{x}^{(2)}, \ldots, \mathbf{x}^{(k)}\}$ are R-linearly independent. If there exists $\{\lambda_i = (\lambda_{i1}, \lambda_{i2}, \ldots, \lambda_{it}) \in R : i = 1, 2, \ldots, k\}$ such that

$$\lambda_1 \mathbf{x}^{(1)} + \lambda_2 \mathbf{x}^{(2)} + \cdots + \lambda_k \mathbf{x}^{(k)} = 0,$$

then

$$\lambda_{1i}\alpha_{i1} + \lambda_{2i}\alpha_{i2} + \cdots + \lambda_{ki}\alpha_{ik} = 0 \in \mathbb{Z}_{p_i^{e_i}}, 1 \le i \le t.$$

Since $\{\alpha_{i1}, \alpha_{i2}, \ldots, \alpha_{ik}\}$ is $\mathbb{Z}_{p_i^{e_i}}$-linearly independent, all the $\lambda_{1i}, \lambda_{2i}, \cdots, \lambda_{ki}$ must be 0. Hence $\lambda_1 = \lambda_2 = \cdots = \lambda_k = 0$, that is, $\{\mathbf{x}^{(1)}, \mathbf{x}^{(2)}, \ldots, \mathbf{x}^{(k)}\}$ is R-linearly independent. Let $A = \langle \mathbf{x}^{(1)}, \mathbf{x}^{(2)}, \ldots, \mathbf{x}^{(k)} \rangle$ be the R-module spanned by $\{\mathbf{x}^{(1)}, \mathbf{x}^{(2)}, \ldots, \mathbf{x}^{(k)}\}$, then A is a rank k free R-submodule of R^r and $\varphi(A) = (A_1, A_2, \ldots, A_t)$.

The proof is finished. □

By Lemma 3.7 we just need to calculate $F_{p^e,r}(k)$, where p is a prime and $e \ge 1$. Ahead of the calculation, we still need some lemmas.

Lemma 3.8. *Assume p is a prime, $e \ge 1$ an integer, and $m = p^e$. Suppose $\{\mathbf{x}^{(1)}, \mathbf{x}^{(2)}, \ldots, \mathbf{x}^{(k)}\} \subset \mathbb{Z}_m^r$ is $\mathbb{Z}_m$-linearly independent, then it can be extended to a $\mathbb{Z}_m$-basis $\{\mathbf{x}^{(1)}, \mathbf{x}^{(2)}, \ldots, \mathbf{x}^{(k)}, \mathbf{x}^{(k+1)}, \ldots, \mathbf{x}^{(r)}\}$ of $\mathbb{Z}_m^r$.*

Proof. We use induction on k.

(1) Assume $k = 1$.

Suppose $\mathbf{x}^{(1)} = (x_1^{(1)}, x_2^{(1)}, \ldots, x_r^{(1)})^T$. Since $\{\mathbf{x}^{(1)}\}$ is $\mathbb{Z}_m$-linearly independent, $(\lambda x_1^{(1)}, \lambda x_2^{(1)}, \ldots, \lambda x_r^{(1)}) \ne 0$ for all $\lambda \ne 0 \in \mathbb{Z}_m$. So $g.c.d(x_1^{(1)}, x_2^{(1)}, \ldots, x_r^{(1)})$ can not be divided by p. Without loss of generality, we may assume $p \nmid x_1^{(1)}$, i.e., $x_1^{(1)}$ is invertible in $\mathbb{Z}_m$.

Take the standard $\mathbb{Z}_m$-basis $\{\mathbf{e}_i = (0,\dots,0,1,0,\dots,0)^T : i = 1,2,\dots,r\}$ of $\mathbb{Z}_m^r$, where $\mathbf{e}_i$ has only one nonzero coordinate 1 at the ith position. Obviously,

$$\mathbf{x}^{(1)} = x_1^{(1)}\mathbf{e}_1 + x_2^{(1)}\mathbf{e}_2 + \cdots + x_r^{(1)}\mathbf{e}_r.$$

Then

$$\mathbf{e}_1 = -{x_1^{(1)}}^{-1}(x_2^{(1)}\mathbf{e}_2 + \cdots + x_r^{(1)}\mathbf{e}_r - \mathbf{x}^{(1)}).$$

Thus $\{\mathbf{e}_1, \mathbf{e}_2, \dots, \mathbf{e}_r\}$ can be $\mathbb{Z}_m$-linearly represented by $\{\mathbf{x}^{(1)}, \mathbf{e}_2, \dots, \mathbf{e}_r\}$, then $\{\mathbf{x}^{(1)}, \mathbf{e}_2, \dots, \mathbf{e}_r\}$ is a $\mathbb{Z}_m$-basis of $\mathbb{Z}_m^r$ as well.

(2) Suppose this lemma holds for $k-1$.

Since $\{\mathbf{x}^{(1)}, \mathbf{x}^{(2)}, \dots, \mathbf{x}^{(k-1)}\}$ is also $\mathbb{Z}_m$-linearly independent, by our assumption it can be extended to a $\mathbb{Z}_m$-basis $\{\mathbf{x}^{(1)}, \mathbf{x}^{(2)}, \dots, \mathbf{x}^{(k-1)}, \mathbf{y}^{(k)}, \dots, \mathbf{y}^{(r)}\}$ of $\mathbb{Z}_m^r$. So we can write $\mathbf{x}^{(k)}$ in the form

$$\mathbf{x}^{(k)} = \lambda_1\mathbf{x}^{(1)} + \lambda_2\mathbf{x}^{(2)} + \cdots + \lambda_{k-1}\mathbf{x}^{(k-1)} + \lambda_k\mathbf{y}^{(k)} + \cdots + \lambda_r\mathbf{y}^{(r)}.$$

There must exist at least one λ_n in $\{\lambda_k, \lambda_{k+1}, \dots, \lambda_r\}$ that is invertible in $\mathbb{Z}_m$. Otherwise, if $p|g.c.d(\lambda_k, \lambda_{k+1}, \dots, \lambda_r)$, we have

$$\begin{aligned} p^{e-1}\mathbf{x}^{(k)} &= p^{e-1}\sum_{n=1}^{k-1}\lambda_n\mathbf{x}^{(n)} + p^{e-1}\sum_{n=k}^{n}\lambda_n\mathbf{y}^{(n)} \\ &= p^{e-1}\sum_{n=1}^{k-1}\lambda_n\mathbf{x}^{(n)}, \end{aligned}$$

which contradicts with the assumption that $\{\mathbf{x}^{(1)}, \mathbf{x}^{(2)}, \dots, \mathbf{x}^{(k)}\}$ is $\mathbb{Z}_m$-linearly independent.

We may assume $p \nmid \lambda_k$, then

$$\mathbf{y}^{(k)} = -{\lambda_k}^{-1}(\sum_{n=1}^{k-1}\lambda_n\mathbf{x}^{(n)} + \sum_{n=k+1}^{r}\lambda_n\mathbf{y}^{(n)} - \mathbf{x}^{(k)}),$$

Thus $\{\mathbf{x}^{(1)}, \mathbf{x}^{(2)}, \dots, \mathbf{x}^{(k)}\}$ can be extended to a $\mathbb{Z}_m$-basis $\{\mathbf{x}^{(1)},\mathbf{x}^{(2)},\dots,\mathbf{x}^{(k)},\mathbf{y}^{(k+1)},\dots,\mathbf{y}^{(r)}\}$ of $\mathbb{Z}_m^r$.

The result follows. □

Corollary 3.9. *On the same condition as Lemma* 3.8, $A = \langle\mathbf{x}^{(1)}, \mathbf{x}^{(2)}, \dots, \mathbf{x}^{(k)}\rangle$ *is a free* $\mathbb{Z}_m$*-module of rank* k*, and* $\mathbb{Z}_m^r/A$ *is a free* $\mathbb{Z}_m$*-module of rank* $r-k$.

Proof. $A = \bigoplus_{n=1}^k \mathbb{Z}_m\mathbf{x}^{(n)} \cong \mathbb{Z}_m^k$ is of course a free $\mathbb{Z}_m$-module of rank k. By Lemma 3.8, $\{\mathbf{x}^{(1)}, \mathbf{x}^{(2)}, \dots, \mathbf{x}^{(k)}\}$ can be extended to a $\mathbb{Z}_m$-basis $\{\mathbf{x}^{(1)}, \mathbf{x}^{(2)}, \dots, \mathbf{x}^{(k)}, \mathbf{x}^{(k+1)}, \dots, \mathbf{x}^{(r)}\}$ of $\mathbb{Z}_m^r$. Therefore

$$\mathbb{Z}_m^r/A \cong \bigoplus_{n=k+1}^{r} \mathbb{Z}_m\mathbf{x}^{(n)} \cong \mathbb{Z}_m^{r-k}.$$

□

Now we can begin to calculate $F_{p^e,r}(k)$.

Lemma 3.10. *Assume p is a prime, $e \geq 1$ and $m = p^e$. Then*

$$F_{m,r}(k) = p^{k(r-k)(e-1)} \begin{bmatrix} r \\ k \end{bmatrix}_p, \quad 1 \leq k \leq r. \tag{3.18}$$

Proof. Let $N_{m,r}(k)$ denote the number of $\mathbb{Z}_m$-linearly independent sets of cardinality k in $\mathbb{Z}_m^r$, then $F_{m,r}(k) = N_{m,r}(k)/N_{m,k}(k)$. We go on using induction on k.

(1) Assume $k = 1$.

An element $\mathbf{x} = (x_1, x_2, \ldots, x_r)^T \in \mathbb{Z}_m^r$ can be taken as a $\mathbb{Z}_m$-basis to span a rank 1 free $\mathbb{Z}_m$-module if and only if $\lambda\mathbf{x} \neq 0$ for all $\lambda \neq 0 \in \mathbb{Z}_m$, that is, $p \nmid g.c.d(x_1, x_2, \ldots, x_r)$. So there are $p^{re} - p^{r(e-1)}$ elements in $\mathbb{Z}_m^r$ that can span a rank 1 free $\mathbb{Z}_m$-module.

On the other hand, in any rank 1 free $\mathbb{Z}_m$-module, there are $\phi(m) = p^e - p^{e-1}$ elements that can be taken as a $\mathbb{Z}_m$-basis. Thus,

$$\begin{aligned} F_{m,r}(1) &= \frac{p^{re} - p^{r(e-1)}}{p^e - p^{e-1)}} \\ &= \frac{p^{(r-1)(e-1)}(p^r - 1)}{p - 1} \\ &= p^{(r-1)(e-1)} \begin{bmatrix} r \\ 1 \end{bmatrix}_p. \end{aligned}$$

(2) Suppose this lemma holds for $k - 1$.

For any $\mathbb{Z}_m$-linearly independent set of cardinality k $\{\mathbf{x}^{(1)}, \mathbf{x}^{(2)}, \ldots, \mathbf{x}^{(k)}\}$, put $A = \langle \mathbf{x}^{(1)}, \mathbf{x}^{(2)}, \ldots, \mathbf{x}^{(k-1)} \rangle$. By Corollary 3.9, A is a free $\mathbb{Z}_m$-module of rank $k-1$, and $\{\overline{\mathbf{x}^{(k)}}\}$ is $\mathbb{Z}_m$-linearly independent in $\mathbb{Z}_m^r/A \cong \mathbb{Z}_m^{r-k+1}$.

On the other hand, to choose a $\mathbb{Z}_m$-linearly independent set $\{\mathbf{x}^{(1)}, \mathbf{x}^{(2)}, \ldots, \mathbf{x}^{(k-1)}\}$ in $\mathbb{Z}_m^r$, we have $N_{m,r}(k-1)$ choices. To make $\{\mathbf{x}^{(1)}, \mathbf{x}^{(2)}, \ldots, \mathbf{x}^{(k)}\}$ $\mathbb{Z}_m$-linearly independent, $\{\overline{\mathbf{x}^{(k)}}\}$ must be $\mathbb{Z}_m$-linearly independent in $\mathbb{Z}_m^r/A$. Thus, $\overline{\mathbf{x}^{(k)}}$ has $N_{m,r-k+1}(1)$ choices in $\mathbb{Z}_m^r/A$. A coset $\overline{\mathbf{x}^{(k)}}$ has m^{k-1} elements in $\mathbb{Z}_m^r$, so $\mathbf{x}^{(k)}$ has $m^{k-1}N_{m,r-k+1}(1)$ choices in $\mathbb{Z}_m^r$. Then

$$N_{m,r}(k) = m^{k-1}N_{m,r-k+1}(1)N_{m,r}(k-1)/k.$$

Therefore,

$$\begin{aligned} F_{m,r}(k) &= N_{m,r}(k)/N_{m,k}(k) \\ &= \frac{N_{m,r-k+1}(1)N_{m,r}(k-1)}{N_{m,1}(1)N_{m,k}(k-1)} \\ &= \frac{F_{m,r-k+1}(1)F_{m,r}(k-1)}{F_{m,k}(k-1)} \\ &= \frac{p^{k(r-k)(e-1)}\prod_{n=r-k+1}^{r}(p^n-1)}{\prod_{n=1}^{k}(p^n-1)} \\ &= p^{k(r-k)(e-1)}\begin{bmatrix} r \\ k \end{bmatrix}_p \end{aligned}$$

as we needed. □

Finally, we get a formula to calculate $F_{m,r}(k)$.

Theorem 3.11. *Let m be a positive integer. Write m in the form $m = p_1^{e_1}p_2^{e_2}\cdots p_t^{e_t}$, where p_i's are different primes and $e_i > 0$ for all i. Then*

$$F_{m,r}(k) = \prod_{i=1}^{t} p_i^{k(r-k)(e_i-1)}\begin{bmatrix} r \\ k \end{bmatrix}_{p_i}, 1 \le k \le r. \tag{3.19}$$

Proof. By Lemma 3.10 and Lemma 3.7. □

Hitherto, we have got a formula to calculate $F_{m,r}(k)$, and hence determined the parameters of $\mathcal{C}$. But unfortunately, for fixed $\theta < t-1$, we have not found a way to obtain some rank t free submodules of $\mathbb{Z}_m^r$: $B_1, B_2, \ldots, B_M$, such that M is as large as possible and $\operatorname{rank}(B_i \cap B_j) \le \theta$ for any $1 \le i \ne j \le M$. Here is a particular result for $\theta = t-1$:

Theorem 3.12. *Let m be a positive integer. Then for all $1 \le s \le t \le r$,*

$$A(F_{m,r}(s), 2F_{m,t}(s) - 2F_{m,t-1}(s), F_{m,t}(s)) \ge F_{m,r}(t). \tag{3.20}$$

4. Examples

In this section, we give some explicit examples from our constructions in Section 3. These examples improve some lower bounds [7] on binary constant-weight codes, and hence show the significance of our constructions.

Example. According to Theorem 3.1, $A(n,d,w) \ge N$ will deduce $A(\binom{n}{2}, 2\binom{w}{2} - 2\binom{\frac{2w-d}{2}}{2}, \binom{w}{2}) \ge N$. We choose some lower bounds on $A(n,d,w)$ from [7], and give some deduced lower bounds in Table 1.

Note that all of these new lower bounds in Table 1 can improve known lower bounds in [7].

TABLE 1

Lower bounds from [7]	Deduced lower bounds
$A(8,4,6) \geq 4$	$A(28,18,15) \geq 4$
$A(9,4,6) \geq 12$	$A(36,18,15) \geq 12$
$A(10,4,6) \geq 30$	$A(45,18,15) \geq 30$
$A(11,4,6) \geq 66$	$A(55,18,15) \geq 66$
$A(12,4,6) \geq 132$	$A(66,18,15) \geq 132$

TABLE 2

q	$m = t = 2, l = s = 1$
8	$64 \leq A(512,16,8) \leq 65$
11	$121 \leq A(1331,22,11) \leq 122$
13	$169 \leq A(2197,26,13) \leq 170$
17	$289 \leq A(4913,34,17) \leq 290$
19	$361 \leq A(6859,38,19) \leq 362$

TABLE 3

q	$m = t = l = 2, s = 1$
8	$4096 \leq A(512,14,8) \leq 4745$
11	$14641 \leq A(1331,20,11) \leq 16226$
13	$28561 \leq A(2197,24,13) \leq 31110$
17	$83521 \leq A(4913,32,17) \leq 89030$
19	$130321 \leq A(6859,36,19) \leq 137922$

Example. Let $1 \leq l \leq t \leq m$ be positive integers. Taking specific s that is close to l, we calculate the lower and upper bounds on some $A(n,d,w)$, and list them in Table 2 and Table 3.

The lower and upper bounds in Table 2 and Table 3 are very close, which illustrates that our construction in Section 3.2 can really produce constant-weight codes with good parameters.

Example. Let $\mathbb{Z}_m = \mathbb{Z}_6$. Assume $r = 4$, $t = 3$, and $s = 1$, we get $F_{6,4}(1) = F_{6,4}(3) = 600$, $F_{6,3}(1) = 91$ and $F_{6,2}(1) = 12$, so there exists a $(600, 600, 156, 91)$ constant-weight code by our construction from free $\mathbb{Z}_m$-module, i.e., $A(600, 156, 91) \geq 600$.

Acknowledgment

The authors would like to thank Prof. Chaoping Xing for helpful comments and corrections.

References

[1] E. Agrell, A. Vardy and K. Zeger, "Upper Bounds for Constant-Weight Codes," *IEEE Trans. Inform. Theory*, vol. 46, No. 7, Nov. 2000.

[2] J.H. Conway and N.J.A. Sloane, *Sphere Packings, Lattices and Groups*, 3rd ed. New York: Springer, 1999.

[3] E.M. Gabidulin, "Theory of Codes with Maximum Rank Distance," *Problems of Information Transmission*, 21(1), 1985.

[4] T.W. Hungerford, *Algebra*, Springer-Verlag, 1974.

[5] T. Johansson, "Authentication Codes for Nontrusting Parties Obtained from Rank Metric Codes," *Designs, Codes and Cryptography*, vol. 6, pp. 205–218, 1995.

[6] F.J. MacWilliams and N.J.A. Sloane, *The Theory of Error-Correcting Codes*, Amsterdam, The Netherlands: North-Holland, 1977.

[7] E.M. Rains, Table of Constant Weight Binary Codes[Online], Available: http://www.research.att.com/ njas/codes/Andw/index.html.

[8] S. Roman, *Coding and Information Theory*, Springer-Verlag, 1992.

[9] R. Safavi-Naini, H. Wang and C. Xing, "Linear Authentication Codes: Bounds and Constructions," Indocrypt'01, *Lecture Notes in Computer Science*, Vol. 2247, 2001, pp. 127–135.

Lei Li and Shoulun Long
Department of Mathematics
University of Science and Technology of China
Hefei, Anhui 230026
P.R. China
e-mail: `harrylee@mail.ustc.edu.cn`
e-mail: `lsl@mail.ustc.edu.cn`

Progress in Computer Science and Applied Logic, Vol. 23, 223–226

Good Self-Dual Quasi-Cyclic Codes over $\mathbf{F}_q$, q Odd

San Ling and Patrick Solé

Abstract. We show that there are long self-dual q-ary quasi-cyclic codes above the Gilbert-Varshamov bound for odd q. We use Hughes's $(\mathbf{u}+\mathbf{v}|\mathbf{u}-\mathbf{v})$ construction.

Mathematics Subject Classification (2000). Primary 94B15.

Keywords. Self-dual codes, quasi-cyclic codes, Gilbert-Varshamov bound, $(\mathbf{u}+\mathbf{v}|\mathbf{u}-\mathbf{v})$ construction.

1. Introduction

It has been known for more than a quarter of a century that long self-dual q-ary codes exist [8].

Recently Hughes introduced the $(\mathbf{u}+\mathbf{v}|\mathbf{u}-\mathbf{v})$ construction [2, 3] for codes over fields of odd characteristic. In [5] the present authors show that all q-ary quasi-cyclic codes of index half the length over a field of odd characteristic can be obtained in that way. In a companion paper [6], building on the results of [1], the authors study asymptotically good self-dual binary quasi-cyclic codes. Here, following [8], we study the existence of asymptotically good self-dual q-ary quasi-cyclic codes, where q is an odd prime power.

2. Combinatorics

Throughout this paper, we assume q to be an odd prime power and that all the codes over $\mathbf{F}_q$ are equipped with the Euclidean inner product.

The following collection of preliminary results may be found in [8, p. 37]:

Proposition 2.1. *Let ℓ be a positive even integer if $q \equiv 1 \bmod 4$ and let ℓ be a positive multiple of* 4 *if $q \equiv 3 \bmod 4$.*

The research of the first-named author is partially supported by MOE-ARF research grant R-146-000-029-112 and DSTA research grant R-394-000-011-422.

(i) *The number of self-dual q-ary codes of length ℓ is given by* $N(q,\ell) = 2\prod_{i=1}^{\frac{\ell}{2}-1}(q^i+1)$.

(ii) *Let* $\mathbf{v}$ *be a nonzero self-orthogonal vector of length ℓ and of weight less than ℓ. The number of self-dual q-ary codes of length ℓ containing* $\mathbf{v}$ *is given by* $M(q,\ell) = 2\prod_{i=1}^{\frac{\ell}{2}-2}(q^i+1)$.

Let q be a power of an odd prime. Consider two q-ary codes C_1 and C_2, each of length ℓ. To this pair of codes we attach an ℓ-quasi-cyclic code C of length 2ℓ by the recent (see [2]) $(\mathbf{u}+\mathbf{v}|\mathbf{u}-\mathbf{v})$ construction. More specifically we define C as

$$C = \{(\mathbf{u}+\mathbf{v}|\mathbf{u}-\mathbf{v}) \mid \mathbf{u} \in C_1,\ \mathbf{v} \in C_2\}. \tag{2.1}$$

Indeed, if $(\mathbf{u}+\mathbf{v}|\mathbf{u}-\mathbf{v})$ is in C, by changing $\mathbf{v}$ into $-\mathbf{v}$ we observe that its shift by ℓ places $(\mathbf{u}-\mathbf{v}|\mathbf{u}+\mathbf{v})$ is also in C. Equivalently, from (2.1), we see readily that

$$C_1 = \{\mathbf{a}+\mathbf{b} \mid (\mathbf{a}|\mathbf{b}) \in C\} \tag{2.2}$$

and

$$C_2 = \{\mathbf{a}-\mathbf{b} \mid (\mathbf{a}|\mathbf{b}) \in C\}. \tag{2.3}$$

The code C is self-dual if and only if so are C_1 and C_2.

Let C be a self-dual q-ary ℓ-quasi-cyclic code of length 2ℓ constructed in that way. (We show in [5] that they can all be constructed in that fashion.) Writing $\mathbf{c} \in C$ as $(\mathbf{a}|\mathbf{b})$, where $\mathbf{a},\mathbf{b} \in \mathbf{F}_q^\ell$, the self-duality of C implies that $\mathbf{a}\cdot\mathbf{a}+\mathbf{b}\cdot\mathbf{b} = 0 \in \mathbf{F}_q$. The ℓ-quasi-cyclicity of C also shows that $\mathbf{a}\cdot\mathbf{b} = 0 \in \mathbf{F}_q$. Suppose that $\mathbf{c}$ corresponds to the pair $(\mathbf{c}_1,\mathbf{c}_2)$, where $\mathbf{c}_1 \in C_1$ and $\mathbf{c}_2 \in C_2$. When $\mathbf{c} \neq \mathbf{0}$ is contained in a self-dual q-ary quasi-cyclic code, there are three possibilities for the pair $(\mathbf{c}_1,\mathbf{c}_2)$:

1. $\mathbf{c}_1 \neq \mathbf{0}$, $\mathbf{c}_2 \neq \mathbf{0}$ and $\mathbf{c}_1,\mathbf{c}_2$ self-orthogonal;
2. $\mathbf{c}_1 = \mathbf{0}$, $\mathbf{c}_2 \neq \mathbf{0}$ and $\mathbf{c}_2$ self-orthogonal; and
3. $\mathbf{c}_1 \neq \mathbf{0}$, $\mathbf{c}_2 = \mathbf{0}$ and $\mathbf{c}_1$ self-orthogonal.

For each of $i = 1,2,3$ and for k even, let $A_i(q,k,d)$ denote the number of nonzero words $\mathbf{c}$ in $\mathbf{F}_q^k$ of the type i that are of weight $< d$. Let $A(q,k,d)$ denote the number of nonzero self-orthogonal vectors of length k and weight $< d$. (For more details on the properties of $A(q,k,d)$, see [8].) Clearly, any word of type i ($i = 1,2,3$) is self-orthogonal. Therefore, $A(q,k,d) \geq \sum_{i=1}^{3} A_i(q,k,d)$.

It follows readily from (2.2) and (2.3) that, if $\mathbf{c}_1 = \mathbf{0}$ and $\mathbf{c}_2 \neq \mathbf{0}$, then in fact the Hamming weight of $\mathbf{c}_2$ is exactly half the Hamming weight of $\mathbf{c}$. Similarly, we can show that, for words of type 3 above, the Hamming weight of $\mathbf{c}_1$ is exactly half that of $\mathbf{c}$. This observation shows in particular that $A_2(q,2\ell,d) = A_3(q,2\ell,d) = A(q,\ell,d/2)$.

Note that if a code contains a nonzero word $\mathbf{c}$, then it also contains all the $q-1$ nonzero multiples of $\mathbf{c}$, all of which have the same weight as $\mathbf{c}$. Using this observation and Proposition 2.1, we see that the number of self-dual q-ary ℓ-quasi-cyclic codes of length 2ℓ whose minimum weight is $< d$ is bounded above by

$$\frac{1}{q-1}\big(A_1(q,2\ell,d)M(q,\ell)^2 + A_2(q,2\ell,d)N(q,\ell)M(q,\ell) \\ + A_3(q,2\ell,d)N(q,\ell)M(q,\ell)\big).$$

On the other hand, the number of distinct self-dual ℓ-quasi-cyclic codes of length 2ℓ over $\mathbf{F}_q$ is given by the following result.

Proposition 2.2 ([5], **Proposition 6.6**). *Suppose $q \equiv 1 \bmod 4$ and ℓ is even, or $q \equiv 3 \bmod 4$ and $\ell \equiv 0 \bmod 4$. The number of distinct self-dual ℓ-quasi-cyclic codes of length 2ℓ over $\mathbf{F}_q$ is $4\prod_{i=1}^{\frac{\ell}{2}-1}(q^i+1)^2$.*

Combining the above two facts, we obtain the following theorem.

Theorem 2.3. *Let ℓ be an even integer if $q \equiv 1 \bmod 4$ and let ℓ be a multiple of 4 if $q \equiv 3 \bmod 4$. Let d be the largest integer such that*

$$A(q,2\ell,d) + q^{\frac{\ell}{2}}\left(A_2(q,2\ell,d) + A_3(q,2\ell,d)\right) \le (q-1)(q^{\frac{\ell}{2}-1}+1)^2. \tag{2.4}$$

Then there exists a self-dual q-ary ℓ-quasi-cyclic code of length 2ℓ with minimum weight at least d.

Remark. In view of the Gleason-Pierce Theorem, the d in Theorem 2.3 can be taken to be the largest even integer that satisfies (2.4) when $q \neq 3$ and the largest multiple of 3 that satisfies (2.4) when $q = 3$.

3. Asymptotic Analysis

With $y < \frac{q-1}{q}$ satisfying the Gilbert-Varshamov bound

$$2y\log(q-1) - 2y\log y - 2(1-y)\log(1-y) = \log q,$$

it can be verified using the usual asymptotic analysis (cf. [8, p. 39]) that

$$A(q,2\ell,2\ell y)/\left((q-1)(q^{\ell-1}+1)\right) \longrightarrow 0 \qquad \text{as} \qquad \ell \to \infty$$

and

$$A(q,\ell,\ell y)/\left((q-1)(q^{\frac{\ell}{2}-1}+1)\right) \longrightarrow 0 \qquad \text{as} \qquad \ell \to \infty.$$

It then follows that, with y satisfying the Gilbert-Varshamov bound as above,

$$\left(A(q,2\ell,2\ell y)+q^{\frac{\ell}{2}}\left(A_2(q,2\ell,2\ell y)+A_3(q,2\ell,2\ell y)\right)\right)/\left((q-1)(q^{\frac{\ell}{2}-1}+1)^2\right) \longrightarrow 0 \quad \text{as } \ell \to \infty.$$

This shows that almost all long self-dual ℓ-quasi-cyclic codes of length 2ℓ enjoy a normalized minimum distance $d/2\ell$ as large as promised by the Gilbert-Varshamov estimate.

Acknowledgments
The authors thank an anonymous referee for his/her helpful comments which improved the exposition of the paper.

References

[1] J.H. Conway, V. Pless & N.J.A. Sloane, *Self-dual codes over* $GF(3)$ *and* $GF(4)$ *of length not exceeding* 16, IEEE Trans. Inform. Theory **25** (1979), 312–322

[2] G. Hughes, *Codes and arrays from cocycles*, Ph.D. thesis, Royal Melbourne Institute of Technology, 2000

[3] G. Hughes, *Constacyclic codes, cocycles and a* $u+v|u-v$ *construction*, IEEE Trans. Inform. Theory **46** (2000), 674–680

[4] T. Kasami, *A Gilbert-Varshamov bound for quasi-cyclic codes of rate* 1/2, IEEE Trans. Inform. Theory **20** (1974), 679

[5] S. Ling & P. Solé, *On the algebraic structure of quasi-cyclic codes I: finite fields*, IEEE Trans. Inform. Theory **47** (2001), 2751–2760

[6] S. Ling & P. Solé, *Good self-dual quasi-cyclic codes exist*, IEEE Trans. Inform. Theory **49** (2003), 1052–1053

[7] F.J. MacWilliams, N.J.A. Sloane & J.G. Thompson, *Good self-dual codes exist*, Discrete Math **3** (1972), 153–162

[8] V. Pless & J.N. Pierce, *Self-dual codes over* $GF(q)$ *satisfy a modified Varshamov-Gilbert bound*, Information and Control **23** (1973), 35–40

[9] E.M. Rains & N.J.A. Sloane, *Self-dual codes*, in *Handbook of Coding Theory*, Vol. II, V.S. Pless & W.C. Huffman eds., Elsevier Science, Amsterdam, 1998, pp. 177–294

San Ling
Department of Mathematics
National University of Singapore
2 Science Drive 2
Singapore 117543, Republic of Singapore
e-mail: `matlings@nus.edu.sg`

Patrick Solé
CNRS, I3S,
ESSI, BP 145
Route des Colles
F-06 903 Sophia Antipolis, France
e-mail: `ps@essi.fr`

Progress in Computer Science and Applied Logic, Vol. 23, 227–235

Linear Complexity and k-Error Linear Complexity for p^n-Periodic Sequences

Wilfried Meidl

Abstract. The k-error linear complexity of an N-periodic sequence with terms in the finite field $\mathbb{F}_q$ is defined to be the smallest linear complexity that can be obtained by changing k or fewer terms of the sequence per period. For the case that $N = p^n$, p is an odd prime, and q is a primitive root modulo p^2, we show a relationship between the linear complexity and the minimum value of k for which the k-error linear complexity is strictly less than the linear complexity.

Mathematics Subject Classification (2000). Primary 94A55; Secondary 68W40.

Keywords. Linear complexity, k-error linear complexity, periodic sequences, stream ciphers.

1. Introduction

Let $S = s_0, s_1, s_2, \dots$ be a sequence with terms in the finite field $\mathbb{F}_q$. Then S is said to be *N-periodic* if $s_i = s_{i+N}$ for all $i \geq 0$. Since an N-periodic sequence is determined by the terms of one period, we can completely describe S by the notation $S = (s_0, s_1, \dots, s_{N-1})^\infty$. The *linear complexity* $L(S)$ of an N-periodic sequence $S = (s_0, s_1, \dots, s_{N-1})^\infty$ with terms in $\mathbb{F}_q$ is the smallest nonnegative integer L for which there exist coefficients $d_1, d_2, \dots, d_L$ in $\mathbb{F}_q$ such that

$$s_i + d_1 s_{i-1} + \dots + d_L s_{i-L} = 0 \qquad \text{for all } i \geq L.$$

For an N-periodic sequence $S = (s_0, s_1, \dots, s_{N-1})^\infty$ we define $S_N(x)$ to be the polynomial

$$S_N(x) = s_0 + s_1 x + s_2 x^2 + \dots + s_{N-1} x^{N-1} \in \mathbb{F}_q[x].$$

Then we have (cf. [7], [15, p.31])

$$L(S) = N - \deg(\gcd(x^N - 1, S_N(x))). \tag{1}$$

The linear complexity is of fundamental importance as a complexity measure for periodic sequences (see [9], [11], [12]). Motivated by security issues of stream ciphers, in [14] Stamp and Martin proposed a new measure of the complexity of periodic sequences, the *k-error linear complexity*, which is defined by

$$L_k(S) = \min_T L(T),$$

where the minimum is taken over all N-periodic sequences $T = (t_0, t_1, \ldots, t_{N-1})^\infty$ over $\mathbb{F}_q$ for which the Hamming distance of the vectors $(s_0, s_1, \ldots, s_{N-1})$ and $(t_0, t_1, \ldots, t_{N-1})$ is at most k.

In this paper we cover the case of N-periodic sequences over a finite prime field $\mathbb{F}_q$, where $N = p^n$ for an odd prime p with the property that q is a primitive root modulo p^2. In all further considerations q will be a prime, and we will denote the set of odd primes p with the property that q is a primitive root modulo p^2 by P_q. We remark that a conjecture of Artin suggests that approximately 3/8 of all primes have q as a primitive root (cf. [13, p.81]), and that it is very seldom that a primitive root modulo the prime p is not also a primitive root modulo p^2.
Our results generalize results of [6] where for the binary case a fast algorithm that computes the k-error linear complexity of a p^n-periodic sequence, $p \in P_2$, has been proposed, and a relationship between the linear complexity and the minimum value of k for which the k-error linear complexity is strictly less than the linear complexity has been established. Very recently an algorithm that calculates the k-error linear complexity of sequences over $\mathbb{F}_q$ with period length p^n, q prime, $p \in P_q$, has been introduced in [16] and summarized in [18]. Thus we only briefly describe and commentate that algorithm in Section 2. The algorithm in [6] can be seen as a special case of that algorithm. In Section 3 we establish lower and upper bounds on the minimal value $m(S)$ of terms that have to be changed within one period to decrease the linear complexity of a p^n-periodic sequence over $\mathbb{F}_q$ with given linear complexity L. We will utilize the algorithm described in Section 2 to show that both bounds are tight.

2. Preliminaries

In this section we firstly summarize some basics on the linear complexity of p^n-periodic sequences over F_q, $p \in P_q$. Secondly, we shortly present the algorithm of [16].

2.1. The Linear Complexity of p^n-Periodic Sequences

Because of the connection between the linear complexity of an N-periodic sequence S and the greatest common divisor of the corresponding polynomial $S_N(x)$ and $x^N - 1$, we will need the canonical factorization of $x^{p^n} - 1$ in $\mathbb{F}_q[x]$. Let Φ denote the p^nth cyclotomic polynomial (see [2], [4]), which is explicitly given by

$$\Phi_{p^n}(x) = 1 + x^{p^{n-1}} + \ldots + x^{(p-1)p^{n-1}}. \tag{2}$$

If $p \in P_q$ then by a lemma in [17] the p^nth cyclotomic polynomial is irreducible in $\mathbb{F}_q[x]$ for all $n \geq 1$, and the canonical factorization of $x^{p^n} - 1$ in $\mathbb{F}_q[x]$ is given by (see also [6], [10])

$$x^{p^n} - 1 = (x-1)\prod_{s=1}^{n} \Phi_{p^s}.$$

Thus the term $\gcd(x^N - 1, S_N(x))$ is of the form

$$\gcd(x^N - 1, S_N(x)) = (x-1)^\epsilon \prod_{t\in T} \Phi_{p^t}, \quad T \subseteq \{1,\dots,n\},\ \epsilon \in \{0,1\},$$

and by equation (1) and (2) the linear complexity $L(S)$ of a p^n-periodic sequence S over $\mathbb{F}_q$ can be written in the form

$$L = \epsilon + (p-1)\sum_{r\in R} p^{r-1}, \quad R \subseteq \{1,\dots,n\},\ \epsilon \in \{0,1\}, \tag{3}$$

2.2. An Algorithm for the k-Error Linear Complexity

The algorithm in [16] is based on the algorithm in [17] which determines the linear complexity of a sequence over $\mathbb{F}_q$ with period length p^n, $p \in P_q$. Given a p^n-periodic sequence $S = (s_0, s_1, \dots, s_{p^n-1})^\infty$ over $\mathbb{F}_q$, we define

$$A_j = (s_{jp^{n-1}}, \dots, s_{(j+1)p^{n-1}-1}), \qquad j = 0,\dots,p-1.$$

If $A_0 = A_1 = \cdots = A_{p-1}$ then $L(S)$ equals the linear complexity of the p^{n-1}-periodic sequence with first period A_0. Else we have $L(S) = (p-1)p^{n-1} + L(B)$, where B is the p^{n-1}-periodic sequence with first period $b = A_0 + A_1 + \cdots + A_{p-1}$, where the addition is elementwise modulo q (see [17]). The algorithm of [17] is obtained by applying these results recursively. The basic idea of the algorithm in [16] is to try to force $A_0 = A_1 = \cdots = A_{p-1}$ in each recursive step by as few term changes as possible. In [18] the algorithm has been described as follows.

Initial values: $a \leftarrow S, N \leftarrow p^n, L \leftarrow 0, \text{cost}[i, a_i] \leftarrow 0, \text{cost}[i,h] \leftarrow 1, 0 \le h \le q-1$ and $h \ne a_i, i = 0,1,\dots,N-1, K \leftarrow k$.

(I) If $N = 1$, then go to (II); else $N \leftarrow N/p, A_j = (s_{jp^{n-1}}, \dots, s_{(j+1)p^{n-1}-1})$, $0 \le j \le p-1, T_{ih} = \sum_{j=0}^{p-1} \text{cost}[i+jN, h], 0 \le i \le N-1, 0 \le h \le q-1, T_i = \min_{0\le h\le q-1}\{T_{ih}\}, T = \sum_{i=0}^{N-1} T_i$, go to (IV).

(II) If $a = 0$, then stop; else go to (III).

(III) If $\text{cost}[0,0] \le K$, then stop; else $L \leftarrow L+1$, stop.

(IV) If $T \le K$, then $K \leftarrow K - T, \text{cost}[i,h] \leftarrow T_{ih} - T_i, 0 \le i \le N-1$, go to (V); else, $a \leftarrow A_0 + A_1 + \cdots + A_{p-1}, L \leftarrow L + (p-1)N$, $\text{cost}[i,h] \leftarrow \min_{e_0+\cdots+e_{p-1}=h-a_i} \{\sum_{j=0}^{p-1} \text{cost}[i+jN, a_{i+jN} + e_j]\}, 0 \le i \le N-1$, go to (I).

(V) For $i = 0,1,\dots,N-1$, if $T_{ih} = T_i$ then $a_i = h$. $a \leftarrow A_0$, go to (I).

In (I) the number T of necessary term changes in the original sequence in order to force $A_0 = A_1 = \cdots = A_{p-1}$ without affecting any previous result is calculated. (IV) calculates $\text{cost}[i,h]$ for the next step, i.e., the number of necessary term changes in the original sequence in order to change the ith element of the sequence to be considered in the next step into h, again without affecting any previous result for both cases, for the case that we were able to force $A_0 = A_1 = \cdots = A_{p-1}$ (i.e., $T \leq K$), and for the case that we had $T > K$. For the latter case (IV) also provides the sequence to be used in the next step. (V) provides this sequence for the first case, and (II) and (III) perform the final step. (See [16].)

3. The Minimum k with $L_k(S) < L(S)$

In [3] for binary sequences S with period 2^n an explicit formula for the minimum value $m(S)$ of k for which the k-error linear complexity is strictly less than the linear complexity $L(S)$ of S was derived. In this section we establish tight lower and upper bounds on $m(S)$ of a given p^n-periodic sequence S with terms in $\mathbb{F}_q$, $p \in P_q$. The results generalize the results for the binary case in [6].
Since $m(S)$ is reasonable if and only if $L(S) > 0$ (i.e., S is not the zero sequence), in the further considerations S will always be a sequence with $L(S) > 0$. Given an N-periodic sequence over $\mathbb{F}_q$, then by equation (1), $m(S)$ is the minimal weight $\omega(e(x))$ of a polynomial $e(x)$ with $\deg(e(x)) < N$, i.e., $\omega(e(x))$ is the number of the nonzero coefficients of $e(x)$, such that

$$\deg(\gcd(x^N - 1, S_N(x) + e(x))) > \deg(\gcd(x^N - 1, S_N(x))). \tag{4}$$

For instance let $L(S) = N$, i.e., $\gcd(x^N - 1, S_N(x)) = 1$. Then with $e(x) = -S_N(1)$ we get $\gcd(x^N - 1, S_N(x) + e(x)) = (x - 1)g(x)$ for a polynomial $g(x) \in \mathbb{F}_q[x]$. Thus $L_1(S) \leq N - 1$ and $m(S) = 1$. To show further results on the relationship between $m(S)$ and $L(S)$ for the case that $N = p^n$, $p \in P_q$, we will need some definitions. Let L be a nonnegative integer of the form (3) for a subset $R \subseteq \{1, \ldots, n\}$. Then with $W_H(L)$ we denote the cardinality of the subset R, i.e., the Hamming weight of the vector $(l_1, \ldots, l_n) \in \mathbb{F}_2^n$ with $l_r = 1$ if and only if $r \in R$. We will utilize the following lemma (see [6]).

Lemma 3.1. *Let $L = (p-1)\sum_{r\in R} p^{r-1}$, $R \subseteq \{1, \ldots, n\}$, and let $e(x) = \prod_{r\in R} \Phi_{p^r}$. Then we have*

$$\omega(e(x)) = p^{W_H(L)}.$$

Suppose that μ is the smallest element of R and let $g(x)$ be a polynomial of degree smaller than $p^{\mu-1}$. Then

$$\omega(e(x)g(x)) = \omega(e(x))\omega(g(x)).$$

3.1. Lower Bound on $m(S)$

In the following theorem we establish a lower bound κ on $m(S)$. Moreover we construct a sequence S with $m(S) = \kappa$ and additionally with κ-error linear complexity $L_\kappa(S)$ being as small as possible.

Theorem 3.2. *Suppose that S is a periodic sequence over $\mathbb{F}_q$ with period length p^n, $n \geq 1$, $p \in P_q$, and let r, $0 \leq r \leq n-1$, be the unique integer such that $(p-1)p^{n-1-r} \leq L(S) \leq p^{n-r}$. Then we have*

$$m(S) \geq p^r \quad \text{and} \quad L_{p^r}(S) \geq p^{n-r} - L(S).$$

Moreover for any given linear complexity L, $(p-1)p^{n-1-r} \leq L \leq p^{n-r}$ there exists a p^n-periodic sequence $\bar{S}$ over $\mathbb{F}_q$ with $L(\bar{S}) = L$ and $L_{p^r}(\bar{S}) = p^{n-r} - L$.

Proof. First we notice, that r is the least integer such that $\Phi_{p^{n-r}}$ does not divide the polynomial $S_{p^n}(x)$ corresponding to the sequence S, i.e.,

$$S_{p^n}(x) = \Phi_{p^n}\Phi_{p^{n-1}} \ldots \Phi_{p^{n-r+1}} \prod_{t\in T} \Phi_{p^t}(x-1)^{\epsilon} s(x),$$

$T \subseteq \{1, \ldots, n-r-1\}, \epsilon \in \{0,1\}$ and $\gcd(s(x), \Phi_{p^i}) = 1$ for $1 \leq i \leq n-r$ and $i \notin T$.

Suppose that $e(x) \in \mathbb{F}_q[x]$ is a polynomial of degree at most $p^n - 1$ such that $\deg(\gcd(x^{p^n} - 1, S_{p^n}(x) + e(x))) > \deg(\gcd(x^{p^n} - 1, S_{p^n}(x)))$, then

$$e(x) = \Phi_{p^n}\Phi_{p^{n-1}} \ldots \Phi_{p^{n-r+1}} e_1(x)$$

for a polynomial $e_1(x) \in \mathbb{F}_q[x]$ of degree at most $p^{n-r} - 1$. Thus Lemma 3.1 yields

$$\omega(e(x)) = p^r \omega(e_1(x)) \geq p^r.$$

Now suppose $\omega(e(x)) = p^r$, i.e., $e_1(x) = ax^j$, $a \neq 0$, for an integer $0 \leq j < p^{n-r}$. Then

$$S_{p^n}(x) + e(x) = \Phi_{p^n}\Phi_{p^{n-1}} \ldots \Phi_{p^{n-r+1}}(\prod_{t\in T} \Phi_{p^t}(x-1)^{\epsilon} s(x) + ax^j).$$

Since the polynomials Φ_{p^t}, $t \in T$, and if $\epsilon = 1$ the polynomial $x - 1$, do not divide $\prod_{t\in T} \Phi_{p^t}(x-1)^{\epsilon} s(x) + ax^j$, we have

$$N - \deg(\gcd(x^{p^n} - 1, S_{p^n}(x) + e(x))) \geq \epsilon + (p-1)\sum_{t\in T} p^{t-1} = p^{n-r} - L(S).$$

Finally let L be any given linear complexity with $(p-1)p^{n-1-r} \leq L \leq p^{n-r}$. Then L is of the form

$$L = p^n - [\epsilon + (p-1)(p^{n-1} + p^{n-2} + \ldots + p^{n-r} + \sum_{t\in T} p^{t-1})]$$

with $T \subseteq \{1, \ldots, n-r-1\}, \epsilon \in \{0,1\}$. Let $\Omega(x) = \Phi_{p^n}\Phi_{p^{n-1}} \ldots \Phi_{p^{n-r+1}}$, $H(x) = \prod_{t\in T} \Phi_{p^t}(x-1)^{\epsilon}$ and $G(x) = (x^{p^{n-r}} - 1)/H(x)$. Then $\gcd(H(x), G(x)) = 1$, and with the Euclidean algorithm we can find unique polynomials $s(x), t(x) \in \mathbb{F}_q[x]$ such that $s(x)H(x) + t(x)G(x) = 1$ and $\deg(s(x)) < \deg(G(x))$, $\deg(t(x)) < \deg(H(x))$ (see [1, p.71]). The polynomial $\bar{S}_{p^n}(x) := \Omega(x)s(x)H(x)$ has degree at most $p^n - 1$. By (1), the corresponding sequence $\bar{S}$ satisfies $L(\bar{S}) = L$. Moreover with $e(x) = -\Omega(x)$ – note that $\omega(-\Omega(x)) = p^r$ – we get $\bar{S}_{p^n}(x) + e(x) = -t(x)\Omega(x)G(x)$. Again with (1) we see that $L_{p^r}(\bar{S}) = p^{n-r} - L$. □

Note that a p^n-periodic sequence S over $\mathbb{F}_q$ has least period length p^n if and only if the linear complexity $L(S)$ of S satisfies $L(S) \geq (p-1)p^{n-1}$. In this case Theorem 3.2 yields the following corollary.

Corollary 3.3. *Let L be an integer of the form* (3) *for a prime $p \in P_q$, and suppose that $L \geq (p-1)p^{n-1}$, i.e., $n \in R$. Then there exists a sequence S over $\mathbb{F}_q$ with (least) period length p^n such that $L(S) = L$ and $L_1(S) = p^n - L$.*

Remark 3.4. If S has maximal possible linear complexity p^n then, as Theorem 3.2 shows, a few term changes can cause a significant decrease of the linear complexity. But the existence of sufficiently many p^n-periodic sequences over $\mathbb{F}_q$ with linear complexity p^n and large k-error linear complexity if k does not exceed a certain percentage of the period length, is guaranteed:
Let $\mathcal{Q}_{p^n}(\gamma)$ be the proportion of the p^n-periodic sequences with $L_{\lfloor \gamma p^n \rfloor}(S) \geq \phi(p^n) = (p-1)p^{n-1}$ for a given γ, $0 < \gamma < 1$, and let $\mathcal{H}_q(\gamma)$ denote the entropy function defined by (cf. [5, p.55])

$$\mathcal{H}_q(\gamma) = \gamma \log(q-1) - \gamma \log \gamma - (1-\gamma)\log(1-\gamma), \quad 0 < \gamma < 1,$$

where $\log z$ denotes the logarithm to the base q. Then in [8] it was shown that $lim_{n\to\infty} \mathcal{Q}_{p^n}(\gamma) = 1$ as long as γ satisfies $p - 1 > p\mathcal{H}_q(\gamma)$.

3.2. Upper Bound on $m(S)$

In the following theorem $W_H(L)$ shall be defined as in Lemma 3.1 for integers L of the form (3). Notice that if $L(S)$ is the linear complexity of an arbitrary sequence S over $\mathbb{F}_q$ with period length $N = p^n$, $p \in P_q$, then $L(S)$ as well as $N - L(S)$ is of the form (3). Thus $W_H(N - L(S))$ is well defined.

Theorem 3.5. *Suppose that S is a sequence over $\mathbb{F}_q$ with period length $N = p^n$, $n \geq 1$, $p \in P_q$, and let $L(S) = \epsilon + (p-1)\sum_{r\in R} p^{r-1}$, $R \subseteq \{1,\ldots,n\}, \epsilon \in \{0,1\}$, be the linear complexity of S. Then with $\delta = (\epsilon + 1) \bmod 2$ and $\Upsilon = \lceil \frac{(p-1)(q-1)}{q} \rceil^{\delta}$, $m(S)$ satisfies*

$$m(S) \leq \Upsilon p^{W_H(N-L(S))}.$$

Proof. First we suppose that $\epsilon = 1$ and define $T = \{1,\ldots,n\} \setminus R$. Then the polynomial $S_{p^n}(x)$ corresponding to the sequence S is of the form

$$S_{p^n}(x) = \prod_{t\in T} \Phi_{p^t} s(x),$$

where $s(x) \in \mathbb{F}_q[x]$ with $\gcd(s(x), x-1) = \gcd(s(x), \Phi_{p^r}) = 1$ for all $r \in R$. Let

$$e(x) = -s(1) \prod_{t\in T} \Phi_{p^t},$$

and consequently

$$S_{p^n}(x) + e(x) = \prod_{t\in T} \Phi_{p^t}(s(x) - s(1)).$$

Then $x - 1$ divides $s(x) - s(1)$ and we have

$$\deg(\gcd(x^{p^n} - 1, S_{p^n}(x) + e(x))) \geq 1 + (p-1)\sum_{t\in T} p^{t-1} > \deg(\gcd(x^{p^n} - 1, S_{p^n}(x))).$$

Hence by (4) and Lemma 3.1 we have

$$m(S) \le \omega(e(x)) = p^{W_H((p-1)\sum_{t\in T} p^{t-1})} = p^{W_H(N-L(S))}.$$

Now let $\epsilon = 0$ and suppose that $1 \le \mu \le n$ is the least integer such that Φ_{p^μ} does not divide $S_{p^n}(x)$. Then $S_{p^n}(x)$ is of the form

$$S_{p^n}(x) = (x-1)\prod_{i=1}^{\mu-1}\Phi_{p^i}\prod_{u\in U}\Phi_{p^u}s(x), \quad U = \{\mu+1,\ldots,n\}\setminus R,$$

where $s(x) \in \mathbb{F}_q[x]$ with $\gcd(s(x),\Phi_{p^r}) = 1$ for all $r \in R$. Let $k(x) = s(x)(x-1)\prod_{i=1}^{\mu-1}\Phi_{p^i}$ and let $g_1(x) = \sum_{j=0}^{(p-1)p^{\mu-1}-1} a_j x^j$, where $-g_1(x)$ is the remainder of $k(x)$ after division by Φ_{p^μ}. Then Φ_{p^μ} divides $k(x)+g_1(x)+h(x)\Phi_{p^\mu}$ for all polynomials $h(x) \in \mathbb{F}_q[x]$. Let us write $g_1(x)$ in the form

$$\begin{aligned} g_1(x) = {} & (a_0 + a_{p^{\mu-1}}x^{p^{\mu-1}} + \cdots + a_{(p-2)p^{\mu-1}}x^{(p-2)p^{\mu-1}}) \\ & + (a_1 + a_{p^{\mu-1}+1}x^{p^{\mu-1}} + \cdots + a_{(p-2)p^{\mu-1}+1}x^{(p-2)p^{\mu-1}})x + \cdots \\ & \cdots + (a_{p^{\mu-1}-1} + a_{2p^{\mu-1}-1}x^{p^{\mu-1}} + \ldots + a_{(p-1)p^{\mu-1}-1}x^{(p-2)p^{\mu-1}})x^{p^{\mu-1}-1} \end{aligned}$$

and consider the term

$$A_i = (a_i + a_{p^{\mu-1}+i}x^{p^{\mu-1}} + \cdots + a_{(p-2)p^{\mu-1}+i}x^{(p-2)p^{\mu-1}})x^i, \quad 0 \le i \le p^{\mu-1}-1.$$

Suppose that at least $\lfloor (p-1)/q \rfloor$ coefficients in A_i are zero. Then the weight $\omega(A_i)$ of A_i satisfies

$$\omega(A_i) \le p - 1 - \left\lfloor \frac{p-1}{q} \right\rfloor = \left\lceil \frac{(p-1)(q-1)}{q} \right\rceil.$$

In this case we define $\alpha_i = 0$.
Else, the most frequent coefficient $\alpha_i \in \mathbb{F}_q \setminus \{0\}$ in A_i occurs at least $\lfloor (p-1)/q + 1 \rfloor$ times, and the weight of

$$A_i - \alpha_i x^i \Phi_{p^\mu} = [a_i - \alpha_i + \cdots + (a_{(p-2)p^{\mu-1}+i} - \alpha_i)x^{(p-2)p^{\mu-1}} - \alpha_i x^{(p-1)p^{\mu-1}}]x^i$$

satisfies

$$\omega(A_i - \alpha_i x^i \Phi_{p^\mu}) \le p - \left\lfloor \frac{p-1}{q} + 1 \right\rfloor = \left\lceil \frac{(p-1)(q-1)}{q} \right\rceil.$$

Thus we can find a polynomial

$$g(x) = g_1(x) - \sum_{i=0}^{p^{\mu-1}-1} \alpha_i x^i \Phi_{p^\mu},$$

such that

$$\omega(g(x)) \le \left\lceil \frac{(p-1)(q-1)}{q} \right\rceil p^{\mu-1} \quad \text{and} \quad \deg(g(x)) \le p^\mu - 1.$$

If we put

$$e(x) = \prod_{u\in U}\Phi_{p^u}g(x)$$

then

$$S_{p^n}(x) + e(x) = \prod_{u \in U} \Phi_{p^u}(k(x) + g(x))$$

and

$$\begin{aligned} \deg(\gcd(x^{p^n} - 1, S_{p^n}(x) + e(x))) &\geq (p-1)\left(p^{\mu-1} + \sum_{u \in U} p^{u-1}\right) \\ &> \deg(\gcd(x^{p^n} - 1, S_{p^n}(x))). \end{aligned}$$

Thus with (4) and Lemma 3.1 we get

$$\begin{aligned} m(S) \leq \omega(e(x)) &\leq \left\lceil \frac{(p-1)(q-1)}{q} \right\rceil p^{\mu-1} p^{W_H((p-1)\sum_{u \in U} p^{u-1})} \\ &= \left\lceil \frac{(p-1)(q-1)}{q} \right\rceil p^{W_H(N-L(S))}. \end{aligned}$$

□

Remark 3.6. By the discussion at the beginning of Section 3 we know that $L(S) = N$ yields $m(S) = 1$. Indeed for this case by Theorem 3.5 we have $m(S) \leq 1$. Thus if $L(S) = N$ the upper bound established in Theorem 3.5 is always attained.

We close this section with some examples of sequences over $\mathbb{F}_3$ (S_1, S_2, S_4) and $\mathbb{F}_5$ (S_3) with linear complexity less than N.

$$\begin{aligned} S_1 = (&2202000112010121201002120012202112021202002122210021020102100002 \\ &2111122010200202011202101021200121102012210210012210121020110 2)^\infty \end{aligned}$$

$$\begin{aligned} S_2 = (&21022001211000102000222211012022211002102102210120012012102201 1 \\ &0022011012020122010220102001102010221100212001022010202010012 0)^\infty \end{aligned}$$

$$\begin{aligned} S_3 = (&4230410032222033323210322303210440014103021034210230 21 \\ &01234143012432100401432123 3)^\infty \end{aligned}$$

$$S_4 = (1110221102012001120210202011202112121100021021020)^\infty$$

Using the algorithm [16] described in Section 2 we can determine the k-error linear complexity of S_i, $i = 1, 2, 3, 4$, for any given k – and thus the value of $m(S_i)$. The results tabulated in the following schedule show that at the sequences S_1, S_2, S_3, S_4 the upper bound established in Theorem 3.5 is attained.

i	N	$L(S_i)$	$W_H(N - L(S_i))$	Υ	$p^{W_H(N-L(S_i))}$	$m(S_i)$	$L_{m(S_i)}(S_i)$
1	125	101	2	1	5^2	25	100
2	125	104	1	3	5	15	101
3	81	62	1	2	3	6	60
4	49	48	0	4	1	4	42

References

[1] J. von zur Gathen and J. Gerhard, *Modern Computer Algebra.* Cambridge University Press, Cambridge, 1999.

[2] D. Jungnickel, *Finite Fields: Structure and Arithmetics.* Bibliographisches Institut, Mannheim, 1993.

[3] K. Kurosawa, F. Sato, T. Sakata and W. Kishimoto, *A relationship between linear complexity and k-error linear complexity.* IEEE Trans. Inform. Theory **46** (2000), 694–698.

[4] R. Lidl and H. Niederreiter, *Finite Fields.* Reading, MA: Addison-Wesley, 1983.

[5] J.H. van Lint, *Introduction to coding theory.* 3rd Edition, Graduate texts in Mathematics, 86. Springer-Verlag, Berlin, 1999.

[6] W. Meidl, *How many bits have to be changed to decrease the linear complexity?* Des. Codes Cryptogr., to appear.

[7] W. Meidl and H. Niederreiter, *Linear complexity, k-error linear complexity, and the discrete Fourier transform.* J. Complexity **18** (2002), 87–103.

[8] W. Meidl and H. Niederreiter, *Periodic sequences with maximal linear complexity and large k-error linear complexity.* AAECC **14** (2003), 273–286.

[9] H. Niederreiter, *Some computable complexity measures for binary sequences.* Sequences and Their Applications (C. Ding, T. Helleseth and H. Niederreiter, Eds.), 67–78, Springer, London, 1999.

[10] K.H Rosen, *Elementary Number Theory and its Applications.* Reading, MA: Addison-Wesley, 1988.

[11] R.A. Rueppel, *Analysis and Design of Stream Ciphers*, Springer, Berlin, 1986.

[12] R.A. Rueppel, *Stream ciphers.* Contemporary Cryptology: The Science of Information Integrity (G.J. Simmons, Ed.), 65–134, IEEE Press, New York, 1992.

[13] D. Shanks, *Solved and Unsolved Problems in Number Theory.* 2nd Edition, Chelsea Publishing Company, New York, 1978.

[14] M. Stamp and C. F. Martin, *An algorithm for the k-error linear complexity of binary sequences with period* 2^n. IEEE Trans. Inform. Theory **39** (1993), 1398–1401.

[15] H.C.A. van Tilborg, *An Introduction to Cryptology.* Kluwer Acad. Publ., Boston, 1988.

[16] S. Wei, G. Xiao and Z. Chen, *An Efficient Algorithm for k-Error Linear Complexity.* Chinese Journal of Electronics **11(2)** (2002), 265–267.

[17] G. Xiao, S. Wei, K.Y. Lam and K. Imamura, *A fast algorithm for determining the linear complexity of a sequence with period* p^n *over* $GF(q)$. IEEE Trans. Inform. Theory **46** (2000), 2203–2206.

[18] G. Xiao, S. Wei, *Fast algorithms for determining the linear complexity of period sequences.* Progress in Cryptology – INDOCRYPT 2002 (A. Menezes and P. Sarkar, Eds.), Lecture Notes in Computer Science **2551**, 12–21, Springer-Verlag, 2002.

Wilfried Meidl
Temasek Laboratories, National University of Singapore
Engineering Drive 3, 10 Kent Ridge Crescent, 119260 Singapore
e-mail: `tslmw@nus.edu.sg`

Progress in Computer Science and Applied Logic, Vol. 23, 237–246

HFE and BDDs: A Practical Attempt at Cryptanalysis

Jean-Francis Michon, Pierre Valarcher and Jean-Baptiste Yunès

Abstract. HFE (Hidden Field Equations) is a public key cryptosystem using univariate polynomials over finite fields. It was proposed by J. Patarin in 1996. Well chosen parameters during the construction produce a system of quadratic multivariate polynomials over $\mathbb{F}_2$ as the public key. An enclosed trapdoor is used to decrypt messages. We propose a ciphertext-only attack which mainly consists in satisfying a boolean formula. Our algorithm is based on BDDs (Binary Decision Diagrams), introduced by Bryant in 1986, which allow to represent and manipulate, possibly efficiently, boolean functions. This paper is devoted to some experimental results we obtained while trying to solve the Patarin's challenge. This approach was not successful, nevertheless it provided some interesting information about the security of HFE cryptosystem.

1. Introduction

1.1. BDDs

Binary Decision Diagrams (BDDs), introduced in [1], are now commonly used in CAD design or verification as tools to represent and manipulate efficiently boolean functions. They have been recently used in cryptography to cryptanalyze some keystream generators (see [5]).

The construction of the BDD of a boolean function f over n-variables labelled $x_1, \ldots, x_n$, noted $BDD(f)$, is obtained using the following Shannon's decomposition: $f(x_1, \ldots, x_n) = x_1.f(1, x_2, \ldots, x_n) + \neg x_1.f(0, x_2, \ldots, x_n)$.

Iterating this decomposition leads to a tree decision diagram in which inner nodes are labelled by variables, leaf nodes by boolean constants and directed edges are rooted from variables and labelled by a boolean assignment (see Figure 1(a)). Each boolean assignment of the variables defines a unique path from the root to a boolean constant in the tree, and that constant gives the valuation of the function for the given assignment.

This work was supported by the "ACI Cryptologie" program of the French "Ministère de la Recherche".

By choosing a graph isomorphism and by iterating it to the subgraphs of the tree, one obtains a unique acyclic directed graph, called OBDD (Ordered BDD or Oblivious Free BDD in [10]). Figure 1(b) is an example of such a graph obtained by: decomposing the function (x first, then y and then z), merging identical subgraphs and finally deleting useless nodes.

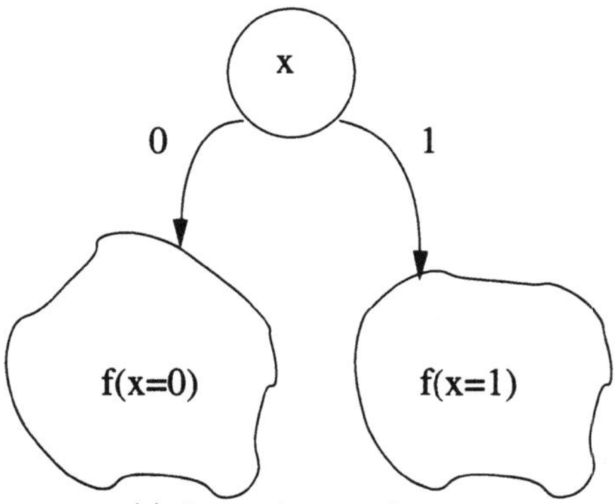

(a) A tree decision diagram

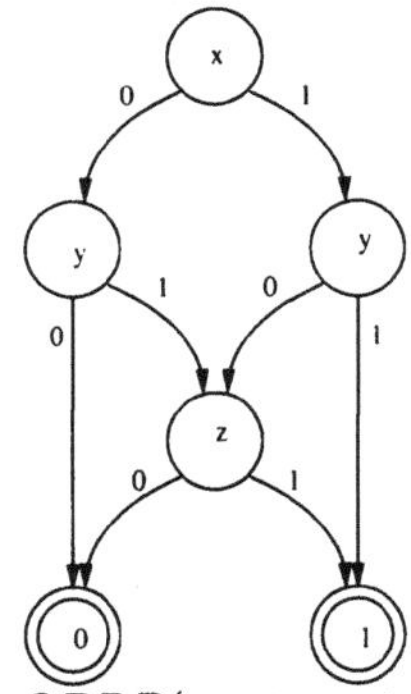

(b) $OBDD(x.y + x.z + y.z)$

FIGURE 1. Decision diagrams

The size of $OBDD(f)$ is the number of its nodes and is denoted by $|OBDD(f)|$, and this number is also the OBDD-complexity of the boolean function, denoted by $C_{obdd}(f)$.

One must note that the resulting graphs (form and size) are variable orderings dependent (in the Shannon's decomposition). Given a function, finding the variable ordering that gives the minimal OBDD size is an NP-complete problem.

We define a boolean hard function of n variables as a function whose OBDD-complexity $C(n) = \max_{f \in \mathbb{B}_n} C_{obdd}(f)$ is maximal over the set $\mathbb{B}_n$ of all n-ary boolean functions. We have:

$$\begin{cases} \underline{\lim}_{n\to\infty} \frac{n.C(n)}{2^n} & = & 1 \\ \overline{\lim}_{n\to\infty} \frac{n.C(n)}{2^n} & = & 2 \end{cases}$$

The practical construction of an $OBBD(f)$ where $f = g \oplus h$ (where $\oplus$ represents any boolean operator) is commonly obtained by application of an algorithm based on a parallel DAG traversal method (called APPLY, see [1]) over $OBDD(g)$ and $OBDD(h)$:

$$OBBD(f) = OBDD(g \oplus h) = APPLY(OBDD(g), \oplus, OBDD(h))$$

1.2. HFE Cryptosystems

The HFE (Hidden Field Equations) public key cryptosystem is based on polynomials over finite fields. This system has been proposed by J. Patarin (see [8]) who showed how insecure Matsumoto & Imai scheme was (see [6]) and then reinforced it. This method can be described as follows.

Let $\mathbb{K}$ be an extension of degree n over the finite field $\mathbb{F}_2$. The field $\mathbb{K}$ can be viewed as an n-dimensional vector space over $\mathbb{F}_2$. Any basis $\{\omega_1, \ldots, \omega_n\}$ of $\mathbb{K}$ defines the following mapping from $\mathbb{K}$ to $(\mathbb{F}_2)^n$:

$$\sum_{i=1}^{n} x_i\omega_i \longleftrightarrow (x_1, \ldots, x_n), \text{ where } x_i \in \mathbb{F}_2. \tag{1}$$

Any quasi-quadratic univariate polynomial $P(X)$ over $\mathbb{K}$

$$P(X) = \sum_{i,j}^{1..r} \alpha_{i,j} X^{2^i+2^j} + \gamma, \text{ with } \alpha_{i,j} \in \mathbb{K} \text{ and } \gamma \in \mathbb{K},$$

can be viewed in $(\mathbb{F}_2)^n$ as the collection of n quasi-quadratic multivariate polynomials over $\mathbb{F}_2$:

$$\left\{\begin{array}{lcl} P_1(x_1, \ldots, x_n) & = & \sum_{j,k}^{1..n} \beta_{1,j,k} x_j x_k + \delta_1 \\ & \vdots & \\ P_n(x_1, \ldots, x_n) & = & \sum_{j,k}^{1..n} \beta_{n,j,k} x_j x_k + \delta_n \end{array}\right. \quad \text{with } \beta_{i,j,k} \in \mathbb{F}_2 \text{ and } \delta_i \in \mathbb{F}_2$$

Such a system is the public key of HFE.

Encryption is done applying the P_i's on cleartext bits. Decryption is suspected to be as hard as the problem of solving multivariate quadratic boolean equations system which is known to be NP-hard (see [2]). But it is also known to be efficiently tractable using Berlekamp's algorithm to extract the roots of $P(X)$ (only if its degree is not too large, see [3]).

Moreover, two linear permutations S and T of $\mathbb{K}$ are introduced as trapdoors. It is important to note at that point that S and T are polynomials over $\mathbb{K}$ and that their expressions over $\mathbb{F}_2$ are invertible linear transformations. The public key is then defined as the expression of $T(P(S(X)))$ over $\mathbb{F}_2$ and it is still a quasi-quadratic multivariate polynomial as before. But now, the polynomial TPS has a very high degree (comparable to the order of the field), the number of its coefficients is much greater than those of P, and is now practically unsolvable if S and T are not known.

It is possible to define such cryptosystems on different fields, but using $\mathbb{F}_2$ allows to encrypt text using simple digital devices as smart cards. The parameter r, which controls the degree of P, must actually be less than 12 (see [4]) allowing Berlekamp's algorithm to run efficiently. To improve his cryptosystem, Patarin published a challenge in which $\mathbb{K} = \mathbb{F}_{2^{80}}$, such a public key size is about 32KB.

1.3. Attack

We chose a ciphertext-only attack, which consists in retrieving the cleartext given only the ciphertext and the public key. We decided not to use algebraic properties of the system. The problem is then reduced to satisfy a boolean formula.

Given a cleartext $X_0 \in \mathbb{K}$, encryption gives $Y_0 = TPS(X_0)$, so breaking the cipher corresponds to solving $Y_0 = TPS(X)$. Using the mapping defined in (1), this can be rewritten as:

$$\begin{cases} y_{0,1} & = & P_1(x_1, \dots, x_n) \\ & \vdots & \\ y_{0,n} & = & P_n(x_1, \dots, x_n) \end{cases}$$

then as:

$$\begin{cases} 1 & = & 1 + y_{0,1} + P_1(x_1, \dots, x_n) \\ & \vdots & \\ 1 & = & 1 + y_{0,n} + P_n(x_1, \dots, x_n). \end{cases}$$

So, satisfying the formula:

$$R(x_1, \dots, x_n) = \bigwedge_{i=1}^{n} E_i(x_1, \dots, x_n)$$

where $E_i(x_1, \dots, x_n) = 1 + y_i + P_i(x_1, \dots, x_n)$, solves the problem and gives a possible cleartext.

We decided to use OBBDs to manipulate the boolean functions involved in the process and the algorithm is the following:

1. Construct $B_{E_i} = OBDD(E_i)$, for all $i \in [1, n]$,
2. Construct $B_{R_1} = B_{E_1}$,
3. Construct $B_{R_i} = APPLY(B_{R_{i-1}}, \wedge, B_{E_i})$, for all $i \in [2, n]$,
4. Extract the set of all satisfying paths in $B_R = B_{R_n}$.

2. The Library: LiBDD

We first tried to use CMU's library (see [11]) to solve the system. But as performances were not satisfactory, we coded our own library to tune up parameters of the implementation (this be shown later in this paper). Written in C, the library has less than 2,000 lines of source code and nothing but data layout is machine-dependent, however porting onto different platforms is quite easy (and was experimentally done). We then focused on some critical points of practical algorithmics: space and time.

2.1. Memory Management

The general-purpose BDD libraries we looked at commonly use a 96 or 128 bits long structure to code a single BDD-node. One can remark that we only need a small set of variables (actually 80 in the challenge), so we were able to pack a node into 64 bits using the following structure:

```
struct node {
 UINT64 mark : 1; // mark and sweep garbage collector
 UINT64 idx  : 7; // variable's index (0<=idx<127)
 UINT64 pthen:28; // 'true' sub-BDD
 UINT64 pelse:28; // 'false' sub-BDD
};
```

The fields named `pthen` and `pelse` contain the significant parts of the "real" memory pointers of the respectives true and false sub-BDDs. As each of these fields is 28 bits long, a realtime expansion must be done on each reference to produce a full 32 bits pointer. A generic coding of this is possible, however a machine-dependent representation is preferable because computation can be faster.

This implementation gave us the ability to manage boolean functions of at most 128 variables, and BDD size less than 2^{28} nodes. One can note that 2^{28} nodes of 2^3 bytes each need 2 giga-bytes to be stored (a common computer maximum memory size at this time).

The overall memory management has been kept as simple as possible, hence a single continuous pool of nodes is allocated at startup time from which, nodes are picked as needed. Clearing an index frees the node but nodes of BDDs are freed when the BDD is freed and the synchronous mark and sweep garbage collector is called later.

2.2. Algorithms and BDD Variant

We used a mixed form of Bryant's algorithms `Apply` and `Reduce` (see [1]) which allows to allocate only useful nodes of the resulting BDD when applying an operator on two BDDs (see [10]). The main difficulty was the tricky management of some auxiliary structures (basically matrices) involved. As such matrices are mostly sparse, we used a mix of hashed and sorted tables.

As we constructed really huge BDDs, we decided to use a variant of BDDs. Recall that the subgraph identification operation in the `Reduce` step of the algorithm can use any suitable graph isomorphism. So we choose to identify two graphs either if they are denoted by the same function f, or if one is denoted by f and the other by $\neg f$. This is known as "BDD with output inverters" (see [7]) and only needs some slight modifications in algorithms and structures. One of the 28 bits devoted to pointers is now used to encode the "negation" mark. The maximum number of nodes decreased to 2^{27} but we expected to identify many more subgraphs and reduce the overall needed memory size.

2.3. Performances

We compared three different BDD libraries to solve the HFE problems we generate. As we can see in Figure 2, the library LiBDD is slightly better than CuDD (VLSI/CAD Laboratory of Colorado University – see [12]), which is far better than CMU's.

3. Experimental Results

As we needed many different HFE cryptosystems, we designed a program to generate such systems that was written in `C++` and uses Shoup's NTL library (Number Theory Library – see [14]).

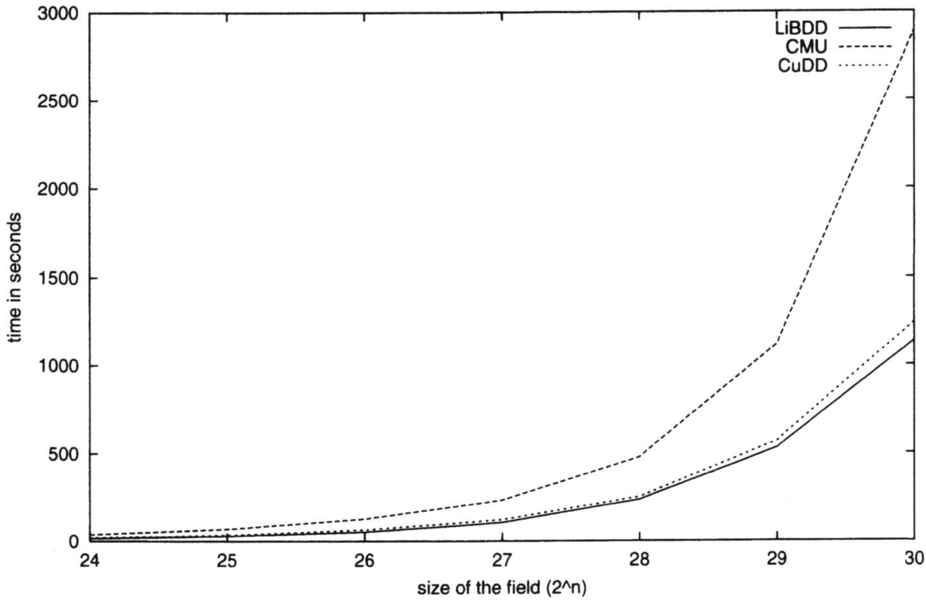

FIGURE 2. Comparing libraries on HFE systems

The input of the main program is a HFE public key in challenge's format. The following is one equation taken from a system generated from a quasi-quadratic polynomial of degree 2,304=2^{11}+2^8 over $\mathbb{F}_{2^{24}}$:

$$y_1 = 1+x_1+x_3+x_4+x_7+x_{10}+x_{11}+x_{12}+x_{17}+x_1x_2+x_1x_3+x_1x_5+x_1x_{10}+x_1x_{11}+x_1x_{13}+x_1x_{14}+ x_1x_{17}+x_1x_{21}+x_1x_{22}+x_1x_{23}+x_1x_{24}+x_2x_3+x_2x_6+x_2x_7+x_2x_9+x_2x_{13}+x_2x_{14}+x_2x_{15}+x_2x_{19}+ x_2x_{20}+x_2x_{22}+x_2x_{24}+x_3x_6+x_3x_7+x_3x_8+x_3x_9+x_3x_{15}+x_3x_{16}+x_3x_{18}+x_3x_{21}+x_3x_{23}+x_3x_{24}+x_4x_9+ x_4x_{10}+x_4x_{11}+x_4x_{16}+x_4x_{17}+x_4x_{18}+x_4x_{20}+x_4x_{21}+x_4x_{22}+x_4x_{24}+x_5x_7+x_5x_8+x_5x_{10}+x_5x_{11}+ x_5x_{13}+x_5x_{15}+x_5x_{17}+x_5x_{19}+x_6x_7+x_6x_8+x_6x_{12}+x_6x_{13}+x_6x_{16}+x_6x_{19}+x_6x_{22}+x_6x_{24}+x_7x_{10}+ x_7x_{12}+x_7x_{15}+x_7x_{16}+x_7x_{18}+x_8x_{12}+x_8x_{15}+x_8x_{18}+x_8x_{19}+x_8x_{22}+x_8x_{23}+x_9x_{12}+x_9x_{13}+x_9x_{14}+ x_9x_{22}+x_9x_{24}+x_{10}x_{15}+x_{10}x_{18}+x_{10}x_{19}+x_{10}x_{20}+x_{10}x_{22}+x_{10}x_{24}+x_{11}x_{13}+x_{11}x_{19}+x_{11}x_{21}+x_{11}x_{24}+ x_{12}x_{13}+x_{12}x_{17}+x_{12}x_{18}+x_{12}x_{19}+x_{12}x_{22}+x_{12}x_{24}+x_{13}x_{14}+x_{13}x_{15}+x_{13}x_{18}+x_{13}x_{19}+x_{13}x_{20}+ x_{13}x_{21}+x_{13}x_{22}+x_{13}x_{23}+x_{13}x_{24}+x_{14}x_{15}+x_{14}x_{21}+x_{14}x_{22}+x_{14}x_{23}+x_{15}x_{16}+x_{15}x_{17}+x_{15}x_{20}+ x_{15}x_{21}+x_{15}x_{22}+x_{15}x_{23}+x_{16}x_{20}+x_{16}x_{21}+x_{16}x_{22}+x_{16}x_{23}+x_{17}x_{19}+x_{17}x_{20}+x_{17}x_{22}+x_{17}x_{23}+ x_{17}x_{24}+x_{18}x_{19}+x_{18}x_{21}+x_{18}x_{24}+x_{19}x_{23}+x_{20}x_{22}+x_{20}x_{23}+x_{20}x_{24}+x_{21}x_{23}+x_{21}x_{24}+x_{22}x_{23}+x_{22}x_{24}$$

A cleartext x is then randomly chosen and encoded with the system, giving the ciphertext y which is then used to solve the boolean system as described before: first, equations E_i are BDD encoded, then they are combined to produce functions R_i. At last, vectors that satisfy R_n are produced. Figure 3 is a typical output.

Figure 4 shows that a computation with different parameters can produce more than one cleartext (a satisfying assignment of the corresponding boolean function).

```
GF2(24)
Degree 2304
X : 110101100010111011001110
Y : 111111011001111111100111
E1  :  0s 12283 nodes
E2  :  0s  9179 nodes
E3  :  0s 10235 nodes
E4  :  0s  8185 nodes
E5  :  0s 12283 nodes
E6  :  0s  7161 nodes
E7  :  0s 10229 nodes
E8  :  0s  6901 nodes
E9  :  0s  9205 nodes
E10 :  1s 10235 nodes
E11 :  1s 12279 nodes
E12 :  1s 12285 nodes
E13 :  1s 12281 nodes
E14 :  1s  7675 nodes
E15 :  1s 10235 nodes
```

```
E16 :  1s 12285 nodes
E17 :  1s 10235 nodes
E18 :  1s 10233 nodes
E19 :  2s  7285 nodes
E20 :  2s 11773 nodes
E21 :  2s  9209 nodes
E22 :  2s 12283 nodes
E23 :  2s  9213 nodes
E24 :  2s 10231 nodes

R1  :  2s  12283 nodes
R2  :  3s 210735 nodes
R3  :  6s 589695 nodes
R4  :  9s 490549 nodes
R5  : 11s 318192 nodes
R6  : 12s 199306 nodes
R7  : 13s 123517 nodes
R8  : 14s  78076 nodes
R9  : 14s  46515 nodes
```

```
R10 : 14s  26830 nodes
R11 : 14s  16579 nodes
R12 : 14s  11242 nodes
R13 : 14s   8080 nodes
R14 : 14s   5524 nodes
R15 : 14s   3591 nodes
R16 : 14s   2307 nodes
R17 : 14s   1359 nodes
R18 : 14s    883 nodes
R19 : 14s    546 nodes
R20 : 14s    332 nodes
R21 : 14s    234 nodes
R22 : 14s    121 nodes
R23 : 14s     67 nodes
R24 : 14s     25 nodes
X     : 110101100010111011001110
Sol 1: 110101100010111011001110 <--
Biggest bdd       :     589695 nodes
Pure CPU time     :     14 seconds
```

FIGURE 3. A typical output

```
GF2(29)
Degree 33
X : 11010110001011101100111001101
Y : 11101011001111110010011011000
E1 : 0s 57339 nodes
E2 : 1s 57339 nodes
...
R27 : 532s 172 nodes
R28 : 532s 126 nodes
R29 : 532s 101 nodes
X     : 11010110001011101100111001101
Sol 1: 00010101101001110110000110011
Sol 2: 10101010101010001011001100011
Sol 3: 11010110001011101100111001101 <--
Sol 4: 11101000000101110000000101101
Biggest bdd       :  11809274 nodes
Pure CPU time     :  532 seconds
```

FIGURE 4. Another output

3.1. Analysis

When we first tried to solve Patarin's challenge, we quickly saw that it would exceed any today computer capability. We then decided to try our method on several different HFE systems.

Figure 5 shows how the size of the BDD of R_i typically evolves. As we combined a few E_i, the size of the result R_i grew exponentially up to a very high peak, then the size decreased down to a very small number. This suggests that not only it is difficult to build a BDD for each E_i, but combining a few is even harder.

Figure 6 shows how this phenomenon is persistent even if we use different fields.

One should be careful in interpreting these results: we measured the size of BDDs, which is not easily related to the number of solutions satisfying the associated boolean formula. What BDD size reveals is in a certain sense the "complexity"

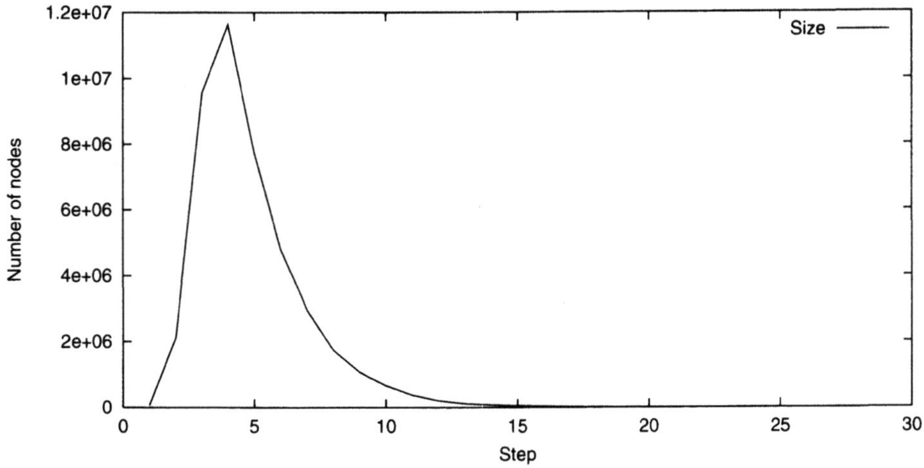

FIGURE 5. Partial results BDD size

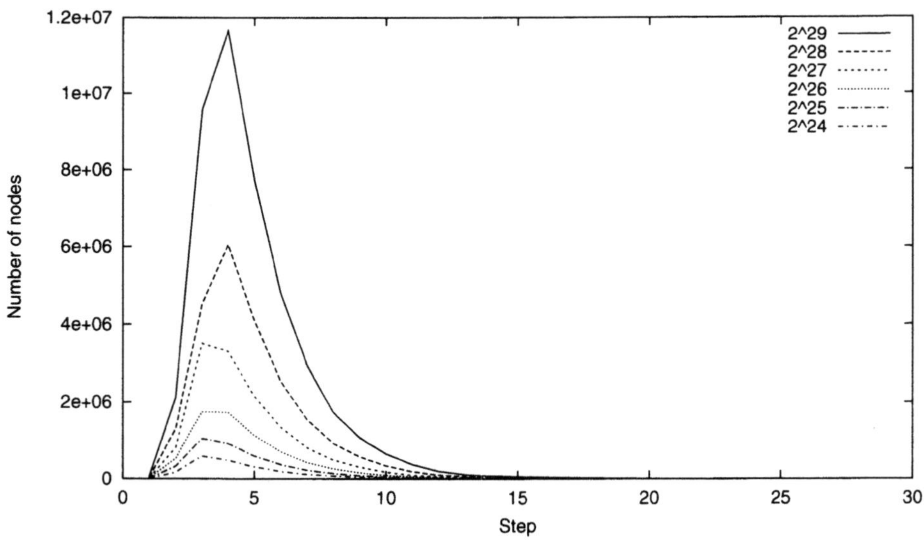

FIGURE 6. Partial results BDD size (different fields)

of the boolean function; remember also that this complexity is variable ordering dependent. Moreover, what our algorithm builds is a set of boolean functions R_i where the sets of solutions $\cdots \subset S_{i+1} \subset S_i \subset \cdots$ are all included one into the other. Thus, even if the number of solutions decreases as i grows, the complexity of the characteristic function of the satisfying set evolves as the previous figures indicate.

Figure 7 illustrates how the size of the peak evolves as the size of the field grows and how the size of the biggest generated BDD is close to the function $2^x/x$.

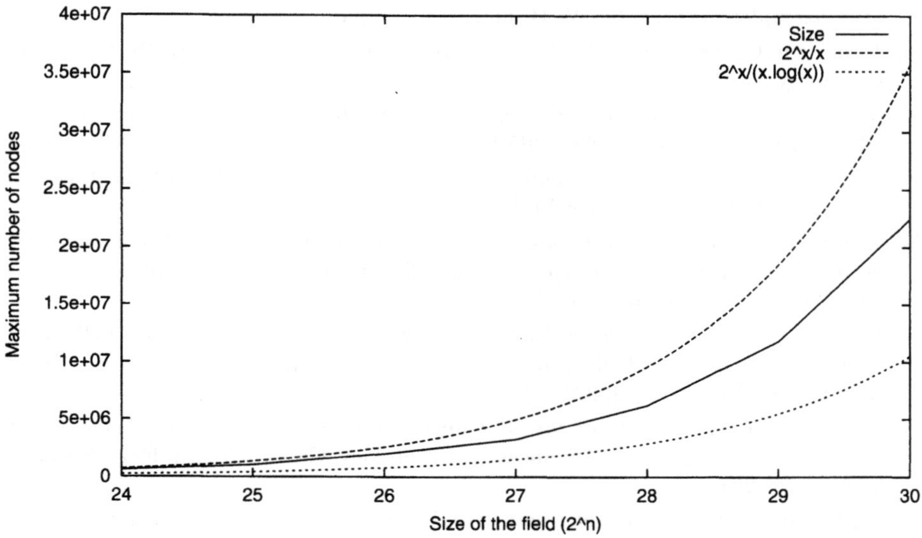

FIGURE 7. Peak size

Some more experiments also showed that the observed complexities are independent of the different variables or equations orderings. More surprisingly, there was nothing to note when we modified P, S or T, even when trying a polynomial as simple as $P(X) = X^3$ with $S(X) = T(X) = X$.

Many more experimental results can be found in [13].

4. Conclusion

All those experiments suggest that this method as it stands is not suitable to break the HFE cryptosystem. However, it is worth noting that although this method is not really tractable, it does work and its complexity is smaller than the exhaustive one.

A side effect is that our algorithm can be used to extract roots of any quasi-quadratic polynomial over a finite field. Our experiments showed that a standard computer with 1Gb of memory can break any HFE system defined in $\mathbb{F}_{2^{30}}$ in about 20 minutes. It will be easy to extend it to extract roots of any polynomial over a finite field.

References

[1] R.E. Bryant. *Graph-based algorithms for boolean function manipulation.* In IEEE Transaction on Computers. C-35 volume 8 (1986). 677–691.

[2] M.R. Garey and D.S. Johnson. *Computers and intractability: a guide to the theory of NP-completeness.* (1979). W.H. Freeman Company.

[3] J. von zur Gathen and J. Gerhard. *Modern computer algebra.* (1999). Cambridge University Press.

[4] A. Kipnis and A. Shamir. *Cryptanalysis of the HFE public key cryptosystem by relinearization.* In LNCS 1966, CRYPTO'99 (1999). Springer-Verlag. 19–30.

[5] M. Krause. *BDD-based cryptanalysis of keystram generators.* In LNCS 2332, EUROCRYPT'2002 (2002). Springer-Verlag. 222–237.

[6] T. Matsumoto and H. Imai. *Public quadratic polynomial-tuples for efficient signature-verification and message encryption.* In LNCS 330, Advances in Cryptology EUROCRYPT'88 (1988). Springer-Verlag. 419–453.

[7] S.I. Minato, N. Ishiura, and S. Yajima. *Shared binary decision diagram with attributed edges for efficient boolean function manipulation.* In 27-th ACM/IEEE Design Automaton Conference (1990). 52–57.

[8] J. Patarin. *Cryptanalysis of the Matsumoto and Imai public key scheme of Eurocrypt'88.* In LNCS 963, Advances in Cryptology CRYPTO'95 (1995). Springer-Verlag. 248–261.

[9] J. Patarin. *Hidden fields equations (HFE) and isomorphism of polynomials (IP): two new families of asymetric algorithms.* In LNCS 1070, EUROCRYPT'96 (1996). Springer-Verlag. 33–48.

[10] I. Wegener. *BDDs – design, analysis, complexity and applications.* `http://ls2-www.cs.uni-dortmund.de/~wegener/papers/BDDdesign.ps`

[11] *CMU package.* `http://www.cs.cmu.edu/afs/cs/project/modck/pub/www/bdd.html`

[12] *CuDD Package.* `ftp://vlsi.colorado.edu/pub/`

[13] *HFE experiments web page.* `http://www.liafa.jussieu.fr/~yunes/HFE/`

[14] *NTL Library.* `http://www.shoup.net/`

J.F. Michon
LIFAR, Université de Rouen
F-76821 Mont Saint-Aignan, France
e-mail: `Jean-Francis.Michon@univ-rouen.fr`

P. Valarcher
LIFAR, Université de Rouen
F-76821 Mont Saint-Aignan, France
e-mail: `Pierre.Valarcher@univ-rouen.fr`

J.B. Yunès
LIAFA, Université Paris 7
F-75251 Paris, France
e-mail: `Jean-Baptiste.Yunes@liafa.jussieu.fr`

Progress in Computer Science and Applied Logic, Vol. 23, 247–257

Digital Nets and Coding Theory

Harald Niederreiter

Abstract. Recent research has established close links between digital nets and coding theory. In fact, the problem of constructing good digital nets can now be viewed as the problem of constructing good linear codes in metric spaces that are more general than Hamming spaces. In this paper we report on the fascinating connections between digital nets and linear codes. In particular, we describe the duality theory for digital nets, the asymptotics of digital-net parameters, and the new concept of cyclic digital nets.

Mathematics Subject Classification (2000). Primary 11K38, 11K45; Secondary 65C05, 94B05.

Keywords. Digital net, linear code, duality theory.

1. Introduction

Digital nets, the main topic of this article, arise in multidimensional numerical integration as point configurations with excellent uniform distribution properties (see [8], [14, Chapter 4]). Despite this provenance in numerical analysis, digital nets are firmly embedded in the area of discrete mathematics in terms of structure theory and construction methods. Indeed, the construction of good digital nets can be viewed as a problem of combinatorial linear algebra over finite fields, and the classical constructions use tools such as number theory and the theory of finite fields.

For an integer $s \geq 1$ let $I^s := [0,1]^s$ denote the s-dimensional unit cube. Our starting point is the following definition, first given in [12], of point configurations in I^s with a very regular uniform distribution behavior. We note that by a point set we mean a "multiset" in the sense of combinatorics, i.e., a set in which multiplicities of elements are allowed and taken into account.

Definition 1.1. Let $0 \leq t \leq m$ and $b \geq 2$ be integers. A (t,m,s)-*net in base* b is a point set P of b^m points in I^s such that every subinterval J of I^s of the form

$$J = \prod_{i=1}^{s} [a_i b^{-d_i}, (a_i + 1) b^{-d_i})$$

with $a_i, d_i \in \mathbb{Z}$, $d_i \geq 0$, $0 \leq a_i < b^{d_i}$ for $1 \leq i \leq s$ and with $\mathrm{Vol}(J) = b^{t-m}$ contains exactly b^t points of the point set P.

It should be observed that $b^t = b^{t-m} \cdot b^m$ is the expected number of points of P in an interval J of volume b^{t-m}, in the case of perfect uniform distribution. It is important to note that the smaller the value of t, the larger the family of intervals J for which the property in Definition 1.1 is requested, and so the stronger the uniform distribution property. Thus, the primary interest is in (t, m, s)-nets with a small value of t. The number t is often called the *quality parameter* of the net.

The standard construction of nets proceeds by the digital method which will be explained in Section 2. Nets obtained by the digital method are, in an obvious terminology, called *digital nets.* The first connections between digital nets and coding theory will be explored in Section 3. Further connections arise in the context of the duality theory for digital nets which will be delineated in Section 4. Sections 5 and 6 are devoted to cyclic digital nets – the analogs of cyclic codes – and to the asymptotic theory of digital nets, respectively.

2. The Digital Method

The digital method was introduced in [12] and provides a general framework for the construction of (t, m, s)-nets. We describe this method for the case of a prime-power base q. Suppose we are given an integer $m \geq 1$ and the dimension $s \geq 1$. Then we have to define q^m points in I^s for a (t, m, s)-net in base q.

The first step is to take a finite field $\mathbb{F}_q$ with q elements. The next step is to choose $m \times m$ matrices $C^{(1)}, \ldots, C^{(s)}$ over $\mathbb{F}_q$, that is, one matrix for each of the s coordinate directions of points in I^s. For a fixed column vector $\mathbf{a} \in \mathbb{F}_q^m$ of length m, we compute the matrix-vector products

$$C^{(i)}\mathbf{a} \in \mathbb{F}_q^m \qquad \text{for } 1 \leq i \leq s.$$

Next we define the map $T : \mathbb{F}_q^m \to [0, 1)$ by

$$T(\mathbf{h}) = \sum_{j=1}^{m} \psi(h_j) q^{-j}$$

for $\mathbf{h} = (h_1, \ldots, h_m) \in \mathbb{F}_q^m$, where $\psi : \mathbb{F}_q \to \{0, 1, \ldots, q-1\}$ is a fixed bijection from $\mathbb{F}_q$ onto the set $\{0, 1, \ldots, q-1\}$ of q-adic digits. Then we put

$$p^{(i)}(\mathbf{a}) = T(C^{(i)}\mathbf{a}) \in [0, 1) \qquad \text{for } 1 \leq i \leq s.$$

In this way we get the point

$$P(\mathbf{a}) = (p^{(1)}(\mathbf{a}), \ldots, p^{(s)}(\mathbf{a})) \in I^s.$$

By letting $\mathbf{a}$ range over all q^m possibilities in $\mathbb{F}_q^m$, we arrive at the desired q^m points in I^s.

If the point set constructed above is a (t, m, s)-net in base q, then it is called a *digital* (t, m, s)*-net over* $\mathbb{F}_q$. Note that these q^m points in I^s always form a

(t, m, s)-net in base q for some value of t, since for instance $t = m$ always satisfies Definition 1.1. The interesting question is, of course, whether we can choose the parameters in the above construction in such a way that we can obtain nontrivial values $t < m$ (or even rather small values) for the quality parameter.

It turns out that the quality parameter t of a digital (t, m, s)-net over $\mathbb{F}_q$ depends only on the choice of the matrices $C^{(1)}, \ldots, C^{(s)}$ in the above construction. These matrices are called the *generating matrices* of the digital net. For $1 \leq i \leq s$ let $\mathbf{c}_j^{(i)} \in \mathbb{F}_q^m$, $1 \leq j \leq m$, be the row vectors of $C^{(i)}$. Then the quality parameter t can be determined by the following result (see [14, Theorem 4.28], [20, Theorem 8.2.4]).

Theorem 2.1. *Let $d \leq m$ be such that for any nonnegative integers $d_1, \ldots, d_s$ with $\sum_{i=1}^{s} d_i = d$ the vectors*

$$\mathbf{c}_j^{(i)} \in \mathbb{F}_q^m, \quad 1 \leq j \leq d_i, \ 1 \leq i \leq s,$$

are linearly independent over $\mathbb{F}_q$. Then the point set constructed above is a digital (t, m, s)-net over $\mathbb{F}_q$ with $t = m - d$.

By maximizing d, we get the minimal value of the quality parameter t for this point set. Similar construction principles are available for arbitrary bases, but then one has to replace the finite field $\mathbb{F}_q$ by a finite commutative ring with identity.

Recent surveys of constructions of (digital) nets are given in Clayman *et al.* [3], Larcher [8], Niederreiter [15], and Xing and Niederreiter [27]. For expository accounts of the theory of digital nets we refer to the books of Niederreiter [14, Chapter 4] and Niederreiter and Xing [20, Chapter 8].

3. Digital Nets and Parity-Check Matrices

In order to make the digital method work concretely and effectively, we have to solve the problem of combinatorial linear algebra arising from Theorem 2.1. In other words, we have to find systems $\{\mathbf{c}_j^{(i)} \in \mathbb{F}_q^m : 1 \leq j \leq m, 1 \leq i \leq s\}$ of vectors for which the number d in Theorem 2.1 is as large as possible.

This problem is reminiscent of that of finding good linear codes over $\mathbb{F}_q$ by means of their parity-check matrices. However, instead of an $s \times m$ array of vectors $\mathbf{c}_j^{(i)}$, $1 \leq i \leq s$, $1 \leq j \leq m$, as above, the coding-theory problem just requires us to find a list of s vectors, i.e., an $s \times 1$ array, with a suitable linear independence property.

The rigorous way of viewing the digital-net problem and the coding-theory problem from a common perspective is based on the following definition from [18].

Definition 3.1. Let k, m, and s be positive integers and let d be an integer with $0 \leq d \leq \min(k, ms)$. The system $\{\mathbf{c}_j^{(i)} \in \mathbb{F}_q^k : 1 \leq j \leq m, 1 \leq i \leq s\}$ of vectors is called a *(d, k, m, s)-system over* $\mathbb{F}_q$ if for any integers $d_1, \ldots, d_s$ with $0 \leq d_i \leq m$ for $1 \leq i \leq s$ and $\sum_{i=1}^{s} d_i = d$ the system $\{\mathbf{c}_j^{(i)} \in \mathbb{F}_q^k : 1 \leq j \leq d_i, 1 \leq i \leq s\}$ is linearly independent over $\mathbb{F}_q$ (the empty system is considered linearly independent).

On account of Theorem 2.1, it is now clear that the construction of a digital (t, m, s)-net over $\mathbb{F}_q$ requires the construction of an $(m-t, m, m, s)$-system over $\mathbb{F}_q$. On the other hand, if we can construct a $(d, k, 1, s)$-system $\{\mathbf{c}^{(i)} \in \mathbb{F}_q^k : 1 \leq i \leq s\}$ over $\mathbb{F}_q$, then we obtain a linear code of length s, dimension $\geq s-k$, and minimum distance $\geq d+1$ if we use $\mathbf{c}^{(1)}, \ldots, \mathbf{c}^{(s)}$ as the columns of a parity-check matrix of the linear code. This follows from a standard result in coding theory (see [2, Corollary 3.2.3], [10, Lemma 8.14]).

To emphasize this important point, the step of going from the construction of good linear codes to the construction of good digital nets is that of going from a list of s vectors to an $s \times m$ array with a suitable linear independence property. In this sense, the problem of constructing good digital nets can be considered more difficult than that of constructing good linear codes.

This connection between linear codes and digital nets suggests the following intuitive procedure for constructing good digital nets:

(i) start from a construction of good linear codes in terms of parity-check matrices;
(ii) write the columns of the parity-check matrix as a single column of vectors, thus obtaining the first column of an $s \times m$ array of vectors;
(iii) fill up the remaining $m-1$ columns of an $s \times m$ array of vectors in a "canonical" way;
(iv) determine the value of d in Definition 3.1 that you get.

Example. Let β be a primitive element of $\mathbb{F}_q$ and let $1 \leq s \leq q-1$. Put

$$\mathbf{c}_1^{(i)} = (1, \beta^i, \beta^{2i}, \ldots, \beta^{(m-1)i}) \in \mathbb{F}_q^m \qquad \text{for } 1 \leq i \leq s.$$

Note that these vectors occur as columns of parity-check matrices of certain Reed-Solomon codes over $\mathbb{F}_q$ (compare with [10, p. 324]). A "canonical" extension to an $s \times m$ array of vectors is obtained by defining the vectors $\mathbf{c}_j^{(i)} \in \mathbb{F}_q^m$, $1 \leq i \leq s$, $2 \leq j \leq m$, by

$$\mathbf{c}_j^{(i)} = (\underbrace{0, \ldots, 0}_{j-1}, 1, \binom{j}{j-1}\beta^i, \binom{j+1}{j-1}\beta^{2i}, \ldots, \binom{m-1}{j-1}\beta^{(m-j)i}).$$

This yields a familiar construction of digital nets, first given by Faure [6] in the special case where q is a prime and then by Niederreiter [12] for the general case of an arbitrary prime power q (compare also with [14, Remark 4.52]).

Further information on the general theory of (d, k, m, s)-systems over $\mathbb{F}_q$ can be found in the papers of Niederreiter [13] and Niederreiter and Pirsic [18], [19].

4. Duality Theory for Digital Nets

In the previous section we described the connection between digital nets and parity-check matrices of linear codes. This raises the question of whether one can construct digital nets in analogy with the construction of linear codes via generator matrices, or equivalently via linear subspaces of the Hamming space. This is indeed the case, but to make it work one has to develop a duality theory for digital nets.

The viewpoint of duality was introduced into the theory of digital nets by Niederreiter and Pirsic [18]; see also Skriganov [26] for a more specialized setting. In this viewpoint, the problem of constructing digital (t, m, s)-nets over $\mathbb{F}_q$ is reduced to that of constructing certain $\mathbb{F}_q$-linear subspaces of $\mathbb{F}_q^{ms}$. The vector space $\mathbb{F}_q^{ms}$ is endowed with a weight function which generalizes the classical Hamming weight in coding theory. Then there is a known relationship between the quality parameter t of the digital net and the minimum distance (or minimum weight) of the corresponding $\mathbb{F}_q$-linear subspace. Small values of t correspond to large values of the minimum distance in this relationship.

The appropriate weight function V_m on $\mathbb{F}_q^{ms}$ was first introduced by Niederreiter [11] and later used in an equivalent form in coding theory by Rosenbloom and Tsfasman [24]. First, we define a weight function v on $\mathbb{F}_q^m$ by putting $v(\mathbf{a}) = 0$ if $\mathbf{a} = \mathbf{0} \in \mathbb{F}_q^m$, and for $\mathbf{a} = (a_1, \ldots, a_m) \in \mathbb{F}_q^m$ with $\mathbf{a} \neq \mathbf{0}$ we set

$$v(\mathbf{a}) = \max\{j : a_j \neq 0\}.$$

Then we extend this definition to $\mathbb{F}_q^{ms}$ by writing a vector $\mathbf{A} \in \mathbb{F}_q^{ms}$ as the concatenation of s vectors of length m, i.e.,

$$\mathbf{A} = (\mathbf{a}^{(1)}, \ldots, \mathbf{a}^{(s)}) \in \mathbb{F}_q^{ms} \quad \text{with } \mathbf{a}^{(i)} \in \mathbb{F}_q^m \text{ for } 1 \leq i \leq s,$$

and putting

$$V_m(\mathbf{A}) = \sum_{i=1}^{s} v(\mathbf{a}^{(i)}).$$

Note that in the case $m = 1$, the weight $V_m(\mathbf{A})$ reduces to the Hamming weight of the vector $\mathbf{A}$. If for any $m \geq 1$ we define the distance $d_m(\mathbf{A}, \mathbf{B})$ of $\mathbf{A}, \mathbf{B} \in \mathbb{F}_q^{ms}$ by $d_m(\mathbf{A}, \mathbf{B}) = V_m(\mathbf{A} - \mathbf{B})$, then $\mathbb{F}_q^{ms}$ turns into a metric space, which for $m = 1$ is the Hamming space in coding theory. As in coding theory, the concept of minimum distance relative to d_m, or equivalently V_m, plays a crucial role.

Definition 4.1. For any nonzero $\mathbb{F}_q$-linear subspace $\mathcal{N}$ of $\mathbb{F}_q^{ms}$ we define the *minimum distance*

$$\delta_m(\mathcal{N}) = \min_{\mathbf{A} \in \mathcal{N} \setminus \{\mathbf{0}\}} V_m(\mathbf{A}).$$

The following result from [18] is an analog of the well-known Singleton bound in coding theory.

Proposition 4.2. *For any nonzero $\mathbb{F}_q$-linear subspace $\mathcal{N}$ of $\mathbb{F}_q^{ms}$ we have*

$$\delta_m(\mathcal{N}) \leq ms + 1 - \dim(\mathcal{N}).$$

The basic step in [18] is to set up a duality between digital (t, m, s)-nets over $\mathbb{F}_q$ and certain $\mathbb{F}_q$-linear subspaces of $\mathbb{F}_q^{ms}$. We assume $s \geq 2$ to avoid the trivial one-dimensional case. Let the $m \times m$ matrices $C^{(1)}, \ldots, C^{(s)}$ over $\mathbb{F}_q$ be the generating matrices of the digital net P. We set up an $m \times ms$ matrix M as follows: for $1 \leq j \leq m$, the jth row of M is obtained by concatenating the jth columns of $C^{(1)}, \dots, C^{(s)}$. Let $\mathcal{M} \subseteq \mathbb{F}_q^{ms}$ be the row space of M and let $\mathcal{M}^\perp \subseteq \mathbb{F}_q^{ms}$ be its dual space as in coding theory, i.e., $\mathcal{M}^\perp$ is the orthogonal complement of $\mathcal{M}$ relative to the standard inner product on $\mathbb{F}_q^{ms}$. With this notation, we can now formulate the following result from [18].

Theorem 4.3. *The digital net P is a digital (t, m, s)-net over $\mathbb{F}_q$ if and only if*

$$\delta_m(\mathcal{M}^\perp) \geq m + 1 - t.$$

This leads to the following procedure for constructing good digital nets on the basis of duality theory:

(i) construct an $\mathbb{F}_q$-linear subspace $\mathcal{N}$ of $\mathbb{F}_q^{ms}$ with $\dim(\mathcal{N}) \geq ms - m$ and a large value of $\delta_m(\mathcal{N})$ (compatible with Proposition 4.2);

(ii) dualize $\mathcal{N}$ to get $\mathcal{M}$, which determines the digital net.

In the notation of Theorem 4.3 we have $\mathcal{N} = \mathcal{M}^\perp$. Note that trivially $\dim(\mathcal{M}) \leq m$, and so we have

$$\dim(\mathcal{N}) = ms - \dim(\mathcal{M}) \geq ms - m,$$

as stated in (i). In the end, this procedure yields a digital (t, m, s)-net over $\mathbb{F}_q$ with $t = m + 1 - \delta_m(\mathcal{N})$. We summarize this as follows.

Corollary 4.4. *Let q be a prime power and let $m \geq 1$ and $s \geq 2$ be integers. Then from any $\mathbb{F}_q$-linear subspace $\mathcal{N}$ of $\mathbb{F}_q^{ms}$ with $\dim(\mathcal{N}) \geq ms - m$ we can construct a digital (t, m, s)-net over $\mathbb{F}_q$ with $t = m + 1 - \delta_m(\mathcal{N})$.*

Example. Based on this construction principle, we can obtain an analog of algebraic-geometry codes for digital nets. For a smooth, projective, absolutely irreducible algebraic curve $\mathcal{A}/\mathbb{F}_q$ of genus g, choose s distinct $\mathbb{F}_q$-rational points $P_1, \dots, P_s$ on $\mathcal{A}$ and a suitable divisor D of $\mathcal{A}$. Recall that an algebraic-geometry code is the image of the $\mathbb{F}_q$-linear map

$$f \in \mathcal{L}(D) \mapsto (f(P_1), \dots, f(P_s)) \in \mathbb{F}_q^s,$$

where $\mathcal{L}(D)$ denotes the Riemann-Roch space of D. Given integers $m \geq \max(1, g)$ and $s \geq 2$, we now replace each $f(P_i)$ – which can be viewed as the constant term in the local expansion of f at P_i – by the m-tuple of the first m coefficients in the local expansion of f at P_i. This yields an $\mathbb{F}_q$-linear map into $\mathbb{F}_q^{ms}$. By letting $\mathcal{N}$ in Corollary 4.4 be the image of this map, we obtain a digital (g, m, s)-net over $\mathbb{F}_q$. In the simplest case where $\mathcal{A}$ is the projective line over $\mathbb{F}_q$, we obtain a digital $(0, m, s)$-net over $\mathbb{F}_q$ whenever $s \leq q + 1$. For the details of this construction and generalizations we refer to Niederreiter and Özbudak [16].

We mention now several more constructions of digital nets that have been carried out with the help of duality theory. The general flavor of the constructions of digital nets based on Corollary 4.4 is to exploit the analogy with coding theory and transfer known constructions of good linear codes to digital nets. The paper [18], which introduced the duality theory for digital nets, contained already one such application, namely an analog of the $(u, u + v)$ construction of codes.

An improved version of the $(u, u+v)$ construction for digital nets was given by Bierbrauer, Edel, and Schmid [1]. This construction was recently generalized by Niederreiter and Özbudak [17] who used an analog of the matrix-product construction of codes. Analogs for digital nets of the Kronecker-product construction of linear codes are due to Bierbrauer, Edel, and Schmid [1] and Niederreiter and Pirsic [19]. Niederreiter and Xing [22] used Corollary 4.4 to obtain a new propagation rule for digital nets.

5. Cyclic Digital Nets

Cyclic codes form a well-known class of linear codes. In this section we introduce an analog of cyclic codes for digital nets. We adopt the viewpoint that cyclic codes can be defined by prescribing roots of polynomials (compare with [2, Chapter 5], [10, Section 8.2]).

Let the integers $m \geq 1$ and $s \geq 2$ and the finite field $\mathbb{F}_q$ be given. Our starting point is the vector space

$$\mathcal{P} = \{f \in \mathbb{F}_{q^m}[x] : \deg(f) < s\}$$

of polynomials. Note that $\dim(\mathcal{P}) = ms$ as a vector space over $\mathbb{F}_q$. Now we fix an element $\alpha \in \mathbb{F}_{q^m}$ and define

$$\mathcal{P}_\alpha = \{f \in \mathcal{P} : f(\alpha) = 0\}.$$

It is clear that $\mathcal{P}_\alpha$ is an $\mathbb{F}_q$-linear subspace of $\mathcal{P}$ with $\dim(\mathcal{P}_\alpha) = ms - m$ as a vector space over $\mathbb{F}_q$.

For each $i = 1, \ldots, s$, we choose an ordered basis B_i of $\mathbb{F}_{q^m}$ over $\mathbb{F}_q$. Now we set up a map from $\mathcal{P}$ to $\mathbb{F}_q^{ms}$ in the following way. Take $f \in \mathcal{P}$ and write this polynomial explicitly as

$$f(x) = \sum_{i=1}^{s} \gamma_i x^{i-1}$$

with $\gamma_i \in \mathbb{F}_{q^m}$ for $1 \leq i \leq s$. For each $i = 1, \ldots, s$, let $\mathbf{c}_i(f) \in \mathbb{F}_q^m$ be the coordinate vector of γ_i with respect to the ordered basis B_i. Then we define

$$\phi : f \in \mathcal{P} \mapsto (\mathbf{c}_1(f), \ldots, \mathbf{c}_s(f)) \in \mathbb{F}_q^{ms}.$$

It is obvious that ϕ is an $\mathbb{F}_q$-linear isomorphism from $\mathcal{P}$ onto $\mathbb{F}_q^{ms}$.

Now let $\mathcal{N}_\alpha$ be the image of the subspace $\mathcal{P}_\alpha$ under ϕ. Since ϕ is an isomorphism, we have

$$\dim(\mathcal{N}_\alpha) = \dim(\mathcal{P}_\alpha) = ms - m$$

as a vector space over $\mathbb{F}_q$. Thus, we can apply Corollary 4.4 to the $\mathbb{F}_q$-linear subspace $\mathcal{N}_\alpha$ of $\mathbb{F}_q^{ms}$. The resulting digital net is called a *cyclic digital net* over $\mathbb{F}_q$.

A complete analogy with cyclic codes would require that α be a root of $x^s - 1$. However, this would imply a restriction on s, and so we have not imposed this additional condition on α. The advantage of the construction of digital nets in this section is that it works for any prime power q and any integers $m \geq 1$ and $s \geq 2$.

6. The Asymptotics of Digital Nets

In this section we discuss the existence of digital $(t, t+d, s)$-nets over $\mathbb{F}_q$ for a fixed integer $d \geq 0$ and a fixed prime power q. Since it is trivial that for $d = 0$ and $d = 1$ such digital nets always exist, we assume $d \geq 2$ in the remainder of the section. The following result from [21] shows that in any sequence of such digital nets with the dimension s tending to ∞, the quality parameter t must have a certain minimal rate of growth. We include the proof for the sake of completeness.

Proposition 6.1. *Let q be an arbitrary prime power and $d \geq 2$ a fixed integer. Then for any sequence of digital $(t_r, t_r + d, s_r)$-nets over $\mathbb{F}_q$ with $s_r \to \infty$ as $r \to \infty$, we have*

$$\liminf_{r \to \infty} \frac{t_r}{\log_q s_r} \geq \left\lfloor \frac{d}{2} \right\rfloor,$$

where $\log_q$ denotes the logarithm to the base q.

Proof. We use the following result of Schmid and Wolf [25]: if there exists a digital $(t, t+d, s)$-net over $\mathbb{F}_q$ with $d \geq 2$, then necessarily

$$\sum_{u=1}^{\lfloor d/2 \rfloor} \sum_{l=1}^{u} \binom{s}{l} \binom{u-1}{l-1} (q-1)^l q^{u-l} < q^{t+d}.$$

This implies, in particular, that

$$\binom{s}{\lfloor d/2 \rfloor} < q^{t+d},$$

and so for $s \geq d$ we get

$$c_d s^{\lfloor d/2 \rfloor} < q^{t+d}$$

with a constant $c_d > 0$ depending only on d. Consequently,

$$\frac{t}{\log_q s} > \left\lfloor \frac{d}{2} \right\rfloor + \frac{C_d(q)}{\log_q s}$$

with a constant $C_d(q)$ depending only on d and q, which immediately implies the result of the proposition. □

An important question is then whether one can construct such sequences of digital nets with the minimal growth rate $t_r = O(\log s_r)$ for the quality parameter. The answer is affirmative, and the currently best result was established in [22] by combining BCH codes with the construction of digital nets from linear codes that is due to Lawrence *et al.* [9]. This yields the following result.

Theorem 6.2. *For every prime power q and every integer $d \geq 2$, we can construct a sequence of digital $(t_r, t_r + d, s_r)$-nets over $\mathbb{F}_q$ with $s_r \to \infty$ as $r \to \infty$ and*

$$\lim_{r \to \infty} \frac{t_r}{\log_q s_r} \leq d - 1 - \left\lfloor \frac{d-1}{q} \right\rfloor.$$

Corollary 6.3. *For every integer $d \geq 2$ there exists a sequence of digital $(t_r, t_r + d, s_r)$-nets over $\mathbb{F}_2$ with $s_r \to \infty$ as $r \to \infty$ and*

$$\lim_{r\to\infty} \frac{t_r}{\log_2 s_r} = \left\lfloor \frac{d}{2} \right\rfloor,$$

and the constant $\lfloor d/2 \rfloor$ is best possible.

Proof. We use Theorem 6.2 with $q = 2$ and note that

$$d - 1 - \left\lfloor \frac{d-1}{2} \right\rfloor = \left\lfloor \frac{d}{2} \right\rfloor \qquad \text{for all } d \geq 2.$$

The rest follows from Proposition 6.1. □

A comparison with Proposition 6.1 shows that Theorem 6.2 is best possible also in two other cases. An obvious case is $d = 2$. Another special case in which Theorem 6.2 is best possible is $(q, d) = (3, 4)$.

For $(q, d) = (2, 4)$ and $(q, d) = (3, 4)$, the result of Theorem 6.2 can be deduced also from the constructions of Edel and Bierbrauer [4], [5]. These constructions use BCH codes as well, but in a way that is different from the proof of Theorem 6.2. For $(q, d) = (2, 4)$ the result of Theorem 6.2 (and Corollary 6.3) can also be obtained by combining the work of Helleseth, Kløve, and Levenshtein [7] and Özbudak [23].

Acknowledgment

This research was partially supported by the grant R-394-000-011-422 with Temasek Laboratories in Singapore.

References

[1] Bierbrauer, J., Edel, Y., Schmid, W.Ch.: Coding-theoretic constructions for (t, m, s)-nets and ordered orthogonal arrays. J. Combin. Designs **10**, 403–418 (2002)

[2] Blahut, R.E.: Theory and Practice of Error Control Codes. Addison-Wesley, Reading, MA (1983)

[3] Clayman, A.T., Lawrence, K.M., Mullen, G.L., Niederreiter, H., Sloane, N.J.A.: Updated tables of parameters of (t, m, s)-nets. J. Combin. Designs **7**, 381–393 (1999)

[4] Edel, Y., Bierbrauer, J.: Construction of digital nets from BCH-codes. In: Niederreiter, H., *et al.* (eds.) Monte Carlo and Quasi-Monte Carlo Methods 1996, Lecture Notes in Statistics, Vol. 127, pp. 221–231. Springer, New York (1998)

[5] Edel, Y., Bierbrauer, J.: Families of ternary (t, m, s)-nets related to BCH-codes. Monatsh. Math. **132**, 99–103 (2001)

[6] Faure, H.: Discrépance de suites associées à un système de numération (en dimension s). Acta Arith. **41**, 337–351 (1982)

[7] Helleseth, T., Kløve, T., Levenshtein, V.I.: Hypercubic 4- and 5-designs from double-error-correcting BCH codes. Designs Codes Cryptogr. **28**, 265–282 (2003)

[8] Larcher, G.: Digital point sets: analysis and application. In: Hellekalek, P., Larcher, G. (eds.) Random and Quasi-Random Point Sets, Lecture Notes in Statistics, Vol. 138, pp. 167–222. Springer, New York (1998)

[9] Lawrence, K.M., Mahalanabis, A., Mullen, G.L., Schmid, W.Ch.: Construction of digital (t, m, s)-nets from linear codes. In: Cohen, S., Niederreiter, H. (eds.) Finite Fields and Applications, London Math. Soc. Lecture Note Series, Vol. 233, pp. 189–208. Cambridge University Press, Cambridge (1996)

[10] Lidl, R., Niederreiter, H.: Introduction to Finite Fields and Their Applications, rev. ed. Cambridge University Press, Cambridge (1994)

[11] Niederreiter, H.: Low-discrepancy point sets. Monatsh. Math. **102**, 155–167 (1986)

[12] Niederreiter, H.: Point sets and sequences with small discrepancy. Monatsh. Math. **104**, 273–337 (1987)

[13] Niederreiter, H.: A combinatorial problem for vector spaces over finite fields. Discrete Math. **96**, 221–228 (1991)

[14] Niederreiter, H.: Random Number Generation and Quasi-Monte Carlo Methods. SIAM, Philadelphia (1992)

[15] Niederreiter, H.: Constructions of (t, m, s)-nets. In: Niederreiter, H., Spanier, J. (eds.) Monte Carlo and Quasi-Monte Carlo Methods 1998, pp. 70–85. Springer, Berlin (2000)

[16] Niederreiter, H., Özbudak, F.: Constructions of digital nets using global function fields. Acta Arith. **105**, 279–302 (2002)

[17] Niederreiter, H., Özbudak, F.: Matrix-product constructions of digital nets. Finite Fields Appl. (to appear)

[18] Niederreiter, H., Pirsic, G.: Duality for digital nets and its applications. Acta Arith. **97**, 173–182 (2001)

[19] Niederreiter, H., Pirsic, G.: A Kronecker product construction for digital nets. In: Fang, K.-T., Hickernell, F.J., Niederreiter, H. (eds.) Monte Carlo and Quasi-Monte Carlo Methods 2000, pp. 396–405. Springer, Berlin (2002)

[20] Niederreiter, H., Xing, C.P.: Rational Points on Curves over Finite Fields: Theory and Applications. London Math. Soc. Lecture Note Series, Vol. 285. Cambridge University Press, Cambridge (2001)

[21] Niederreiter, H., Xing, C.P.: A construction of digital nets with good asymptotic behavior. Technical Report, Temasek Laboratories, National University of Singapore (2001)

[22] Niederreiter, H., Xing, C.P.: Constructions of digital nets. Acta Arith. **102**, 189–197 (2002)

[23] Özbudak, F.: Elements of prescribed order, prescribed traces and systems of rational functions over finite fields. Designs Codes Cryptogr. (to appear)

[24] Rosenbloom, M.Yu., Tsfasman, M.A.: Codes for the m-metric. Problems of Inform. Transmission **33**, 45–52 (1997)

[25] Schmid, W.Ch., Wolf, R.: Bounds for digital nets and sequences. Acta Arith. **78**, 377–399 (1997)

[26] Skriganov, M.M.: Coding theory and uniform distributions (Russian). Algebra i Analiz **13**, 191–239 (2001)

[27] Xing, C.P., Niederreiter, H.: Digital nets, duality, and algebraic curves. In: Niederreiter, H. (ed.) Monte Carlo and Quasi-Monte Carlo Methods 2002. Springer, Berlin (to appear)

Harald Niederreiter
Department of Mathematics
National University of Singapore
2 Science Drive 2
Singapore 117543
Republic of Singapore
e-mail: nied@math.nus.edu.sg

Progress in Computer Science and Applied Logic, Vol. 23, 259–275

Constructive Asymptotic Codes with an Improvement on the Tsfasman-Vlăduţ-Zink and Xing Bounds

Harald Niederreiter and Ferruh Özbudak

Abstract. This paper is a contribution to the asymptotic theory of algebraic codes and considers the well-known function α_q in this theory. The lower bound on α_q due to Tsfasman, Vlăduţ, and Zink was recently improved by Xing in a nonconstructive manner. We improve on the Xing bound by using a constructive approach.

Mathematics Subject Classification (2000). Primary 94B27, 94B65; Secondary 11R58, 11T71.

Keywords. Asymptotic theory of codes, algebraic-geometry code, global function field, digital net.

1. Introduction

Let $\mathbb{F}_q$ be the finite field of order q, where q is an arbitrary prime power. For a (not necessarily linear) code C over $\mathbb{F}_q$, we denote by $n(C)$ its length and by $d(C)$ its minimum distance. We write $|M|$ for the cardinality of a finite set M.

An important function in the asymptotic theory of algebraic codes is α_q, which is defined by

$$\alpha_q(\delta) = \sup\{R \in [0,1] : (\delta, R) \in U_q\} \qquad \text{for} \quad 0 \le \delta \le 1. \tag{1.1}$$

Here U_q is the set of all ordered pairs $(\delta, R) \in [0,1]^2$ for which there exists a sequence $\{C_i\}_{i=1}^{\infty}$ of (not necessarily linear) codes C_i over $\mathbb{F}_q$ such that $n(C_i) \to \infty$ as $i \to \infty$ and

$$\delta = \lim_{i\to\infty} \frac{d(C_i)}{n(C_i)}, \qquad R = \lim_{i\to\infty} \frac{\log_q |C_i|}{n(C_i)},$$

where $\log_q$ is the logarithm to the base q. We refer to [10, Section 1.3.1] for some basic properties of the function α_q. In particular, we note that $\alpha_q(0) = 1$ and $\alpha_q(\delta) = 0$ for $(q-1)/q \le \delta \le 1$.

A central problem in algebraic coding theory is to find lower bounds on $\alpha_q(\delta)$ for $0 < \delta < (q-1)/q$. A classical lower bound is the asymptotic Gilbert-Varshamov bound which says that

$$\alpha_q(\delta) \geq 1 - \delta \log_q(q-1) + \delta \log_q \delta + (1-\delta)\log_q(1-\delta) \quad \text{for} \quad 0 < \delta < \frac{q-1}{q}.$$

It was widely believed that this bound is best possible, but in an important breakthrough Tsfasman, Vlăduţ, and Zink [11] showed that one can beat the asymptotic Gilbert-Varshamov bound by using Goppa's algebraic-geometry codes (see [3], [8, Chapter 6], [9], [10] for the latter codes).

Before we can introduce the Tsfasman-Vlăduţ-Zink bound, we have to recall the quantity

$$A(q) = \limsup_{g \to \infty} \frac{N_q(g)}{g} \tag{1.2}$$

from the theory of global function fields. Here $N_q(g)$ denotes the maximum number of rational places that a global function field of genus g with full constant field $\mathbb{F}_q$ can have (compare with [8, Chapter 1] and [9, Chapter V]). Now the Tsfasman-Vlăduţ-Zink bound in [11] says that

$$\alpha_q(\delta) \geq 1 - \delta - \frac{1}{A(q)} \qquad \text{for} \quad 0 \leq \delta \leq 1. \tag{1.3}$$

The bound (1.3) was used by Tsfasman, Vlăduţ, and Zink [11] and later by Niederreiter and Xing [7] to beat the asymptotic Gilbert-Varshamov bound in various cases. We refer also to [8, Section 6.2] for an expository account of these results.

It took over 20 years for the bound (1.3) to get improved. This was recently achieved by Xing [12] who showed that for any $\delta \in (0,1)$ we have

$$\alpha_q(\delta) \geq 1 - \delta - \frac{1}{A(q)} + \sum_{i=2}^{\infty} \log_q \left(1 + \frac{q-1}{q^{2i}}\right). \tag{1.4}$$

The proof of (1.4) given in [12] proceeds in a nonconstructive manner.

In the present paper we establish an improvement on the Xing bound (1.4) and thus also on the Tsfasman-Vlăduţ-Zink bound (1.3). Moreover, the proof of our bound is obtained constructively in a certain range for δ (see Corollary 4.8). However, if we want to extend the range for δ to the full range in which our new bound is meaningful, then we have to use nonconstructive arguments (see Corollary 5.4). Our main results are based on the theory of global function fields (see [8] and [9] for the necessary background). The construction of our (not necessarily linear) codes yielding the improved lower bound on $\alpha_q(\delta)$ takes its cue from the construction of digital nets given by the authors in [5].

2. Preliminaries

In this section we give some definitions and preliminary results that we use later in the paper. Throughout the paper we assume that m is an integer with $m \geq 2$.

Let $n \geq 1$ be an integer. For $\boldsymbol{a} = (a_1^{(1)}, \ldots, a_m^{(1)}, \ldots, a_1^{(n)}, \ldots, a_m^{(n)}) \in \mathbb{F}_q^{mn}$, we define the subsets $I_m(\boldsymbol{a}), I_{m-1}(\boldsymbol{a}), \ldots, I_1(\boldsymbol{a})$ of $\{1, \ldots, n\}$ as

$$\begin{aligned}
I_m(\boldsymbol{a}) &= \left\{i \in \{1, \ldots n\} : a_m^{(i)} \neq 0\right\}, \\
I_{m-1}(\boldsymbol{a}) &= \left\{i \in \{1, \ldots, n\} : a_m^{(i)} = 0,\ a_{m-1}^{(i)} \neq 0\right\}, \\
&\ \vdots \\
I_1(\boldsymbol{a}) &= \left\{i \in \{1, \ldots, n\} : a_m^{(i)} = \cdots = a_2^{(i)} = 0,\ a_1^{(i)} \neq 0\right\}.
\end{aligned}$$

For given positive real numbers $x_1, \ldots, x_m$ with $x_1 + \cdots + x_m < 1$, let $M_{q,n}(x_1, \ldots, x_m)$ be the subset of $\mathbb{F}_q^{mn}$ defined as

$$M_{q,n}(x_1, \ldots, x_m) = \{\boldsymbol{a} \in \mathbb{F}_q^{mn} : |I_1(\boldsymbol{a})| = \lfloor x_1 n \rfloor, \ldots, |I_m(\boldsymbol{a})| = \lfloor x_m n \rfloor\}. \tag{2.1}$$

Note that, in general, $M_{q,n}(x_1, \ldots, x_m) \subseteq \mathbb{F}_q^{mn}$ is not in the form of a cartesian product $W_1 \times \cdots \times W_m$ with subsets $W_1, \ldots, W_m \subseteq \mathbb{F}_q^n$.

We observe that for the cardinality of the set $M_{q,n}(x_1, \ldots, x_m)$ we have

$$\begin{aligned}
&|M_{q,n}(x_1, \ldots, x_m)| \\
&= \binom{n}{\lfloor x_m n \rfloor}(q-1)^{\lfloor x_m n \rfloor} q^{(m-1)\lfloor x_m n \rfloor} \\
&\times \binom{n - \lfloor x_m n \rfloor}{\lfloor x_{m-1} n \rfloor}(q-1)^{\lfloor x_{m-1} n \rfloor} q^{(m-2)\lfloor x_{m-1} n \rfloor} \\
&\times \cdots \\
&\times \binom{n - (\lfloor x_m n \rfloor + \lfloor x_{m-1} n \rfloor + \cdots + \lfloor x_2 n \rfloor)}{\lfloor x_1 n \rfloor}(q-1)^{\lfloor x_1 n \rfloor}.
\end{aligned} \tag{2.2}$$

Now we introduce some weight functions that we use later. For $\boldsymbol{\alpha} = (\alpha_1, \ldots, \alpha_{m+1}) \in \mathbb{F}_q^{m+1}$, let

$$\begin{aligned}
v_{m+1}(\boldsymbol{\alpha}) &= m+1 && \text{if } \alpha_{m+1} \neq 0, \\
v_{m+1}(\boldsymbol{\alpha}) &= m && \text{if } \alpha_{m+1} = 0 \text{ and } \alpha_m \neq 0, \\
& && \vdots \\
v_{m+1}(\boldsymbol{\alpha}) &= 1 && \text{if } \alpha_{m+1} = \cdots = \alpha_2 = 0 \text{ and } \alpha_1 \neq 0, \\
v_{m+1}(\boldsymbol{\alpha}) &= 0 && \text{if } \boldsymbol{\alpha} = \boldsymbol{0}.
\end{aligned} \tag{2.3}$$

For any $\boldsymbol{a} = (a_1^{(1)}, a_2^{(1)}, \ldots, a_{m+1}^{(1)}, \ldots\ldots, a_1^{(n)}, a_2^{(n)}, \ldots, a_{m+1}^{(n)}) \in \mathbb{F}_q^{(m+1)n}$, we denote $\boldsymbol{a}^{(i)} = (a_1^{(i)}, a_2^{(i)}, \ldots, a_{m+1}^{(i)}) \in \mathbb{F}_q^{m+1}$ for $i = 1, \ldots, n$. We define the weight function $V_{m+1}(\cdot)$ on $\mathbb{F}_q^{(m+1)n}$ as

$$V_{m+1}(\boldsymbol{a}) = \sum_{i=1}^{n} v_{m+1}(\boldsymbol{a}^{(i)}) \tag{2.4}$$

(see also [6]).

For $\boldsymbol{\beta} = (\beta_1, \ldots, \beta_m) \in \mathbb{F}_q^m$, let

$$\begin{array}{ll} v_{(2,\ldots,m+1)}(\boldsymbol{\beta}) = m+1 & \text{if } \beta_m \neq 0, \\ v_{(2,\ldots,m+1)}(\boldsymbol{\beta}) = m & \text{if } \beta_m = 0 \text{ and } \beta_{m-1} \neq 0, \\ & \vdots \\ v_{(2,\ldots,m+1)}(\boldsymbol{\beta}) = 2 & \text{if } \beta_m = \cdots = \beta_2 = 0 \text{ and } \beta_1 \neq 0, \\ v_{(2,\ldots,m+1)}(\boldsymbol{\beta}) = 0 & \text{if } \boldsymbol{\beta} = \mathbf{0}. \end{array} \tag{2.5}$$

For any $\boldsymbol{b} = (b_1^{(1)}, b_2^{(1)}, \ldots, b_m^{(1)}, \ldots\ldots, b_1^{(n)}, b_2^{(n)}, \ldots, b_m^{(n)}) \in \mathbb{F}_q^{mn}$, we denote $\boldsymbol{b}^{(i)} = (b_1^{(i)}, b_2^{(i)}, \ldots, b_m^{(i)}) \in \mathbb{F}_q^m$ for $i = 1, \ldots, n$. We define the weight function $V_{(2,\ldots,m+1)}(\cdot)$ on $\mathbb{F}_q^{mn}$ as

$$V_{(2,\ldots,m+1)}(\boldsymbol{b}) = \sum_{i=1}^{n} v_{(2,\ldots,m+1)}(\boldsymbol{b}^{(i)}). \tag{2.6}$$

For a nonempty set $W \subseteq \mathbb{F}_q^{mn}$, let

$$R_{(2,\ldots,m+1)}(W) = \max\{V_{(2,\ldots,m+1)}(\boldsymbol{a}) : \boldsymbol{a} \in W\}$$

and

$$D_{(2,\ldots,m+1)}(W) = \max\{V_{(2,\ldots,m+1)}(\boldsymbol{a} - \boldsymbol{b}) : \boldsymbol{a}, \boldsymbol{b} \in W\}.$$

We prove a simple lemma.

Lemma 2.1. *For any nonempty set $W \subseteq \mathbb{F}_q^{mn}$, we have*

$$D_{(2,\ldots,m+1)}(W) \leq 2R_{(2,\ldots,m+1)}(W).$$

Proof. Let

$$\boldsymbol{a} = (a_1^{(1)}, \ldots, a_m^{(1)}, \ldots, a_1^{(n)}, \ldots, a_m^{(n)}) \in W \quad \text{and}$$
$$\boldsymbol{b} = (b_1^{(1)}, \ldots, b_m^{(1)}, \ldots, b_1^{(n)}, \ldots, b_m^{(n)}) \in W$$

be such that $D_{(2,\ldots,m+1)}(W) = V_{(2,\ldots,m+1)}(\boldsymbol{a} - \boldsymbol{b})$. For $i = 1, \ldots, n$, let

$$\boldsymbol{a}^{(i)} = (a_1^{(i)}, \ldots, a_m^{(i)}) \quad \text{and} \quad \boldsymbol{b}^{(i)} = (b_1^{(i)}, \ldots, b_m^{(i)}).$$

Then we have

$$D_{(2,\ldots,m+1)}(W) = \sum_{i=1}^{n} v_{(2,\ldots,m+1)}(\boldsymbol{a}^{(i)} - \boldsymbol{b}^{(i)}). \tag{2.7}$$

First we prove that for each $1 \leq i \leq n$,

$$v_{(2,\ldots,m+1)}(\boldsymbol{a}^{(i)} - \boldsymbol{b}^{(i)}) \leq v_{(2,\ldots,m+1)}(\boldsymbol{a}^{(i)}) + v_{(2,\ldots,m+1)}(\boldsymbol{b}^{(i)}). \tag{2.8}$$

Let $1 \leq i \leq n$ be a fixed index. If $\boldsymbol{a}^{(i)} = \boldsymbol{b}^{(i)} = \boldsymbol{0}$, then (2.8) holds trivially. Assume that either $\boldsymbol{a}^{(i)} \neq \boldsymbol{0}$ or $\boldsymbol{b}^{(i)} \neq \boldsymbol{0}$. Let l be the largest integer such that either $a_l^{(i)} \neq 0$ or $b_l^{(i)} \neq 0$. Then $a_t^{(i)} - b_t^{(i)} = 0$ for each $t \geq l+1$ and hence $v_{(2,\dots,m+1)}(\boldsymbol{a}^{(i)} - \boldsymbol{b}^{(i)}) \leq l+1$. Since either $a_l^{(i)} \neq 0$ or $b_l^{(i)} \neq 0$, we have $v_{(2,\dots,m_1)}(\boldsymbol{a}^{(i)}) \geq l+1$ or $v_{(2,\dots,m+1)}(\boldsymbol{b}^{(i)}) \geq l+1$. This proves (2.8).

Hence using (2.7) and (2.8), we obtain

$$D_{(2,\dots,m+1)}(W) \leq \sum_{i=1}^{n} v_{(2,\dots,m+1)}(\boldsymbol{a}^{(i)}) + \sum_{i=1}^{n} v_{(2,\dots,m+1)}(\boldsymbol{b}^{(i)}). \tag{2.9}$$

By the definition of $R_{(2,\dots,m+1)}(W)$ we also have

$$\begin{aligned} V_{(2,\dots,m+1)}(\boldsymbol{a}) &= \textstyle\sum_{i=1}^{n} v_{(2,\dots,m+1)}(\boldsymbol{a}^{(i)}) \leq R_{(2,\dots,m+1)}(W) \quad \text{and} \\ V_{(2,\dots,m+1)}(\boldsymbol{b}) &= \textstyle\sum_{i=1}^{n} v_{(2,\dots,m+1)}(\boldsymbol{b}^{(i)}) \leq R_{(2,\dots,m+1)}(W). \end{aligned} \tag{2.10}$$

Combining (2.9) and (2.10), we obtain

$$D_{(2,\dots,m+1)}(W) \leq 2R_{(2,\dots,m+1)}(W).$$

□

Throughout the paper we denote the Hamming weight on $\mathbb{F}_q^n$ as $V_1(\cdot)$.

3. The Basic Code Construction

In this section we give our basic code construction. We note that the crucial idea of the definition of the map Φ below stems from the construction of digital nets in [5].

Let $F/\mathbb{F}_q$ be a global function field over $\mathbb{F}_q$ with at least $n+1$ rational places, where $n \geq 1$. Assume that the genus g of F satisfies $2g \leq n$. Let $m \geq 2$ and r be integers with

$$mn + 2g - 1 \leq r \leq (m+1)n - 1. \tag{3.1}$$

Let $x_1, \dots, x_m$ be positive real numbers with $x_1 + \cdots + x_m < 1$ satisfying

$$(m+1)n - r - 2\left(2\lfloor x_1 n \rfloor + 3\lfloor x_2 n \rfloor + \cdots + (m+1)\lfloor x_m n \rfloor\right) \geq 1. \tag{3.2}$$

Let $P_0, P_1, \dots, P_n$ be distinct rational places of F. Let G be the divisor of F defined as $G = rP_0$; more generally, G could be any divisor of F of degree r with $\operatorname{supp}(G) \cap \{P_1, \dots, P_n\} = \emptyset$. For each $i = 1, \dots, n$, let t_i be a local parameter at P_i. For each $i = 1, \dots, n$ and f in the Riemann-Roch space $\mathcal{L}(G)$, let $f^{(0)}(P_i) \in \mathbb{F}_q$ be the evaluation $f(P_i)$ and for $l = 1, \dots, m$, let $f^{(l)}(P_i) \in \mathbb{F}_q$ be the recursively defined evaluation

$$\left(\frac{f - (f^{(0)}(P_i) + f^{(1)}(P_i)t_i + \cdots + f^{(l-1)}(P_i)t_i^{l-1})}{t_i^l}\right)(P_i)$$

(compare also with [5, Section 3] and [8, pp. 5–6]).

For each $i = 1, \ldots, n$, let

$$\begin{array}{rcl} \varphi_i : \mathcal{L}(G) & \to & \mathbb{F}_q^m \\ f & \mapsto & \left(f^{(m-1)}(P_i), \ldots, f^{(1)}(P_i), f^{(0)}(P_i)\right), \end{array}$$

and let Φ be the linear map defined as

$$\begin{array}{rcl} \Phi : \mathcal{L}(G) & \to & \mathbb{F}_q^{mn} \\ f & \mapsto & (\varphi_1(f), \ldots, \varphi_n(f)). \end{array}$$

Note that $\operatorname{Ker} \Phi = \mathcal{L}\left(G - m(P_1 + \cdots + P_n)\right)$. Using the Riemann-Roch theorem, we get

$$\dim \operatorname{Ker} \Phi = r - mn + 1 - g, \tag{3.3}$$

since $r - mn \geq 2g - 1$ by (3.1). Again using the Riemann-Roch theorem, we obtain

$$\dim \mathcal{L}(G) = r + 1 - g$$

and hence Φ is surjective.

Let $N_{q,n}(x_1, \ldots, x_m) = \Phi^{-1}\left(M_{q,n}(x_1, \ldots, x_m)\right)$ be the inverse image of the set $M_{q,n}(x_1, \ldots, x_m) \subseteq \mathbb{F}_q^{mn}$ under the linear map Φ. By (3.3) we have

$$|N_{q,n}(x_1, \ldots, x_m)| = q^{r+1-mn-g} |M_{q,n}(x_1, \ldots, x_m)|.$$

Let ϕ be the map from $N_{q,n}(x_1, \ldots, x_m) \subseteq \mathcal{L}(G)$ to $\mathbb{F}_q^n$ defined as

$$\begin{array}{rcl} \phi : N_{q,n}(x_1, \ldots, x_m) & \to & \mathbb{F}_q^n \\ f & \mapsto & \left(f^{(m)}(P_1), \ldots, f^{(m)}(P_n)\right). \end{array}$$

Note that ϕ depends on the real numbers $x_1, \ldots, x_m$. Let $C \subseteq \mathbb{F}_q^n$ be the code defined as the image of ϕ. Note that C is a nonlinear code in general.

Theorem 3.1. *Under the notation and assumptions as above,*

$$C \text{ is a } q\text{-ary } (n, q^{r+1-mn-g} |M_{q,n}(x_1, \ldots, x_m)|, d) \text{ code}$$

with

$$d \geq (m+1)n - r - 2 \sum_{j=1}^{m} (j+1) \lfloor x_j n \rfloor.$$

Proof. Let $f,\ g \in N_{q,n}(x_1, \ldots, x_m) \subseteq \mathcal{L}(G)$ be distinct functions. For $i = 1, \ldots, n$, let $\boldsymbol{a}_i,\ \boldsymbol{b}_i \in \mathbb{F}_q^{m+1}$ be the vectors defined as

$$\boldsymbol{a}_i = \left(f^{(m)}(P_i), \ldots, f^{(0)}(P_i)\right),$$

$$\boldsymbol{b}_i = \left(g^{(m)}(P_i), \ldots, g^{(0)}(P_i)\right).$$

Using these we define the vectors $\boldsymbol{a},\ \boldsymbol{b} \in \mathbb{F}_q^{(m+1)n}$ as

$$\boldsymbol{a} = (\boldsymbol{a}_1, \ldots, \boldsymbol{a}_n), \quad \boldsymbol{b} = (\boldsymbol{b}_1, \ldots, \boldsymbol{b}_n).$$

Since $f,\ g \in \mathcal{L}(G)$, from the definition (2.3) we obtain

$$f - g \in \mathcal{L}\Big(G - \sum_{i=1}^{n} \left(m + 1 - v_{m+1}\left(\boldsymbol{a}_i - \boldsymbol{b}_i\right)\right) P_i\Big). \tag{3.4}$$

As $f \neq g$, the degree of the divisor in (3.4) is nonnegative and hence using (2.4) we get

$$V_{m+1}(\boldsymbol{a}-\boldsymbol{b}) \geq (m+1)n - r. \tag{3.5}$$

On the other hand, the definitions (2.4) and (2.6) and the definitions of ϕ and Φ yield

$$V_{m+1}(\boldsymbol{a}-\boldsymbol{b}) \leq V_1(\phi(f)-\phi(g)) + V_{(2,\dots,m+1)}(\Phi(f)-\Phi(g)). \tag{3.6}$$

By the definition of $M_{q,n}(x_1,\dots,x_m)$ we have

$$\begin{aligned} &R_{(2,\dots,m+1)}(M_{q,n}(x_1,\dots,x_m)) \\ &\leq 2\lfloor x_1 n\rfloor + 3\lfloor x_2 n\rfloor + \cdots + (m+1)\lfloor x_m n\rfloor. \end{aligned} \tag{3.7}$$

Using Lemma 2.1 and (3.7), we get

$$\begin{aligned} &V_{(2,\dots,m+1)}(\Phi(f)-\Phi(g)) \\ &\leq 2\left(2\lfloor x_1 n\rfloor + 3\lfloor x_2 n\rfloor + \cdots + (m+1)\lfloor x_m n\rfloor\right). \end{aligned} \tag{3.8}$$

From (3.5), (3.6), and (3.8) we obtain

$$\begin{aligned} &V_1(\phi(f)-\phi(g)) \\ &\geq (m+1)n - r - 2\left(2\lfloor x_1 n\rfloor + 3\lfloor x_2 n\rfloor + \cdots + (m+1)\lfloor x_m n\rfloor\right). \end{aligned} \tag{3.9}$$

Using (3.2) and (3.9), it follows also that ϕ is injective, and so the number of codewords in C is $|N_{q,n}(x_1,\dots,x_m)|$. This completes the proof. □

4. Asymptotic Construction

In this section we present a constructive class of asymptotic codes leading to an improvement on the Tsfasman-Vlăduţ-Zink and Xing bounds.

First we give some definitions and we prove some technical lemmas that we use later in this section.

Let $E(x)$ be the real-valued function defined on the open interval $0 < x < 1$ by

$$E(x) = -x\log_q x - (1-x)\log_q(1-x).$$

Lemma 4.1. *For any real number $0 < x < 1$, we have*

$$\lim_{n\to\infty} \frac{\log_q \binom{n}{\lfloor xn\rfloor}}{n} = E(x).$$

Proof. This follows immediately from [4, Chapter 10, Lemma 7]. □

Let $P_m(x_1,\dots,x_m)$ be the real-valued function defined on the open domain $D=\{(x_1,\dots,x_m)\in\mathbb{R}^m : x_1>0,\dots,x_m>0, \text{ and } x_1+\cdots+x_m<1\}$ by

$$\begin{aligned}
&P_m(x_1,\dots,x_m)\\
&= E(x_m)+\sum_{j=1}^{m-1}(1-x_m-x_{m-1}-\cdots-x_{j+1})E\left(\frac{x_j}{1-x_m-x_{m-1}-\cdots-x_{j+1}}\right)\\
&\quad+\left(\sum_{j=1}^{m}x_j\right)\log_q(q-1)+\sum_{j=2}^{m}(j-1)x_j.
\end{aligned}$$

Lemma 4.2. *Under the notation and assumptions as above, for each $(x_1,\dots,x_m)$ in the domain D of the function $P_m(x_1,\dots,x_m)$, we have*

$$\lim_{n\to\infty}\frac{\log_q|M_{q,n}(x_1,\dots,x_m)|}{n}=P_m(x_1,\dots,x_m).$$

Proof. This follows from (2.2) and Lemma 4.1. □

Next we note that, by straightforward manipulations, the expression for $P_m(x_1,\dots,x_m)$ can be simplified to

$$\begin{aligned}
P_m(x_1,\dots,x_m) &= -\textstyle\sum_{j=1}^m x_j\log_q x_j-\left(1-\sum_{j=1}^m x_j\right)\log_q\left(1-\sum_{j=1}^m x_j\right)\\
&\quad+\left(\textstyle\sum_{j=1}^m x_j\right)\log_q(q-1)+\sum_{j=2}^m(j-1)x_j.
\end{aligned}\tag{4.1}$$

Let $Q_m(x_1,\dots,x_m)$ be the real-valued function defined on the same domain D as above by

$$Q_m(x_1,\dots,x_m)=P_m(x_1,\dots,x_m)-2\sum_{j=1}^{m}(j+1)x_j.\tag{4.2}$$

We deduce from (4.1) and (4.2) that for each $1\le l\le m$, the partial derivative of $Q_m(x_1,\dots,x_m)$ with respect to x_l is given by

$$\begin{aligned}
&\frac{\partial Q_m(x_1,\dots,x_m)}{\partial x_l}\\
&= \log_q\left(\frac{1-(x_1+\cdots+x_m)}{x_l}\right)+\log_q(q-1)-(l+3).
\end{aligned}\tag{4.3}$$

Let

$$\alpha_m=\frac{q-1}{q^{m+3}+q^m-1}\tag{4.4}$$

and for $1\le l\le m-1$ let

$$\alpha_{m-l}=q^l\alpha_m=\frac{q^l(q-1)}{q^{m+3}+q^m-1}.\tag{4.5}$$

Lemma 4.3. *Under the notation and assumptions as above, the critical point of the function $Q_m(x_1,\dots,x_m)$ is $(\alpha_1,\dots,\alpha_m)$.*

Proof. By (4.3), it suffices to show that $(\alpha_1,\dots,\alpha_m) \in D \subseteq \mathbb{R}^m$ is the unique solution of the system of equations

$$\frac{x_l}{1-(x_1+\cdots+x_m)} = \frac{q-1}{q^{l+3}} \qquad \text{for } l = 1,\dots,m. \tag{4.6}$$

Adding the equations in (4.6), we get

$$\frac{x_1+\cdots+x_m}{1-(x_1+\cdots+x_m)} = \frac{q^m-1}{q^{m+3}}.$$

Hence

$$x_1+\cdots+x_m = \frac{q^m-1}{q^{m+3}+q^m-1}$$

and

$$1-(x_1+\cdots+x_m) = \frac{q^{m+3}}{q^{m+3}+q^m-1}. \tag{4.7}$$

We complete the proof using (4.6) and (4.7). □

Lemma 4.4. *Under the notation and assumptions as above, we have*

$$Q_m(\alpha_1,\dots,\alpha_m) = \log_q\left(1+\frac{q^m-1}{q^{m+3}}\right).$$

Proof. By (4.1) and (4.2) we have

$$\begin{aligned}
Q_m(\alpha_1,\dots,\alpha_m) &= Q_m(q^{m-1}\alpha_m, q^{m-2}\alpha_m,\dots,\alpha_m) \\
&= -\sum_{j=0}^{m-1} q^j\alpha_m \log_q(q^j\alpha_m) - \left(1-\frac{q^m-1}{q-1}\alpha_m\right)\log_q\left(1-\frac{q^m-1}{q-1}\alpha_m\right) \\
&\quad +\frac{q^m-1}{q-1}\alpha_m\log_q(q-1) - \sum_{j=1}^{m}(j+3)q^{m-j}\alpha_m \\
&= -\alpha_m\sum_{j=0}^{m-1} jq^j - \frac{q^m-1}{q-1}\alpha_m\log_q\alpha_m + \frac{q^{m+3}}{q^{m+3}+q^m-1}\log_q\left(1+\frac{q^m-1}{q^{m+3}}\right) \\
&\quad +\frac{q^m-1}{q-1}\alpha_m\log_q(q-1) - \alpha_m\sum_{j=0}^{m-1}(m-j+3)q^j \\
&= \frac{q^m-1}{q-1}\alpha_m(\log_q(q-1)-\log_q\alpha_m - m-3) \\
&\quad +\frac{q^{m+3}}{q^{m+3}+q^m-1}\log_q\left(1+\frac{q^m-1}{q^{m+3}}\right) \\
&= \frac{q^m-1}{q^{m+3}+q^m-1}\log_q\left(1+\frac{q^m-1}{q^{m+3}}\right) + \frac{q^{m+3}}{q^{m+3}+q^m-1}\log_q\left(1+\frac{q^m-1}{q^{m+3}}\right),
\end{aligned}$$

which yields the desired result. □

Now we describe our asymptotic construction. Let $\{F_i/\mathbb{F}_q\}_{i=1}^{\infty}$ be a sequence of global function fields over $\mathbb{F}_q$, where F_i is of genus g_i with n_i rational places for $i = 1, 2, \ldots$, $\lim_{i\to\infty} g_i = \infty$, and $\lim_{i\to\infty} \frac{n_i}{g_i} = \lambda$ such that

$$\lambda > \frac{2}{1 - 2\left(2\alpha_1 + \cdots + (m+1)\alpha_m\right)}.$$

It is easily seen that

$$2\alpha_1 + \cdots + (m+1)\alpha_m < \frac{1}{2},$$

so we have, in particular, $\lambda > 0$. Let μ be any real number with

$$0 \leq \mu < 1 - \frac{2}{\lambda} - 2\left(2\alpha_1 + \cdots + (m+1)\alpha_m\right).$$

Then for sufficiently large i, we have $2g_i \leq n_i - 1$ and we can choose integers r_i such that the inequalities

$$mn_i + 2g_i - 1 \leq r_i,$$

$$(m+1)n_i - r_i - 2\left(2\lfloor\alpha_1 n_i\rfloor + \cdots + (m+1)\lfloor\alpha_m n_i\rfloor\right) \geq 1,$$

are satisfied and

$$\lim_{i\to\infty} \frac{r_i}{n_i} = m + \frac{2}{\lambda} + \mu. \tag{4.8}$$

For sufficiently large i, let C_i be the code obtained from Theorem 3.1, using F_i together with r_i as determined above, with $x_j = \alpha_j$ for $j = 1, \ldots, m$. Let $d(C_i)$ be the minimum distance of the code C_i.

Theorem 4.5. *Under the notation and assumptions as above, we have*

$$\lim_{i\to\infty} \frac{d(C_i)}{n_i} \geq 1 - \frac{2}{\lambda} - \mu - 2\sum_{j=1}^{m}(j+1)\alpha_j$$

and

$$\lim_{i\to\infty} \frac{\log_q |C_i| + d(C_i)}{n_i} \geq 1 - \frac{1}{\lambda} + \log_q\left(1 + \frac{q^m - 1}{q^{m+3}}\right).$$

Proof. For sufficiently large i, by Theorem 3.1 we have

$$\log_q |C_i| = r_i + 1 - mn_i - g_i + \log_q |M_{q,n_i}(\alpha_1, \ldots, \alpha_m)| \tag{4.9}$$

and

$$d(C_i) \geq (m+1)n_i - r_i - 2\sum_{j=1}^{m}(j+1)\lfloor\alpha_j n_i\rfloor. \tag{4.10}$$

As $\lim_{i\to\infty} \frac{\lfloor\alpha_j n_i\rfloor}{n_i} = \alpha_j$ for each $j = 1, \ldots, m$, from (4.8) and (4.10) we obtain by passing to a suitable subsequence of $\{F_i/\mathbb{F}_q\}_{i=1}^{\infty}$,

$$\lim_{i\to\infty} \frac{d(C_i)}{n_i} \geq 1 - \frac{2}{\lambda} - \mu - 2\sum_{j=1}^{m}(j+1)\alpha_j.$$

By (4.9) and (4.10) we have

$$\lim_{i\to\infty} \frac{\log_q |C_i| + d(C_i)}{n_i}$$
$$\geq 1 - \frac{1}{\lambda} + \lim_{i\to\infty} \frac{\log_q |M_{q,n_i}(\alpha_1, \ldots, \alpha_m)|}{n_i} - 2\sum_{j=1}^{m}(j+1)\alpha_j.$$

Using Lemma 4.2, (4.2), and Lemma 4.4 we obtain

$$\begin{aligned} &\lim_{i\to\infty} \frac{\log_q |M_{q,n_i}(\alpha_1, \ldots, \alpha_m)|}{n_i} - 2\sum_{j=1}^{m}(j+1)\alpha_j \\ &= P_m(\alpha_1, \ldots, \alpha_m) - 2\sum_{j=1}^{m}(j+1)\alpha_j \\ &= Q_m(\alpha_1, \ldots, \alpha_m) = \log_q\left(1 + \frac{q^m - 1}{q^{m+3}}\right). \end{aligned}$$

This completes the proof. □

Remark 4.6. We recall that the codes C_i constructed above are nonlinear in general. We note that the codes C_i are constructive. Moreover, in the notation of (1.1) and (1.2), with $\lambda = A(q)$ in Theorem 4.5 we obtain a constructive class of asymptotic codes such that

$$\alpha_q(\delta) \geq 1 - \delta - \frac{1}{A(q)} + \log_q\left(1 + \frac{q^m - 1}{q^{m+3}}\right) \tag{4.11}$$

for each $m \geq 2$ and for any δ in the range

$$\delta \in \left(0, 1 - \frac{2}{A(q)} - \frac{2q^m(q-1)}{q^{m+3} + q^m - 1}\sum_{j=1}^{m}\frac{j+1}{q^j}\right).$$

Here we used the definition of the α_j in (4.4) and (4.5).

Remark 4.7. From Lemma 4.3 we observe that the choice $x_j = \alpha_j$ for $j = 1, \ldots, m$ is optimal in order to get the maximal improvement in (4.11).

Corollary 4.8. *We have*

$$\alpha_q(\delta) \geq 1 - \delta - \frac{1}{A(q)} + \log_q\left(1 + \frac{1}{q^3}\right)$$

for any δ in the range

$$\delta \in \left(0, 1 - \frac{2}{A(q)} - \frac{4q - 2}{(q-1)(q^3+1)}\right].$$

Proof. It is straightforward to show that the sequence of numbers

$$\frac{2q^m(q-1)}{q^{m+3} + q^m - 1}\sum_{j=1}^{m}\frac{j+1}{q^j}, \quad m = 2, 3, \ldots,$$

is increasing. Furthermore, we have

$$\begin{aligned}\lim_{m\to\infty}\frac{2q^m(q-1)}{q^{m+3}+q^m-1}\sum_{j=1}^{m}\frac{j+1}{q^j} &= \frac{2(q-1)}{q^3+1}\sum_{j=1}^{\infty}\frac{j+1}{q^j}\\ &= \frac{2(q-1)}{q^3+1}\left(\frac{q}{(q-1)^2}+\frac{1}{q-1}\right)=\frac{4q-2}{(q-1)(q^3+1)}.\end{aligned}$$

Therefore, for any δ in the range

$$\delta\in\left(0,1-\frac{2}{A(q)}-\frac{4q-2}{(q-1)(q^3+1)}\right]$$

the bound (4.11) can be applied for all $m\geq 2$. Letting $m\to\infty$ in (4.11), we obtain the desired result. □

Remark 4.9. If q is a square, then $A(q)=\sqrt{q}-1$, and so Corollary 4.8 yields

$$\alpha_q(\delta)\geq 1-\delta-\frac{1}{\sqrt{q}-1}+\log_q\left(1+\frac{1}{q^3}\right)$$

for any δ in the range in Corollary 4.8. Sequences of global function fields achieving $A(q)=\sqrt{q}-1$ in (1.2) were constructed explicitly by Garcia and Stichtenoth [1], [2].

Corollary 4.8 improves on the Xing bound in (1.4). This follows from Lemma 4.10 below and a comparison of the bounds in Corollary 4.8 and (1.4).

Lemma 4.10. *For any $q\geq 2$ we have*

$$\log_q\left(1+\frac{1}{q^3}\right)>\sum_{i=2}^{\infty}\log_q\left(1+\frac{q-1}{q^{2i}}\right).$$

Proof. We prove the equivalent inequality

$$\prod_{i=2}^{\infty}\left(1+\frac{q-1}{q^{2i}}\right)<1+\frac{1}{q^3}.$$

With $\exp(x)=e^x$ for $x\in\mathbb{R}$ we have

$$\begin{aligned}\prod_{i=2}^{\infty}\left(1+\frac{q-1}{q^{2i}}\right) &< \prod_{i=2}^{\infty}\exp\left(\frac{q-1}{q^{2i}}\right)=\exp\left(\sum_{i=2}^{\infty}\frac{q-1}{q^{2i}}\right)\\ &= \exp\left(\frac{1}{q^2(q+1)}\right).\end{aligned}$$

Using the simple bound

$$\exp(x)\leq 1+x+2x^2 \qquad \text{for } 0\leq x\leq 1,$$

we obtain

$$\begin{aligned}\exp\left(\frac{1}{q^2(q+1)}\right) &\leq 1+\frac{1}{q^2(q+1)}+\frac{2}{q^4(q+1)^2}\\ &< 1+\frac{1}{q^2(q+1)}+\frac{1}{q^3(q+1)}=1+\frac{1}{q^3},\end{aligned}$$

and the proof is complete. □

5. A Nonconstructive Extension

We now extend the validity of the lower bound on $\alpha_q(\delta)$ in Corollary 4.8 to a wider range for δ (see Corollary 5.4 below) by using a nonconstructive method.

For each $\boldsymbol{c} \in \mathbb{F}_q^{mn}$ and positive real numbers $x_1, \ldots, x_m$ with $x_1+\cdots+x_m < 1$, let $\widetilde{M}_{q,n}(x_1, \ldots, x_m; \boldsymbol{c})$ be the subset of $\mathbb{F}_q^{mn}$ defined as

$$\widetilde{M}_{q,n}(x_1,\ldots,x_m;\boldsymbol{c}) = \{\boldsymbol{a} \in \mathbb{F}_q^{mn} : |I_1(\boldsymbol{a}-\boldsymbol{c})| \leq \lfloor x_1 n\rfloor, \ldots, |I_m(\boldsymbol{a}-\boldsymbol{c})| \leq \lfloor x_m n\rfloor\}.$$

A comparison with (2.1) shows that for each $\boldsymbol{c} \in \mathbb{F}_q^{mn}$ we have

$$|\widetilde{M}_{q,n}(x_1,\ldots,x_m;\boldsymbol{c})| = |\widetilde{M}_{q,n}(x_1,\ldots,x_m;\boldsymbol{0})| \geq |M_{q,n}(x_1,\ldots,x_m)|. \tag{5.1}$$

Let $F/\mathbb{F}_q$ be a global function field over $\mathbb{F}_q$ with at least $n+1$ rational places, where $n \geq 1$. Let $m \geq 2$ and r be integers and $x_1, \ldots, x_m$ be positive real numbers with $x_1 + \cdots + x_m < 1$ satisfying

$$r \leq (m+1)n - 1 - 2\left(2\lfloor x_1 n\rfloor + 3\lfloor x_2 n\rfloor + \cdots + (m+1)\lfloor x_m n\rfloor\right). \tag{5.2}$$

Let $P_0, P_1, \ldots, P_n$ be distinct rational places of F. Let G be the divisor of F defined as $G = rP_0$. For each $i = 1, \ldots, n$, let

$$\begin{aligned}\psi_i : \mathcal{L}(G) &\rightarrow \mathbb{F}_q^m\\ f &\mapsto \left(f^{(m-1)}(P_i), \ldots, f^{(1)}(P_i), f^{(0)}(P_i)\right),\end{aligned}$$

and let Ψ be the linear map defined as

$$\begin{aligned}\Psi : \mathcal{L}(G) &\rightarrow \mathbb{F}_q^{mn}\\ f &\mapsto (\psi_1(f), \ldots, \psi_n(f)).\end{aligned}$$

Note that Ψ is not necessarily surjective.

Let S be the subset of the cartesian product $\mathcal{L}(G) \times \mathbb{F}_q^{mn}$ defined as

$$S = \{(f, \boldsymbol{c}) \in \mathcal{L}(G) \times \mathbb{F}_q^{mn} : f \in \mathcal{L}(G),\ \Psi(f) \in \widetilde{M}_{q,n}(x_1, \ldots, x_m; \boldsymbol{c})\}.$$

From (5.1) we obtain

$$|S| = |\mathcal{L}(G)| \cdot |\widetilde{M}_{q,n}(x_1, \ldots, x_m; \boldsymbol{0})|.$$

For each $\boldsymbol{c} \in \mathbb{F}_q^{mn}$, let $N_{\boldsymbol{c}} \subseteq \mathcal{L}(G)$ and $S_{\boldsymbol{c}} \subseteq S$ be the subsets defined as

$$N_{\boldsymbol{c}} = \{f \in \mathcal{L}(G) : \Psi(f) \in \widetilde{M}_{q,n}(x_1, \ldots, x_m; \boldsymbol{c})\}$$

and

$$S_{\boldsymbol{c}} = \{(f, \boldsymbol{c}) \in S : f \in N_{\boldsymbol{c}}\}.$$

Note that

$$S = \bigcup_{\boldsymbol{c} \in \mathbb{F}_q^{mn}} S_{\boldsymbol{c}}$$

and that for each $\boldsymbol{c} \in \mathbb{F}_q^{mn}$ we have

$$|S_{\boldsymbol{c}}| = |N_{\boldsymbol{c}}|.$$

Hence there exists $\boldsymbol{c} \in \mathbb{F}_q^{mn}$ such that

$$|N_{\boldsymbol{c}}| = |S_{\boldsymbol{c}}| \geq \frac{|S|}{q^{mn}} = \frac{|\mathcal{L}(G)| \cdot |\widetilde{M}_{q,n}(x_1, \ldots, x_m; \mathbf{0})|}{q^{mn}}. \tag{5.3}$$

We assume that

$$|\mathcal{L}(G)| \cdot |\widetilde{M}_{q,n}(x_1, \ldots, x_m; \mathbf{0})| > q^{mn} \tag{5.4}$$

and we fix an element $\boldsymbol{c} \in \mathbb{F}_q^{mn}$ satisfying (5.3).

Let ψ be the map from $N_{\boldsymbol{c}} \subseteq \mathcal{L}(G)$ to $\mathbb{F}_q^n$ defined as

$$\begin{array}{rcl} \psi : N_{\boldsymbol{c}} & \rightarrow & \mathbb{F}_q^n \\ f & \mapsto & \left(f^{(m)}(P_1), \ldots, f^{(m)}(P_n)\right). \end{array}$$

Let $\widetilde{C} \subseteq \mathbb{F}_q^n$ be the code defined as the image of ψ.

Theorem 5.1. *Under the notation and assumptions as above, $\widetilde{C}$ is a q-ary $(n, |N_{\boldsymbol{c}}|, d)$ code with*

$$|N_{\boldsymbol{c}}| \geq \left\lceil \frac{|\mathcal{L}(G)| \cdot |\widetilde{M}_{q,n}(x_1, \ldots, x_m; \boldsymbol{c})|}{q^{mn}} \right\rceil$$

and

$$d \geq (m+1)n - r - 2\sum_{j=1}^{m} (j+1)\lfloor x_j n \rfloor.$$

Proof. From (5.3) and (5.4), we have $|N_{\boldsymbol{c}}| \geq 2$. Let $f, g \in N_{\boldsymbol{c}}$ such that $f \neq g$. We have $\Psi(f), \Psi(g) \in \widetilde{M}_{q,n}(x_1, \ldots, x_m; \boldsymbol{c})$ and hence

$$\Psi(f) - \boldsymbol{c},\ \Psi(g) - \boldsymbol{c} \in \widetilde{M}_{q,n}(x_1, \ldots, x_m; \mathbf{0}).$$

By the definition of $\widetilde{M}_{q,n}(x_1, \ldots, x_m; \mathbf{0})$ we have

$$R_{(2,\ldots,m+1)}(\widetilde{M}_{q,n}(x_1, \ldots, x_m; \mathbf{0})) \leq 2\lfloor x_1 n \rfloor + \cdots + (m+1)\lfloor x_m n \rfloor. \tag{5.5}$$

Using Lemma 2.1 and (5.5), we get

$$\begin{aligned} V_{(2,\ldots,m+1)}(\Psi(f) - \Psi(g)) &= V_{(2,\ldots,m+1)}((\Psi(f) - \boldsymbol{c}) - (\Psi(g) - \boldsymbol{c})) \\ &\leq 2(2\lfloor x_1 n \rfloor + \cdots + (m+1)\lfloor x_m n \rfloor). \end{aligned}$$

We complete the proof as in the proof of Theorem 3.1. □

Remark 5.2. Note that $\widetilde{C}$ is nonconstructive. Moreover $\widetilde{C}$ is a nonlinear code in general. We also observe that

$$|\mathcal{L}(G)| \cdot |M_{q,n}(x_1, \ldots, x_m)| > q^{mn} \tag{5.6}$$

implies the assumption in (5.4).

Now we describe our nonconstructive asymptotic codes. Let $\{F_i/\mathbb{F}_q\}_{i=1}^{\infty}$ be a sequence of global function fields over $\mathbb{F}_q$, where F_i is of genus g_i with n_i rational places for $i = 1, 2 \ldots$, $\lim_{i\to\infty} g_i = \infty$, and $\lim_{i\to\infty} \frac{n_i}{g_i} = \gamma$ such that

$$\gamma > \frac{1}{1 + \log_q \left(1 + \frac{q^m - 1}{q^{m+3}}\right)}.$$

Let ν be any real number with

$$\frac{1}{\gamma} - \log_q \left(1 + \frac{q^m - 1}{q^{m+3}}\right) < \nu < 1.$$

Using Lemma 4.2, (4.2), and Lemma 4.4, for sufficiently large i, we can choose integers r_i such that the inequalities

$$r_i \leq (m+1)n_i - 1 - 2(2\lfloor \alpha_1 n_i \rfloor + 3\lfloor \alpha_2 n_i \rfloor + \cdots + (m+1)\lfloor \alpha_m n_i \rfloor) \tag{5.7}$$

and

$$r_i > mn_i - 1 + g_i - \log_q |M_{q,n_i}(\alpha_1, \ldots, \alpha_m)| \tag{5.8}$$

are satisfied and

$$\lim_{i\to\infty} \frac{r_i}{n_i} = m - 2\sum_{j=1}^{m} (j+1)\alpha_j + \nu.$$

From (5.4), (5.6), (5.7), and (5.8), we observe that for sufficiently large i, the conditions of Theorem 5.1 using F_i and r_i, with $x_j = \alpha_j$ for $j = 1, \ldots, m$, are satisfied. For sufficiently large i, let $\widetilde{C}_i$ be the code obtained from Theorem 5.1 and $d(\widetilde{C}_i)$ be the minimum distance of the code $\widetilde{C}_i$.

As in the proof of Theorem 4.5, we prove the following theorem.

Theorem 5.3. *Under the notation and assumptions as above, we have*

$$\lim_{i\to\infty} \frac{d(\widetilde{C}_i)}{n_i} \geq 1 - \nu$$

and

$$\lim_{i\to\infty} \frac{\log_q |\widetilde{C}_i| + d(\widetilde{C}_i)}{n_i} \geq 1 - \frac{1}{\gamma} + \log_q \left(1 + \frac{q^m - 1}{q^{m+3}}\right).$$

Proof. For sufficiently large i, by Theorem 5.1 and (5.1) we have

$$\begin{aligned} \log_q |\widetilde{C}_i| &\geq \log_q |\mathcal{L}(G)| + \log_q |M_{q,n_i}(\alpha_1, \ldots, \alpha_m)| - mn_i \\ &\geq r_i + 1 - g_i - mn_i + \log_q |M_{q,n_i}(\alpha_1, \ldots, \alpha_m)| \end{aligned} \tag{5.9}$$

and

$$d(\widetilde{C}_i) \geq (m+1)n_i - r_i - 2\sum_{j=1}^{m}(j+1)\lfloor \alpha_j n_i \rfloor. \tag{5.10}$$

From Lemma 4.2, (4.2), Lemma 4.4, (5.9), and (5.10) we obtain by passing to a suitable subsequence of $\{F_i/\mathbb{F}_q\}_{i=1}^{\infty}$,

$$\lim_{i\to\infty} \frac{d(\widetilde{C}_i)}{n_i} \geq 1 - \nu$$

and

$$\lim_{i\to\infty} \frac{\log_q |\widetilde{C}_i| + d(\widetilde{C}_i)}{n_i} \geq 1 - \frac{1}{\gamma} + \log_q\left(1 + \frac{q^m - 1}{q^{m+3}}\right).$$

□

Corollary 5.4. *We have*

$$\alpha_q(\delta) \geq 1 - \delta - \frac{1}{A(q)} + \log_q\left(1 + \frac{1}{q^3}\right)$$

for any δ in the range

$$\delta \in \left(0, 1 - \frac{1}{A(q)} + \log_q\left(1 + \frac{1}{q^3}\right)\right).$$

Remark 5.5. If q is a square, then $A(q) = \sqrt{q} - 1$, and so Corollary 5.4 yields

$$\alpha_q(\delta) \geq 1 - \delta - \frac{1}{\sqrt{q}-1} + \log_q\left(1 + \frac{1}{q^3}\right)$$

for any δ in the range in Corollary 5.4.

Acknowledgments

The research of the first author is partially supported by the grant R-394-000-011-422 with Temasek Laboratories in Singapore. The second author is partially supported by the Turkish Academy of Sciences in the framework of the Young Scientists Award Programme (F.Ö./TÜBA-GEBIP/2003-13).

References

[1] Garcia, A., Stichtenoth, H.: A tower of Artin-Schreier extensions of function fields attaining the Drinfeld-Vladut bound. Invent. Math. **121**, 211–222 (1995)

[2] Garcia, A., Stichtenoth, H.: On the asymptotic behaviour of some towers of function fields over finite fields. J. Number Theory **61**, 248–273 (1996)

[3] Goppa, V.D.: Codes on algebraic curves (in Russian). Dokl. Akad. Nauk SSSR **259**, 1289–1290 (1981)

[4] MacWilliams, F.J., Sloane, N.J.A.: The Theory of Error-Correcting Codes. North-Holland, Amsterdam (1977)

[5] Niederreiter, H., Özbudak, F.: Constructions of digital nets using global function fields. Acta Arith. **105**, 279–302 (2002)

[6] Niederreiter, H., Pirsic, G.: Duality for digital nets and its applications. Acta Arith. **97**, 173–182 (2001)

[7] Niederreiter, H., Xing, C.P.: Towers of global function fields with asymptotically many rational places and an improvement on the Gilbert-Varshamov bound. Math. Nachr. **195**, 171–186 (1998)

[8] Niederreiter, H., Xing, C.P.: Rational Points on Curves over Finite Fields: Theory and Applications. Cambridge University Press, Cambridge (2001)

[9] Stichtenoth, H.: Algebraic Function Fields and Codes. Springer, Berlin (1993)

[10] Tsfasman, M.A., Vlăduţ, S.G.: Algebraic-Geometric Codes. Kluwer, Dordrecht (1991)

[11] Tsfasman, M.A., Vlăduţ, S.G., Zink, T.: Modular curves, Shimura curves, and Goppa codes, better than Varshamov-Gilbert bound. Math. Nachr. **109**, 21–28 (1982)

[12] Xing, C.P.: Nonlinear codes from algebraic curves improving the Tsfasman-Vlăduţ-Zink bound. IEEE Trans. Inform. Theory **49**, 1653–1657 (2003)

Harald Niederreiter
Department of Mathematics
National University of Singapore
2 Science Drive 2
Singapore 117543, Republic of Singapore
e-mail: `nied@math.nus.edu.sg`

Ferruh Özbudak
Department of Mathematics
Middle East Technical University
İnönü Bulvarı
06531 Ankara, Turkey
e-mail: `ozbudak@math.metu.edu.tr`

Progress in Computer Science and Applied Logic, Vol. 23, 277–288

Malleability Attacks on Multi-Party Key Agreement Protocols

Josef Pieprzyk and Huaxiong Wang

Abstract. Multi-party key agreement protocols indirectly assume that each principal equally contributes to the final form of the key. In this paper we consider three malleability attacks on multi-party key agreement protocols. The first attack, called *strong key control*, allows a dishonest principal (or a group of principals) to fix the key to a pre-set value. The second attack is *weak key control* in which the key is still random, but the set from which the key is drawn is much smaller than expected. The third attack is named *selective key control* in which a dishonest principal (or a group of dishonest principals) is able to remove a contribution of honest principals to the group key. The paper discusses the above three attacks on several key agreement protocols, including DH (Diffie-Hellman), BD (Burmester-Desmedt) and JV (Just-Vaudenay). We show that dishonest principals in all three protocols can weakly control the key, and the only protocol which does not allow for strong key control is the DH protocol. The BD and JV protocols permit to modify the group key by any pair of neighboring principals. This modification remains undetected by honest principals.

Mathematics Subject Classification (2000). Primary 94A60, 68P25, 68M10.

Keywords. Public-key cryptography, key agreement protocols, malleability attacks, key control.

1. Introduction

Multi-party key establishment protocols are necessary cryptographic tools whenever a group of principals (people or processes) would like to create a secure channel among themselves (either to provide confidential or authenticated communication or both). Traditionally, key establishment protocols are divided into two broad categories: key transport and key agreement protocols. In key transport protocols, the common key is generated by the so-called trusted authority and distributed via pre-set secure communication channels. The most prominent example of such a

The work was in part supported by Australian Research Council Discovery grants DP0345366 and DP0344444.

protocol is the Needham-Schroeder protocol [12]. In key agreement protocols, however, the common key is composed from information provided by the co-operating principals. Diffie and Hellman in their revolutionary paper [6] proposed the first key agreement protocol. It allows two parties to agree on a secret via a public discussion. The protocol, however, is subject to the so called "meet-in-the-middle" attack. The reason for this is the lack of mutual authentication of the parties. They certainly can establish a common secret key, but they are not sure with whom. Later Diffie, van Oorschot and Wiener [7] rectified the Diffie-Hellman (DH) protocol in which both parties are sure of the identity of each other. This is the well-known Station-to-Station (STS) protocol [7].

The DH 2-party protocol was generalized by Burmester and Desmedt [4] for multi-party. Although elegant, the original BD protocol did not consider the entity authentication and so suffered from the man-in-the-middle attack as well. A number of protocols for authenticated multi-party key agreement protocols have been then suggested [9, 3, 1, 2, 14], unfortunately, none of them gave rigorous security proofs. Recently, Katz and Yung [8] presented the first constant-round authenticated multi-party key agreement protocol that is provably secure under known cryptographic assumptions.

The cryptographic protocols are evaluated using well-defined security goals (see [10] for details). Among them the following goals are considered as the basic ones.

- Key freshness.
- Key confidentiality.
- Mutual entity authentication.
- Key confirmation.

One of the few security goals which has not been fully studied and discussed is key control. This goal does not apply to key transport protocols as the trusted authority takes the full responsibility for key generation and the participants have no control over it. The situation is different for key agreement protocols as the secret key is negotiated among the principals.

The paper is structured as follows. Key control, its properties and its relation to other security goals are discussed in Section 2. Section 3 studies key control in the DH protocol. The BD protocol is examined in Section 4. Section 5 considers key control in the JV protocol. We conclude the paper in Section 6.

2. Properties of Key Control

The issue of key control could be ignored all together if all collaborating principals are honest and follow the protocol. However, even when the principals are honest, they may be tempted to choose their private elements in a non-random way, perhaps from elements generated previously in order to save time and computing resources. In practice, it is reasonable to assume that principals will follow the protocol in order to achieve the common goal (such as an agreement on the common secret key) especially when the discussion among principals is being done via public channel. It is, however, unreasonable to expect from principals to behave "honestly" whenever they can get some advantage over other principals while their dishonest behavior is not going to be detected.

The problem of key control in key agreement protocols was first discussed by Mitchell, Ward and Wilson [11]. They observed that in two-party key agreement protocols, always the party who sends a "partial key" first is disadvantaged as the second party may exert some control over the jointly agreed key. To make key agreements "fair" for both parties, the authors proposed to use commitments based, for instance, on collision intractable hash functions.

Multi-party cryptography was defined to enable groups of people to perform cryptographic operations. Again the belief about the group honesty was replaced by a more realistic assumption that in the group of n principals, there are at least t honest ones $(t < n)$.

Intuitively, principals who interact to jointly and equally influence the final form of the key must not be able to force other principals to accept a key which has been prefixed. It is desired that no principal or group of conspiring participants are able to fix the secret key to some (non-random) value. One can argue that the name key agreement means that all parties equally contribute to the final value of the key. Ideally, as long as there is a single honest principal who selects her contribution at random, the agreed secret is also random. In other words, key control should be impossible in an ideal key agreement protocol.

The way principals can exercise their influence over the key can be classified into

- strong key control – a dishonest principal (or group of principals) can fix the secret key to any value of their choice (i.e., the principal waits until others have published their public information and adjusts their contribution in order to get the requested value of the key),
- weak key control – a dishonest principal (or group of principals) can restrict the value of the key so it belongs to a smaller set of elements,
- selective key control – a dishonest principal (or group of principals) can remove a key contribution of a single or group of honest principals,
- no key control – a dishonest principal (or group of principals) can neither fix nor restrict the value of the key.

Note that strong key control implies weak key control. Also strong key control implies selective key control. If the protocol allows for selective key control, then a big enough collection of dishonest principals may be able to strongly control the key. However, weak key control and selective key control seem to be independent (to some degree).

Key control can be exercised by

- a single principal,
- a subgroup of principals,
- other party involved in the protocol (typically a trusted registry who generates public elements used during a run of the protocol).

Principals who wish to control the key may need to behave differently from the other principals. In other words, protocols may allow to

- hide any attempt to control the key,
- detect an attempt to control the key,
- identify the principal who wants to control the key.

If a principal or a subgroup of principals may control the key, then some other security goals may not be achievable. In particular,

- key freshness cannot be guaranteed – honest principals cannot be sure that the group key is fresh. Strong key control allows a dishonest principal to fix the key to any value including those used in the past runs of the protocol. Weak key control increases the probability that the new key has been used before (as the key belongs to a relatively small set of elements),
- key confidentiality is likely to fail – if dishonest principals force the group to accept a key known to some outsiders. The key can be compromised before the execution of the protocol in the case of strong key control. When weak key control is exercised by dishonest principals, then the exhaustive search of key space may be possible.

Clearly, key control (or lack of key freshness) makes the protocol susceptible to the replay attack. If the dishonest principals force the group to use a previously used key, then all communication can be replayed and the principals will not be able to distinguish current multi-party dialog from the replayed one.

3. Key Control in DH Protocols

The DH protocol is one of the basic cryptographic tools and in fact, it was the very first protocol used by Diffie and Hellman to demonstrate viability of public key cryptography. It can be used by any two principals A, B who wish to establish a common secret key via a public discussion. Let p be a long enough prime so that the discrete logarithm problem in $GF(p)$ is intractable and g be a primitive element randomly chosen from $GF(p)$. Both p and g are public. Now A chooses at random an integer $\alpha \in GF(p)$ and calculates

$$a = g^{\alpha}$$

Similarly, the principal B selects at random β and computes

$$b = g^{\beta}$$

Now A communicates a to B while B sends b to A. The communication channel can be public. Both principals compute the common key as

$$K = b^{\alpha} = a^{\beta} = g^{\alpha\beta}$$

From the last equality, one could conclude that both parties contribute equally to the final value of the key. It is not difficult to see that the protocol can be subject to the well-known small subgroup attack described by the following theorem.

Theorem 3.1. *Given the DH protocol with parameters as described above. Let $p - 1 = 2p_1 \dots p_t$. Then any principal can force the key K to assume a value from the short cyclic group.*

Proof. Consider that the principal A selects α deterministically. Let it be

$$\alpha = \frac{p-1}{p_i}$$

then the value $a = g^\alpha$ is an element of cyclic group with p_i elements. The principal B can only force the random choice within the group with elements $(a, a^2, \dots, a^{p_i})$. □

The following corollaries can be derived.

- If both principals attempt to control the value of the key, then the resulting key belongs to the cyclic group which is an intersection of two cyclic groups selected by A and B.
- Principals can detect an attempt of key fixing by checking whether the key K is a primitive element of $GF(p)$, or in other words whether

$$K^{\frac{p-1}{p_i}} \stackrel{?}{=} 1$$

  for some p_i. If there is such i for which the above equality holds, then the element is not primitive. The protocol allows principals to identify potential cheaters.
- There is also a version of the man-in-the-middle attack in which the attacker forces the two parties to accept the final key from a short subgroup. Assume that the attacker C sits between A and B and is able to intercept and modify messages communicated between the two parties. So the communication from A to B, i.e., $A \to B : g^\alpha$ is replaced by

$$A \to C : g^\alpha \text{ and } C \to B : (g^\alpha)^{\frac{p-1}{p_i}}$$

  Similarly, the communication $B \to A : g^\beta$ is intercepted and modified by C as follows:

$$B \to C : g^\beta \text{ and } C \to A : \left(g^\beta\right)^{\frac{p-1}{p_i}}$$

  Clearly, both parties recover the key which is $K = \left(g^{\alpha\beta}\right)^{\frac{p-1}{p_i}}$.
- To exclude the possibility of key control by using elements of short cyclic groups, it is enough to apply $GF(2^n)$ for which $2^n - 1$ is a Mersenne prime as any element different from 1 (and -1) generates the full cyclic group. Alternatively, for $GF(p)$, the prime p is chosen as a strong one, i.e., $p-1 = 2q$ where q is prime.

Recall the DH problem.

<u>Instance.</u> Given $GF(p)$, a primitive element g and a pair of integers g^α, g^β.

<u>Question.</u> What is $g^{\alpha\beta}$?

The DH problem is considered to be computationally intractable.

Theorem 3.2. *Given the DH key agreement protocol in $GF(p)$ where p, q are primes and $p-1 = 2 \times q$. If instances of the DH problem are intractable, then principals are not able to fix the common key to the pre-defined value K_F where $K_F \notin \{-1, 1\}$.*

Proof. By contradiction. Assume that the principal B after getting the partial key g^α from A can force K_F to be the agreed key. In other words, there is a polynomial-time probabilistic algorithm $\mathbb{A}$ which accepts the pair $(K_F = g^{\alpha\beta}, g^\alpha)$ and returns the proper g^β so it can be communicated to A. The algorithm may not work if the cyclic group generated by g^α does not include the element K_F. In this case, it is enough to choose K_F appropriately, i.e., from the cyclic group generated by g^α.

Note that the algorithm $\mathbb{A}$ also solves the DH problem. Given two integers (g^α, g^β). It is easy to verify that if we put the pair $(g^\alpha, g^{\beta^{-1}})$ as the input to our algorithm, then it outputs $g^{\alpha\beta}$ or

$$\mathbb{A}(g^\alpha, g^{\beta^{-1}}) = g^{\alpha\beta}$$

Clearly, to get this result we have to generate $g^{\beta^{-1}}$ first. Observe that two applications of the algorithm $\mathbb{A}$ generates it because

$$\mathbb{A}(\mathbb{A}(g^\alpha, g^\beta), g^\alpha) = g^{\beta^{-1}}$$

That is, the DH problem is easy and we have obtained the contradiction which proves the theorem. □

An attempt of key control can be detected by checking the order of two partial keys g^α and g^β. Equivalently, it is enough to check the order of $K = g^{\alpha\beta}$ as if K is a primitive element then there is no way to fix or restrict the value of the key unless the DH problem is easy.

4. The Burmester-Desmedt Protocol

The Burmester-Desmedt (BD) protocol is described in [4]. The arithmetics is performed in $GF(p)$ with g as a primitive element where $p - 1 = 2 \times q$. The protocol is an extension of the DH protocol and allows a group of n principals $P_0, \ldots, P_{n-1}$ to agree on a common and secret key. The BD protocol runs as follows.

- P_i; $i = 0, \ldots, n-1$, picks up at random $r_i \in_R GF(p)$, computes $z_i = g^{r_i}$ and broadcasts

$$P_i \rightarrow \star : z_i$$

- Each P_i checks the order of g or $ord(g) = q$. Then P_i computes

$$X_i = \left(\frac{z_{i+1}}{z_{i-1}}\right)^{r_i}$$

and broadcasts

$$P_i \rightarrow \star : X_i$$

Note that subscripts are counted modulo n.

- Principal P_i computes the common secret key

$$K_i = z_{i-1}^{nr_i} X_i^{n-1} X_{i+1}^{n-2} \dots X_{i-2}$$

Note that if all principals follow the protocol, the common key is

$$K = K_i = g^{r_0 r_1 + r_1 r_2 + \cdots + r_{n-1} r_0}$$

The BD protocol consists of two phases. In the first phase, principals broadcast their z_i. In the second one, they announce their X_i. In both phases, principals are bound to use the same r_i. Assume that principals are going to use different elements and $z_i = g^{r_i}$ while

$$X_i = \left(\frac{z_{i+1}}{z_{i-1}}\right)^{R_i}$$

and r_i, R_i are not necessarily the same. After some transformations, we can get an explicit expression for the keys $K_i = g^{e_i}$ computed by particular principal P_i where

$$\begin{aligned} e_i &= (nr_{i-1}r_i - (n-1)r_{i-1}R_i) + ((n-1)R_i r_{i+1} - (n-2)r_i R_{i+1}) \\ &\quad + \cdots + (2R_{i-3}r_{i-2} - r_{i-3}R_{i-2}) + R_{i-2}r_{i-1} \end{aligned}$$

Assume that $r_i = R_i$ for all i except $i = 1$. Then principals will recover keys with their exponents $(r_1 \neq R_1)$

$$\begin{aligned} e_1 &= (nr_0 r_1 - (n-1)r_0 R_1) + ((n-1)R_1 r_2 - (n-2)r_1 r_2) + r_2 r_3 + \cdots + r_{n-1}r_0 \\ e_2 &= (nr_1 r_2 - (n-1)r_1 r_2) + r_2 r_3 + \cdots + r_0 r_1 \\ e_3 &= r_2 r_3 + \cdots + (2r_0 r_1 - r_0 R_1) + R_1 r_2 \\ &\vdots \\ e_0 &= r_{n-1}r_0 + ((n-1)r_0 r_1 - (n-2)r_0 R_1) \\ &\quad + ((n-2)R_1 r_2 - (n-3)r_1 r_2) + r_2 r_3 + \cdots + r_{n-2}r_{n-1} \end{aligned}$$

Clearly, every key will be different so the protocol fails. The BD protocol allows to generate the same key only if all principals consistently apply the same r_i for z_i and X_i generation.

Consider the BD protocol executed by n principals with parameters defined as above. Is it possible for a group of conspiring principals to force a pre-determined key?

Theorem 4.1. *Given the BD protocol with parameters as above. A group of $(n-1)$ conspirators can force the honest principal to accept the key to the pre-defined value K_F of their choosing, i.e., the strong key control is possible by the group of $(n-1)$ principals.*

Proof. Assume that the group of conspirators is $\{P_1, \dots, P_{n-1}\}$ and the honest principal is P_0. The BD protocol is executed as prescribed to the point when the values X_i are to be announced. The dishonest principals stop there and wait until the honest one broadcasts her X_0. Now they can calculate the correct key $K = K_i$

for $i = 1, \ldots, n-1$. One of the conspirators, say P_{n-2}, now modifies his X_{n-2} to a new false value

$$X'_{n-2} = \frac{K_F X_{n-2}}{K}.$$

This value together with other (correct) ones are broadcast as requested by the protocol. The honest principal computes

$$K_0 = z_{n-1}^{nr_0} X_0^{n-1} X_1^{n-1} \ldots X'_{n-2} = K_F.$$

□

Note that a single dishonest principal may force a single honest principal to obtain the false key K_F. However, other honest principals will end up with keys different from K and K_F. This will be detected when the principals start using the keys causing their communication to fail.

Theorem 4.2. *Given the BD protocol as described above. A pair of two dishonest principals, say P_{n-2}, P_{n-1}, is able to strongly control the key. Other principals $P_0, P_1, \ldots, P_{n-3}$ are unable to detect the cheating.*

Proof. Assume that the dishonest principals wish to replace the original key K by a false one of their choice K_F. When the BD protocol is being executed, the cheaters follow it until the second stage when X_i are about to be broadcast. They wait until all honest principals have announced their X_i; $i = 0, 1, \ldots, n-3$. Having their X_{n-2}, X_{n-1}, the dishonest principals compute the key K as in the BD protocol and broadcast

$$P_{n-2} \rightarrow \star : \quad X'_{n-2} = \tfrac{K_F}{K} X_{n-2}$$
$$P_{n-1} \rightarrow \star : \quad X'_{n-1} = \tfrac{K}{K_F} X_{n-1}$$

where X'_{n-2}, X'_{n-1} are modified values of the original X_{n-2}, X_{n-1} obtained in the protocol. The key derived by P_0 is K_F and it is computed as in the proof of Theorem 4.1. Other honest principals compute their keys using their correct values X_i for $i = 0, 1, \ldots, n-3$ and false values X'_{n-2}, X'_{n-1}. Therefore

$$K_i = z_{i-1}^{nr_i} X_i^{n-1} \ldots {X'}_{n-2}^{i+1} {X'}_{n-1}^{i} \ldots X_{i-2}$$

We now evaluate

$${X'}_{n-2}^{i+1} {X'}_{n-1}^{i} = \frac{K_F}{K} X_{n-2}^{i+1} X_{n-1}^{i}$$

So all honest principals calculate $K_i = K_F$ for $i = 0, 1, \ldots, n-3$. There is no facility in the protocol to detect the manipulation by the dishonest principals. □

Now we investigate the case of weak key control.

Theorem 4.3. *Given the BD protocol with parameters as described above. Assume that P_{i-1} and P_{i+1} are cheaters who wish to annihilate P_i's contribution to the key. Then it is enough for them to choose their r_{i-1} and r_{i+1} such that*

$$g^{r_{i-1}+r_{i+1}}$$

generate a short cyclic group.

Proof. Consider the extreme case when P_{i-1} and P_{i+1} collectively select their r_{i-1} and r_{i+1} such that

$$r_{i-1} + r_{i+1} = p - 1,$$

then the agreed key is

$$K = g^{(r_{i-1}+r_{i+1})r_i} g^{r_{i+1}r_{i+3}+\cdots+r_{i-2}r_{i-1}}.$$

As $g^{(r_{i-1}+r_{i+1})r_i} = 1$, the key depends on contributions of other principals – the contribution of P_i has been removed. Clearly, the neighbors P_{i-1} and P_{i+1} can choose their contributions so $g^{r_{i-1}+r_{i+1}}$ generates a short cyclic group – the contribution of P_i can be restricted to a random element from a short cyclic group. □

Note that a single principal, either P_{i-1} or P_{i+1}, cannot weakly control the key. Assume that one of them, say P_{i-1}, wish to control the key. First P_{i-1} waits until the other neighbor P_{i+1} broadcasts their z_{i+1} and then selects an element from a short cyclic group $g^c = z_{i-1}z_{i+1}$ and computes z_{i-1}. P_{i-1}, however, is unable to compute the exponent r_{i-1} which correspond to it unless the discrete logarithm problem is easy.

The good news for P_i is that they can detect this kind of attack by checking whether the element $z_{i-1}z_{i+1}$ is primitive. If it is not, then P_i may decide to abort the protocol.

5. Key Control in the Just-Vaudenay Protocol

We give a brief description of the protocol and the reader is encouraged to refer to the paper [9] for details. It is assumed that the group of principals $\mathbb{P} = \{P_0, \ldots, P_{n-1}\}$ is already arranged into a cycle and each pair of principals has run a two-party key agreement protocol. The two party key agreement protocol is a version of the STS protocol. For the sake of clarity we assume that this protocol provides the standard collection of goals: key freshness, key confidentiality, explicit key authentication and mutual entity authentication. As the result of the execution of the protocol for pairs (P_0, P_1), ..., (P_{n-1}, P_0), the pair (P_i, P_{i+1}) holds the session key K_i. Note that the indices are computed modulo n.

JV Multi-party Key Agreement Protocol [9]

1. Each pair of principals (P_i, P_{i+1}) holds a session key K_i.
2. Each P_i computes and broadcasts $W_i = \frac{K_i}{K_{i-1}}$.
3. After receiving the values W_j, P_i computes the group key

$$K = K_{i-1}^n W_i^{n-1} W_{i+1}^{n-2} \ldots W_{i-1} = K_1 K_2 \ldots K_n.$$

The authors analyzed their protocol and proved that their protocol was secure against passive attacks. Also they considered two active attacks (the shielding and middle-person attacks). Unfortunately, the key derived in the JV protocol can be manipulated by principals.

Theorem 5.1. *The JV multi-party key agreement protocol allows a pair of principals to strongly control the key.*

Proof. The proof describes an attack in which two dishonest principals, say (P_1, P_2), force the other principals to accept a key K_F of their choice. We assume that each pair has completed their two-party key agreement protocol and each pair (P_i, P_{i+1}) holds the session key K_i. The cheaters (P_1, P_2) wait until other principals have announced their W_i. Then knowing $W_0, W_3, \ldots, W_{n-1}$, both compute

$$k = K_0 K_2, K_3, \ldots, K_{n-1}.$$

P_2 computes keys in the following order:

$$\begin{aligned} (K_2, W_3) &\rightarrow K_3, \\ (K_3, W_4) &\rightarrow K_4, \\ &\ldots \\ (K_{n-2}, W_{n-1}) &\rightarrow K_{n-1}. \end{aligned}$$

P_1 obtains the same collection of keys differently:

$$\begin{aligned} (K_n, W_n) &\rightarrow K_{n-1}, \\ (K_{n-1}, W_{n-1}) &\rightarrow K_{n-2}, \\ &\ldots \\ (K_4, W_4) &\rightarrow K_3. \end{aligned}$$

The cheaters compute their new session key (different from K_1)

$$k_1 = K_F k^{-1}$$

and they broadcast $W_1 = \frac{k_1}{K_0}$ and $W_2 = \frac{K_2}{k_1}$. Now every principal can calculate their final group key which is

$$K = K_0 k_1 K_2, K_3, \ldots, K_{n-1} = K_F.$$

The group key is K_F as intended by the cheaters. □

It seems that key control can be removed by adding a step in the protocol requiring principals to commit themselves to their true values W_i before broadcasting them. This could be done by sending signed hash values of W_i. Clearly, the attack demonstrated in the proof will no longer work. This way of ensuring of key-control freeness has been first suggested in [11]. This method of prevention does not work if manipulation of keys can be done before W_i are broadcast. The next theorem shows how a subgroup of principals can selectively remove key contributions of some victim principals.

Theorem 5.2. *Given* 5 *principals* $P_{i-2}, P_{i-1}, P_i, P_{i+1}, P_{i+2}$ *in the JV protocol holding their pairwise keys* $K_{i-2}, K_{i-1}, K_i, K_{i+1}$, *where* $n \geq 5$. *Then* (P_{i+1}, P_{i+2}) *and* (P_{i-1}, P_{i-2}) *can jointly remove the contribution of* P_i *by modifying their keys* K_{i+1}, K_{i-2} *so their keys satisfy the following equation*

$$K'_{i+1} K'_{i-2} = K_{i+1} K_{i-2} (K_i K_{i-1})^{-1}.$$

The group key is

$$K = \frac{K_0 K_1 \cdots K_{n-1}}{K_i K_{i-1}}.$$

Proof. Unlike in the previous cheating scenarios, the dishonest principals P_{i-2}, P_{i-1}, P_{i+1}, P_{i+2} do not wait for broadcast of W_i. They modify their keys using the knowledge of the keys K_i and K_{i-1} they agreed with the honest P_i. Note that P_{i-1} and P_i share with P_i keys K_{i-1} and K_i, respectively. Now the modification of their keys

$$K'_{i+1} K'_{i-2} = K_{i+1} K_{i-2} (K_i K_{i-1})^{-1}$$

removes the contribution of P_i. □

Interestingly enough the theorem is also true for $n = 3$ when any coalition of two principals can set the key to the value of their choice before broadcasting values W_i.

6. Conclusion

In this paper we have considered key control problems in multi-party key agreement protocols and presented malleability attacks on several existing key agreement protocols in which malicious principals can send "bad" protocol messages in order to bias the probability distribution of the final key or even force the session key to some desired value. Such an attack by the principals (insiders) can allow collaborating outsiders to derive the group key, even if the communication between the malicious insiders and outsiders is not allowed after the key agreement protocol begins.

While previous work has been mainly concerned with security (privacy) against (active) attacks by group outsiders, we believe that security (non-malleability) against attacks by the group insiders should be considered as a basic security requirement as well. As we have shown for DH, BD and JV protocols, other existing key agreement protocols (DH-type based) also suffer from the key control problems under the malleability attacks, thus it is desirable to design key agreement protocols that are non-malleable against active insiders. After this paper was submitted, we continued to work on this problem and made some positive progress. In the coming paper [5], we give the first multi-party agreement protocol that simultaneously achieves *key privacy* against outsiders and *non-malleability* against insiders.

Acknowledgement

The authors thank Yvo Desmedt, Keith Martin and Ron Steinfeld for inspiring discussions about key agreement protocols. We also would like to thank the anonymous reviewer for critical comments and suggested improvements to the work.

References

[1] G. Ateniese, M. Steiner and G. Tsudik. Authenticated Group Key Agreement and Friends. ACM CCCS '98.

[2] G. Ateniese, M. Steiner and G. Tsudik. New Multi-Party Authentication Services and Key Agreement Protocols. *IEEE Journal on Selected Areas in Communications*, 18(4):628–639, 2000.

[3] C. Boyd. On key agreement and conference key agreement. ACISP97.

[4] M. Burmester and Y. Desmedt. A secure and efficient conference key distribution system. In A. De Santis, editor, *Advances in Cryptology - EUROCRYPT'94*, pages 275–286. Springer, 1995. Lecture Notes in Computer Science No. 950.

[5] Y. Desmedt, J. Pieprzyk, R. Steinfeld and H. Wang. A Non-Malleable Group Key Exchange Protocol Robust Against Active Insiders. Preprint, 2004 (31 pages).

[6] W. Diffie and M.E. Hellman. New directions in cryptography. *IEEE Transactions on Information Theory*, 22:644–654, 1976.

[7] W. Diffie, P. Van Oorschot, and M. Wiener. Authentication and authenticated key exchanges. *Designs, Codes, and Cryptography*, 2:107–125, 1992.

[8] J. Katz and M. Yung. Scalable Protocols for Authenticated Group Key Exchange. *Advances in Cryptology - CRYPTO'03*, pages 110–125, Springer, 2003. Lecture Notes in Computer Science, No. 2729.

[9] M. Just and S. Vaudenay. Authenticated multi-party key agreement. In K. Kim and T. Matsumoto, editors, *Advances in Cryptology - ASIACRYPT'96*, pages 36–49. Springer, 1996. Lecture Notes in Computer Science No. 1163.

[10] A. Menezes, P. van Oorschot, and S. Vanstone. *Handbook of Applied Cryptography*. CRC Press, Boca Raton, 1997.

[11] C.J. Mitchell, M. Ward, and P. Wilson. Key control in key agreement protocols. *Electronics Letters*, 34(10):980–981, 1998.

[12] R.M. Needham and M.D. Schroeder. Using encryption for authentication in large networks of computers. *Communications of the ACM*, 21(12):993–999, December 1978.

[13] T.P. Pedersen. Non-interactive and information-theoretic secure verifiable secret sharing. In J. Feigenbaum, editor, *Advances in Cryptology - CRYPTO'91*, pages 129–140. Springer, 1992. Lecture Notes in Computer Science No. 576.

[14] W.-G. Tzeng. A Practical and Secure Fault-Tolerant Conference Key Agreement Protocol. PKC 2000.

Josef Pieprzyk and Huaxiong Wang
Centre for Advanced Computing – Algorithms and Cryptography
Department of Computing
Macquarie University
Australia
e-mail: `josef@ics.mq.edu.au`
e-mail: `hwang@ics.mq.edu.au`

Progress in Computer Science and Applied Logic, Vol. 23, 289–301

Combinatorial Tableaux in Isoperimetry

Charles C. Pinter

Abstract. The purpose of this paper is to introduce a new tool for the study of isoperimetry in hypercubes. With its help, new interconnections are found between isoperimetric quantities, and some traditional isoperimetric inequalities may be refined.

Mathematics Subject Classification (2000). 05D05.

Keywords. Extremal combinatorics, isoperimetry.

1. Overview

The purpose of this paper is to introduce a new tool for the study of isoperimetry in hypercubes. Let Q_n be the graph of the n-cube. We introduce a sequence $\{\Delta_n : n = 1, 2, 3, \dots\}$ of tableaux such that for each positive integer n, Δ_n is Q_n together with a labelling of its vertices. The sequence $\{\Delta_n\}$ is defined by a very simple recursion: Level k of Δ_n is the concatenation of levels $k-1$ and k of Δ_{n-1}. The initial tableau is $\Delta_1 = \binom{1}{0}$. Thus, the following are tableaux for $n = 2, 3, 4$.

$$\Delta_2 = \left(\begin{array}{ccc} & 2 & \\ 0 & & 1 \\ & 0 & \end{array}\right), \quad \Delta_3 = \left(\begin{array}{ccc} & 3 & \\ 0 & 1 & 2 \\ 0 & 0 & 1 \\ & 0 & \end{array}\right) \quad \text{and} \quad \Delta_4 = \left(\begin{array}{cccccc} & & \multicolumn{2}{c}{4} & & \\ & 0 & 1 & 2 & 3 & \\ 0 & 0 & 1 & 0 & 1 & 2 \\ & 0 & 0 & 0 & 1 & \\ & & \multicolumn{2}{c}{0} & & \end{array}\right)$$

From the tableau Δ_n one may read off exact values of major isoperimetric quantities: The maximum size of the interior of an m-element set in Q_n, the minimum size of the boundary of such a set, the minimum size of the upper and lower shadow, and other quantities related to these. The tableaux reveal a rich structure of interconnections between these quantities, and bring to light patterns which are otherwise inaccessible. Some new isoperimetric inequalities emerge from the patterns discerned in the tableaux.

2. Structure of Boolean Hypercubes

Consider the n-dimensional hypercube whose sides are of length 1. The set of all its vertices is the set $\{0,1\}^n$ consisting of all n-tuples of 0s and 1s. Let the set of vertices of the n-cube be denoted by Q_n.

If $X = \{x_1, x_2, \ldots, x_n\}$ is the set of axes of the n-cube, then every vertex u is a function $u : X \to \{0,1\}$. Thus, one may identify every vertex u with a subset $\mathbf{u} \subseteq X$; u is the characteristic function of $\mathbf{u}$. The vertices $a_1 = (1,0,\ldots,0)$, $a_2 = (0,1,0,\ldots,0), \ldots, a_n = (0,\ldots,0,1)$ are called *generators* and correspond to all the one-element sets. Finally, the operation $\oplus$ of symmetric difference on sets leads to an analogous operation on vertices, which will be denoted multiplicatively.

Example 1. *In the* 3*-dimensional boolean cube,* $a_1 = (1,0,0), a_2 = (0,1,0)$ *and* $a_3 = (0,0,1)$*. Then* $a_1a_2 = (1,1,0)$*,* $a_1a_3 = (1,0,1)$*, and so on. Also, if* $u = a_1a_2$ *and* $v = a_2a_3$*, then* $uv = a_1a_3$ *because* $\{a_1,a_2\} \oplus \{a_2,a_3\} = \{a_1,a_3\}$.

Under the operation $\oplus$, the set of all the vertices of the n-cube forms an abelian group in which every element is equal to its inverse, and the empty set, to be denoted by e, is the identity element.

Aside from its algebraic structure, Q_n has an order structure which will play a dominant role in the arguments of this Section. We begin by observing that the set of vertices of Q_n is naturally partitioned into $n+1$ *levels* where, for any $u \in \{0,1\}^n$, u belongs to level k iff there are k ones among the coordinates of u.

Definition 1. *The following relation* $<$ *is called the standard ordering of* Q_n*:*

(i) e *is the minimum element of* Q_n.

(ii) *The set of generators (level 1) is ordered by:* $a_1 < a_2 < \ldots < a_n$.

(iii) *The elements of levels* $2,\ldots,n$ *are ordered lexicographically.*

(iv) *If* $h < k$*, every element of level* h *precedes every element of level* k.

It is helpful to represent Q_n in the form of a graph (where edges connect pairs of vertices which differ in only one component). In this form, Q_n can be explicitly pictured as a graph whose vertices are partitioned into $n+1$ levels, and the vertices in each level are ordered lexicographically.

Throughout this paper, we use the symbol $u_{k,j}$ to denote the vertex of Q_n which lies in the jth position (in lex order) of level k of Q_n. Moreover, we sometimes denote vertices of Q_n by the letter u with a single subscript: Our convention is that $u_1, u_2, \ldots, u_{2^n}$ is the set of all the vertices of Q_n in standard order.

If u is a vertex of Q_n, say u is the product of the generators $a_{i_1}, a_{i_2}, \ldots, a_{i_s}$, then the *complement* of u, denoted by u', is defined to be the product of all the generators which are not among $a_{i_1}, a_{i_2}, \ldots, a_{i_s}$. It is known that if a set of vertices $v_1, v_2 \ldots, v_q$ is ordered lexicographically, then the set $v_1', v_2' \ldots, v_q'$ is in inverse lex order. In fact, the mapping $u \to u'$ is a bijection $Q_n \to Q_n$ which preserves adjacency and has the property

$$u < v \quad \text{in lex order} \quad \Leftrightarrow \quad u' < v' \quad \text{in inverse lex order.}$$

It follows that

$$u = u_{k,j} \quad \text{iff} \quad u' = u_{n-k,\binom{n}{k}-j} \tag{1}$$

and

$$u = u_m \quad \text{iff} \quad u' = u_{2^n-m}. \tag{2}$$

3. Splitting Cubes

In order to use induction to prove properties of subsets of Q_n, we shall resort to the simple device of splitting Q_n, along any one of its axes, into two cubes of dimension $n-1$. Specifically, here is what it means to split $Q = Q_n$ along the x_1-axis. As stated earlier, every vertex u in Q_n is a function $u : X \to \{0,1\}$, where $X = \{x_1, x_2, \ldots, x_n\}$ are the axes of the cube. Then $Q = Q^{(0)} \cup Q^{(1)}$, where $Q^{(0)} = \{u \in Q : u(x_1) = 0\}$ and $Q^{(1)} = \{u \in Q : u(x_1) = 1\}$.

Concerning $Q^{(1)}$, a simple observation turns out to be quite important:

If a vertex u is in level k of $Q^{(1)}$, there are $k+1$ ones among the coordinates of u. (*)

The proof of this remark is an easy induction on k. Level 0 of $Q^{(1)}$ contains only one vertex, namely $a_1 = (1, 0, \ldots, 0)$. Assume (*) is true up to level $k-1$. Then every vertex in level $k-1$ of $Q^{(1)}$ has k ones, and therefore every vertex in level k of $Q^{(1)}$ has $k+1$ ones.

As a consequence of (*), level k of $Q^{(1)}$ is a subset of level $k+1$ of Q. Moreover, from Definition 1, the first $\binom{n-1}{k}$ entries of level $k+1$ of Q are those vertices which have a_1 as a factor (that is, they correspond to n-tuples whose first component is a 1). We conclude that for $k = 0, \ldots, n-1$:

Level k of $Q^{(1)}$ is an initial segment of level $k+1$ of Q.

By an analogous (but simpler) argument, every element of level k of $Q^{(0)}$ lies in level k of Q. The conclusion to be drawn from these observations follows:

Let ρ_k be the sequence of all the vertices in level k of Q_n, in standard order. We shall say simply that ρ_k is level k of Q_n. Likewise, let σ_k be level k of $Q^{(1)}$ and let τ_k be level k of $Q^{(0)}$. Then for $k = 0, \ldots, n$,

$$\rho_k = \sigma_{k-1} \cdot \tau_k \tag{3}$$

where the dot signifies concatenation.

4. Isoperimetry of the n-Cube

In this Section we shall examine certain properties of subsets $C \subseteq Q_n$. The positive integer n is arbitrary, but is assumed to remain fixed throughout our discussion in this Section. For $k = 0, \ldots, n$, the symbol ρ_k will designate the set of vertices in level k of Q_n, ordered by the standard ordering. (A helpful mnemonic is to read ρ_k as "row k".)

Definition 2. *Let $C \subseteq Q_n$. We say that a vertex $v \in C$ is an interior vertex of C if every vertex $u \in Q_n$ adjacent to v is in C. It is equivalent to say, for any $v \in C$, that v is an interior vertex of C iff $va_i \in C$ for $i = 1, \dots, n$. The interior of C, denoted by $I(C)$, is the set of all the interior vertices of C.*

Definition 3. *A set $C \subseteq Q_n$ is called a level set in Q_n if*

$$[v \in C \ \& \ u < v] \ \Rightarrow \ u \in C$$

In the literature, levels sets are sometimes called Hamming spheres.

From the Kruskal-Katona theorem, if $C \subseteq Q_n$ is a level set, then $I(C) \geq I(D)$ for any $D \subseteq Q_n$ such that $|D| = |C|$.
Remark 1. If C is a level set in Q_n, then for some integers k and r, C consists of the first k levels of Q_n (that is, levels 0 through $k-1$), together with an initial segment of level k, consisting of the first r elements of level k of Q_n. Then the cardinality of C is given by:

$$|C| = \sum_{i=0}^{k-1} \binom{n}{i} + r \quad , \quad r < \binom{n}{k}.$$

Remark 2. If m is any integer in the range $0 \leq m \leq 2^n$, there are integers $k \geq 0$ and $r > 0$ such that

$$m = \sum_{i=0}^{k-1} \binom{n}{i} + r \quad , \quad r < \binom{n}{k}.$$

We shall abbreviate this by writing $m = \langle k, r \rangle_n$. Informally, this means that a level set of cardinality m in Q_n may be divided into k disjoint subsets ρ_i, $i = 0, \dots, k-1$, where $|\rho_i| = \binom{n}{i}$, leaving a "remainder" $r < \binom{n}{k}$.

5. Number of Interior Elements in a Level Set

Let C be a level subset of Q_n with $|C| = m = \langle k, r \rangle_n$. Then C consists of the following vertices:

(a) All the vertices in levels 0 through $k-1$ of Q_n , as well as

(b) An initial segment R of level k of Q_n , where $|R| = r$.

From (a), $I(C)$ contains all the vertices in levels 0 through $k-2$ of Q_n. From (b), $I(C)$ contains a subset T of the vertices in level $k-1$ of Q_n, where $T = \{u \in \rho_{k-1} : (v \text{ is adjacent to } u \text{ and } v \in \rho_k) \Rightarrow v \in R\}$. Since every vertex in ρ_{k-2} is in C, every element of T is in the interior of C.

Definition 4. *Let R be an initial segment of row k of Q_n. The set*

$$v(R) = \{u \in \rho_{k-1} : (v \text{ is adjacent to } u \text{ and } v \in \rho_k) \Rightarrow v \in R\}$$

will be called the **umbra** *of R. (Informally, u is in the umbra of R iff every level-k vertex adjacent to u is in R.)*

Comment. We use the umbra rather than the shadow here, because the notion of umbra is a more comfortable fit in the kind of arguments needed in this paper.

Lemma 1. *If R is an initial segment of ρ_k, then $v(R)$ is an initial segment of ρ_{k-1}.*

Though easy to prove, Lemma 1 is obviously a consequence of the Kruskal-Katona theorem.

Lemma 2. *For any $k = 1, 2, \ldots, n-2$, let u be an element of level k of Q_n. Suppose that $ua_i \in Q^{(1)}$ for every $ua_i \in \rho_{k+1}$. Then $u \in Q^{(1)}$.*

Proof. Suppose $u \notin Q^{(1)}$. Since $k \leq n-2$, there is a generator $a_j \neq a_1$ such that a_j is not a factor of u. Thus, $ua_j \in \rho_{k+1}$ and $ua_j \notin Q^{(1)}$. □

Definition 5. *The function $\gamma_n(,)$ is defined by the following induction on n:*
(1) *$\gamma_1(0,1) = 0$ and $\gamma_1(1,1) = 1$.*
(2) *Let k and r be integers where $1 \leq k \leq n-1$ and $1 \leq r \leq \binom{n}{k}$.*

For $r = 1, \ldots, \binom{n-1}{k-1}$: $\quad \gamma_n(k,r) = \gamma_{n-1}(k-1,r)$.
For $r = \binom{n-1}{k-1} + 1, \ldots, \binom{n}{k}$: $\quad \gamma_n(k,r) = \gamma_{n-1}(k, r - \binom{n-1}{k-1}) + \binom{n-1}{k-2}$.

Lastly, $\gamma_n(0,1) = 0$ and $\gamma_n(n,1) = n$.

Theorem 1. *Let R be an initial segment of level k of Q_n, with $|R| = r$. Then $|v(R)| = \gamma_n(k,r)$.*

Proof. The theorem is assumed to be true for $n-1$, and will be proved for n. Let k be any integer in the range $1 \leq k \leq n-1$.

Case I: $r \leq \binom{n-1}{k-1}$. In this case, all the elements of R belong to $Q^{(1)}$, hence by Lemma 2, $v(R)$ contains only elements of $Q^{(1)}$. Thus, $R \subseteq Q^{(1)}$ and $v(R) \subseteq Q^{(1)}$. By the hypothesis of induction applied to $Q^{(1)}$, the number of elements in $v(R)$ is equal to $\gamma_{n-1}(k-1,r) = \gamma_n(k,r)$.

Case II: $r > \binom{n-1}{k-1}$. In this case, R contains the totality of level $k-1$ of $Q^{(1)}$, as well as an initial segment of level k of $Q^{(0)}$ numbering $r - \binom{n-1}{k-1}$ elements. So the number of vertices in $v(R)$ is equal to $\binom{n-1}{k-2} + \gamma_{n-1}(k, r - \binom{n-1}{k-1}) = \gamma_n(k,r)$. □

Definition 6. *For any two positive integers n and $m < 2^n$, the symbol $\kappa_n(m)$ will denote the number of interior elements in the level subset $C \subseteq Q_n$ where $|C| = m$.*

By the previous theorem and Remark 1, if $m = \langle k,r \rangle_n$, then

$$\kappa_n(m) = \sum_{i=0}^{k-2} \binom{n}{i} + \gamma_n(k,r).$$

6. Combinatorial Tableaux

The concept presented in this section is a useful tool for relating and unifying several notions from the study of isoperimetry.

Definition 7. *A combinatorial tableau of dimension n is a set A of positive integers together with a graph isomorphism $\mu : Q_n \to A$. Informally, the combinatorial tableau A is an array of integers consisting of $n+1$ rows, (corresponding to the $n+1$ levels of Q_n), where for $i = 0, \ldots, n$, the ith row is a sequence of $\binom{n}{i}$ entries. Thus, every entry of A occupies a position which is a vertex of Q_n. The set of all the entries of A is considered to be linearly ordered, having inherited the standard ordering of the vertices of Q_n whose positions they occupy. The entry of A located in the jth position of row k will be denoted by a_{kj}.*

There is an especially important example of a combinatorial tableau which will concern us in the sequel.

Example 2. *The array of this example is a combinatorial tableau of dimension n to be denoted by Δ_n. We define it by specifying the entry d_{kj} in the (k, j) position of the array. For the purposes of this definition, let S_j stand for the initial segment, of length j, of row k of Q_n. Then*

$$d_{kj} = |v(S_j) - v(S_{j-1})|.$$

Let C be a level subset of Q_n whose final term lies in position $(k, j-1)$ of Q_n. In practical terms, d_{kj} is the number of interior vertices added to C when one additional element is adjoined to C. Moreover, from Theorem 1,

$$d_{k,j} = \gamma_n(k, j) - \gamma_n(k, j-1).$$

The sequence $\{\Delta_n : n = 1, 2, 3, \ldots\}$ of combinatorial tableaux has a property which makes it especially valuable in the present context. This property is given next:

Theorem 2. *Let the successive rows of Δ_n be denoted by $\delta_0, \delta_1, \ldots, \delta_n$ and the successive rows of Δ_{n-1} by $\epsilon_0, \epsilon_1, \ldots, \epsilon_{n-1}$. Then for $k = 1, 2, \ldots, n-1$,*

$$\delta_k = \epsilon_{k-1} \cdot \epsilon_k. \tag{4}$$

Remark: For every positive integer n, and for k in the range $0 \leq k \leq n-1$, the first entry of level k of Δ_n is equal to 0. This is easy to understand: If R is an initial segment of ρ_k and R contains only one vertex, then the umbra of R is empty. The only exception is level n.

Proof. Let the entry in the (k, r) position of Δ_n be $d_{k,r}$ and let the entry in the (k, r) position of Δ_{n-1} be $e_{k,r}$. Consider an arbitrary row of Δ_n (say row k), and an arbitrary entry $d_{k,r}$ in the row. There are two cases:

Case I: $2 \leq r \leq \binom{n-1}{k-1}$. In this case

$$d_{k,r} = \gamma_n(k, r) - \gamma_n(k, r-1) = \gamma_{n-1}(k-1, r) - \gamma_{n-1}(k-1, r-1) = e_{k-1,r}.$$

Case II: $\binom{n-1}{k-1}+2 \le r \le \binom{n-1}{k}$. Set $t = r - \binom{n-1}{k-1}$. Then for $t = 1,\ldots,\binom{n-1}{k-1}$,

$$d_{k,r} = \gamma_n(k,r) - \gamma_n(k,r-1) = \gamma_{n-1}(k,t) - \gamma_{n-1}(k,t-1) = e_{k,t}.$$

This proves the claim of the theorem for all but two of the entries of δ_k. The two remaining entries of δ_k are: First, $d_{k,1}$. From Remark 1, $d_{k,1} = 0 = e_{k,1}$. Next, $d_{k,r}$, where $r = \binom{n-1}{k-1}+1$. The following fact must first be noted:

$$\gamma_{n-1}(k-1,\binom{n-1}{k-1}) = \binom{n-1}{k-2}. \tag{5}$$

Indeed, if R is the entire level $k-1$, its umbra is all of level $k-2$. In this case,

$$d_{k,r} = \gamma_n(k,r) - \gamma_n(k,r-1) = \gamma_{n-1}(k,1) + \binom{n-1}{k-2} - \gamma_{n-1}(k-1,\binom{n-1}{k-1})$$

and from Definition 5, this is equal to $\gamma_{n-1}(k,1) = 0$. □

Remark. For every $n > 0$, the first and last levels of Δ_n each have only one entry. That entry is as follows:

$$\delta_0 = (0) \quad \text{and} \quad \delta_n = (n). \tag{6}$$

Remark. *It is an interesting and important fact that the sequence of combinatorial tableaux* $\{\Delta_n : n = 1,2,\ldots\}$ *is generated from* $\Delta_1 = \binom{1}{0}$ *by equations* (4) *and* (6). *This fact turns out to be a useful tool for proving some assertions about isoperimetric quantities by induction on the dimension of the ambient cube.*

Let R be the initial segment of level k of Q_n such that $|R| = j$. Note that $|v(R)|$ is equal to the sum of the first j entries of level k of Δ_n. Moreover, the cumulative sum of the m first entries of Δ_n is the size of the interior of an m-element level set.

7. The Exact Structure of Δ_n

We begin by defining certain sequences of integers, to be called *elementary sequences*, given by the following recursion: For all integers k and n, $(k \le n)$,

(a) $\omega_n^1 = 012\ldots n$.
(b) $\omega_n^k = \omega_0^{k-1}\omega_1^{k-1}\ldots\omega_n^{k-1}$.

Lastly, $\omega_0^k = 0$ and $\omega_n^0 = n$.

Example 3. *From (a),* $\omega_0^1 = 0$, $\omega_1^1 = 01$, $\omega_2^1 = 012$ *and* $\omega_3^1 = 0123$. *So from (b),* $\omega_1^2 = \omega_0^1\omega_1^1 = 001$ *and* $\omega_2^2 = \omega_0^1\omega_1^1\omega_2^1 = 001012$. *Also,* $\omega_1^3 = \omega_0^2\omega_1^2 = 0001$. *Then one observes that the levels of* Δ_4*, from top to bottom, are:*

$$\omega_4^0 = 4, \qquad \omega_3^1 = 0123, \qquad \omega_2^2 = 001012, \qquad \omega_1^3 = 0001, \qquad \omega_0^4 = 0$$

The pattern exhibited in this example is perfectly general. First, it follows immediately from Condition (b), above, that

$$\omega_{n-1}^k\omega_n^{k-1} = \omega_n^k. \tag{7}$$

Combining (7) with Theorem 2, we conclude as follows:

Theorem 3. *For every positive integer n, the levels of Δ_n, from bottom to top, are*

$$\omega_0^n, \omega_1^{n-1}, \ldots, \omega_n^0.$$

Theorem 3 completely describes the structure of the tableaux Δ_n. In order to prove properties of the tableaux Δ_n by reasoning with elementary sequences, the following two formulas are very useful:

$$|\omega_n^k| = \binom{n+k}{k}, \qquad \sum |\omega_n^k| = \binom{n+k}{k+1} = |\omega_{n-1}^{k+1}| \tag{8}$$

where $|\sigma|$ denotes the length of a string σ. The proof of (8) follows from the definition of ω_n^k by using $\binom{a-1}{b-1} + \binom{a-1}{b} = \binom{a}{b}$.

The connection between tableaux Δ_n and quantities such as the interior and boundary of subsets of Q_n is described in the next Section.

8. Δ_n and Shadows

The following simple remark concerning the ordering of Q_n is of importance here. Suppose we reverse the order of the generators of Q_n, so now they are ordered by $a_n < a_{n-1} < \cdots < a_1$. The elements of level k of Q_n – listed in their order prior to the reversal of the generators – are now in anti-lex order. So if the elements of level k are read from right to left, they are in reverse anti-lex (that is, squashed) order. This is important, because it shows that every final segment of Q_n in lex order (with $a_1 < a_2 < \cdots < a_n$), is equal to an initial segment of Q_n in squashed order if read from right to left and $a_n < a_{n-1} < \cdots < a_1$.

If R is an initial segment of ρ_k in lex order, then its complement in ρ_k, to be denoted by R', is a final segment of ρ_k in lex order. Thus, R' is an initial segment of ρ_k in squashed order if read right to left and $a_n < a_{n-1} < \cdots < a_1$.

Theorem 4. *Let R be an initial segment of ρ_k and U an initial segment of ρ_{k-1}.*

$$U = v(R) \quad \textit{iff} \quad U' = sh(R')$$

where $sh(R')$ denotes the shadow of R'.

Proof. Let $U = v(R)$:

$$x \notin U \iff \text{at least one upper neighbor of } x \text{ is in } R' \iff x \in sh(R').$$

Let $U' = sh(R')$:

$$x \in U \iff \text{no upper neighbor of } x \text{ is in } R' \iff x \in v(R).$$

□

Theorem 4 shows that Δ_n yields the same information about shadows as it does about umbras: Informally first: let F be a final segment of row k of Q_n. If the rows of Δ_n are read from right to left, each entry of row k gives the marginal increment in the size of the shadow of F, as F is increased by one element. More formally, let F be a final segment of row k of Q_n, whose initial element is in position (k,j). Then $d_{k,j-1}$ is the marginal increment of the shadow of F following the addition of one element to F. The following is now self-evident:

The size of the shadow of the first m elements of level k of Q_n, in squashed order, is equal to the sum of the last m elements of level k of Δ_n.

Let $\{d_1, d_2, \ldots, d_{2^n}\}$ be the entries of Δ_n in standard order and let Δ_n^{-1} be the combinatorial tableau of dimension n whose entries, in standard order, are $\{d_{2^n}, \ldots, d_2, d_1\}$. Let $e_{k,j}$ denote that element of Δ_n^{-1} which lies in position (k,j). That is, $e_{k,j} = d_{n-k,\binom{n}{k}-j}$.

Theorem 5. *Let S be a level set in Q_n, say $S = \rho_0 \ldots \rho_{k-1} R$, where R is an initial segment of ρ_k with $|R| = j-1$. If one element is joined to R, the upper shadow of R is increased by $e_{k,j}$.*

Comment. *It follows immediately from the above assertion that the sum of the first j entries of level k of Δ_n^{-1} is the size of the upper shadow of R, $|R| = j$.*

Proof. Let $R = \{v_1, v_2, \ldots, v_{j-1}\}$, and $\hat{R} = \{v_1', v_2', \ldots, v_{j-1}'\}$ the set of complements of $v_1, v_2, \ldots, v_{j-1}$. (Note that $\hat{R}$ is a final segment of ρ_{n-k}.) For any vertex u, u is a lower neighbor of an element of $\hat{R}$ iff u' is an upper neighbor of an element of R. If the addition of one element to $\hat{R}$ increases the lower shadow of $\hat{R}$ by a, then the addition of one element to R increases the upper shadow of R by a. □

9. Δ_n and the Boundary

The tableau Δ_n is also useful when reasoning about the boundary of a level set. By the *inner boundary* of a set $S \subseteq Q_n$ we mean the set of those elements of S which are not in the interior of S. The *outer boundary* of S is the inner boundary of the complement of S in Q_n. It is immediate that if S is a level set whose cardinality is m, the size of the inner boundary of S is $m - \sum_{i=1}^{m} d_i$. Likewise, the size of the outer boundary of S is

$$\sum\nolimits_{i=m+1}^{2^n} d_i - m + 1.$$

If $d_{k,j}$ is the element of Δ_n in the (k,j) position, let $g_{k,j} = 1 - d_{k,j}$. Let Γ_n be the combinatorial tableau of dimension n whose entry in the (k,j) position is $g_{k,j}$.

Theorem 6. *Let S be a level subset of Q_n, say $S = \rho_0 \rho_1 \ldots \rho_{k-1} R$ where R is an initial segment of ρ_k, $|R| = j-1$. The inner boundary of S is increased by $g_{k,j}$ when one element is joined to R.*

Remark. *As a consequence of this theorem, the size of the inner boundary of S is equal to the sum of all the entries of Γ_n preceding (and including) $g_{k,j}$.*

Proof. If one element is joined to R then $|S|$ is increased by 1, and at the same time, $I(S)$ is increased by $d_{k,j}$. Thus, the inner boundary of S is increased by $1 - d_{k,j}$. □

Actually, there is an isolated exception to Theorem 6. If S is the set of all but the last element of Q_n, and the last element of Q_n is joined to S the boundary of S increases by $-n$. Thus, we define $\hat{\Gamma}_n$ to be the tableau which has the same entries as Γ_n except for the final entry which, in $\hat{\Gamma}_n$, is equal to $-n$.

Let Γ_n^{-1} denote the combinatorial tableau of dimension n whose entries are those of Γ_n, in reverse order. If the entry in the (k, j) position of Γ_n^{-1} is denoted by $h_{k,j}$, then $h_{k,j} = 1 - e_{k.j}$.

Theorem 7. *Let S be a level subset of Q_n, say $S = \rho_0\rho_1 \dots \rho_{k-1}R$ where $|R| = j-1$. If one element is joined to R, the outer boundary of S is incremented by $1 - e_{k.j}$.*

Proof. The outer boundary of S is the union of $\rho_k \backslash R$ with the upper shadow of R. Thus, when one element is joined to R, the outer boundary of S first decreases by 1, then increases by the amount of the upper shadow of R, which is equal to $e_{k,j}$. So finally, the outer boundary of S increases by $1 - e_{k,j}$. □

Remark. *It follows from the theorem that the size of the outer boundary of S is equal to the cumulative sum of the entries of Γ_n^{-1} preceding (and including) $h_{k.j}$.*

Remark The sequence of tableaux $\{\Gamma_n : n = 1, 2, 3, \dots\}$ can be generated recursively from the initial tableau $\Gamma_1 = \binom{0}{1}$ by using Equation (4) with the boundary conditions

$$\delta_0 = 1 \qquad \text{and} \qquad \delta_n = -n$$

where δ_i refers to level i of Γ_n.

10. Some Properties of the Tableaux Δ_n

In this final Section, we prove several properties of Δ_n which may be used to refine some of the classical isoperimetric inequalities. Throughout this discussion, the entries of Δ_n, as well as Δ_{n-1}, are assumed to be ordered by the standard order. The symbol σ will denote any sequence of entries of either Δ_n or Δ_{n-1}, and $\sum \sigma$ will denote the sum of the entries in σ. The symbol $|\sigma|$ will denote the length of σ (that is, the number of terms of σ). It is convenient, too, to refer to a sequence of length q as a "*q-tuple*".

A Notational Convention. For any appropriate positive integer q, the symbols ϕ_q and ν_q will be used as follows:

$\phi_q(\sigma)$ = the final q-tuple of σ.

$\nu_q(\sigma)$ = the initial q-tuple of σ.

Thus, for example, $\nu_p(\phi_q(\sigma))$ = the p initial entries of the final q-tuple of σ.

Theorem 8. *Let n be an arbitrary positive integer, and let ρ_i denote level i of Δ_n. For $i = 1, \dots, n-1$,*

(a) $\sum \nu_j(\rho_i) \leq \sum \nu_j(\rho_{i+1})$ (c) $\sum \phi_j(\rho_{i-1}) \leq \sum \phi_j(\rho_i)$

(b) $\sum \nu_j(\rho_i) \leq \sum \nu_j(\phi_q(\rho_i))$ (d) $\sum \nu_j(\phi_q(\rho_i)) \leq \phi_j(\rho_i)$

It is assumed in (b) *and* (d) *that $j < q < |\rho_i|$ and $j \leq |\rho_{i+1}|$.*

Proof. The proof is straightforward but comprises a number of cases, distinguished by the relative sizes of j, $|r_i|$ and $|r_{i+1}|$. The following is a typical case: Suppose $|r_{i-1}| \leq j \leq |r_i|$. Let $g = |r_{i-1}|$ and $h = j - g$.

Proof of (a). The proof is by induction on n. Recall that $\rho_i = r_{i-1} \cdot r_i$ and $\rho_{i+1} = r_i \cdot r_{i+1}$. Then $\nu_j(\rho_i) = r_{i-1} \cdot \nu_h(r_i)$ and $\nu_j(\rho_{i+1}) = \nu_g(r_i) \cdot \sigma_h$, where $\sigma_h = \nu_h(\phi_{|r_i|-g}(r_i)$.

From the hypothesis of induction,

$$\sum r_{i-1} \leq \sum \nu_g(r_i) \quad \text{and} \quad \sum \nu_h(r_i) \leq \sum \sigma_h.$$

Thus,

$$\sum \nu_j(\rho_i) = \sum r_{i-1} + \sum \nu_h(r_i) \leq \sum \nu_g(r_i) + \sum \sigma_h = \sum \nu_j(\rho_{i+1}).$$

Proof of (b). Let $\sigma_j(\rho_i)$ be an arbitrary sequence of ρ_i of length j. Then $\sigma_j(\rho_i)$ has one of the following forms:

(i) $\sigma_j(\rho_i)$ is a j-tuple of r_{i-1};
(ii) $\sigma_j(\rho_i)$ is a j-tuple of r_i.
(iii) $\sigma_j(\rho_i) = \phi_{j_1}(r_{i-1}) \cdot \nu_{j_2}(r_i)$, $j = j_1 + j_2$.

In Cases (i) and (ii), the result is a straightforward application of the hypothesis of induction. For Case (iii), let $g = |r_{i-1}|$, $h = j - g$ and $k = j_2 - h$. From the hypothesis of induction,

$$\sum \nu_k(r_{i-1}) \leq \sum \nu_k(r_i) \leq \sum \nu_k(\phi_{|r_i|-h}(r_i)).$$

But

$$\nu_j(\rho_i) = r_{i-1} \cdot \nu_h(r_i) = \nu_k(r_{i-1}) \cdot \phi_{g-k}(r_{i-1}) \cdot \nu_h(r_i).$$

Thus

$$\begin{aligned}\sum \nu_j(\rho_i) &\leq \sum \phi_{g-q}(r_{i-1}) + \sum \nu_h(r_i) + \sum \nu_k(\phi_{|r_i|-h}(r_i)) \\ &= \sum \phi_{j_1}(r_{i-1}) + \sum \nu_{j_2}(r_i) = \sum \sigma_j(\rho_i).\end{aligned}$$

Proof of (c). Let g, h and j be as defined above. We have $\phi_j(\rho_{i-1}) = \phi_h(r_{i-2}) \cdot r_{i-1}$ and $\phi_j(\rho_i) = \phi_j(r_i)$. By the hypothesis of induction,

$$\sum \phi_h(r_{i-2}) \leq \sum \phi_h(r_{i-1}) \leq \sum \phi_h(r_i)$$

and

$$\sum r_{i-1} \leq \sum \nu_g(r_i) \leq \sum \nu_g(\phi_j(r_i)).$$

Thus,

$$\sum \phi_j(\rho_{i-1}) \leq \sum \phi_h(r_i) + \sum \nu_g(\phi_j(r_i)) = \sum \phi_j(\rho_i).$$

Proof of (d). Let σ_j denote $\nu_j(\phi_q(\rho_i))$. There are three cases to consider:

Case (i): σ_j is a j-tuple of r_i. This is given by the hypothesis of induction (H.I.).

Case (ii): σ_j is a j-tuple of r_{i-1}. From the H.I., $\sum \sigma_j \leq \sum \phi_j(r_{i-1}) \leq \sum \phi_j(r_i)$.

Case (iii): $\sigma_j = \phi_{j_1}(r_{i-1}) \cdot \nu_{j_2}(r_i), \quad j_1 + j_2 = p$. Then,

$$\sum \phi_{j_1}(r_{i-1}) \leq \sum \phi_{j_1}(r_i) \text{ and } \sum \nu_{j_2}(r_i) \leq \sum \nu_{j_2}(\phi_j(r_i)).$$

Thus,

$$\sum \sigma_j = \sum \phi_{j_1}(r_{i-1}) + \sum \nu_{j_2}(r_i) \leq \sum \phi_{j_1}(r_i) + \sum \nu_{j_2}(\phi_j(r_i)) = \sum \phi_j(\rho_i).$$

□

In the remainder of this section, the following notation is convenient. If d_{kj} is an entry of Δ_n, and $i = \langle k, j \rangle_n$, then $d_{kj} = d_i$. (The single subscript distinguishes the two notations.) So the entries of Δ_n in standard order are $d_0, d_1, \ldots, d_{2^n-1}$.

Now we wish to use Theorem 8 to prove the following important fact about the tableaux Δ_n: Suppose $d_h, d_{h+1}, \ldots, d_{h+r}$ is a sequence of $r+1$ consecutive entries of Δ_n. Then the sequence $d_{h+j}, d_{h+j+1}, \ldots, d_{h+j+r}$ satisfies $\sum_{i=h+j}^{h+j+r} d_i > \sum_{i=h}^{h+r} d_i$, on condition that d_{h+j+r} is the last entry of a level.

In the theorem which follows, δ_j is level j of Δ_n. Also, χ, σ and τ denote sequences of consecutive entries of Δ_n. For $n = 2p$ or $2p+1$, we shall refer to levels 0 to p as the *"lower half"* of Δ_n, and levels $p+1$ to n as the *"upper half"* of Δ_n.

Theorem 9. *Let χ be a sequence of ℓ consecutive entries of Δ_n, say $\chi = (d_{a+1}, \ldots, d_{a+\ell})$. We may write χ in the form*

$$\chi = \sigma \cdot \delta_i \cdots \delta_{i+r-1} \cdot \tau$$

where $\sigma = \phi_s(\delta_{i-1})$ and $\tau = \nu_t(\delta_{i+r})$. Let $\chi*$ be a sequence of ℓ consecutive entries of Δ_n such that the last term of $\chi*$ is the last term of level δ_{i+r}. Then

$$|\chi *| = |\chi| \quad \textit{and} \quad \sum \chi* \geq \sum \chi. \tag{9}$$

Proof. We shall assume, in our proof, that $d_{a+\ell}$ lies in the "upper half" of Δ_n: This assumption simplifies the proof, and is the only case to be used in this paper. Note that in the "lower half" the length of successive levels increases, while in the "upper half" it decreases.

For now, we assume that $s, t \neq 0$. If $p = \binom{n}{i+r} - t$, we consider two cases depending on whether or not $s < p$.

Case 1: $s < p$, say $p = s + q$. Let $\sigma* = \phi_s(\delta_{i+r})$ and $\lambda = \nu_q(\delta_{i+1})$. Finally, let $\lambda* = \nu_q(\phi_p(\delta_{i+r}))$.

From Theorem 8, $\sum \sigma* \geq \sum \sigma$ and $\sum \lambda* \geq \sum \lambda$. So if we let

$$\chi* = \phi_{\binom{n}{i+1}-q}(\delta_{i+1}) \cdot \delta_{i+2} \cdots \delta_{i+r}$$

then (9) is satisfied.

Case 2: $s > p$, say $s = p + u$. Let $\alpha* = \phi_p(\delta_{i+r})$ and $\beta* = \phi_u(\delta_i)$. From Theorem 3, $\sum \alpha* \geq \sum \phi_p(\delta_i)$ and $\sum \beta* \geq \sum \nu_u(\phi_s(\delta_i))$.

If we let $\chi* = \beta * \cdot \delta_{i+1} \cdots \delta_{i+r}$, then (9) is satisfied. □

In addition to the inequalities discussed above, a number of tighter and more complex inequalities prevail between the consecutive levels of Δ_n and of Γ_n. These are perhaps worth investigating.

References

[1] Anderson, Ian. *Combinatorics of Finite Sets*, Oxford University Press, 1987.

[2] Bollobas, Bela. *Combinatorics*, Cambridge University Press, 1986.

[3] Jukna, S. *Extremal Combinatorics*, Springer Verlag, 2003.

Charles C. Pinter
Department of Mathematics
Bucknell University
Lewisburg, PA 17837, USA
e-mail: cpinter@bucknell.edu

Progress in Computer Science and Applied Logic, Vol. 23, 303–317

On the Error Exponents of Reliability-Order-Based Decoding Algorithms for Linear Block Codes

Yuansheng Tang

Abstract. In this paper, for any reliability-order-based decoding algorithm of linear block codes, we prove that the error exponent and the squared error-correction radius are equal to each other. A known method for computing the squared error-correction radius is improved. The improved method can also be used further to compute the effective error coefficient of the reliability-order-based decoding algorithms.

Mathematics Subject Classification (2000). 94B05; 94B35.

Keywords. Linear block code, reliability-order-based decoding algorithm, error exponent, squared error-correction radius, optimization problem.

1. Introduction

Many soft-decision decoding algorithms have been proposed to decode linear binary codes over the *additive white Gaussian noise* channel. If all codewords are transmitted with the same probability, some of the proposed algorithms, called *maximum-likelihood* (ML) decodings, minimize the probability of decoding error and thus are optimum. Since the computational complexity of the ML decodings remains very high for long codes, some sub-optimum soft-decision decoding algorithms, which provide an efficient tradeoff between error performance and decoding complexity, are highly attractive as well for many coding theorists and practicing communication engineers. By a reliability-order-based decoding algorithm (ROBDA), we mean a soft-decision decoding algorithm which decodes to the best (most likely) codeword of form that is the sum of the hard-decision tuple and an error pattern in a set, which is determined only in the order of the reliabilities of the hard-decisions.

This work was completed with the support of NUS-ARF research grants R-252-000-015-112, R-252-000-015-303 and DSTA project (POD 0103223).

To evaluate the error performance of a soft-decision decoding algorithm, since tight bounds on the probability of decoding error are quite difficult to obtain, in literature it is a common way to give an accurate approximation for the probability of decoding error when the *signal-to-noise ratio* (SNR) becomes large. For many soft-decision decoding algorithms, the natural logarithm of the probability of decoding error is approximatively in inverse ratio to SNR for high SNRs. These ratios are negative numbers and their absolute values are called the *error exponents* of the decoding algorithms. In this paper, we show that the error exponent of any ROBDA is equal to its *squared error-correction radius* (SECR), which is the largest positive number ρ such that the received sequence must be decoded correctly if it is within squared Euclidean distance ρ of the bipolar sequence corresponding to the transmitted codeword. For computing the SECRs of ROBDAs, Fossorier and Lin proposed in [6] a unified method, which is improved in this paper. The improved method can be used further to compute the effective error coefficient, which is also an important measure for the asymptotic property of the probability of decoding error [4]–[7].

The paper is organized as follows. Section 2 gives the definition of ROBDAs. By a detailed investigation on the decision regions, the error performance of ROBDAs is studied in Section 3. For any ROBDA, the error exponent is shown to be equal to the SECR. In Section 4, the method proposed in [6] for computing the SECRs of ROBDAs is improved. Some examples for computing the error exponents or SECRs of ROBDAs are also given in Section 4.

2. Definition of ROBDAs

Suppose that a binary $(N, K, d_{\min})$ linear block code C is used for error control over the additive white Gaussian noise (AWGN) channel with BPSK signaling. If the transmitted codeword is $\boldsymbol{c} = (c_1, c_2, \ldots, c_N) \in C$, then the received sequence $\boldsymbol{r}$ is an N-tuple in $\mathbb{R}^N$ which can be written as $\boldsymbol{s}(\boldsymbol{c}) + \boldsymbol{w}$, where $\boldsymbol{s}(\boldsymbol{c}) \triangleq ((-1)^{c_1}, (-1)^{c_2}, \ldots, (-1)^{c_N})$ is the bipolar sequence corresponding to $\boldsymbol{c}$ and the components of $\boldsymbol{w}$ are independent Gaussian random variables with common density function:

$$g(w) \triangleq \frac{1}{(\pi N_0)^{1/2}} e^{-w^2/N_0}.$$

The density function of $\boldsymbol{r}$ is

$$p(\boldsymbol{r}|\boldsymbol{c}) = \frac{1}{(\pi N_0)^{N/2}} e^{-d_{\mathrm{E}}(\boldsymbol{r}.\boldsymbol{s}(\boldsymbol{c}))/N_0}, \tag{2.1}$$

where $d_{\mathrm{E}}(\boldsymbol{r}, \boldsymbol{s}(\boldsymbol{c}))$ is the *squared Euclidean distance* (SED) between $\boldsymbol{r}$ and $\boldsymbol{s}(\boldsymbol{c})$. For each received sequence $\boldsymbol{r}$, the *maximum-likelihood* (ML) decoding outputs the codeword $\boldsymbol{c}_{\mathrm{opt},\boldsymbol{r}}$ which satisfies

$$p(\boldsymbol{r}|\boldsymbol{c}_{\mathrm{opt}.\boldsymbol{r}}) = \max_{\boldsymbol{c} \in C} p(\boldsymbol{r}|\boldsymbol{c}).$$

Let V^N denote the set of binary N-tuples. For a received sequence $\boldsymbol{r} = (r_1, r_2, \dots, r_N)$, let $\boldsymbol{z_r} = (z_1, z_2, \dots, z_N) \in V^N$ denote the hard-decision sequence defined by: $z_i = 0$ for $r_i > 0$ and $z_i = 1$ for $r_i \leq 0$. A tuple $\boldsymbol{v} \in V^N$ is said to be *better* than another tuple $\boldsymbol{v'} \in V^N$ if $d_{\mathrm{E}}(\boldsymbol{r}, \boldsymbol{s}(\boldsymbol{v})) \leq d_{\mathrm{E}}(\boldsymbol{r}, \boldsymbol{s}(\boldsymbol{v'}))$. Thus, $\boldsymbol{z_r}$ is the best tuple in V^N and $\boldsymbol{c}_{\mathrm{opt},\boldsymbol{r}}$ is the best codeword in C, respectively.

Let Λ_N be the set of permutations of $(1, 2, \dots, N)$. A permutation $\boldsymbol{\lambda} = (\lambda_1, \lambda_2, \dots, \lambda_N) \in \Lambda_N$ permutes any N-tuple $\boldsymbol{v} = (v_1, v_2, \dots, v_N)$ into $\boldsymbol{\lambda}(\boldsymbol{v}) = (v_{\lambda_1}, v_{\lambda_2}, \dots, v_{\lambda_N})$. For a set $\mathcal{E} = \{E_{\boldsymbol{\lambda}} \subseteq V^N : \boldsymbol{\lambda} \in \Lambda_N\}$, let $\mathcal{A}(\mathcal{E})$ denote the *reliability-order-based decoding algorithm* (ROBDA) which decodes the received sequence $\boldsymbol{r}$ to the best codeword in the *search region*

$$\mathcal{S}_{\mathcal{E}}(\boldsymbol{r}) \triangleq \{\boldsymbol{z_r} + \boldsymbol{\lambda_r}(\boldsymbol{e}) : \boldsymbol{e} \in E_{\boldsymbol{\lambda_r}}\}$$

if it contains some codewords, where $\boldsymbol{\lambda_r} \in \Lambda_N$ is a permutation whose reverse $(\boldsymbol{\lambda_r})^{-1} = (\lambda_1, \lambda_2, \dots, \lambda_N)$ satisfies

$$|r_{\lambda_1}| \leq |r_{\lambda_2}| \leq \cdots \leq |r_{\lambda_N}|.$$

If $\mathcal{S}_{\mathcal{E}}(\boldsymbol{r})$ contains no codeword, the decoding algorithm $\mathcal{A}(\mathcal{E})$ declares failure. In general, the subsets $E_{\boldsymbol{\lambda}}$ are big enough such that $\mathcal{S}_{\mathcal{E}}(\boldsymbol{r})$ contains at least one codeword for almost all received sequence $\boldsymbol{r}$. The tuples in $E_{\boldsymbol{\lambda}}$ are also called *error patterns.*

Though, in the definition of search region $\mathcal{S}_{\mathcal{E}}(\boldsymbol{r})$, $\boldsymbol{\lambda_r}(\boldsymbol{e})$ and $E_{\boldsymbol{\lambda_r}}$ can be replaced by $\boldsymbol{e}$ and $E'_{\boldsymbol{\lambda_r}} = \boldsymbol{\lambda_r}^{-1}(E_{\boldsymbol{\lambda_r}})$, respectively, the present form is convenient for the description of many known algorithms, such as the generalized minimum distance (GMD) decoding algorithm proposed in [1] and the Chase decoding algorithms proposed in [2]. For example, for the Chase-3 algorithm [2], the subset $E_{\boldsymbol{\lambda}}$ is independent of the permutation $\boldsymbol{\lambda} \in \Lambda_N$ and consists of the tuples $\boldsymbol{e} \in V^N$ which have at most i nonzero entries in the last $N - d_{\min} + 2i + 1$ bit positions for every integer i with $0 \leq i \leq \lfloor (d_{\min} - 1)/2 \rfloor$ (cf. [11]). A detailed list of references about ROBDAs can be found in [6].

3. Error Performance of ROBDAs

3.1. Decision Regions of ROBDAs

A soft-decision decoding algorithm A divides $\mathbb{R}^N$ into $2^K + 1$ disjoint regions: $\mathcal{F}_A$ and $D_A(\boldsymbol{c})$ for $\boldsymbol{c} \in C$. If the received sequence $\boldsymbol{r}$ belongs to $D_A(\boldsymbol{c})$ for some codeword $\boldsymbol{c}$ in C, the soft-decision decoding algorithm A declares that $\boldsymbol{c}$ is the transmitted codeword. If the received sequence $\boldsymbol{r}$ belongs to $\mathcal{F}_A$, the soft-decision decoding algorithm A makes no decision on the transmitted codeword, i.e., a decoding failure occurs in this case. $D_A(\boldsymbol{c})$ is called the *decision region* associated with $\boldsymbol{c}$. The error performance analysis of the soft-decision decoding algorithm A is then to estimate the probability of decoding error:

$$\Pr(e_A) \triangleq \sum_{\boldsymbol{c} \in C} \Pr(\boldsymbol{r} \in D_A^c(\boldsymbol{c}) | \boldsymbol{c}) \Pr(\boldsymbol{c}), \tag{3.1}$$

where $D_A^c(\boldsymbol{c}) \triangleq \mathbb{R}^N \setminus D_A(\boldsymbol{c})$ and $\Pr(\boldsymbol{c})$ is the probability that $\boldsymbol{c}$ is transmitted.

For two sequences $\boldsymbol{x} = (x_1, x_2, \ldots, x_N)$ and $\boldsymbol{x}' = (x'_1, x'_2, \ldots, x'_N)$ in $\mathbb{R}^N$, write

$$\boldsymbol{x} \circ \boldsymbol{x}' \triangleq (x_1 x'_1, x_2 x'_2, \ldots, x_N x'_N).$$

Let Δ_N denote the set of sequences $\boldsymbol{x} = (x_1, x_2, \ldots, x_N) \in \mathbb{R}^N$ which satisfy

$$\sum_{i=1}^{N} a_i |x_i| \neq 0, \text{ for all } (a_1, a_2, \ldots, a_N) \in \{1, 0, -1\}^N \setminus \{\boldsymbol{0}\}. \tag{3.2}$$

Lemma 3.1. *Assume that $\boldsymbol{x}$ is a sequence in Δ_N. For any codeword $\boldsymbol{c} \in C$, the sequence $\boldsymbol{x}$ belongs to $D_{\mathcal{A}(\mathcal{E})}(\boldsymbol{0})$ if and only if the sequence $\boldsymbol{x} \circ \boldsymbol{s}(\boldsymbol{c})$ belongs to $D_{\mathcal{A}(\mathcal{E})}(\boldsymbol{c})$.*

Proof. Let $C_{\boldsymbol{x}}$ be the set of codewords in $\mathcal{S}_{\mathcal{E}}(\boldsymbol{x})$. For any codeword $\boldsymbol{c}$, from $\boldsymbol{x} \in \Delta_N$ and (3.2), $\boldsymbol{z}_{\boldsymbol{x} \circ \boldsymbol{s}(\boldsymbol{c})} = \boldsymbol{z}_{\boldsymbol{x}} + \boldsymbol{c}$ and $\boldsymbol{\lambda}_{\boldsymbol{x} \circ \boldsymbol{s}(\boldsymbol{c})} = \boldsymbol{\lambda}_{\boldsymbol{x}}$. Hence,

$$\mathcal{S}_{\mathcal{E}}(\boldsymbol{x} \circ \boldsymbol{s}(\boldsymbol{c})) = \{\boldsymbol{v} + \boldsymbol{c} : \boldsymbol{v} \in \mathcal{S}_{\mathcal{E}}(\boldsymbol{x})\}. \tag{3.3}$$

Then, from $\boldsymbol{c} \in C$ and (3.3),

$$C_{\boldsymbol{x} \circ \boldsymbol{s}(\boldsymbol{c})} = \{\boldsymbol{c}' + \boldsymbol{c} : \boldsymbol{c}' \in C_{\boldsymbol{x}}\}. \tag{3.4}$$

If $\boldsymbol{x}$ belongs to $D_{\mathcal{A}(\mathcal{E})}(\boldsymbol{0})$, then $\boldsymbol{0} \in C_{\boldsymbol{x}}$ and, from (3.2),

$$d_{\mathrm{E}}(\boldsymbol{x}, \boldsymbol{s}(\boldsymbol{0})) < d_{\mathrm{E}}(\boldsymbol{x}, \boldsymbol{s}(\boldsymbol{c}')), \text{ for } \boldsymbol{c}' \in C_{\boldsymbol{x}} \setminus \{\boldsymbol{0}\}. \tag{3.5}$$

From $\boldsymbol{0} \in C_{\boldsymbol{x}}$ and (3.4), $\boldsymbol{c}$ belongs to $C_{\boldsymbol{x} \circ \boldsymbol{s}(\boldsymbol{c})}$. Since $d_{\mathrm{E}}(\boldsymbol{x} \circ \boldsymbol{s}(\boldsymbol{c}), \boldsymbol{s}(\boldsymbol{c}' + \boldsymbol{c})) = d_{\mathrm{E}}(\boldsymbol{x}, \boldsymbol{s}(\boldsymbol{c}'))$, from (3.4) and (3.5),

$$d_{\mathrm{E}}(\boldsymbol{x} \circ \boldsymbol{s}(\boldsymbol{c}), \boldsymbol{s}(\boldsymbol{c})) < d_{\mathrm{E}}(\boldsymbol{x} \circ \boldsymbol{s}(\boldsymbol{c}), \boldsymbol{s}(\boldsymbol{c}'')), \text{ for } \boldsymbol{c}'' \in C_{\boldsymbol{x} \circ \boldsymbol{s}(\boldsymbol{c})} \setminus \{\boldsymbol{c}\}.$$

Hence, $\boldsymbol{c}$ is the unique best codeword in $C_{\boldsymbol{x} \circ \boldsymbol{s}(\boldsymbol{c})}$ and thus $\boldsymbol{x} \circ \boldsymbol{s}(\boldsymbol{c})$ belongs to $D_{\mathcal{A}(\mathcal{E})}(\boldsymbol{c})$.

Similarly, one can also show that $\boldsymbol{x}$ belongs to $D_{\mathcal{A}(\mathcal{E})}(\boldsymbol{0})$ if $\boldsymbol{x} \circ \boldsymbol{s}(\boldsymbol{c})$ belongs to $D_{\mathcal{A}(\mathcal{E})}(\boldsymbol{c})$. □

Since almost all of the sequences $\boldsymbol{x}$ in $\mathbb{R}^N$ belong to Δ_N, Lemma 3.1 shows that the decision regions $D_{\mathcal{A}(\mathcal{E})}(\boldsymbol{c})$ are almost symmetrical to each other. Hence, from (2.1) and Lemma 3.1,

$$\begin{aligned} \Pr(\boldsymbol{r} \in D^c_{\mathcal{A}(\mathcal{E})}(\boldsymbol{c})|\boldsymbol{c}) &= \frac{1}{(\pi N_0)^{N/2}} \int_{D^c_{\mathcal{A}(\mathcal{E})}(\boldsymbol{c})} e^{-d_{\mathrm{E}}(\boldsymbol{x}, \boldsymbol{s}(\boldsymbol{c}))/N_0} d\boldsymbol{x} \\ &= \frac{1}{(\pi N_0)^{N/2}} \int_{D^c_{\mathcal{A}(\mathcal{E})}(\boldsymbol{0})} e^{-d_{\mathrm{E}}(\boldsymbol{y}, \boldsymbol{s}(\boldsymbol{0}))/N_0} d\boldsymbol{y} \\ &= \Pr(\boldsymbol{r} \in D^c_{\mathcal{A}(\mathcal{E})}(\boldsymbol{0})|\boldsymbol{0}), \end{aligned} \tag{3.6}$$

where $\boldsymbol{y} = \boldsymbol{x} \circ \boldsymbol{s}(\boldsymbol{c})$. Then, it follows from (3.1) and (3.6) that

$$\Pr(e_{\mathcal{A}(\mathcal{E})}) = \Pr(\boldsymbol{r} \in D^c_{\mathcal{A}(\mathcal{E})}(\boldsymbol{0})|\boldsymbol{0}). \tag{3.7}$$

To simplify the analysis of error performance of some decoding algorithms, in many published papers it is a common method to consider only a particular case that a specified codeword is transmitted without strict proof. The equality (3.7) shows that the commonly used method is correct for the ROBDAs. We notice that some decoding algorithms, say the decoding algorithms which use only quantized reliabilities, do not possess a symmetry similar to that shown in Lemma 3.1 and consequently the equalities (3.6) and (3.7) may fail.

3.2. Squared Error-Correction Radii of ROBDAs

The *squared error-correction radius* (SECR) of a decoding algorithm A is defined as the largest number $\rho(A)$ such that A decodes correctly whenever the received sequence is within SED $\rho(A)$ of the bipolar sequence corresponding to the transmitted codeword. It is well known that the SECR is not greater than $d_{\min}$.

For any tuple $\boldsymbol{e} = (e_1, e_2, \ldots, e_N) \in V^N$, let $\mathcal{H}(\boldsymbol{e})$ denote the set of sequences $(x_1, x_2, \ldots, x_N) \in \mathbb{R}^N$ with

$$|x_1| \leq |x_2| \leq \cdots \leq |x_N|, \text{ and}$$
$$x_i(-1)^{e_i} \geq 0, \text{ for } i = 1, 2, \ldots, N.$$

For a ROBDA $\mathcal{A}(\mathcal{E})$ and a codeword $\boldsymbol{c}$, assume that $\boldsymbol{x}$ is a sequence in $D^c_{\mathcal{A}(\mathcal{E})}(\boldsymbol{c})$ such that the SED between $\boldsymbol{x}$ and the bipolar sequence $\boldsymbol{s}(\boldsymbol{c})$ is less than $d_{\min}$. Then $\boldsymbol{x}$ belongs to $D_{\mathrm{ML}}(\boldsymbol{c}) \setminus D_{\mathcal{A}(\mathcal{E})}(\boldsymbol{c})$ and thus $\boldsymbol{c}$ is not in the search region $\mathcal{S}_{\mathcal{E}}(\boldsymbol{x})$. Hence, there is an error pattern $\boldsymbol{e}$ in $V^N \setminus E_{\boldsymbol{\lambda}_{\boldsymbol{x}}}$ such that $\boldsymbol{z}_{\boldsymbol{x}} + \boldsymbol{\lambda}_{\boldsymbol{x}}(\boldsymbol{e}) = \boldsymbol{c}$, which implies $\boldsymbol{x} \in \boldsymbol{\lambda}_{\boldsymbol{x}}(\mathcal{H}(\boldsymbol{\lambda}_{\boldsymbol{x}}^{-1}(\boldsymbol{c}) + \boldsymbol{e}))$. Clearly, for any $\boldsymbol{x}' \in \boldsymbol{\lambda}_{\boldsymbol{x}}(\mathcal{H}(\boldsymbol{\lambda}_{\boldsymbol{x}}^{-1}(\boldsymbol{c}) + \boldsymbol{e})) \cap \Delta_N$, $\boldsymbol{\lambda}_{\boldsymbol{x}'} = \boldsymbol{\lambda}_{\boldsymbol{x}}$ and

$$\boldsymbol{z}_{\boldsymbol{x}'} = \boldsymbol{\lambda}_{\boldsymbol{x}}(\boldsymbol{\lambda}_{\boldsymbol{x}}^{-1}(\boldsymbol{c}) + \boldsymbol{e}) = \boldsymbol{c} + \boldsymbol{\lambda}_{\boldsymbol{x}'}(\boldsymbol{e}). \tag{3.8}$$

Since $\boldsymbol{e} \in V^N \setminus E_{\boldsymbol{\lambda}_{\boldsymbol{x}}} = V^N \setminus E_{\boldsymbol{\lambda}_{\boldsymbol{x}'}}$, the equality (3.8) implies that $\boldsymbol{c}$ is not in the search region $\mathcal{S}_{\mathcal{E}}(\boldsymbol{x}')$ and thus $\boldsymbol{x}'$ belongs to $D^c_{\mathcal{A}(\mathcal{E})}(\boldsymbol{c})$. Then, according to Lemma 3.1, $\boldsymbol{s}(\boldsymbol{c}) \circ \boldsymbol{x}' \in D^c_{\mathcal{A}(\mathcal{E})}(\boldsymbol{0})$ and thus

$$\boldsymbol{s}(\boldsymbol{c}) \circ \left(\boldsymbol{\lambda}_{\boldsymbol{x}}(\mathcal{H}(\boldsymbol{\lambda}_{\boldsymbol{x}}^{-1}(\boldsymbol{c}) + \boldsymbol{e})) \cap \Delta_N\right) \subset D^c_{\mathcal{A}(\mathcal{E})}(\boldsymbol{0}). \tag{3.9}$$

Since the sequence $\boldsymbol{x}$ belongs to $\boldsymbol{\lambda}_{\boldsymbol{x}}(\mathcal{H}(\boldsymbol{\lambda}_{\boldsymbol{x}}^{-1}(\boldsymbol{c}) + \boldsymbol{e}))$, which is a polyhedron, and almost all sequences in $\mathbb{R}^N$ belong to Δ_N, the sequence $\boldsymbol{x}$ can be approached infinitely by sequences in $\boldsymbol{\lambda}_{\boldsymbol{x}}(\mathcal{H}(\boldsymbol{\lambda}_{\boldsymbol{x}}^{-1}(\boldsymbol{c}) + \boldsymbol{e})) \cap \Delta_N$ and thus $\boldsymbol{s}(\boldsymbol{c}) \circ \boldsymbol{x}$ can also be approached infinitely by sequences in $\boldsymbol{s}(\boldsymbol{c}) \circ \left(\boldsymbol{\lambda}_{\boldsymbol{x}}(\mathcal{H}(\boldsymbol{\lambda}_{\boldsymbol{x}}^{-1}(\boldsymbol{c}) + \boldsymbol{e})) \cap \Delta_N\right)$. Thus, from (3.9), $\boldsymbol{s}(\boldsymbol{c}) \circ \boldsymbol{x}$ can be approached infinitely by sequences in $D^c_{\mathcal{A}(\mathcal{E})}(\boldsymbol{0})$. Consequently, according to $d_{\mathrm{E}}(\boldsymbol{x}, \boldsymbol{s}(\boldsymbol{c})) = d_{\mathrm{E}}(\boldsymbol{s}(\boldsymbol{c}) \circ \boldsymbol{x}, \boldsymbol{1}_N)$, where $\boldsymbol{1}_N$ is the all-one sequence of length N, one can conclude that the SECR of a ROBDA is the squared radius of the smallest sphere centered at the bipolar sequence $\boldsymbol{s}(\boldsymbol{0}) = \boldsymbol{1}_N$ of the codeword $\boldsymbol{0}$ that touches the region $D^c_{\mathcal{A}(\mathcal{E})}(\boldsymbol{0})$. Then, we have proved the following lemma, which is also given in [6] without strict proof.

Lemma 3.2. *For any $\mathcal{E} = \{E_{\boldsymbol{\lambda}} \subseteq V^N : \boldsymbol{\lambda} \in \Lambda_N\}$, the SECR of the ROBDA $\mathcal{A}(\mathcal{E})$ is equal to the smaller one between $d_{\min}$ and*

$$\begin{aligned}\delta(\mathcal{E}) &\triangleq \min_{\mathbf{e}\in\mathbb{R}^N\setminus E_{\boldsymbol{\lambda}},\boldsymbol{\lambda}\in\Lambda_N} d_{\mathrm{E}}(\mathbf{1}_N, \boldsymbol{\lambda}(\mathcal{H}(\mathbf{e}))) \qquad (3.10)\\ &= \min_{\mathbf{e}\in\mathbb{R}^N\setminus E_{\boldsymbol{\lambda}},\boldsymbol{\lambda}\in\Lambda_N} \sigma(\mathbf{e}),\end{aligned}$$

where $\sigma(\mathbf{e})$ is the minimal SED between $\mathbf{1}_N$ and the sequences in $\mathcal{H}(\mathbf{e})$.

3.3. Error Exponents of ROBDAs

For many decoding algorithms A, the following limit

$$\lim_{N_0\to 0}(-N_0 \ln(\Pr(e_A))) \qquad (3.11)$$

is a positive number, which is called the *error exponent* of A. For ROBDA decoding algorithms, we have the following lemma.

Lemma 3.3. *For any ROBDA decoding algorithm $\mathcal{A}(\mathcal{E})$, the error exponent is equal to the SECR, i.e.,*

$$\lim_{N_0\to 0}\left(-N_0 \ln(\Pr(e_{\mathcal{A}(\mathcal{E})}))\right) = \min\{d_{\min}, \delta(\mathcal{E})\}. \qquad (3.12)$$

Proof. Assume that the codeword $\mathbf{0}$ is transmitted and $\boldsymbol{r} = (r_1, r_2, \ldots, r_N)$ is the received sequence. From (3.7) and Lemma 3.2,

$$\begin{aligned}\Pr(e_{\mathcal{A}(\mathcal{E})}) &= \Pr\left(\boldsymbol{r} \in D^c_{\mathcal{A}(\mathcal{E})}(\mathbf{0})|\mathbf{0}\right) \qquad (3.13)\\ &\le \Pr\left(d_{\mathrm{E}}(\boldsymbol{r}, \mathbf{1}_N) \ge \min\{d_{\min}, \delta(\mathcal{E})\}\big|\mathbf{0}\right).\end{aligned}$$

Since $\{r_i - 1\}_{j=1}^N$ are independent Gaussian random variables each with mean 0 and variance $N_0/2$,

$$\zeta \triangleq \frac{2}{N_0}\sum_{i=1}^N (r_i - 1)^2$$

is a standard chi-squared random variable with N degrees of freedom and has density function

$$f(\zeta) = \frac{\zeta^{(N/2)-1}e^{-\zeta/2}}{\Gamma(N/2)2^{N/2}},$$

where $\Gamma(\cdot)$ is the gamma function. Thus, for any positive number δ,

$$\Pr\left(d_{\mathrm{E}}(\boldsymbol{r}, \mathbf{1}_N) \ge \delta|\mathbf{0}\right) = \int_{2\delta/N_0}^{\infty} \frac{\zeta^{(N/2)-1}e^{-\zeta/2}}{\Gamma(N/2)2^{N/2}} d\zeta. \qquad (3.14)$$

By L'Hospital rule, one can show easily that

$$\lim_{N_0\to 0} \frac{\int_{2\delta/N_0}^{\infty} \zeta^{(N/2)-1}e^{-\zeta/2}d\zeta}{(2\delta/N_0)^{(N/2)-1}e^{-\delta/N_0}} = 1. \qquad (3.15)$$

It follows from (3.13)–(3.15) that

$$\liminf_{N_0\to 0}\left(-N_0 \ln(\Pr(e_{\mathcal{A}(\mathcal{E})}))\right) \ge \min\{d_{\min}, \delta(\mathcal{E})\}. \qquad (3.16)$$

To complete the proof, we claim now, for any positive number ϵ, that the volume, denoted $\nu(\epsilon)$, of the part of $D^c_{\mathcal{A}(\mathcal{E})}(\mathbf{0})$ that is within SED $\min\{d_{\min}, \delta(\mathcal{E})\}+\epsilon$ of the sequence $\mathbf{1}_N$ is a positive number.

If $\delta(\mathcal{E}) < d_{\min}$, there is at least one sequence $\boldsymbol{x} \in D^c_{\mathcal{A}(\mathcal{E})}(\mathbf{0})$ that is within SED $\min\{\delta(\mathcal{E}) + \epsilon/2, d_{\min}\}$ of the sequence $\mathbf{1}_N$. For such a sequence $\boldsymbol{x}$, there is an error pattern $\boldsymbol{e} \in V^N \setminus E_{\boldsymbol{\lambda}_{\boldsymbol{x}}}$ such that $\boldsymbol{z}_{\boldsymbol{x}} + \boldsymbol{\lambda}_{\boldsymbol{x}}(\boldsymbol{e}) = \mathbf{0}$, which implies the sequence $\boldsymbol{x}$ belongs to the polyhedron $\boldsymbol{\lambda}_{\boldsymbol{x}}(\mathcal{H}(\boldsymbol{e}))$. Since almost all of the sequences in $\boldsymbol{\lambda}_{\boldsymbol{x}}(\mathcal{H}(\boldsymbol{e}))$ are in $G^c_{\mathcal{A}(\mathcal{E})}(\mathbf{0}) \subseteq D^c_{\mathcal{A}(\mathcal{E})}(\mathbf{0})$, the claim is true for this case.

For the remaining case $\delta(\mathcal{E}) \geq d_{\min}$, the claim is also true since for any codeword $\boldsymbol{c}$ of Hamming weight $d_{\min}$ the sphere of squared radius $d_{\min}$ centered at $\boldsymbol{s}(\boldsymbol{c})$ is a subset of $D_{\mathcal{A}(\mathcal{E})}(\boldsymbol{c}) \subset D^c_{\mathcal{A}(\mathcal{E})}(\mathbf{0})$.

Hence, from (2.1) and (3.7),

$$\begin{aligned}\Pr(e_{\mathcal{A}(\mathcal{E})}) &= \Pr\left(\boldsymbol{r} \in D^c_{\mathcal{A}(\mathcal{E})}(\mathbf{0})|\mathbf{0}\right) \\ &\geq \Pr\left(\boldsymbol{r} \in D^c_{\mathcal{A}(\mathcal{E})}(\mathbf{0}), d_{\mathrm{E}}(\boldsymbol{r}, \mathbf{1}_N) \leq \min\{d_{\min}, \delta(\mathcal{E})\} + \epsilon \,\middle|\mathbf{0}\right) \\ &\geq \frac{\nu(\epsilon)}{(\pi N_0)^{N/2}} e^{-(\min\{d_{\min},\delta(\mathcal{E})\}+\epsilon)/N_0}.\end{aligned} \tag{3.17}$$

It follows from (3.17) that

$$\limsup_{N_0\to 0} \left(-N_0 \ln(\Pr(e_{\mathcal{A}(\mathcal{E})}))\right) \leq \min\{d_{\min}, \delta(\mathcal{E})\} + \epsilon. \tag{3.18}$$

Then, (3.12) follows from (3.16), (3.18) and the arbitrariness of ϵ. □

4. Computation of Error Exponents for ROBDAs

To compute the error exponents of ROBDAs, according to Lemmas 3.2 and 3.3, it is necessary to compute the minimal SED $\sigma(\boldsymbol{e})$ for given error pattern $\boldsymbol{e} \in V^N$. In [6], Fossorier and Lin transform the computation of $\sigma(\boldsymbol{e})$ into an optimization problem.

Assume that the tuple $\boldsymbol{e} = (e_1, e_2, \ldots, e_N)$ has nonzero components at the indices $n_1, n_2, \ldots, n_{l_0}$ with $1 \leq n_1 < n_2 < \cdots < n_{l_0} \leq N$. Assume that $\boldsymbol{x}^* = (x_1^*, x_2^*, \ldots, x_N^*)$ is a sequence in $\mathcal{H}(\boldsymbol{e})$ such that

$$d_{\mathrm{E}}(\boldsymbol{x}^*, \mathbf{1}_N) = \sigma(\boldsymbol{e}). \tag{4.1}$$

If $x^*_{i_0} < -1$ for some i_0, then the sequence $\boldsymbol{x}' = (x_1', x_2', \ldots, x_N')$ defined by

$$x_i' \triangleq \begin{cases} -1, & \text{if } x_i^* \leq -1, \\ 1, & \text{if } x_i^* \geq 1, \\ x_i^*, & \text{otherwise,} \end{cases} \qquad i = 1, 2, \ldots, N,$$

is also in $\mathcal{H}(\boldsymbol{e})$ and satisfies

$$d_{\mathrm{E}}(\boldsymbol{x}', \mathbf{1}_N) < d_{\mathrm{E}}(\boldsymbol{x}^*, \mathbf{1}_N),$$

which contradicts (4.1). Hence, we have

$$0 \geq x^*_{n_1} \geq x^*_{n_2} \geq \cdots \geq x^*_{n_{l_0}} \geq -1. \tag{4.2}$$

Clearly, we also have

$$x_i^* = 1, \text{ for } i > n_{l_0}, \tag{4.3}$$

$$x_i^* = -x_{n_j}^*, \text{ for } n_{j-1} < i < n_j, j = 1, 2, \ldots, l_0, \tag{4.4}$$

where $n_0 \triangleq 0$.

Let $P(\boldsymbol{e})$ denote the following optimization problem:

$$\text{Minimize } d(\boldsymbol{y}) \triangleq \sum_{j=1}^{l_0} \left((1+y_j)^2 + (n_j - n_{j-1} - 1)(1-y_j)^2\right)$$

$$\text{where } \boldsymbol{y} = (y_1, \ldots, y_{l_0}) \text{ subject to } 0 \le y_1 \le \cdots \le y_{l_0} \le 1.$$

Then, we have proved the following lemma [6].

Lemma 4.1. *A sequence $\boldsymbol{x}^* = (x_1^*, x_2^*, \ldots, x_N^*) \in \mathcal{H}(\boldsymbol{e})$ satisfies (4.1) if and only if $\boldsymbol{x}^*$ satisfies (4.2)–(4.4) and the sequence $(-x_{n_1}^*, -x_{n_2}^*, \ldots, -x_{n_{l_0}}^*)$ is a solution of the optimization problem $P(\boldsymbol{e})$. In particular, $\sigma(\boldsymbol{e})$ is equal to the optimal value of $P(\boldsymbol{e})$.*

A method for solving the optimization problem $P(\boldsymbol{e})$ is also proposed in [6]. In Section 4.1, we will give a new method to solve this optimization problem. Our algorithm is much simpler than the one given in [6], and can be modified easily to generate all the boundary points of $D_{\mathcal{A}(\mathcal{E})}(\mathbf{0})$ which are nearest to the sequence $\mathbf{1}_N$. These nearest boundary points play important roles in the analysis of the asymptotic error performance of the decoding algorithm, the computation of the effective error coefficients [4]–[7].

4.1. New Method for Solving the Optimization Problem $P(\boldsymbol{e})$

Assume that the tuple $\boldsymbol{e} = (e_1, e_2, \ldots, e_N)$ has nonzero components at the indices $n_1, n_2, \ldots, n_{l_0}$ with $1 \le n_1 < n_2 < \cdots < n_{l_0} \le N$. Let

$$t(\boldsymbol{e}) \triangleq \max\left\{\frac{n_{l_0} - 2l_0}{n_{l_0}}, 0\right\}. \tag{4.5}$$

The tuple $\boldsymbol{e}$ is called *singular* if

$$\frac{n_{l_0} - n_j - 2(l_0 - j)}{n_{l_0} - n_j} \le t(\boldsymbol{e}), \text{ for } 1 \le j < l_0. \tag{4.6}$$

Lemma 4.2. *If $\boldsymbol{e}$ is a singular tuple, then $t(\boldsymbol{e})\mathbf{1}_{l_0}$ is the unique solution of $P(\boldsymbol{e})$ and the optimal value is equal to*

$$\begin{cases} \dfrac{4l_0(n_{l_0} - l_0)}{n_{l_0}}, & \textit{if } n_{l_0} > 2l_0, \\ n_{l_0}, & \textit{otherwise.} \end{cases} \tag{4.7}$$

Proof. Assume that $\boldsymbol{y} = (y_1, y_2, \ldots, y_{l_0})$ is an arbitrary sequence of real numbers with $y_0 = 0 \le y_1 \le y_2 \le \cdots \le y_{l_0} \le y_{l_0+1} = 1$.

At first we consider the case $t(\boldsymbol{e}) = 0$. From (4.6),

$$2(l_0 - i) - n_{l_0} + n_i \geq 0, \text{ for } 0 \leq i \leq l_0. \tag{4.8}$$

It follows from (4.8) that

$$\begin{aligned}
&d(\boldsymbol{y}) - d(t(\boldsymbol{e})\mathbf{1}_{l_0}) \\
&= \sum_{i=1}^{l_0} \left((n_i - n_{i-1}) y_i^2 + 2(2 - n_i + n_{i-1}) y_i \right) \\
&= \sum_{i=1}^{l_0} (n_i - n_{i-1}) y_i^2 + 2 \sum_{i=0}^{l_0 - 1} (y_{i+1} - y_i)(2(l_0 - i) - n_{l_0} + n_i) \\
&\geq \sum_{i=1}^{l_0} (n_i - n_{i-1}) y_i^2.
\end{aligned}$$

Thus, the sequence $t(\boldsymbol{e})\mathbf{1}_{l_0}$ is the unique solution of $P(\boldsymbol{e})$ for the case $t(\boldsymbol{e}) = 0$.

Next we consider the remaining case $t(\boldsymbol{e}) > 0$. From (4.5),

$$t(\boldsymbol{e}) = \frac{n_{l_0} - 2l_0}{n_{l_0}}. \tag{4.9}$$

From (4.6), we have

$$(n_{l_0} - n_i)t(\boldsymbol{e}) + (2(l_0 - i) - n_{l_0} + n_i) \geq 0, \text{ for } 0 \leq i \leq i_0. \tag{4.10}$$

If there is an integer i with $0 < i < l_0$ such that

$$n_i t(\boldsymbol{e}) + (2i - n_i) > 0,$$

from (4.9) and (4.10), we have

$$t(\boldsymbol{e}) = \frac{(n_{l_0} - n_i - 2(l_0 - i)) + (n_i - 2i)}{n_{l_0}} < \frac{(n_{l_0} - n_i)t(\boldsymbol{e}) + n_i t(\boldsymbol{e})}{n_{l_0}} = t(\boldsymbol{e}),$$

which is a contradiction. Hence

$$n_i t(\boldsymbol{e}) + (2i - n_i) \leq 0, \text{ for } 0 \leq i \leq l_0. \tag{4.11}$$

Let i_0 be the index with $0 \leq i_0 \leq l_0$ and

$$y_{i_0} \leq t(\boldsymbol{e}) \leq y_{i_0+1}. \tag{4.12}$$

It follows from (4.10)–(4.12) that

$$\begin{aligned}
&d(\boldsymbol{y}) - d(t(\boldsymbol{e})\mathbf{1}_{l_0})\\
=&\sum_{i=1}^{l_0}\Big((n_i-n_{i-1})(y_i-t(\boldsymbol{e}))^2+2(y_i-t(\boldsymbol{e}))\big((n_i-n_{i-1})t(\boldsymbol{e})+(2-n_i+n_{i-1})\big)\Big)\\
=&\sum_{i=1}^{l_0}(n_i-n_{i-1})\,(y_i-t(\boldsymbol{e}))^2+2\sum_{i=1}^{i_0-1}(y_i-y_{i+1})\big(n_it(\boldsymbol{e})+(2i-n_i)\big)\\
&\quad+2\,(y_{i_0}-t(\boldsymbol{e}))\,\big(n_{i_0}t(\boldsymbol{e})+(2i_0-n_{i_0})\big)\\
&\quad+2\,(y_{i_0+1}-t(\boldsymbol{e}))\,\big((n_{l_0}-n_{i_0})t(\boldsymbol{e})+(2(l_0-i_0)-n_{l_0}+n_{i_0})\big)\\
&\quad+2\sum_{i=i_0+1}^{l_0-1}(y_{i+1}-y_i)\big((n_{l_0}-n_i)t(\boldsymbol{e})+(2(l_0-i)-n_{l_0}+n_i)\big)\\
\geq&\sum_{i=1}^{l_0}(n_i-n_{i-1})\,(y_i-t(\boldsymbol{e}))^2,
\end{aligned}$$

where the following equivalences are applied:

$$\sum_{i=1}^{i_0}a_ib_i\equiv\sum_{i=1}^{i_0-1}(a_i-a_{i+1})\sum_{j=1}^{i}b_j+a_{i_0}\sum_{j=1}^{i_0}b_j,$$

$$\sum_{i=i_0+1}^{l_0}a_ib_i\equiv\sum_{i=i_0+1}^{l_0-1}(a_{i+1}-a_i)\sum_{j=i+1}^{l_0}b_j+a_{i_0+1}\sum_{j=i_0+1}^{l_0}b_j.$$

Thus, the sequence $t(\boldsymbol{e})\mathbf{1}_{l_0}$ is also the unique solution of $P(\boldsymbol{e})$ for the case $t(\boldsymbol{e})>0$.

Obviously, (4.7) gives the optimal value of $P(\boldsymbol{e})$. □

For $0\leq j<j'\leq n$ and a tuple $\boldsymbol{v}=(v_1,v_2,\ldots,v_n)$, let $\gamma_{j,j'}(\boldsymbol{v})$ denote the subtuple $(v_{j+1},v_{j+2},\ldots,v_{j'})$ of $\boldsymbol{v}$.

Let l_1 be the smallest integer such that $\gamma_{n_{l_1},N}(\boldsymbol{e})$ is a singular tuple. Clearly, $0\leq l_1<l_0$. Furthermore, for $i\geq 1$ and $l_i>0$, let l_{i+1} be the smallest integer such that $\gamma_{n_{l_{i+1}},n_{l_i}}(\boldsymbol{e})$ is a singular tuple. Then, $0\leq l_{i+1}<l_i$. Assume that p is the positive integer satisfying $l_{p+1}=0$. Clearly, for $0\leq i\leq p$ the integer l_{i+1} is the smallest such that

$$t(\gamma_{n_{l_{i+1}},n_{l_i}}(\boldsymbol{e}))=\max_{0\leq j<l_i}t(\gamma_{n_j,n_{l_i}}(\boldsymbol{e})).$$

Let $\boldsymbol{y}(\boldsymbol{e})$ denote the sequence of nonnegative numbers of length l_0 that satisfy

$$\gamma_{l_{i+1},l_i}(\boldsymbol{y}(\boldsymbol{e}))=t(\gamma_{n_{l_{i+1}},n_{l_i}}(\boldsymbol{e}))\mathbf{1}_{l_i-l_{i+1}},\quad\text{for }0\leq i\leq p.$$

Theorem 4.3. *The sequence $\boldsymbol{y}(\boldsymbol{e})$ is the unique solution of the optimization problem $P(\boldsymbol{e})$ and the optimal value is equal to*

$$\begin{cases} \sum_{i=0}^{p} \frac{4(l_i - l_{i+1})(n_{l_i} - n_{l_{i+1}} - (l_i - l_{i+1}))}{n_{l_i} - n_{l_{i+1}}}, & \text{if } n_{l_p} > 2l_p, \\ n_{l_p} + \sum_{i=0}^{p-1} \frac{4(l_i - l_{i+1})(n_{l_i} - n_{l_{i+1}} - (l_i - l_{i+1}))}{n_{l_i} - n_{l_{i+1}}}, & \text{otherwise.} \end{cases}$$

Proof. Clearly, for $1 \leq i \leq p$, from $l_i > 0$,

$$t(\gamma_{n_{l_i}, n_{l_{i-1}}}(\boldsymbol{e})) > t(\gamma_{n_j, n_{l_{i-1}}}(\boldsymbol{e})) \geq 0, \text{ for any } 0 \leq j < l_i, \tag{4.13}$$

and thus

$$n_{l_{i-1}} - n_{l_i} - 2(l_{i-1} - l_i) = (n_{l_{i-1}} - n_{l_i}) t(\gamma_{n_{l_i}, n_{l_{i-1}}}(\boldsymbol{e})). \tag{4.14}$$

If $t(\gamma_{n_{l_{i+1}}, n_{l_i}}(\boldsymbol{e})) \geq t(\gamma_{n_{l_i}, n_{l_{i-1}}}(\boldsymbol{e}))$ for some i with $1 \leq i \leq p$, then $t(\gamma_{n_{l_{i+1}}, n_{l_i}}(\boldsymbol{e}))$ is also positive and

$$\begin{aligned} n_{l_i} - n_{l_{i+1}} - 2(l_i - l_{i+1}) &= (n_{l_i} - n_{l_{i+1}}) t(\gamma_{n_{l_{i+1}}, n_{l_i}}(\boldsymbol{e})) \\ &\geq (n_{l_i} - n_{l_{i+1}}) t(\gamma_{n_{l_i}, n_{l_{i-1}}}(\boldsymbol{e})). \end{aligned} \tag{4.15}$$

From (4.14) and (4.15),

$$\begin{aligned} & t(\gamma_{n_{l_{i+1}}, n_{l_{i-1}}}(\boldsymbol{e})) = \max\left\{ \frac{n_{l_{i-1}} - n_{l_{i+1}} - 2(l_{i-1} - l_{i+1})}{n_{l_{i-1}} - n_{l_{i+1}}}, 0 \right\} \\ \geq & \max\left\{ \frac{(n_{l_{i-1}} - n_{l_i}) t(\gamma_{n_{l_i}, n_{l_{i-1}}}(\boldsymbol{e})) + (n_{l_i} - n_{l_{i+1}}) t(\gamma_{n_{l_i}, n_{l_{i-1}}}(\boldsymbol{e}))}{n_{l_{i-1}} - n_{l_{i+1}}}, 0 \right\} \\ = & t(\gamma_{n_{l_i}, n_{l_{i-1}}}(\boldsymbol{e})), \end{aligned} \tag{4.16}$$

which contradicts (4.13). Hence,

$$1 \geq t(\gamma_{n_{l_1}, n_{l_0}}(\boldsymbol{e})) > t(\gamma_{n_{l_2}, n_{l_1}}(\boldsymbol{e})) > \cdots > t(\gamma_{n_{l_{p+1}}, n_{l_p}}(\boldsymbol{e})) \geq 0. \tag{4.17}$$

Since $\gamma_{n_{l_{i+1}}, n_{l_i}}(\boldsymbol{e})$ are singular tuples for $0 \leq i \leq p$, this theorem follows from Lemma 4.2 and (4.17). □

According to Theorem 4.3, the following Algorithm 1 computes the minimal SED $\sigma(\boldsymbol{e})$ for any $\boldsymbol{e} \in V^N$.

Algorithm 1
Input A tuple $\boldsymbol{e} \in V^N$ with nonzero entries at positions $n_1 < n_2 < \cdots < n_{l_0}$.
Output The minimal SED $\sigma(\boldsymbol{e})$.
Step 1. Set $J = l_0$, $T = 0$ and $n_0 = 0$.
Step 2. Find the smallest index I such that

$$\max\left\{ \frac{(n_J - n_I) - 2(J - I)}{n_J - n_I}, 0 \right\}$$

is maximized. If the maximum is positive, add

$$\frac{4(J-I)(n_J-n_I-J+I)}{n_J-n_I}$$

to T and goto Step 3. Otherwise, output $T+n_J$ and END.
Step 3. If $I>0$, set $J=I$ and goto Step 2. Otherwise, output T and END. □

To compute the SECRs of ROBDAs, the unified method proposed in [6] iteratively constructs a solution of the optimization problem based on the Kuhn-Tucker conditions. Algorithm 1 is more direct and much simpler than this known method. Especially, the Kuhn-Tucker conditions are not used in the proposed method. Furthermore, according to the uniqueness of the solution shown in Theorem 4.3, Algorithm 1 can be modified easily to determine all the nearest boundary points and consequently to compute the effective error coefficient, which is also an important measure for the asymptotic property of the probability of decoding error of the ROBDAs [4]–[7].

4.2. Error Exponents of Some Known Decoding Algorithms

First, we introduce a partial order $\succeq$ in V^N. We write $\boldsymbol{e}\succeq\boldsymbol{e}'$ if the Hamming weight of $\boldsymbol{\gamma}_{i,N}(\boldsymbol{e})$ is not smaller than that of $\boldsymbol{\gamma}_{i,N}(\boldsymbol{e}')$ for all $0\le i<N$. For computing the Error Exponents of ROBDAs, a useful result which is summarized in the following lemma can also be found in [6]. For completeness, we give an alternative proof for it. Our proof is more direct than that given in [6].

Lemma 4.4. *For any two binary N-tuples* $\mathbf{e}$ *and* $\mathbf{e}'$,

$$\sigma(\mathbf{e})\ge\sigma(\mathbf{e}'),\ \ \text{if } \mathbf{e}\succeq\mathbf{e}'. \tag{4.18}$$

Proof. Assume that the nonzero entries of $\mathbf{e}$ and $\mathbf{e}'$ are at the positions $n_1<n_2<\cdots<n_j$ and $n_1'<n_2'<\cdots<n_{j'}'$, respectively. From $\boldsymbol{e}\succeq\boldsymbol{e}'$, one can see easily that $j\ge j'$ and

$$n_{j-i}\ge n'_{j'-i},\ \text{for } 0\le i<j'. \tag{4.19}$$

For $0\le a_1\le a_2\le\cdots\le a_N$, let $\boldsymbol{x}=(x_1,x_2,\ldots,x_N)$ and $\boldsymbol{x}'=(x_1',x_2',\ldots,x_N')$ be the sequences in $\mathcal{H}(\boldsymbol{e})$ and $\mathcal{H}(\boldsymbol{e}')$ which satisfy $|x_i|=|x_i'|=a_i$ for $1\le i\le N$, respectively. Then, from $j\ge j'$ and (4.19),

$$\begin{aligned} d_{\mathrm{E}}(\boldsymbol{x},\mathbf{1})-d_{\mathrm{E}}(\boldsymbol{x}',\mathbf{1}) &= 4\sum_{i=1}^{j}a_{n_i}-4\sum_{i'=1}^{j'}a_{n'_{i'}}\\ &= 4\sum_{i=0}^{j'-1}(a_{n_{j-i}}-a_{n'_{j'-i}})+4\sum_{i=j'}^{j-1}a_{n_{j-i}} \end{aligned}$$

is nonnegative and thus (4.18) follows. □

Example. For $j=1,2,3$, the Chase-j decoding algorithm [2] is the ROBDA with $V^N\setminus E_{\boldsymbol{\lambda}}=\{\boldsymbol{e}\in V^N:\boldsymbol{e}\succeq\boldsymbol{e}_j\}$ for all $\boldsymbol{\lambda}\in V^N$ (cf. [11]), where $\boldsymbol{e}_j$ is the binary

N-tuple whose nonzero entries are at the positions a) $1, 2, \ldots, d_{\min}$ if $j = 1$; b) $\lfloor d_{\min}/2 \rfloor + i$ for $1 \le i \le t_0 \triangleq \lceil d_{\min}/2 \rceil$ if $j = 2$; c) $d_{\min} - 2i$ for $0 \le i < t_0$ if $j = 3$.

Since $\boldsymbol{e}_j$ is a singular tuple with $t(\boldsymbol{e}_j) = 0$ and the rightmost nonzero entry is at the position $d_{\min}$, it follows from Lemma 4.2 that $\sigma(\boldsymbol{e}_j) = d_{\min}$. Hence, from (3.10), Lemmas 3.3 and 4.4, the error exponents of Chase decoding algorithms are equal to $d_{\min}$.

The full GMD [1] is equivalent to the Chase-3 decoding algorithm and thus has error exponent $d_{\min}$. □

Example. The modified GMD decoding algorithm proposed in [3, 4] is the ROBDA with $V^N \setminus E_{\boldsymbol{\lambda}} = \{\boldsymbol{e} \in V^N : \boldsymbol{e} \succeq \boldsymbol{e}^*\}$ for all $\boldsymbol{\lambda} \in V^N$, where $\boldsymbol{e}^*$ is the binary N-tuple whose nonzero entries are at the position $d_{\min} + 1$ and the positions $d_{\min} - 2i$ for $0 < i < t_0$.

From the definition, one can deduce easily that $\gamma_{0,d_{\min}-2}(\boldsymbol{e}^*)$ is singular and $d_{\min} - 2$ is the smallest value of l such that $\gamma_{l,N}(\boldsymbol{e}^*)$ is singular. Thus, from Theorem 4.3,

$$\sigma(\boldsymbol{e}^*) = (d_{\min} - 2) + \frac{4 \cdot 1 \cdot ((d_{\min} + 1) - (d_{\min} - 2) - 1)}{(d_{\min} + 1) - (d_{\min} - 2)} = d_{\min} + 2/3.$$

Hence, the error exponent of the modified GMD decoding algorithm is also equal to $d_{\min}$. □

Example. Some multiple Chase-like decoding algorithms are proposed in [8]. Like the original Chase algorithms, the multiple Chase-like decoding algorithms employ a bounded-distance-$(t_0 - 1)$ binary decoder to generate the candidate codewords. For $0 \le \tau \le N - t_0$ and binary N-tuple $\boldsymbol{u}$, Chase$(\tau, \boldsymbol{u})$ is a 2^τ stage decoding algorithm at each stage the bounded-distance-$(t_0 - 1)$ binary decoder applies to a binary N-tuple obtained by adding $\boldsymbol{u}$ to an error pattern whose non-zero components are confined in the τ least reliable positions. For a positive integer h, the (h, τ)-Chase decoding is a multiple Chase-like decoding algorithm consisting of successive Chase$(\tau, \boldsymbol{u}^{(i)})$ with $1 \le i \le h$, where $\boldsymbol{u}^{(i)}$, called the ith search center, is the "best" binary N-tuple among those which are not yet searched by the $(i-1)$-Chase decoding. The first search center is the hard-decision tuple $\boldsymbol{z_r}$. The second search center is (cf. [9])

$$\boldsymbol{u}^{(2)} = \boldsymbol{z_r} + \boldsymbol{\lambda_r}\big((\underbrace{0, \ldots, 0}_{\tau}, \underbrace{1, \ldots, 1}_{t_0}, \underbrace{0, \ldots, 0}_{N-\tau-t_0})\big).$$

Hence, the $(2, \tau)$-Chase decoding is the ROBDA with $V^N \setminus E_{\boldsymbol{\lambda}} = \{\boldsymbol{e} \in V^N : \boldsymbol{e} \succeq \boldsymbol{f}\}$, where

$$\boldsymbol{f} \triangleq (\underbrace{0, \ldots, 0}_{\tau}, \underbrace{1, \ldots, 1}_{\lfloor t_0/2 \rfloor}, \underbrace{0, \ldots, 0}_{\lceil t_0/2 \rceil}, \underbrace{1, \ldots, 1}_{\lceil t_0/2 \rceil}, \underbrace{0\ , \ \ldots\ , \ 0}_{N-\tau-\lceil 3t_0/2 \rceil}).$$

The error exponent of $(2, \tau)$-Chase decoding is $\min\{d_{\min}, \sigma(\boldsymbol{f})\}$. Since $\boldsymbol{f}$ is singular, it follows from Lemma 4.2 that

$$\sigma(\boldsymbol{f}) = \begin{cases} \tau + \lceil t_0/2 \rceil + t_0, & \text{if } \tau \le \lfloor t_0/2 \rfloor, \\ \dfrac{4t_0(\tau + \lceil t_0/2 \rceil)}{\tau + \lceil t_0/2 \rceil + t_0}, & \text{otherwise.} \end{cases}$$

Hence, the error exponent of $(2, \tau)$-Chase decoding is equal to $d_{\min}$ if and only if

$$\tau \ge \lfloor d_{\min}/2 \rfloor - \lceil t_0/2 \rceil. \qquad \square$$

5. Conclusion

For the reliability-order-based decoding algorithms of linear block codes, we study the error performance in this paper by a detailed investigation on the decision regions. We prove that the error exponent and the squared error-correction radius are equal to each other for any reliability-order-based decoding algorithm. The computation of the squared error-correction radii of reliability-order-based decoding algorithm is transformed in [6] by Fossorier and Lin to an optimization problem. A unified method which iteratively constructs the solution of this optimization problem based on the Kuhn-Tucker conditions is also proposed in [6]. This method is improved in this paper. The improved method is direct and much simpler. Especially, the Kuhn-Tucker conditions are not used in the improved method. We also notice that the improved method can be used further to compute the effective error coefficient, which is also an important measure for the asymptotic property of the probability of decoding error of the reliability-order-based decoding algorithms.

References

[1] G.D. Forney Jr., "Generalized Minimum Distance Decoding," IEEE Trans. Inform. Theory, vol.IT-12, no. 2, pp. 125–131, April 1966.

[2] D. Chase, "A Class of Algorithms for Decoding Block Codes with Channel Measurement Information," IEEE Trans. Inform. Theory, vol. IT-18, no. 1, pp. 170–182, Jan. 1972.

[3] G.D. Forney Jr., "A Bounded-Distance Decoding Algorithm for the Leech Lattice, with Generalizations," IEEE Trans. Inform. Theory, vol. 35, no. 4, pp. 906–909, July 1989.

[4] B. Shen, K.K. Tzeng and C. Wang, "A Bounded-Distance Decoding Algorithm for Binary Linear Block Codes Achieving the Minimum Effective Error Coefficient," IEEE Trans. Inform. Theory, vol. 42, no. 6, pp. 1987–1991, Nov. 1996.

[5] G.D. Forney and A. Vardy, "Generalized Minimum-Distance Decoding of Euclidean Codes and Lattices," IEEE Trans. Inform. Theory, vol. 42, no. 6, pp. 1992–2026, Nov. 1996.

[6] M.P.C. Fossorier and S. Lin, "A Unified Method for Evaluating the Error-Correction Radius of Reliability-Based Soft-Decision Algorithms for Linear Block Codes," IEEE Trans. Inform. Theory, vol. 44, no. 2, pp. 691–700, Mar. 1998.

[7] E. Fishler, O. Amrani and Y. Be'ery, "Geometrical and Performance Analysis of GMD and Chase Decoding Algorithms," IEEE Trans. Inform. Theory, vol. 45, no. 5, pp. 1406–1422, July 1999.

[8] T. Kasami, "On Integer Programming Problems Related to Soft-Decision Iterative Decoding Algorithms," 13th Int. Symp. Proceedings/AAECC 13, Honolulu, Hawaii, USA, pp. 43–54, Nov. 1999.

[9] Y. Tang, T. Kasami and T. Fujiwara, "On the Computation of the Search Centers and the Evaluation of the Testing Conditions for the h-Chase Decoding," 23rd Symp. on Inform. Theory & Its Appl., Aso, Kumamoto, Japan, pp. 77–80, Oct. 2000.

[10] Y. Tang, T. Fujiwara and T. Kasami, "Computation of Nearest Boundary Points for Symmetrical Decoding of Linear Block Codes,"Designs, Codes, Graphs and their Links II, Kyoto, Japan, Aug. 2001.

[11] Y. Tang, T. Fujiwara and T. Kasami, "Asymptotic Optimality of the GMD and Chase Decoding Algorithms", IEEE Trans. Inform. Theory, vol. 48, no. 8, pp. 2401–2405, August 2002.

Yuansheng Tang

Former address
Department of Computer Science
School of Computing
National University of Singapore
3 Science Drive 2
Singapore 117543

Actual address
Temasek Laboratories
National University of Singapore
10 Kent Ridge Crescent
Singapore 119260.
e-mail: `tsltangy@nus.edu.sg`

Progress in Computer Science and Applied Logic, Vol. 23, 319–330

A Construction of Authentication Codes with Secrecy

Xiaojian Tian and Cunsheng Ding

Abstract. In this paper, we construct a class of authentication/secrecy codes which provide ordered perfect L-fold secrecy. The class of authentication codes meets the information-theoretic bounds of deception probabilities. Some of them also meet a lower bound on the number of encoding rules. The authentication codes are thus optimal with respect to the two types of bounds.

Mathematics Subject Classification (2000). Primary 94A60; Secondary 94A62.

Keywords. Authentication codes, cryptography, perfect secrecy.

1. Introduction

Authentication and secrecy of messages are two main security services in computer and communication systems, and two important areas in cryptography. In 1974 Gilbert, MacWilliams and Sloane studied authentication codes [4]. Ten years later Simmons [15] developed a theory of unconditional authentication, which is analogous to Shannon's theory of unconditional secrecy [14]. Since then codes that provide both authentication and secrecy have been considered, and bounds and characterizations of these codes have been established (see, for example, [4, 17, 13, 18, 16, 20, 2]).

The purpose of this paper is to construct a class of authentication/secrecy codes which provide ordered perfect L-fold secrecy. The class of authentication codes meets the information-theoretic bounds of deception probabilities. Some of them also meet a lower bound on the number of encoding rules. The authentication codes are thus optimal with respect to the two types of bounds. Our construction is influenced by and related to the construction by Pei [8], but the two constructions are different.

2. Authentication/Secrecy Codes and Some Bounds

We use the basic model developed by Simmons [15]. In this model, a *transmitter* communicates a sequence of distinct *source states* from a set S to a *receiver* by encoding them using one from a set of *encoding rules* $\mathcal{E}$. Each encoding rule is an injective mapping from $\mathcal{S}$ into a set $\mathcal{M}$ of messages. An *opponent* can intercept the messages transmitted and modify them. We will always assume that the opponent knows the whole system, the only secret is the actual encoding rule shared and used by the transmitter and the receiver.

We follow the notations of [7]. An *authentication/secrecy code* is a 3-tuple $\mathcal{A} = (\mathcal{S}, \mathcal{M}, \mathcal{E})$ where $\mathcal{S}$ is a set of k *source states*, $\mathcal{M}$ is a set of v *messages*, $\mathcal{E}$ is a set of b *encoding rules* such that each $e \in \mathcal{E}$ is an injective mapping from $\mathcal{S}$ to $\mathcal{M}$. $p_{\mathcal{E}}$ is a probability distribution defined on $\mathcal{E}$ and $p_{\mathcal{S}_k}$ is a probability distribution defined on the collection of all sequences of distinct source states

$$\mathcal{S}_k = \{(s_1, s_2, \ldots, s_k) | s_1, \ldots, s_k \in \mathcal{S}, s_i \neq s_j \text{ for } 1 \leq i < j \leq k\}.$$

Define for any fixed l with $1 \leq l \leq k$

$$\mathcal{S}^l = \{s^l = (s_1, \ldots, s_l) | s_i \in \mathcal{S}, 1 \leq i \leq l\}$$

and

$$\mathcal{M}^l = \{m^l = (m_1, \ldots, m_l) | m_i \in \mathcal{M}, 1 \leq i \leq l\}.$$

We use $\mathbf{E}$, $\mathbf{S}$, $\mathbf{M}$, $\mathbf{S}^l$, and $\mathbf{M}^l$ to denote the random variables taking values e, s, m, s^l, m^l in $\mathcal{E}$, $\mathcal{S}$, $\mathcal{M}$, $\mathcal{S}^l$, and $\mathcal{M}^l$, respectively.

If the opponent observes l messages before attempting to deceive the receiver with another message then we say that he is carrying out a spoofing attack of order l. If he adopts an optimal strategy, the probability that he succeeds in deceiving the receiver is denoted by P_{d_l}.

For any $m^l = (m_1, \ldots, m_l) \in \mathcal{M}^l$ and $e \in \mathcal{E}$, let $f_e(m_i)$ be the encoding of m_i under e, and let

$$f_e(m^l) = (f_e(m_1), \ldots, f_e(m_l))$$

denote the unique element $(s_1, \ldots, s_l) \in \mathcal{S}^l$, when it exists, such that $s_i = f_e(m_i)$ $(1 \leq i \leq l)$. For any $m^l = (m_1, \ldots, m_l) \in \mathcal{M}^l$, define

$$\mathcal{E}(m^l) = \left\{ e \in \mathcal{E} : \begin{array}{l} m_i \ (1 \leq i \leq l) \text{ acceptable under } e \text{ and} \\ f_e(m_i) \ (1 \leq i \leq l) \text{ pairwise distinct} \end{array} \right\},$$

i.e., $\mathcal{E}(m^l)$ is the set of all encoding rules under which m^l is valid, and may be empty for some m^l.

For simplicity, we abbreviate by omitting the names of random variables in a probability distribution when this causes no confusion. For instance, we abbreviate $p(\mathbf{S} = s)$ to $p(s)$, and $p(\mathbf{E} = e | \mathbf{M}^l = m^l)$ to $p(e|m^l)$. Let $p(m|m^l)$ denote the probability that m would be a valid $(l+1)$th message when m^l has been observed. It is proved in [7] that the unconditional probability of success in an optimum spoofing attack of order l is

$$P_{d_l} = \sum_{m^l \in \mathcal{M}^l} p(m^l) \max_{m \in \mathcal{M}} p(m|m^l). \tag{1}$$

We shall need this formula later.

It was proved in [7, 10, 12] that

$$P_{d_l} \geq 2^{H(\mathcal{E}|\mathcal{M}^{l+1})-H(\mathcal{E}|\mathcal{M}^l)} \tag{2}$$

for any $l \geq 0$. These are the information-theoretic bounds which hold for authentication codes with and without secrecy. We shall need these bounds later.

Let L be such that $1 \leq L \leq k$. We say that an authentication/secrecy code $(\mathcal{S}, \mathcal{M}, \mathcal{E})$ has *ordered perfect L-fold secrecy (O(L)-secrecy)* if for every $m^L \in \mathcal{M}^L$ with $p(m^L) > 0$ and every $s^L \in \mathcal{S}^L$ with $p(s^L) > 0$ we have $p(s^L|m^L) = p(s^L)$ [3].

Define the space

$$\mathcal{S}_l = \{(s_1, \ldots, s_l)|s_1 \in \mathcal{S}, \ldots, s_l \in \mathcal{S} \text{ are pairwise distinct}\}.$$

We say that $p_{\mathcal{S}_l}$ is *positive* if $p_{\mathcal{S}_l}(s_1, \ldots, s_l) > 0$ for every $(s_1, \ldots, s_l) \in \mathcal{S}_l$.

The following lemma gives a lower bound on the number of encoding rules and will be needed later.

Lemma 1. [2] *Let $\mathcal{A} = (\mathcal{S}, \mathcal{M}, \mathcal{E})$ be an authentication code and let L be such that $1 \leq L \leq k$. If $\mathcal{A}$ has O(L)-secrecy and $p_{\mathcal{S}_L}$ is positive, then*

$$|\mathcal{E}| \geq k!/(k-L)! \prod_{i=0}^{L-1} (1/P_{d_i}).$$

We have also the following bounds [6, 9].

Lemma 2. *In any authentication system without splitting,*

$$P_{d_i} \geq \frac{|\mathcal{S}| - i}{|\mathcal{M}| - i}.$$

3. The Construction of the Authentication Codes

We shall use $\mathbf{F}_q$ to denote the finite field with q elements. Let n be a positive integer, and let Δ be another integer such that $1 \leq \Delta \leq n$.

- The source state space $\mathcal{S}$ of our authentication codes is a set of k nonzero vectors in $\mathbf{F}_q^n$ such that any Δ of them are linearly independent. $p_{\mathcal{S}_\Delta}$ is a uniform distribution on $\mathcal{S}_\Delta$. We will deal with the construction of $\mathcal{S}$ later in Section 7.
- The message space $\mathcal{M}$ of our codes is $\mathbf{F}_q^n \setminus \{0\}$.
- Our encoding rule space $\mathcal{E}$ is defined by $\mathcal{E} = GL_n(\mathbf{F}_q)$, the *general linear group* of degree n over $\mathbf{F}_q$. In other words, $\mathcal{E}$ is the set of $n \times n$ nonsingular matrices over $\mathbf{F}_q$. $\mathcal{E}$ can be regarded as the set of nonsingular linear transformations from $\mathbf{F}_q^n$ to $\mathbf{F}_q^n$. All encoding rules are equally likely.

Given a source state $s = (s_1, s_2, \ldots, s_n) \in \mathcal{S}$ and an encoding rule $e \in \mathcal{E}$, the transmitter will generate the message $m = se$. When the receiver receives a message m', he will recover the source state s' by $s' = m'e^{-1}$, then determine the authenticity of the message by checking whether $s' \in \mathcal{S}$.

4. The Authenticity of the Authentication Codes

Having described the construction of the authentication/secrecy codes in the previous section, we now calculate the deception probabilities P_{d_l}. To this end, we need some auxiliary results.

Lemma 3. [21] *In the construction of Section* 3, *we have*

$$|\mathcal{E}| = |GL_n(\mathbf{F}_q)| = q^{\frac{n(n-1)}{2}} \prod_{i=1}^{n}(q^i - 1).$$

Lemma 4. *In the authentication code of Section* 3, *for* $0 \le i \le \Delta$, $m^i \in \mathcal{M}^i$ *with* $p(m^i) > 0$,

$$|\mathcal{E}(m^i)| = \frac{k!}{(k-i)!} q^{\frac{(n+i-1)(n-i)}{2}} \prod_{j=1}^{n-i}(q^j - 1).$$

Proof. Note that $p((m_1, m_2, \dots, m_i)) > 0$ if and only if $m_1, m_2, \dots, m_i$ are linearly independent. Let $m^i = (m_1, m_2, \dots, m_i) \in \mathcal{M}^i$ and be linearly independent. The m^i can be transformed from any ordered i-set of source states. There are $\frac{k!}{(k-i)!}$ such ordered i-sets. For each ordered i-set $s^i = (s_1, s_2, \dots, s_i)$, we now calculate how many linear transformations e can transform s^i to m^i. Since $s_1, s_2, \dots, s_i$ are linearly independent, we can extend s^i to form a basis of $\mathbf{F}_q^n$, $s^n = (s_1, s_2, \dots, s_n)$. To fully determine T, we only need to determine the image of s_1, s_2, …, s_n under e. These images

$$m^n = (m_1, m_2, \dots, m_n) = (s_1, s_2, \dots, s_n)e$$

form another basis of $\mathbf{F}_q^n$. Since we already know $m_1, \dots, m_i$, m_{i+1} can be any vector of $\mathbf{F}_q^n$ which is linearly independent with $m_1, \dots, m_i$. But since $m_1, \dots, m_i$ are linearly independent, there are $q^n - q^i$ possible choices of m_{i+1}. After m_{i+1} is chosen, m_{i+2} can be any vector of $\mathbf{F}_q^n$ which is linear independent with $m_1, \dots, m_{i+1}$, but since $m_1, \dots, m_{i+1}$ are linearly independent, there are $q^n - q^{i+1}$ possible choices of m_{i+2}. Proceeding this way, there are $q^n - q^{n-1}$ possible choices of m_n. So the number of linear transformations that can transform s^i to m^i is $\prod_{j=i}^{n-1}(q^n - q^j)$. Therefore,

$$\begin{aligned} |\mathcal{E}(m^i)| &= \frac{k!}{(k-i)!} \prod_{j=i}^{n-1}(q^n - q^j) \\ &= \frac{k!}{(k-i)!} q^{\frac{(n+i-1)(n-i)}{2}} \prod_{j=1}^{n-i}(q^j - 1). \end{aligned}$$

Theorem 5. *In the construction of Section* 3, *for* $0 \le l < \Delta$, *the deception probabilities*

$$P_{d_l} = \frac{k-l}{q^n - q^l}.$$

Proof. First, since $l < \Delta \leq k$, $p(m^l) > 0$ if and only if m^l is a sequence of l linearly independent vectors. We now compute the probability $p(m|m^l)$, the probability that m would be a valid $(l+1)$th message when m^l has been observed. By Lemma 4, $|\mathcal{E}(m^l)|$ is constant and independent of m^l. Hence

$$p(m^l * m \text{ valid}) = \frac{|\mathcal{E}(m^{l+1})|}{|\mathcal{E}|}$$

if m is linearly independent of $m_1, \ldots, m_l$, where $m^l * m$ denotes the message sequence $(m_1, \ldots, m_l, m)$. For the same reason, if $p(m^l) > 0$ (i.e., m^l valid), we have

$$p(m^l) = \frac{|\mathcal{E}(m^l)|}{|\mathcal{E}|}.$$

Hence if $m_1, \ldots, m_l, m$ are linearly independent, m^l and $m^l * m$ must be valid, and thus we have

$$p(m|m^l) = \frac{p(m^l * m)}{p(m^l)} = \frac{|\mathcal{E}(m^{l+1})|}{|\mathcal{E}(m^l)|} = \frac{k-l}{q^n - q^l},$$

where $m^l * m$ denotes the message sequence $(m_1, \ldots, m_l, m)$. It then follows from (1) that

$$P_{d_l} = \sum_{m^l \in \mathcal{M}^l} p(m^l) \max_{m \in \mathcal{M}} p(m|m^l) = \frac{k-l}{q^n - q^l} \sum_{m^l \in \mathcal{M}^l} p(m^l) = \frac{k-l}{q^n - q^l}.$$

□

5. The Secrecy of the Authentication Codes

We now prove that this authentication code provides ordered perfect Δ-fold secrecy. First, since every Δ source states are linearly independent, every Δ messages under the same encoding rule are linearly independent. Suppose $m^\Delta = (m_1, m_2, \ldots, m_\Delta)$ is an ordered set of Δ linearly independent messages, and $s^\Delta = (s_1, s_2, \ldots, s_\Delta)$ is an ordered set of Δ source states, from the proof of Lemma 4, we know that there are always $\prod_{j=\Delta}^{n-1}(q^n - q^j)$ encoding rules that map s^Δ to m^Δ. Note that all source states are equally likely and all encoding rules are equally likely by system assumption. Thus given an ordered set of Δ messages, the opponent will have no information about the source states being communicated. So this code provides $O(\Delta)$-secrecy.

6. The Optimality of the Authentication Codes

Theorem 6. *The authentication/secrecy codes of Section* 3 *meet the lower bounds of* (2), *and are thus optimal with respect to these bounds.*

Proof. By Lemma 4, we have

$$H(\mathcal{E}|\mathcal{M}^{l+1}) = \log_2 |\mathcal{E}(m^{l+1})|, \ H(\mathcal{E}|\mathcal{M}^{l}) = \log_2 |\mathcal{E}(m^{l})|,$$

where $m^{l+1} \in \mathcal{M}^{l+1}$ with $p(m^{l+1}) > 0$ and $m^l \in \mathcal{M}^l$ with $p(m^l) > 0$. Hence

$$2^{H(\mathcal{E}|\mathcal{M}^{l+1}) - H(\mathcal{E}|\mathcal{M}^{l})} = \frac{k-l}{q^n - q^l} = P_{d_l}.$$

So the codes are optimal with respect to the information-theoretic lower bounds of (2). □

Theorem 7. *If $\Delta = n$, the code of Section* 3 *meets the lower bound of Lemma* 1 *for $L = n$, and is thus optimal with respect to this bound.*

Proof. The proof is straightforward and omitted. □

Theorem 8. *The authentication/secrecy codes of Section* 3 *meet the lower bound $P_{d_0} = \frac{|S|}{|\mathcal{M}|}$ of Lemma* 2*.*

Furthermore, they also meet the bound $P_{d_1} = \frac{|S|-1}{|\mathcal{M}|-1}$ if $q = 2$.

Proof. The proof is straightforward and omitted. □

7. Specific Constructions of Authentication Codes Within the Framework

The construction of authentication codes presented in the previous sections is quite generic. Different constructions of the source state space $\mathcal{S}$ yield different authentication codes with secrecy. In this section we present several classes of authentication codes within the framework of this generic construction of Section 3.

In the generic construction, we require that $1 \leq \Delta \leq n$. In fact the case $\Delta = 1$ is not interesting. So we shall consider only the cases that $2 \leq \Delta \leq n$. In our construction, the source state space $\mathcal{S}$ is a subset S of $\mathbf{F}_q^n$ such that any Δ of them are linearly independent. For efficiency purpose, we wish to have the size of $\mathcal{S}$ maximal when Δ and n are fixed. We are interested in not only this maximal size, but also the specific constructions of such a set.

Note that $\Delta \geq 2$. Our source state space $\mathcal{S}$ is actually a subset of the projective space $PG(n-1, q)$. In this section we will give several constructions of authentication codes using finite geometries and coding theory.

7.1. With Finite Geometries

Let $PG(N, q)$ be the projective space of N dimensions over the finite field $\mathbf{F}_q$, $q = p^h$ with p prime, and let $|PG(N,q)| = \theta_N = \left(q^{N+1} - 1\right)/(q-1)$. A set of points in $PG(N, q)$ are linearly independent if and only if the vectors representing them are a set of linearly independent vectors in the space $\mathbf{F}_q^{N+1}$. In $PG(N, q)$, subspaces will be denoted by Π_l, where l is the dimension of the subspace. A Π_0 is a *point*, a Π_1 is a *line*, and a Π_2 is a *plane*, a Π_3 is a *solid*, and a Π_{N-1} is a *hyperplane* or *prime*.

A (k,r)*-set* is a set of k points at most r of which lie in Π_{r-1} but some $r+2$ lie in a Π_r; that is, $r+1$ points are always linearly independent, but some $r+2$ points are linearly dependent. We use $M_r(N,q)$ to denote the maximum k such that a (k,r)-set exists in $PG(N,q)$.

Let $e_i = (0,\ldots,0,1,0,\ldots,0)$ for $i = 0,1,\ldots,N$, where 1 is in the $(i+1)$th position and let $e = (1,1,\ldots,1)$. We will need these points later for defining the set S, which can be used to define the source state space $\mathcal{S}$.

Any (k,r)-set could be used as the source state space $\mathcal{S}$ in the construction of Section 3, and the authentication code obtained will still be optimal with respect to the information-theoretic lower bounds of (2). But we are interested more in the lower bound of Lemma 1, and thus the case that $\Delta = n$. The construction of (k,r)-sets is a big area in finite geometries [5]. Here we intend to demonstrate some examples of the (k,r)-sets and thus some constructions of authentication codes within our generic construction of Section 3.

7.2. The First Class of Authentication Codes

We consider the case $\Delta = 2$, and define $\mathcal{S} = PG(n-1,q)$. Then any two elements in $\mathcal{S}$ are linearly independent. Thus $k = |\mathcal{S}| = (q^n-1)/(q-1)$. Hence the generic construction of Section 3 gives an authentication code

$$(\mathcal{S},\mathcal{M},\mathcal{E}) = (PG(n-1,q), \mathbf{F}_q^n \setminus \{0\}, GL_n(\mathbf{F}_q))$$

with

$$|\mathcal{S}| = \frac{q^n-1}{q-1},\ |\mathcal{M}| = q^n-1,\ |\mathcal{E}| = q^{n(n-1)/2}\prod_{i=1}^{n}(q^i-1)$$

and

$$P_{d_l} = \frac{1}{q-1}$$

for $l = 0,1$.

This class of authentication/secrecy codes is optimal with respect to both the information-theoretic lower bounds of (2), but not optimal with respect to the lower bound of Lemma 1. However they also meet the lower bound $P_{d_i} \geq \frac{|\mathcal{S}|-i}{|\mathcal{M}|-i}$ for $i = 0$.

7.3. The Second Class of Authentication Codes

We set $\Delta = n = 3$ and consider the 3-dimensional space $\mathbf{F}_q^3$, and equivalently the 2-dimensional projective space $PG(2,q)$. In this case, it is known that

$$M_2(2,q) = \begin{cases} q+1, & q \text{ odd} \\ q+2, & q \text{ even} \end{cases}$$

An $(M_2(2,q),2)$-set in $PG(2,q)$, q odd, is called an *oval* and an $(M_2(2,q),2)$-set in $PG(2,q)$, q even, is called an *hyperoval*. For the detailed construction of ovals and hyperovals, the reader is referred to [11] and [1].

If we use such an oval and hyperoval as our source state space $\mathcal{S}$ in the construction of Section 3, we obtain an authentication code $(\mathcal{S}, \mathcal{M}, \mathcal{E})$ with

$$|\mathcal{S}| = \begin{cases} q+1, & q \text{ odd}, \\ q+2, & q \text{ even}, \end{cases} \quad |\mathcal{M}| = q^3 - 1, \ |\mathcal{E}| = q^3(q^3-1)(q^2-1)(q-1),$$

and

$$P_{d_l} = \begin{cases} \frac{q+1-l}{q^3-q^l}, & q \text{ odd}, \\ \frac{q+2-l}{q^3-q^l}, & q \text{ even} \end{cases}$$

for $l = 0, 1, 2$.

This class of authentication/secrecy codes is optimal with respect to both the information-theoretic lower bounds of (2), and the lower bound of Lemma 1. They also meet the lower bound $P_{d_0} \geq \frac{|\mathcal{S}|}{|\mathcal{M}|}$.

7.4. The Third Class of Authentication Codes

We set $\Delta = n \geq q$ and consider the n-dimensional space $\mathbf{F}_q^n$, and equivalently the $(n-1)$-dimensional projective space $PG(n-1, q)$. In this case, it is known that $M_{n-1}(n-1, q) = n+1$. The set $\{e_0, \ldots, e_{n-1}, e\}$ is an $(n+1, n-1)$-set.

If we use this $(n+1, n-1)$-set S as our source state space $\mathcal{S}$ in the construction of Section 3, we obtain an authentication code $(\mathcal{S}, \mathcal{M}, \mathcal{E})$ with

$$|\mathcal{S}| = n+1, \quad |\mathcal{M}| = q^n - 1, \ |\mathcal{E}| = q^{n(n-1)/2} \prod_{i=1}^{n} (q^i - 1),$$

and

$$P_{d_l} = \frac{n+1-l}{q^n - q^l}$$

for $l = 0, 1, \ldots, n-1$.

This class of authentication/secrecy codes is optimal with respect to both the information-theoretic lower bounds of (2), and the lower bound of Lemma 1. They also meet the lower bound $P_{d_0} \geq \frac{|\mathcal{S}|}{|\mathcal{M}|}$.

7.5. Other Classes of Authentication Codes

$M_r(r, q)$ is known for $r = 4, 5, 6, 7$ and also in some other cases. Furthermore, constructions of such an $M_r(r, q)$-set are known. We refer to [5] for a survey. Such an $M_r(r, q)$-set can be similarly used as the source state space in the construction of authentication/secrecy codes described before. This justifies that the construction of Section 3 is indeed generic.

Linear error correcting codes can also be used to construct (k, r)-sets. A $[k, n]$ linear code C over $\mathbf{F}_q$ is an n-dimensional subspace of $\mathbf{F}_q^k$. A *generator matrix* G of C is any matrix whose row vectors form a basis of the subspace C. The column vectors of any generator matrix G of a $[k, n]$ linear code form a (k, r)-set, where $r = d - 2$ and d is the minimum distance of the dual code of C. Of course the size of such a (k, r)-set may not be maximal, i.e., it may be smaller than $M_r(N, q)$.

However, such a (k, r)-set can be used as the source state space in the construction of Section 3, and the authentication code is still optimal.

The construction above of (k, r)-sets is very general and effective. As long as the minimum distance of a linear code over $\mathbf{F}_q$ is known, any generator matrix of its dual code gives a (k, r)-set. Almost MDS codes are especially useful [5]. The reader is referred to [5] for a survey.

8. Comparison With Pei's Construction

The idea of our construction of authentication/secrecy codes is influenced by and related to that of the construction by Pei, which is based on rational normal curves over finite fields. Here we demonstrate some similarities and differences of the two constructions.

Let $n \geq 3$ and let $q \geq n+1$ be a prime power. In Pei's construction, the source state space $\mathcal{S}$ is defined by

$$\mathcal{S} = \{(1, a, a^2, \ldots, a^{n-1}) : a \in \mathbf{F}_q\} \cup \{(0, 0, 0, \ldots, 0, 1)\}.$$

Hence $\mathcal{S} = q + 1$. The message space $\mathcal{M} = PG(n-1, q)$. Thus

$$|\mathcal{M}| = (q^n - 1)/(q - 1).$$

Pei proved that

$$|\mathcal{E}| = q^{n(n-1)/2-1} \prod_{i=3}^{n} (q^i - 1).$$

Also the probabilities P_{d_l} of the authentication/secrecy code are given by

$$\begin{cases} p_{d_0} & = \frac{q^2-1}{q^n-1}, \\ p_{d_1} & = \frac{q-1}{q^{n-1}-1}, \\ p_{d_l} & = \frac{(q-1)(q-l+1)}{q^n-q^l}, \quad (2 \leq l \leq n-2) \\ p_{d_{n-1}} & = \frac{q-n}{q^{n-1}}, \\ p_{d_n} & = \frac{q-n+1}{(q-1)^{n-1}}, \\ p_{d_{n+1}} & = \prod_{i=2}^{n-1} (q-i)^{-1}. \end{cases} \tag{3}$$

In our construction, the source state space $\mathcal{S}$ is a subset of $PG(n-1, q)$ such that any Δ elements of $\mathcal{S}$ are linearly independent, $2 \leq \Delta \leq n$. For our authentication/secrecy code we have

$$\begin{aligned} |\mathcal{S}| &= k, \\ |\mathcal{M}| &= q^n - 1, \\ |\mathcal{E}| &= q^{n(n-1)/2} \prod_{i=1}^{n} (q^i - 1) \end{aligned}$$

and for each $0 \leq l \leq \Delta$, the deception probabilities

$$P_{d_l} = \frac{k-l}{q^n - q^l}. \tag{4}$$

Similarities and differences between Pei's construction and ours are the following:

- Pei's construction has the conditions that $n \geq 3$ and $q \geq n+1$, while our construction works for both $q \geq n+1$ and $q < n+1$.
- For any fixed q, in Pei's construction the source state space is fixed and the size is $q+1$ which is independent of n; while in our construction the source state space is variable and the size could be $n+1$, $q+1$, $(q^n-1)/(q-1)$ or any number less than these numbers. Thus our construction contains many subclasses of authentication codes.
- Our encoding rule space is $GL_n(\mathbf{F}_q)$, while that in Pei's construction is a proper subset of $GL_n(\mathbf{F}_q)$. By using more encoding rules our codes offer a better level of secrecy than Pei's codes.
- Pei's codes meet both the bounds of Lemma 2 for $i=0$ and $i=1$, while our codes meet only the bound for $i=0$.
- In both Pei's and our construction, the source state space is a subset of $PG(n-1,q)$ such that any Δ of them are linearly independent.
- In both Pei's and our construction, each encoding rule e is an element from $GL_n(\mathbf{F}_q)$, and the decoding and authentication checking is the same.

A summary of comparisons between Pei's and our codes is given in Table 1.

Item	Pei's codes	Our codes
$\|\mathcal{S}\|$	$q+1$ when $q \geq n+1$	$q+1$, $q+2$ $n+1$, $\frac{q^n-1}{q-1}$ and many others
$\|\mathcal{M}\|$	$\frac{q^n-1}{q-1}$	q^n-1
$\|\mathcal{E}\|$	$q^{n(n-1)/2-1}\prod_{i=3}^{n}(q^i-1)$	$q^{n(n-1)/2}\prod_{i=1}^{n}(q^i-1)$
P_{d_l}	see (3)	See (4)
Level of secrecy	unknown	Δ-fold
Bound of (2)	meet	meet
Bound of Lemma 1	unknown	meet when $\Delta = n$
Bound of Lemma 2	meet both	meet only the first

TABLE 1. Comparison of Pei's and our codes

9. Concluding Remarks

In this paper we presented a construction of authentication/secrecy codes that provide perfect secrecy and is optimal with respect to two types of lower bounds. The encoding of messages is very efficient, as it involves only the multiplication of a matrix with a vector. However, a drawback of the authentication/secrecy codes is that the time complexity of checking the authenticity of messages could be relatively high. This is true when the source state space does not have a good mathematical structure, and in this case checking the membership of the source state space may be time-consuming.

Since the authentication/secrecy codes constructed in this paper are optimal with respect to the two types of bounds, there is no need to compare them with other known authentication/secrecy codes in terms of goodness.

Acknowledgments

This project is supported by the Research Grants Council of the Hong Kong Special Administration Region, China, under Project Nos. HKUST6173/03E and DAG02/03.EG18.

References

[1] R.C. Bose, Mathematical theory of the symmetrical factorial design, *Sankhya* 8 (1947), 107–166.

[2] L.R.A. Casse, K.M. Martin and P.R. Wild, Bounds and characterizations of authentication/secrecy schemes, *Designs, Codes and Cryptography* 13 (1998), 107–129.

[3] P. Godlewski and C. Mitchell, Key-minimal cryptosystems for unconditional secrecy, *J. Cryptology* 3 (1990), 1–25.

[4] E.N. Gilbert, F.J. MacWilliams and N.J.A. Sloane, Codes which detect deception, *Bell System Tech. Journal* 53 (1974), 405–424.

[5] J.W.P. Hirschfeld and L. Storme, The packing problem in statistics, coding theory and finite projective spaces, *J. Statistics Planning and Inference* 72 (1998), 355–380.

[6] J.L. Massey, Cryptography – A selective survey, in: *Digital Communications*, North-Holland (1986), 3–21.

[7] D. Pei, Information-theoretic bounds for authentication codes and block designs, *J. of Cryptology* 8 (1995), 177–188.

[8] D. Pei, A problem of combinatorial designs related to authentication codes, *J. of Combinatorial Designs* 6(6) (1998), 417–429.

[9] R.S. Rees and D.R. Stinson, Combinatorial characterizations of authentication codes, *Designs, Codes and Cryptography* 7 (1996), 239–259.

[10] U. Rosenbaum, A lower bound on authentication after having observed a sequence of messages, *J. Cryptology* 6 (1993), 135–156.

[11] B. Segre, Ovals in a finite projective plane, *Canad. J. Math.* 7 (1955), 414–416.

[12] A. Sgarro, Information-theoretic bounds for authentication frauds, *J. Computer Security* 2 (1993), 53–63.

[13] M.D. Soete, Some constructions for authentication-secrecy codes, in: *Advances in Cryptology: Proceedings of Eurocrypt' 88,* Lecture Notes in Computer Science 330, Springer-Verlag, 1988, 57–75.

[14] C.E. Shannon, Communication theory of secrecy systems, *Bell System Tech. Journal* 28 (1949), 656–715.

[15] G.J. Simmons, Authentication theory/coding theory, in: *Advances in Cryptology: Proceedings of Crypto' 84,* Lecture Notes in Computer Science 196, 411–432. Berlin: Springer-Verlag, 1985.

[16] D.R. Stinson and L. Teirlinck, A construction for authentication/secrecy codes from 3-homogeneous permutation groups, *J. Combinatorics* 11 (1990), 73–79.

[17] D.R. Stinson, A construction for authentication/secrecy codes from certain combinatorial designs, *J. Cryptology* 1 (1988), 37–51.

[18] D.R. Stinson, The combinatorics of authentication and secrecy codes, *J. Cryptology* 2 (1990), 23–49

[19] D.R. Stinson, Combinatorial characterizations of authentication codes, *Designs, Codes, and Cryptography* 2 (1992), 175–187.

[20] T.V. Trung, On the construction of authentication and secrecy codes, *Designs, Codes and Cryptography* 5 (1995), 269–280.

[21] Z. Wan, *Geometry of Classical Groups over Finite Fields*, Studentlitteratur, 1993.

Xiaojian Tian
Department of Computer Science
The Hong Kong University of Science and Technology
Clearwater Bay, Kowloon
Hong Kong, China
e-mail: `xjtian@cs.ust.hk`

Cunsheng Ding
Department of Computer Science
The Hong Kong University of Science and Technology
Clearwater Bay, Kowloon
Hong Kong, China
e-mail: `cding@cs.ust.hk`

Progress in Computer Science and Applied Logic, Vol. 23, 331–343

Selection Method of Test Patterns in Soft-Input and Output Iterative Bounded Distance Decoding Algorithm

Hitoshi Tokushige, Jun Asatani, Marc P.C. Fossorier and Tadao Kasami

Abstract. In this paper, we consider a soft-input and soft-output iterative bounded-distance decoding (SISO-IBDD) algorithm in which the bounded-distance decodings are performed sequentially. And we study a computation method of a soft-output value to approximate Max-Log-MAP. In the decoding algorithm, search centers are generated by the sums of the hard-decision sequence obtained from a received sequence and test patterns which are selected beforehand, and the bounded-distance decodings with respect to the search centers are performed. We have proposed a selection method of test patterns for an iterative bounded-distance decoding (IBDD) algorithm in [5]. Simulation results have shown that the IBDD algorithm whose test patterns are selected by the proposed method with considerably small number of iterations provides better error performance than the Chase II decoding algorithm for several BCH codes over an AWGN channel using BPSK signaling. We adopt the selection method of test patterns for the SISO-IBDD algorithm and a computation method of a soft-output value which is modified based on the proposed method in [2] with respect to reliability factors. Over the AWGN channel using the BPSK signaling, we show bit error rates and average soft-output values given by the SISO-IBDD algorithm for the extended BCH (64, 36, 12) and (64, 45, 8) codes by computer simulation. For the extended BCH codes, the simulation results show that bit error rates of the SISO-IBDD algorithm are relatively smaller than those of a Chase-like decoding algorithm with the same number of iterations of the bounded-distance decodings and for each component bit, average soft-output values provided by the SISO-IBDD algorithm are close to those by the Max-Log-MAP decoding algorithm by computer simulation.

Keywords. Soft-input and output decoding, Chase-like decoding, bounded-distance decoding.

1. Introduction

In [1], [2], [4] and other papers, soft-input and soft-output decoding algorithms for binary product codes have been presented. In [1], the maximum a *posteriori* (MAP) or the Log-MAP decoding algorithm is applied to row and column component codes of the product codes. These two decoding algorithms are optimum, however the decoding complexities are considerably large. To reduce the decoding complexities, a Chase-like and an encoding-based decoding algorithm with soft-output values have been proposed for the component codes in [2] and [4], respectively. A soft-output value for a component bit of a decoded codeword indicates reliability of the component bit of the decoded codeword. In the both algorithms, the soft-output value is computed based on the set D of generated candidate codewords. A computation method of the soft-output value has been presented in [2]. The computation of a soft-output value for the ith component bit of the decoded codeword requires one candidate codeword in D, called a competing codeword, with the ith component bit value different from that of the decoded codeword. If there is no competing codeword at the ith component bit, a value β_i, called a reliability factor, is used to compensate the soft-output value with an approximation in the computation method. Because the quality of a soft-output value depends on D, it is important how a good set of candidate codewords is generated by the decoding algorithm and how suitable reliability factors are prepared beforehand.

In this paper, we consider a soft-input and soft-output iterative bounded-distance decoding (SISO-IBDD) algorithm in which the bounded-distance decodings are performed sequentially. And we study a computation method of a soft-output value to approximate Max-Log-MAP. In the decoding algorithm, sequences, called search centers, are generated by the sums of the hard-decision sequence obtained from a received sequence and test patterns which are selected beforehand, and the bounded-distance decodings with respect to the search centers are performed. Simulation results have shown that the test patterns affect a generation of candidate codewords considerably. We have proposed a selection method of test patterns for an iterative bounded-distance decoding (IBDD) algorithm in [5]. In [5], simulation results have shown that the IBDD decoding algorithm whose test patterns are selected by the proposed method with considerably small number of iterations provides better error performance than the Chase II decoding algorithm for the BCH(127, 64, 21), the BCH(255, 123, 39), the BCH(255, 147, 29), the BCH(255, 163, 25) the BCH(511, 313, 47) and the BCH(511, 349, 39) codes over an AWGN channel using BPSK signaling. The difference between the IBDD and the Chase II algorithms is only a set of test patterns which are generated beforehand. Decoding complexities of the IBDD algorithm and the Chase II decoding algorithm with the same number of test patterns are almost the same. We adopt the selection method of test patterns for the SISO-IBDD algorithm. For computing soft-output values, we modify the proposed method in [2] with respect to reliability factors. We propose a computation method of the reliability factors for each component bit of a decoded codeword. Over the AWGN channel using the BPSK signaling, we show that bit error rates (BER's) of the SISO-IBDD algorithm are relatively smaller than those of a Chase-like decoding algorithm in which the bounded-distance decodings are performed with the same iteration number of bounded-distance decodings in the SISO-IBDD algorithm, for the extended BCH (EBCH)(64, 36, 12) and the EBCH(64, 45, 8) codes by computer simulation. Hereafter, we also show that for each component bit, average soft-output values which are output by the SISO-IBDD algorithm are close to those which are output by the Max-Log-MAP decoding algorithm for the EBCH codes by computer simulation. Also, we show average soft-output values which are output by the SISO-IBDD algorithm for a BCH code by computer simulation.

The paper is organized as follows. Preliminary definitions and notations are introduced in Section 2. A SISO-IBDD algorithm is presented in Section 3. In Section 4, we show a selection method of test patterns for the SISO-IBDD algorithm and the computation of the reliability factors. In Section 5, we show a computation method of soft-output values briefly and present a computation method of the reliability factors. In Section 6, we show error performances of the SISO-IBDD algorithm by computer simulation. Concluding remarks are given in Section 7.

2. Preliminaries

Assume that a binary $(N, K, d_{\min})$ linear block code C of length N, number of information bits K and minimum distance $d_{\min}$ is used over an AWGN channel with BPSK signaling, and each codeword is transmitted with equal probability. For a positive integer n, let V^n denote the vector space of all binary n-tuples and let $\mathbb{R}^n$ denote the vector space of all real value n-tuples. Let $\boldsymbol{r}(\in \mathbb{R}^N)$ be a received sequence obtained from the channel. For a received sequence $\boldsymbol{r} = (r_1, r_2, \ldots, r_N)$ and $1 \leq i \leq N$, the reliability of component r_i is given by the absolute value of r_i and for simplicity, we assume without loss of generality that the order of the components is permuted in increasing order of reliability, i.e.,

$$|r_i| \leq |r_j|, \text{ for } 1 \leq i < j \leq N. \tag{2.1}$$

For integers $1 \leq i \leq j \leq N$, let $[i, j]$ denote the set of integers from i to j. For $\boldsymbol{u} = (u_1, u_2, \ldots, u_N) \in V^N$, and a subset $I = \{i_1, i_2, \ldots, i_m\}$ of $[1, N]$ where $i_s < i_t$ for $1 \leq s < t \leq m$,

$$p_I(\boldsymbol{u}) \triangleq (u_{i_1}, u_{i_2}, \ldots, u_{i_m}). \tag{2.2}$$

For $U(\neq \emptyset) \subseteq V^n$,

$$p_I U \triangleq \{p_I(\boldsymbol{u}) : \boldsymbol{u} \in U\}. \tag{2.3}$$

For a received sequence $\boldsymbol{r} = (r_1, r_2, \ldots, r_N)$, $z(\boldsymbol{r})$ (or simply $\boldsymbol{z}$)$= (z_1, z_2, \ldots, z_N)$ denotes the binary hard-decision sequence for $\boldsymbol{r}$ by using the hard decision function: $z_i = 1$ for $r_i > 0$ and $z_i = 0$ for $r_i \leq 0$. For $\boldsymbol{u} = (u_1, u_2, \ldots, u_N) \in V^N$, the correlation discrepancy of $\boldsymbol{u}$ with respect to $\boldsymbol{r}$ is defined as

$$L(\boldsymbol{u}) \triangleq \sum_{i=1, u_i \neq z_i}^{N} |r_i|. \tag{2.4}$$

For $U \subseteq V^N$, let $\underline{L}(U)$ be defined as

$$\underline{L}(U) \triangleq \begin{cases} \min_{\boldsymbol{u} \in U} L(\boldsymbol{u}), & \text{if } U \neq \emptyset, \\ \infty, & \text{if } U = \emptyset. \end{cases} \tag{2.5}$$

For $U(\neq \emptyset) \subseteq V^N$, let $\boldsymbol{v}(U)$ denote the binary N-tuple in U such that $L(\boldsymbol{v}(U)) = \underline{L}(U)$. For $U(\neq \emptyset) \subseteq V^N$, $\boldsymbol{u} \in U$ such that $L(\boldsymbol{u}) = \underline{L}(U)$ is called the best word in U.

For $J \subseteq [1, N]$ and $\boldsymbol{u} \in p_J V^N$, let $w(\boldsymbol{u})$ denote the Hamming weight of $\boldsymbol{u}$. For $U(\neq \emptyset) \subseteq p_J V^N$ and $\boldsymbol{u} \in p_J V^N$, define

$$w_{\min}(U) \triangleq \min_{\boldsymbol{v} \in U \setminus \{\mathbf{0}_J\}} w(\boldsymbol{v}), \tag{2.6}$$

$$d(U, \boldsymbol{u}) \triangleq \min_{\boldsymbol{v} \in U} w(\boldsymbol{u} + \boldsymbol{v}), \tag{2.7}$$

$$\text{and } \delta(U) \triangleq \max_{\boldsymbol{v} \in p_J V^N} d(U, \boldsymbol{v}), \tag{2.8}$$

where $\mathbf{0}_J \triangleq p_J \mathbf{0}$. The value $\delta(U)$ (or simply δ) is called the covering radius of U. For $\boldsymbol{u} = (u_1, u_2, \ldots, u_N) \in V^N$ and positive integers $1 \leq s \leq e \leq N$,

$$w_{[s,e]}(\boldsymbol{u}) \triangleq |\{u_i = 1 : s \leq i \leq e\}|. \tag{2.9}$$

For an integer $\tau \in [1, N]$, define $I \triangleq [1, \tau]$ and $I' \triangleq [\tau + 1, N]$. For $\boldsymbol{u} = (u_1, u_2, \ldots, u_\tau) \in p_I V^N$, $\boldsymbol{v} = (v_{\tau+1}, v_{\tau+2}, \ldots, v_N) \in p_{I'} V^N$,

$$\boldsymbol{u} \circ \boldsymbol{v} \triangleq (u_1, u_2, \ldots, u_\tau, v_{\tau+1}, v_{\tau+2}, \ldots, v_N), \tag{2.10}$$

and for nonempty sets $U^{(1)} \subseteq p_I V^N$ and $U^{(2)} \subseteq p_{I'} V^N$,

$$U^{(1)} \circ U^{(2)} = \left\{ \boldsymbol{u}^{(1)} \circ \boldsymbol{u}^{(2)} : \boldsymbol{u}^{(1)} \in U^{(1)} \text{ and } \boldsymbol{u}^{(2)} \in U^{(2)} \right\}. \tag{2.11}$$

3. Soft-Input and Soft-Output Decoding Algorithm

Define

$$t_0 \triangleq \lfloor (d_{\min} - 1)/2 \rfloor. \tag{3.1}$$

Bounded distance-t_0 decoding is capable of correcting t_0 errors. For $\boldsymbol{u} \in V^N$, BDD-$t_0(\boldsymbol{u})$ (or simply BDD($\boldsymbol{u}$)) denotes the bounded distance-t_0 decoding with respect to $\boldsymbol{u}$, where $\boldsymbol{u}$ is called a search center. If the BDD($\boldsymbol{u}$) succeeds, then it outputs a unique codeword. Or else, nothing is output. In this paper, we consider the following soft-input and soft-output iterative bounded distance-t_0 decoding algorithm, called a SISO-IBDD algorithm. The decoding algorithm requires a set of test patterns as a parameter. To specify the parameter, it is also denoted

SISO-IBDD(U_{tp}) and define $h \triangleq |U_{\text{tp}}|$ where the U_{tp} is a set of h test patterns $\boldsymbol{t}^{(1)}, \boldsymbol{t}^{(2)}, \ldots, \boldsymbol{t}^{(h)} (\in V^N)$. For a positive integer i and $1 \leq i \leq h$, define

$$\boldsymbol{v}^{(i)} \triangleq z(\boldsymbol{r}) + \boldsymbol{t}^{(i)}, \tag{3.2}$$

where $\boldsymbol{v}^{(i)}$ is called the ith search center of the SISO-IBDD algorithm. The SISO-IBDD(U_{tp}) is an iterative decoding algorithm in which BDD($\boldsymbol{v}^{(1)}$), BDD($\boldsymbol{v}^{(2)}$), $\cdots$, BDD($\boldsymbol{v}^{(h)}$) are performed sequentially. Codewords which are obtained from the bounded distance-t_0 decodings in the SISO-IBDD(U_{tp}) are called candidate codewords and let D be the set of the candidate codewords. Define $\boldsymbol{d} \triangleq \boldsymbol{v}[D]$ as the best codeword in D and the $\boldsymbol{d}$ is called a decoded codeword. The SISO-IBDD(U_{tp}) outputs a sequence of soft-output values which are computed based on the set D as is shown in Section 5. These soft-output values indicate the reliabilities of the component bits of the $\boldsymbol{d}$, respectively.

4. Selection of Test Patterns

In [5], we have presented a selection method of test patterns for an iterative bounded distance-t_0 decoding (IBDD) algorithm. The selection method outputs a set of test patterns for the IBDD algorithm. In this section, we present a selection method of test patterns for the SISO-IBDD algorithm. We modify the selection method proposed in [5]. It outputs two sets of test patterns, one is for the SISO-IBDD algorithm and the other is a generation of reliability factors which are used in computation of soft-output values. The generation method of reliability factors is described in Section 5. This selection method of test patterns is denoted C_{c}-Select($\bar{h}, h, \rho$), where C_{c}, called a covering code, is a binary $(\tau, k; \delta)$ block code of code length τ and the dimension k with the reasonably small covering radius δ, and $\bar{h}$, h with $\bar{h} > h$ and ρ are non-negative integers used as parameters. We generate two sets of reasonably effective test patterns by the following two steps. At the first step, we compose a set of candidate test patterns, denoted U_{ctp}. Every binary sequence in V^N can be a candidate test pattern for a U_{tp}. For a relatively large N, the number of candidate test patterns becomes huge. In this composition method, to reduce a number of candidate test patterns in the U_{ctp} down to a reasonable number, the covering code C_{c} is used for composition of the U_{ctp}. At the second step, we select two subsets from the composed set U_{ctp}. In this step, $\bar{h}$ test patterns in the U_{ctp} whose effectiveness are assigned by computer simulation are selected and set to U'_{ctp}. Then, the best h test patterns in the U'_{ctp} are set to U_{tp}. These sets U_{tp} and U'_{ctp} are output by the selection method. The U'_{ctp} is used for the generation of reliability factors. We describe two steps as follows:

(i) The first step:
We choose a covering code C_{c} for composition of a U_{ctp}. Define $I_{\text{c}} \triangleq [1, \tau]$ and $I'_{\text{c}} \triangleq [\tau + 1, N]$. We initially consider the following set U_{ctp} of candidate test patterns:

$$U_{\text{ctp}} = \left\{ \boldsymbol{u} \circ \boldsymbol{v} : \boldsymbol{u} \in C_{\text{c}} (\subseteq V^{\tau}) \text{ and } \boldsymbol{v} \in p_{I'_{\text{c}}} V^N (= V^{N-\tau}) \right\}.$$

The number of candidate test patterns in U_{ctp} is $2^k \cdot 2^{N-\tau} = 2^{N-(\tau-k)} (\leq 2^N)$. In [8], a table giving tight bounds of covering radius for codes of length ≤ 64 is presented. For the code length τ and the dimension k, if the covering radius δ of C_c is the smallest possible, then C_c is called optimal. In this paper, we consider the following cases:

(1) $C_c = \{0^\tau, 1^\tau\}$ repetition code (τ: odd number). These codes are optimal.
(2) $C_c = (2^m - 1, 2^m - 1 - m; 1)$ Hamming codes (m: positive integer). These codes are optimal.
(3) $C_c = (15, 7; 3)$ double error correcting primitive BCH code. This BCH code is optimal.
(4) $C_c = (23, 12; 3)$ Golay code. The Golay code is optimal.
(5) $C_c = \{0^{\tau_e}, 1^{\tau_e}\} \circ$ (Hamming codes, BCH code or Golay code), where $\tau_e \triangleq \tau - (2^m - 1), 15$ or 23, respectively.

For example, if we choose the Hamming (15, 11, 3) code as a C_c, then the number of candidate test patterns in U_{ctp} is 2^{N-4}. This may remain too large for most applications. In the second step, we select relatively small subsets U'_{ctp} and U_{tp} by computer simulation.

(ii) The second step:
In this step, we describe a selection method to determine two reasonably good subsets $U_{\text{tp}} \subseteq U'_{\text{ctp}} (\subset U_{\text{ctp}})$ according to evaluation of effectiveness of each candidate test pattern in U_{ctp}. At first, we give several definitions. For a covering code C_c, let H_c be a parity check matrix of C_c. Define $\varphi \triangleq \tau - k$, as the number of redundant bits of C_c. For $\boldsymbol{u} \in V^\tau$, $s(\boldsymbol{u}) \triangleq \boldsymbol{u} H_c^T$ is the syndrome of $\boldsymbol{u}$. For $\boldsymbol{s} \in V^\varphi$, let $B(\boldsymbol{s})$ denote the coset in V^τ / C_c associated with the syndrome $\boldsymbol{s}$. Define the weight of a coset leader $w_{\text{cl}}(\boldsymbol{s})$ and the set of coset leaders $U_{\text{cl}}(\boldsymbol{s})$ as follows:

$$\begin{aligned} w_{\text{cl}}(\boldsymbol{s}) &\triangleq w_{\min}(B(\boldsymbol{s})), \\ U_{\text{cl}}(\boldsymbol{s}) &\triangleq \left\{ \boldsymbol{u} \in B(\boldsymbol{s}) : w_{[1,\tau]}(\boldsymbol{u}) = w_{\text{cl}}(\boldsymbol{s}) \right\}. \end{aligned}$$

If C_c is a perfect code, then $|U_{\text{cl}}(\boldsymbol{s})| = 1$. For C_c, a table $S(C_c)$ whose entry for index $\boldsymbol{s}$ is $(w_{\text{cl}}(\boldsymbol{s}), U_{\text{cl}}(\boldsymbol{s}))$ can be constructed if $\sum_{i=0}^{\delta} \binom{\tau}{i}$ remains moderate.

For a given signal to noise ratio (SNR) value, we generate sequentially received sequences of length N for the transmitted zero word by computer simulation. Let R_s denote the multi-set of generated sequences. We assume that $\forall \boldsymbol{r} \in R_s$, the order of components of $\boldsymbol{r}$ is permuted according to increasing order of reliability.

The selection procedure C_c-Select$(\bar{h}, h, \rho)$ is performed as follows:

Selection Procedure
The set $T (\subseteq U_{\text{ctp}})$ represents a current partial set of candidate test patterns. If T is not an empty set, we assume that a positive integer $\sharp(\boldsymbol{t})$ is assigned to each test pattern $\boldsymbol{t}$ in T and it is used as a temporal measure of effectiveness of the test pattern $\boldsymbol{t}$. For $T \neq \emptyset$, all test patterns in T have indices which are

always given by decreasing order of the above assigned integers as follows:

$$\sharp(\boldsymbol{t}_i) \geq \sharp(\boldsymbol{t}_j) \text{ for } 1 \leq i < j \leq |T|.$$

If T is replaced by new one in this selection procedure, these indices are updated according to the above order.

Initialization: $T \leftarrow \{\boldsymbol{t}_1 \triangleq \boldsymbol{0}\}$ and $\sharp(\boldsymbol{t}_1) \leftarrow 0$.

(1) For each newly generated $\boldsymbol{r}$, $\boldsymbol{z} \leftarrow z(\boldsymbol{r})$ and go to (2).

(2) Define $T(\boldsymbol{z}) \triangleq \{\boldsymbol{t} \in T : w_{[1,N]}(\boldsymbol{z}+\boldsymbol{t}) \leq \rho\}$, the set of effective test patterns for $\boldsymbol{z}$ in T.
- If $T(\boldsymbol{z}) \neq \emptyset$, then
  - choose $\boldsymbol{t}_j \in T(\boldsymbol{z})$ with the smallest index,
  - add one credit to $\boldsymbol{t}_j$, that is, $\sharp(\boldsymbol{t}_j) \leftarrow \sharp(\boldsymbol{t}_j) + 1$,
  - permute test patterns in $T \setminus \{\boldsymbol{t}_1\}$ in the decreasing order of $\sharp(\boldsymbol{t})$ and
  - go to (4).
- Else, go to (3).

(3) We calculate a syndrome $\boldsymbol{s}$ of C_{c} for $p_{I_{\mathrm{c}}}(\boldsymbol{z})$, $\boldsymbol{s} \leftarrow p_{I_{\mathrm{c}}}(\boldsymbol{z})H^T$ and retrieve $w_{\mathrm{cl}}(\boldsymbol{s})$ and $U_{\mathrm{cl}}(\boldsymbol{s})$ from Table $S(C_{\mathrm{c}})$.

(A) For each $\boldsymbol{u} \in U_{\mathrm{cl}}(\boldsymbol{s})$,
  - if $w_{\mathrm{cl}}(\boldsymbol{s}) + w_{I'_{\mathrm{c}}}(\boldsymbol{z}) \leq \rho$, then

$$\boldsymbol{t}_{|T|+1} \triangleq (\boldsymbol{u} + p_{I_{\mathrm{c}}}(\boldsymbol{z})) \circ \boldsymbol{0}_{I'_{\mathrm{c}}},$$

  - else

$$\boldsymbol{t}_{|T|+1} \triangleq (\boldsymbol{u} + p_{I_{\mathrm{c}}}(\boldsymbol{z})) \circ \ (z_{\tau+1}, z_{\tau+2}, \ldots, z_l) \circ \boldsymbol{0}_{[l+1,N]},$$

  where for $\tau < l \leq N$, the integer l satisfies $w_{[l+1,N]}(\boldsymbol{z}) + w_{\mathrm{cl}}(\boldsymbol{s}) = \rho$ and $z_l = 1$.
  - $T \leftarrow T \cup \{\boldsymbol{t}_{|T|+1}\}$ and $\sharp(\boldsymbol{t}_{|T|+1}) \leftarrow 1$.
  - If $U_{\mathrm{cl}}(\boldsymbol{s})$ has been exhausted, go to (4).
  Otherwise, go to (A).

(4) If $|T| = \bar{h}$, the T is set to a U'_{ctp}. For a required number $h(\ll \bar{h})$ of test patterns, the first h test patterns in the U'_{ctp} are set to a U_{tp}. Then, output the U'_{ctp} and the U_{tp}, and terminate. Otherwise, go to (1).

5. Computation Method of Soft-Output Values

A computation method of a soft-output value at a component of a decoded codeword has been presented in [2]. To compute a soft-output value at a component of a decoded codeword, a candidate codeword whose component value is different from the component value of the decoded codeword at the component position, is required. This candidate codeword is called a competing codeword. If there is no competing codeword, a value called a reliability factor is used in the computation method. It has been shown by computer simulation that the reliability factor has big influence on error performance. We adopt the computation method proposed in [2] and modify it with respect to the reliability factor. First, we show the computation method briefly and then propose a new generation method of the reliability factor. Let $\boldsymbol{s}' \triangleq (s'_1, s'_2, \ldots, s'_N)(\in \mathbb{R}^N)$ denote a sequence of soft-output values. Define $\boldsymbol{d} = (d_1, d_2, \ldots, d_N)$ as a decoded codeword. For an integer $1 \leq i \leq N$, $D_i \triangleq \{\boldsymbol{c} = (c_1, c_2, \ldots, c_N) \in D : c_i \neq d_i\}$. The set D_i is a set of candidate competing codewords at the ith component. If $D_i \neq \emptyset$, then the ith soft-output value s'_i is given by the following equation:

$$s'_i = \frac{1}{2\sigma^2}(|\boldsymbol{r} - \boldsymbol{d}'|^2 - |\boldsymbol{r} - \boldsymbol{d}|^2)(2d_i - 1), \tag{5.1}$$

where $\boldsymbol{d}' \triangleq \boldsymbol{v}[D_i]$ is a competing codeword. If $D_i = \emptyset$, that is, there is no competing codeword in D, then

$$s'_i = \beta_i(2d_i - 1), \tag{5.2}$$

where β_i is a reliability factor for the ith component d_i. In [2], the reliability factors $\beta_1, \beta_2, \ldots, \beta_N$ are the same constant value which is selected on error performance. Essentially, since the reliability factor should be an estimation value computed under the condition that a set of candidate codewords is large enough to exist a competing codeword at each component position, it is desirable to select a suitable reliability factor for each component position.

We present a generation method of the reliability factors. The generation method has performed beforehand and those reliability factors are stored in a read only memory (ROM). If $D_i = \emptyset$, then a reliability factor is retrieved from the ROM. We assume that the selection method C_c-Select$(\bar{h}, h, \rho)$ with a large $\bar{h}$ described in Section 4 has been performed and a set U'_{ctp} has already output by the selection method. For a given SNR value, let R_s be a multi-set of sampled received sequences generated by computer simulation. To obtain meaningful reliability factors, it is desirable that the R_s is large enough. For all $\boldsymbol{r}_s \in R_s$, we assume that all components of $\boldsymbol{r}_s$ are arranged according to increasing order of the reliability. For each $\boldsymbol{r}_s \in R_s$, the SISO-IBDD(U'_{ctp}) is performed and soft-output values computed by the equation (5.1) are output. For an integer $1 \leq i \leq N$, the average of absolute values of the ith component over all generated sequences of soft-output values is used as the ith reliability factor β_i.

6. Simulation Results

For positive integers $\bar{h} \gg h$ and $1 \leq \rho \leq t_0$, let (h, ρ)-IBDD denote the SISO-IBDD algorithm whose iteration number of the bounded distance-t_0 decodings is h and test patterns are selected by C_c-Select$(\bar{h}, h, \rho)$. In this paper, we consider the following five covering codes C_c's in Section 4: the Hamming(7, 4; 1), the Hamming(15, 11; 1), the Hamming(31, 26; 1), the BCH(15, 7; 3) and the Golay(23, 12; 3) codes. We have compared the above five C_c's together with the values of ρ to each other for the EBCH(64, 36, 12) and the EBCH(64, 45, 8) codes. The simulation results show that for $h = 32, 64$ or 128, the SISO-IBDD algorithm whose test patterns are selected by Hamming(15, 11; 1)-Select$(\bar{h}, h, 2)$ provides the best BER at $E_b/N_0 = 2.0$dB to 5.0dB. Figures 1 and 2 show the BER's for the EBCH(64, 36,

12) and the EBCH(64, 45, 8) codes at $E_b/N_0 = 2.0$dB to 5.0dB, respectively. For an integer $1 \leq l \leq N$, a Chase-like decoding algorithm, denoted Chase(2^l), is defined as the IBDD algorithm whose set of test patterns is $U_{\text{tp}} \triangleq \{\boldsymbol{u} \circ \mathbf{0}_{[l+1,N]} : \boldsymbol{u} \in V^l\}$. In the Chase-like decoding algorithm, the least reliable l components in a received sequence are selected for generating search centers. However, in the IBDD algorithm, the number of selected least reliable components is relatively large. It follows from the computation method presented in Section 5 that BER depends on a decoded codeword $\boldsymbol{d}$ only, and is independent of soft-output values. For comparison, BER's of the Chase-like decoding algorithm without soft-output values are also shown. Figure 1 shows that BER's of Chase(2^l) are almost the same as those of $(2^{l-1}, \rho)$-IBDD. Also, Figure 2 shows that $(2^l, \rho)$-IBDD provides better BER's than Chase(2^l). Figures 3 and 4 show the reliability factors for each component position which are computed by the proposed method at $E_b/N_0 = 2.0$dB to 5.0dB for the above two EBCH codes, respectively. Figures 5 and 6 show the averages of soft-output values which are computed by using the reliability factors shown in Figures 3 and 4 at $E_b/N_0 = 3.0$dB for the above two EBCH codes, respectively. For comparison, the averages of soft-output values generated by the Log-MAP and the Max-Log-MAP decoding algorithms are also shown. From the figures, it is shown that the averages of soft-output values generated by the (h, ρ)-IBDD and the Max-Log-MAP decoding algorithm are relatively close.

7. Conclusion

In this paper, we have studied a soft-input and soft-output iterative decoding algorithm using the bounded distance-t_0 decoding, a SISO-IBDD algorithm. For the SISO-IBDD algorithm, we have proposed a selection method, C_c-Select($\bar{h}, h, \rho$), of test patterns based on the selection method presented in [5]. The proposed selection method outputs two sets of test patterns for the SISO-IBDD algorithm and a generation of reliability factors which are used in a computation of soft-output values, respectively. We also have proposed a new generation method of the reliability factors for each component of a decoded codeword. For the EBCH(64, 36, 12) and the EBCH(64, 45, 8) codes, we have shown bit error rates of the SISO-IBDD algorithm, reliability factors generated by the proposed method at $E_b/N_0 = 2.0$dB to 5.0dB and averages of soft-output values generated by the SISO-IBDD algorithm at $E_b/N_0 = 3.0$dB. Compared with a Chase-like decoding algorithm with the same iteration number of the bounded distance decodings, the SISO-IBDD algorithm provides better bit error rates. Averages of soft-output values generated by the SISO-IBDD algorithm with a relatively small number of iterations are relatively close to those by the Max-Log-MAP decoding algorithm. With increase in the number of iterations, averages of soft-output values of the SISO-IBDD algorithm seem to be away from those of the Max-Log-MAP decoding algorithm. Originally, the selection method of test patterns has been designed to provide good candidate codewords with a small number of iterations. We are studying such a selection method of test patterns that good competing codewords are generated as well as good candidate codewords.

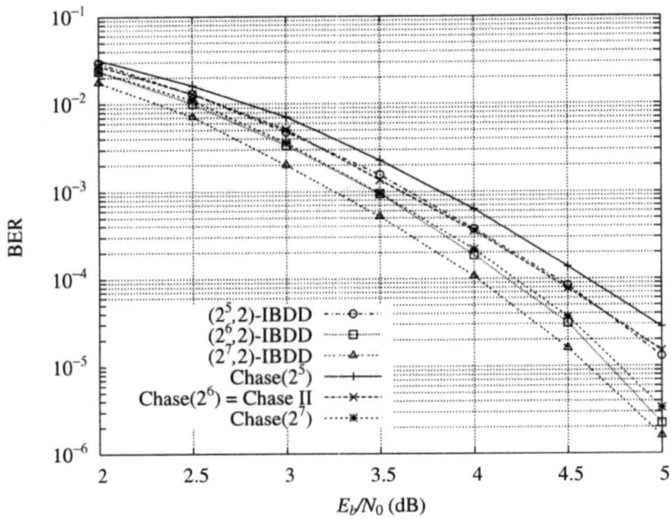

FIGURE 1. BER's for EBCH(64, 36, 12) code.

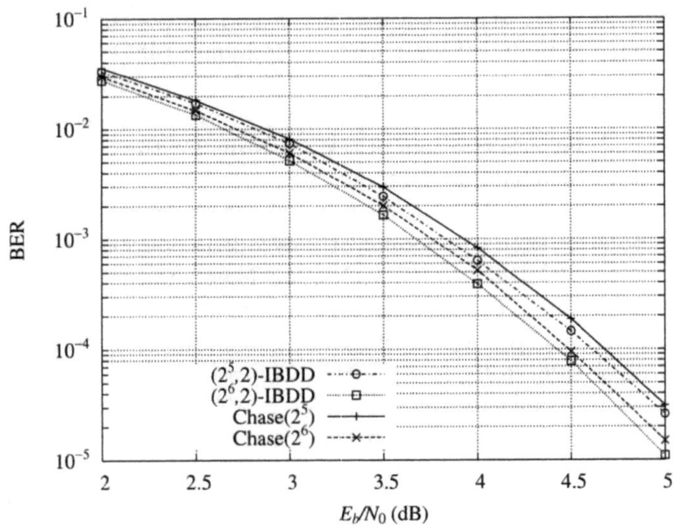

FIGURE 2. BER's for EBCH(64, 45, 8) code.

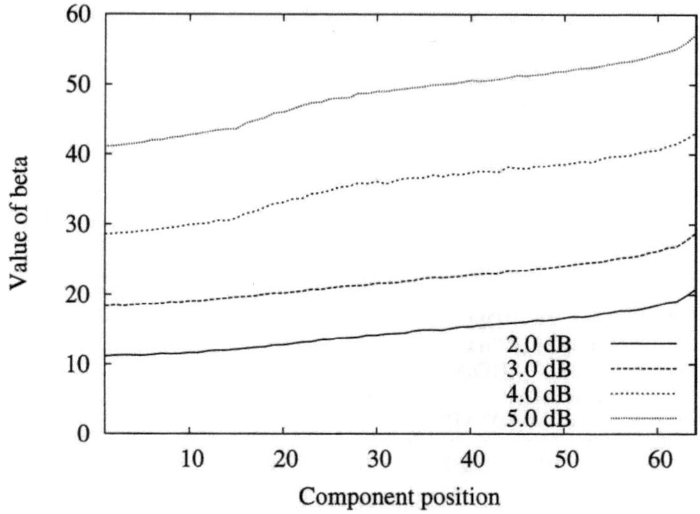

FIGURE 3. Reliability factors for EBCH(64, 36, 12) code.

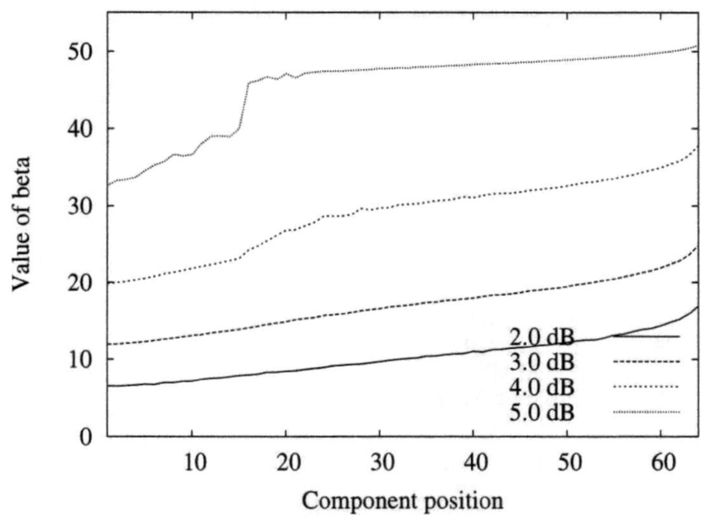

FIGURE 4. Reliability factors for EBCH(64, 45, 8) code.

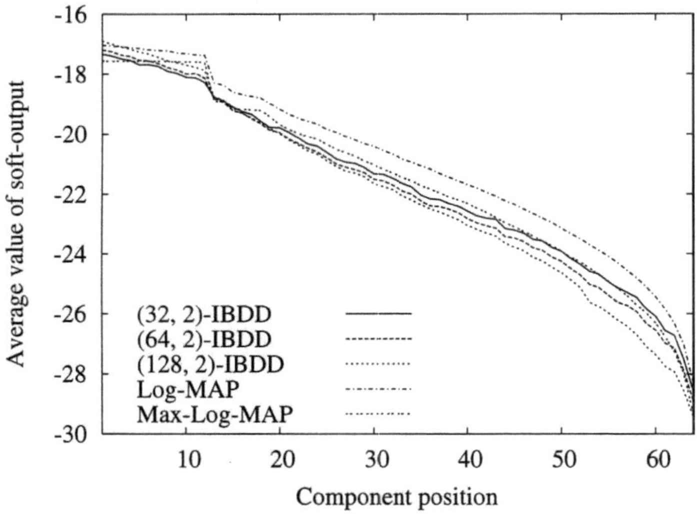

FIGURE 5. Average soft-output values for EBCH(64, 36, 12) code.

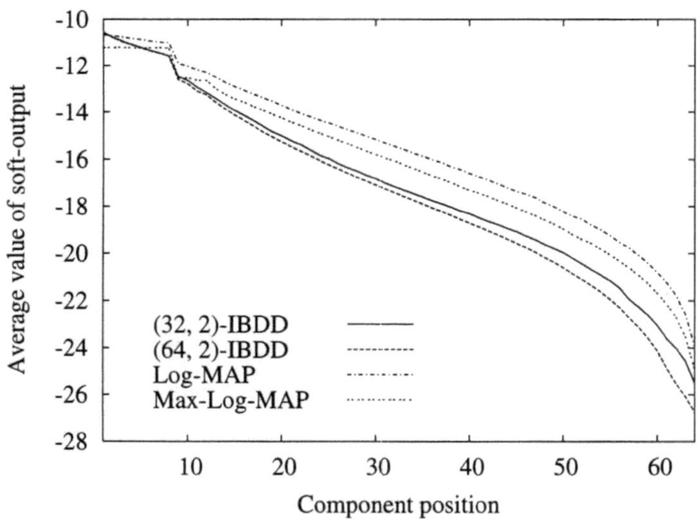

FIGURE 6. Average soft-output values for EBCH(64, 45, 8) code.

References

[1] J. Hagenauer, E. Offer and L. Papke, "Iterative Decoding of Binary Block and Convolutional Codes," *IEEE Trans. Inform. Theory*, vol. 42, No. 2, pp. 429–445, 1996.

[2] R.M. Pyndiah, "Near Optimum Decoding of Product Codes: Block Turbo Codes," *IEEE Trans. on Commun.*, vol. 46, no. 8, pp. 1003–1010, Aug. 1998.

[3] P.A. Martin and D.P. Taylor, "On Multilevel Codes and Iterative Multistage Decoding," *IEEE Trans. Commun.*, vol. 49, no. 11, pp. 1916–1925, Nov. 2001.

[4] P.A. Martin, D.P. Taylor and M.P.C. Fossorier, "Soft-input Soft-output List-based Decoding Algorithm," *Proc. of 2002 IEEE International Symposium on Information Theory*, Lausanne, Switzerland, pp. 339, Jun. 2002.

[5] H. Tokushige, T. Koumoto, M. Fossorier and T. Kasami, "Selection Method of Test Patterns in Soft-decision Iterative Bounded Distance Decoding Algorithms" *IEICE Trans. Fundamentals*, vol. E86-A, no. 10, pp. 2445–2451, Oct. 2003.

[6] D. Chase, "A Class of Algorithms for Decoding Block Codes with Channel Measurement Information," *IEEE Trans. Inform. Theory*, IT-18, pp. 170–182, Jan. 1972.

[7] M. Fossorier and S. Lin, "Error Performance Analysis for Reliability-Based Decoding Algorithms," *IEEE Trans. on Inform. Theory*, vol. IT-48, pp. 287–293, Jan. 2002.

[8] R.L. Graham and N.J.A. Sloane. "On the Covering Radius of Codes," *IEEE Trans. Inform. Theory*, IT-31, pp. 385–401, May 1985.

[9] G. Cohen, I. Honkala, S. Litsyn and A. Lobstein, *Covering Codes*, Amsterdam, The Netherlands: North-Holland, 1997.

[10] M. Fossorier and S. Lin, "Soft-Decision Decoding of Binary Linear Block Codes Based on Ordered Statistics," *IEEE Trans. on Inform. Theory*, vol. 44, No. 5, pp. 1217–1234, 1995.

Hitoshi Tokushige
Hiroshima City University
3-4-1, Ozuka-Higashi
Asaminami, Hiroshima 731-3194, Japan
e-mail: `tokusige@cs.hiroshima-cu.ac.jp`

Jun Asatani
Osaka University
1-3 Machikaneyama
Toyonaka-shi, Osaka, 560-8531, Japan
e-mail: `asatani@ist.osaka-u.ac.jp`

Marc P.C. Fossorier
University of Hawaii, 2540 Dole Street
Room 455
Honolulu, HI 96822, USA
e-mail: `marc@aravis.eng.hawaii.edu`

Tadao Kasami
e-mail: `kasami@empirical.jp`

Progress in Computer Science and Applied Logic, Vol. 23, 345–358

Deletion Correcting Using Generalized Reed-Solomon Codes

Yejing Wang, Luke McAven and Reihaneh Safavi-Naini

Abstract. Deletion correction codes have numerous applications including transmission synchronisation and more recently, tracing traitors. We consider the deletion correcting property of generalized Reed-Solomon codes and describe a class of generalized Reed-Solomon codes which correct one deletion. We also identify other codes that can correct numerous deletion errors including one that can correct the deletion of over half of the components of a codeword.

Mathematics Subject Classification (2000). 94B50, 94B60.

Keywords. Deletion correction, generalized Reed-Solomon codes.

1. Introduction

Error correcting codes have been widely used in correcting substitution and erasure errors. A different, less studied, class of codes are the deletion correcting codes, introduced by Levenshtein [6] to correct synchronisation errors. The applications of deletion correcting codes include packet loss in Internet transmission [11] and more recently, tracing pirate media [9] when the embedded fingerprint is shortened.

The ability to correct deletions is often considered in tandem with the ability to correct insertions [3], since the former implies the latter. Levenshtein constructed binary deletion/insertion correcting codes [6] and a family of non-binary deletion correcting codes [7]. Various other studies of deletion correcting codes are in [2, 5, 11, 12]. Perfect deletion correcting codes also exist [1, 8, 10, 13], for example length 6 codes allowing the correction of 4 deletions, over many alphabets [10].

We consider the deletion correction capabilities of generalized Reed-Solomon (GRS) codes. These codes have been extensively studied for their error and erasure correction capability. We give a method of constructing codes that can correct one deletion. We also give the result of an exhaustive survey of the deletion correcting properties of GRS codes with small parameters, and also selected codes with larger parameters. In addition to deletion correcting codes over fields, we also consider GRS-like codes over rings and show their deletion correction capacities.

The deletion correcting codes obtained from GRS codes are not perfect deletion correcting codes but, as we will show, they can be very effective in correcting deletions. In particular we show codes where deletions of up to half of the length of the code can be corrected. The decoding algorithm for these codes can be formulated as a polynomial reconstruction problem [9]. This results in an efficient decoding algorithm that uses the list decoding algorithm of Guruswami and Sudan [4].

This paper is structured as follows. In Section 2 we introduce basics of deletion correcting codes and generalized Reed-Solomon codes. In Section 3 we construct a class of generalized Reed-Solomon codes which correct one deletion. Section 4 includes our computational results on the deletion correcting properties of generalized Reed-Solomon codes. We include tables of results and a brief discussion of the ramifications. We summarize our work and make some observations in Section 5.

2. Preliminaries

2.1. Deletion Correcting

Let $X = x_1 x_2 \cdots x_m, Y = y_1 y_2 \cdots y_n$ be strings over an alphabet Λ. We say a string X is a subword of Y if X can be obtained from Y by only removing component(s) of Y. For example, 2234 is a subword of 142254364, while 452 is not (since reordering is required).

A code of length ℓ over an alphabet Λ is a subset of Λ^ℓ. A code can correct r deletions if any string of length $\ell - r$ is a subword of at most one codeword. A perfect r-deletion correcting code has any string of length $\ell - r$ as a subword of exactly one codeword. To find the deletion correction capability of a code we need to find the longest common substring of any pair of codewords. Let u and v be two codewords of a code Γ and let $\rho(u, v)$ denote the longest common subword of u and v. We define $\rho(\Gamma) = \max_{u,v\in\Gamma, u\neq v} \rho(u, v)$ and then $\rho(\Gamma) + 1$ is the length of the shortest substring that uniquely identifies a codeword. It follows that Γ is an r-deletion correcting code, where $r = \ell - \rho(\Gamma) - 1$.

A code capable of correcting up to r deletions is also capable of correcting r insertions. This is true because the r deletion correction property implies that substrings of length $\ell - r$ of codewords are unique and so any string of length $\ell + r$ can be obtained from a unique codeword. In general an r-deletion correcting code may correct any combination of up to r deletions and insertions [1, 8]. Given that one may view errors or substitutions as a combination of a deletion and an insertion, r-deletion correcting also implies $r/2$ substitution errors can be corrected.

2.2. Generalized Reed-Solomon Codes

Let F_q be a field of q elements, k be an integer, and

$$F_q[x]_k = \{f(x) : f(x) \text{ is a polynomial over } F_q \text{ with } \deg(f) \leq k\}.$$

Let $\ell \leq q$, $\alpha_1, \alpha_2, \ldots, \alpha_\ell \in F_q$ be distinct elements of F_q and let $v_1, v_2, \ldots, v_\ell \in F_q$ be non-zero elements. Write $\alpha = (\alpha_1, \alpha_2, \ldots, \alpha_\ell)$ and $\mathbf{v} = (v_1, v_2, \ldots, v_\ell)$. A $(k+1)$-dimensional GRS code of length ℓ over F_q is the set of all vectors

$$(v_1 f(\alpha_1),\ v_2 f(\alpha_2),\ \ldots,\ v_\ell f(\alpha_\ell)),\ \forall f \in F_q[x]_k$$

This code is denoted by $\mathrm{GRS}_{k+1}(q, \ell, \alpha, \mathbf{v})$. We refer to α as the *selector* and $\mathbf{v}$ as the *multiplier*. We note that $\mathrm{GRS}_{k+1}(q, \ell, \alpha, \mathbf{v})$ has q^{k+1} codewords.

3. A Class of Single Deletion Correcting GRS Codes

In this section we show a class of $\mathrm{GRS}_{k+1}(q, \ell, \alpha, \mathbf{v})$ codes, for $k = 1$, that correct a single deletion.

We recall a theorem from [9]. Let $K = 2k + 2 < \ell$. Consider two K-subsets of $\{1, 2, \ldots, \ell\}$, denoted by $I = \{i_1, i_2, \ldots, i_K\}$ and $I' = \{i'_1, i'_2, \ldots, i'_K\}$, such that

$$\begin{cases} 1 \leq i_1 < i_2 < \cdots < i_K \leq \ell \\ 1 \leq i'_1 < i'_2 < \cdots < i'_K \leq \ell \\ i_j = i'_j \ \text{ for at most } k \text{ values of } j \in \{1, 2, \ldots, K\} \end{cases} \tag{3.1}$$

Consider the following $K \times K$ matrix M.

$$M = \tag{3.2}$$

$$\begin{pmatrix} v_{i_1} & v_{i_1}\alpha_{i_1} & v_{i_1}\alpha_{i_1}^2 & \cdots & v_{i_1}\alpha_{i_1}^k & v_{i'_1} & v_{i'_1}\alpha_{i'_1} & v_{i'_1}\alpha_{i'_1}^2 & \cdots & v_{i'_1}\alpha_{i'_1}^k \\ v_{i_2} & v_{i_2}\alpha_{i_2} & v_{i_2}\alpha_{i_2}^2 & \cdots & v_{i_2}\alpha_{i_2}^k & v_{i'_2} & v_{i'_2}\alpha_{i'_2} & v_{i'_2}\alpha_{i'_2}^2 & \cdots & v_{i'_2}\alpha_{i'_2}^k \\ v_{i_3} & v_{i_3}\alpha_{i_3} & v_{i_3}\alpha_{i_3}^2 & \cdots & v_{i_3}\alpha_{i_3}^k & v_{i'_3} & v_{i'_3}\alpha_{i'_3} & v_{i'_3}\alpha_{i'_3}^2 & \cdots & v_{i'_3}\alpha_{i'_3}^k \\ \vdots & & & & & \vdots & & & & \\ v_{i_K} & v_{i_K}\alpha_{i_K} & v_{i_K}\alpha_{i_K}^2 & \cdots & v_{i_K}\alpha_{i_K}^k & v_{i'_K} & v_{i'_K}\alpha_{i'_K} & v_{i'_K}\alpha_{i'_K}^2 & \cdots & i'_K\alpha_{i'_K}^k \end{pmatrix}$$

Theorem 3.1. ([9]) *Let* Γ *be a* $GRS_{k+1}(q, \ell, \alpha, \mathbf{v})$ *code, where* $K < \ell < 2K$. *If* α *and* $\mathbf{v}$ *satisfy the property (P), then* $\rho(\Gamma) \leq K - 1$, *where*

(P) *The rank of* M *in* (3.2) *is* K *for any two* K*-sets* I, I' *satisfying* (3.1).

In the following we will use the above theorem to construct 1-deletion correcting codes from $\mathrm{GRS}_{k+1}(q, \ell, \alpha, \mathbf{v})$ with $k = 1$, and hence $\ell = K + 1 = 5$. We will construct α and $\mathbf{v}$ such that matrices M in (3.2) have rank $K = 4$, for all possible I, I'.

Suppose $I, I' \subset \{1, 2, 3, 4, 5\}$ are two 4-sets as defined in (3.1) with $i_{j_0} < i'_{j_0}$, for $j_0 = \min\{j : i_j \neq i'_j, 1 \leq j \leq 4\}$. Then (I, I') will be one of the following 3 possibilities:

$$(I, I') = (\{1,2,3,5\}, \{2,3,4,5\}), (\{1,2,3,4\}, \{1,3,4,5\}), (\{1,2,3,4\}, \{2,3,4,5\})$$

Each pair of (I, I') gives a matrix, which we call M_1, M_2 and M_3, respectively.

$$M_1 = \begin{pmatrix} v_1 & v_1\alpha_1 & v_2 & v_2\alpha_2 \\ v_2 & v_2\alpha_2 & v_3 & v_3\alpha_3 \\ v_3 & v_3\alpha_3 & v_4 & v_4\alpha_4 \\ v_5 & v_5\alpha_5 & v_5 & v_5\alpha_5 \end{pmatrix}, \quad M_2 = \begin{pmatrix} v_1 & v_1\alpha_1 & v_1 & v_1\alpha_1 \\ v_2 & v_2\alpha_2 & v_3 & v_3\alpha_3 \\ v_3 & v_3\alpha_3 & v_4 & v_4\alpha_4 \\ v_4 & v_4\alpha_4 & v_5 & v_5\alpha_5 \end{pmatrix},$$

$$M_3 = \begin{pmatrix} v_1 & v_1\alpha_1 & v_2 & v_2\alpha_2 \\ v_2 & v_2\alpha_2 & v_3 & v_3\alpha_3 \\ v_3 & v_3\alpha_3 & v_4 & v_4\alpha_4 \\ v_4 & v_4\alpha_4 & v_5 & v_5\alpha_5 \end{pmatrix}$$

Let $X = (x_1, x_2, x_3, x_4)^T$. In the following, we show conditions on parameters $\alpha_1, \alpha_2, \ldots, \alpha_5$ and $v_1, v_2, \ldots, v_5$ to ensure systems $M_i X = 0, i = 1, 2, 3$, have only zero solution. Let $M_i^{(m,n)}, i = 1, 2, 3$, denote the submatrix of M_i obtained by considering the first m rows and the first n columns.

3.1. Parameters from M_1

Consider the following sub-matrices of M_1.

$$M_1^{(3,4)} = \begin{pmatrix} v_1 & v_1\alpha_1 & v_2 & v_2\alpha_2 \\ v_2 & v_2\alpha_2 & v_3 & v_3\alpha_3 \\ v_3 & v_3\alpha_3 & v_4 & v_4\alpha_4 \end{pmatrix}, \quad M_1^{(3,3)} = \begin{pmatrix} v_1 & v_1\alpha_1 & v_2 \\ v_2 & v_2\alpha_2 & v_3 \\ v_3 & v_3\alpha_3 & v_4 \end{pmatrix}.$$

Choose $\alpha_1, \alpha_2, \alpha_3 \in F_q$ to be distinct, and $v_1, v_2, v_3 \in F_q^*$ arbitrarily. Then choose $v_4 \in F_q^*$ so that

$$v_4 \neq \frac{(\alpha_3 - \alpha_1)v_1 v_3^2 - (\alpha_3 - \alpha_2)v_2^2 v_3}{(\alpha_2 - \alpha_1)v_1 v_2}. \tag{3.3}$$

Then we have

$$\det(M_1^{(3,3)}) = (\alpha_2 - \alpha_1)v_1 v_2 v_4 - (\alpha_3 - \alpha_1)v_1 v_3^2 + (\alpha_3 - \alpha_2)v_2^2 v_3 \neq 0.$$

Hence equation

$$M_1^{(3,4)} X = 0 \tag{3.4}$$

has a solution $X^{(1)} = (x_1^{(1)}, x_2^{(1)}, x_3^{(1)}, -1)$ obtained by

$$\begin{pmatrix} x_1^{(1)} \\ x_2^{(1)} \\ x_3^{(1)} \end{pmatrix} = (M_1^{(3,3)})^{-1} \begin{pmatrix} v_2\alpha_2 \\ v_3\alpha_3 \\ v_4\alpha_4 \end{pmatrix} \tag{3.5}$$

Now choose α_4 so that

$$\alpha_4 \neq \tag{3.6}$$

$$\frac{(v_3^2 - v_2 v_4)v_1\alpha_1 + (v_1 v_4 + v_2 v_4 - v_2 v_3 - v_3^2)v_2\alpha_2 + (v_2^2 + v_2 v_3 - v_1 v_3 - v_1 v_4)v_3\alpha_3}{(v_2^2 - v_1 v_3)v_4}$$

Then equation $x_2^{(1)} \neq 1$. Finally, choose $v_5 \in F_q^*$ arbitrarily and α_5 such that

$$\alpha_5 \neq -\frac{x_1^{(1)} + x_3^{(1)}}{x_2^{(1)} - 1} \tag{3.7}$$

Then we have $(v_5, v_5\alpha_5, v_5, v_5\alpha_5)X^{(1)} \neq 0$. Therefore $\mathrm{rank}(M_1) = 4$ for v_4, α_4, α_5 satisfying (3.3), (3.6), (3.7), respectively.

3.2. Parameters from M_2

Consider the following sub-matrices of M_2.

$$M_2^{(3,4)} = \begin{pmatrix} v_1 & v_1\alpha_1 & v_1 & v_1\alpha_1 \\ v_2 & v_2\alpha_2 & v_3 & v_3\alpha_3 \\ v_3 & v_3\alpha_3 & v_4 & v_4\alpha_4 \end{pmatrix}, \quad M_2^{(3,3)} = \begin{pmatrix} v_1 & v_1\alpha_1 & v_1 \\ v_2 & v_2\alpha_2 & v_3 \\ v_3 & v_3\alpha_3 & v_4 \end{pmatrix}$$

Choose $v_4 \in F_q^*$ so that

$$v_4 \neq \frac{(\alpha_3 - \alpha_1)v_3^2 - (\alpha_3 - \alpha_2)v_2v_3}{(\alpha_2 - \alpha_1)v_2}. \tag{3.8}$$

Then

$$\det(M_2^{(3,3)}) = (\alpha_2 - \alpha_1)v_1v_2v_4 - (\alpha_3 - \alpha_1)v_1v_3^2 + (\alpha_3 - \alpha_2)v_1v_2v_3 \neq 0,$$

and so equation

$$M_2^{(3,4)}X = 0 \tag{3.9}$$

has a solution $X^{(2)} = (x_1^{(2)}, x_2^{(2)}, x_3^{(2)}, -1)$ obtained by

$$\begin{pmatrix} x_1^{(2)} \\ x_2^{(2)} \\ x_3^{(2)} \end{pmatrix} = (M_2^{(3,3)})^{-1} \begin{pmatrix} v_1\alpha_1 \\ v_3\alpha_3 \\ v_4\alpha_4 \end{pmatrix} \tag{3.10}$$

Finally, we choose the last three parameters; $\alpha_4 \in F_q \setminus \{\alpha_1, \alpha_2, \alpha_3\}$, $v_5 \in F_q^*$ arbitrarily, and $\alpha_5 \in F_q \setminus \{\alpha_1, \alpha_2, \alpha_3, \alpha_4\}$ so that

$$\alpha_5 \neq \frac{v_4x_1^{(2)} + v_4\alpha_4x_2^{(2)} + v_5x_3^{(2)}}{v_5} \tag{3.11}$$

Then $(v_4, v_4\alpha_4, v_5, v_5\alpha_5)X^{(2)} \neq 0$. Hence $\mathrm{rank}(M_2) = 4$ for v_4, α_5 satisfying (3.8), (3.11), respectively.

3.3. Parameters from M_3

Consider the following sub-matrices of M_3.

$$M_3^{(3,4)} = \begin{pmatrix} v_1 & v_1\alpha_1 & v_2 & v_2\alpha_2 \\ v_2 & v_2\alpha_2 & v_3 & v_3\alpha_3 \\ v_3 & v_3\alpha_3 & v_4 & v_4\alpha_4 \end{pmatrix}, \quad M_3^{(3,3)} = \begin{pmatrix} v_1 & v_1\alpha_1 & v_2 \\ v_2 & v_2\alpha_2 & v_3 \\ v_3 & v_3\alpha_3 & v_4 \end{pmatrix}$$

Note that $M_3^{(3,4)} = M_1^{(3,4)}, M_3^{(3,3)} = M_1^{(3,3)}$, so $X^{(1)}$ defined in (3.5) with $x_4^{(1)} = -1$ is a solution to

$$M_3^{(3,4)}X = 0 \tag{3.12}$$

Finally, choose $\alpha_4 \in F_q \setminus \{\alpha_1, \alpha_2, \alpha_3\}$, $v_5 \in F_q^*$ arbitrarily, and α_5 is chosen as

$$\alpha_5 \neq \frac{v_4 x_1^{(1)} + v_4 \alpha_4 x_2^{(1)} + v_5 x_3^{(1)}}{v_5} \tag{3.13}$$

Then $(v_4, v_4\alpha_4, v_5, v_5\alpha_5)X^{(1)} \neq 0$. Therefore $rank(M_3) = 4$ for v_4, α_5 satisfying (3.3), (3.13), respectively. Let

$$V_4 = \left\{ \frac{(\alpha_3-\alpha_1)v_1v_3^2 - (\alpha_3-\alpha_2)v_2^2v_3}{(\alpha_2-\alpha_1)v_1v_2}, \frac{(\alpha_3-\alpha_1)v_3^2 - (\alpha_3-\alpha_2)v_2v_3}{(\alpha_2-\alpha_1)v_2} \right\}$$

$$A_4 =$$

$$\left\{ \frac{(v_3^2 - v_2v_4)v_1\alpha_1 + (v_1v_4 + v_2v_4 - v_2v_3 - v_3^2)v_2\alpha_2 + (v_2^2 + v_2v_3 - v_1v_3 - v_1v_4)v_3\alpha_3}{(v_2^2 - v_1v_3)v_4} \right\}$$

$$A_5 = \left\{ \frac{x_1^{(1)} + x_3^{(1)}}{1 - x_2^{(1)}}, \frac{v_4x_1^{(1)} + v_4\alpha_4x_2^{(1)} + v_5x_3^{(1)}}{v_5}, \frac{v_4x_1^{(2)} + v_4\alpha_4x_2^{(2)} + v_5x_3^{(2)}}{v_5} \right\}$$

From the above discussion we have the following theorem.

Theorem 3.2. *Let $k = 1$, $\ell = 5$, and $\mathbf{v}$ and α be sequentially defined as*

$$\begin{aligned} & v_1, v_2, v_3 \in F_q^* \\ & \alpha_1, \alpha_2, \alpha_3 \in F_q \\ & v_4 \in F_q^* \setminus V_4 \\ & \alpha_4 \in F_q \setminus (\{\alpha_1, \alpha_2, \alpha_3\} \cup A_4) \\ & v_5 \in F_q^* \\ & \alpha_5 \in F_q \setminus (\{\alpha_1, \alpha_2, \alpha_3, \alpha_4\} \cup A_5) \end{aligned}$$

The $GRS_{k+1}(q, \ell, \alpha, \mathbf{v})$ code is capable of correcting a single deletion for $q > 8$.

Proof. By the choices of α and $\mathbf{v}$, the ranks of M_1, M_2, M_3 are 4. When $q > 8$, $\alpha_i, v_i, 1 \le i \le 5$ exist. The theorem is then proved by applying Theorem 3.1. □

3.4. Extending the Theorem

Theorem 3.2 can be extended to $k > 1$. Suppose $\ell = K + 1$. Recall that the aim is to find α and $\mathbf{v}$ such that all matrices M in (3.2) have rank K. In the following we show how these values can be found.

Suppose the first K components of α and $\mathbf{v}$ are chosen such that for any I, I' as defined in (3.1), the following two properties are satisfied.

(L1) The first $K - 1$ rows of (3.2) are linearly independent.

(L2) The system of linear equations $M' \begin{pmatrix} X \\ -X \end{pmatrix} = 0$ has only zero-solution, where M' is the submatrix of M in (3.2) obtained by removing the last row.

Let us consider how we choose α_{K+1} and v_{K+1}. Let $I = \{i_1, i_2, \ldots, i_K\}, I' = \{i'_1, i'_2, \ldots, i'_K\}$ be two K-subsets of $\{1, 2, \ldots, \ell\}$ in (3.1) with $i_{j_0} < i'_{j_0}$ where $j_0 = \min\{j : i_j \neq i'_j\}$. Consider the system of equations

$$M' \begin{pmatrix} X \\ Y \end{pmatrix} = 0$$

with K unknowns and assume $(x_0, \ldots, x_k, y_0, \ldots, y_k)$ is a non-zero solution. Let

$$A^{[I,I']}(x) = x_0 + x_1 x + \cdots + x_k x^k, \; B^{[I,I']}(x) = y_0 + y_1 x + \cdots + y_k x^k$$

From property (L2) we know that $A^{[I,I']}(x) + B^{[I,I']}(x)$ is a non-zero polynomial. Define a set

$$A(I,I') = \tag{3.14}$$
$$\left\{\alpha \in F_q : A^{[I,I']}(\alpha) + B^{[I,I']}(\alpha) = 0, \text{ or } A^{[I,I']}(\alpha) = 0, \text{ or } B^{[I,I']}(\alpha) = 0\right\}.$$

For a given value of α_{K+1}, we define the set

$$V(I,I') = \left\{0, -v_K \frac{A^{[I,I']}(\alpha_K)}{B^{[I,I']}(\alpha_{K+1})}, -v_K \frac{B^{[I,I']}(\alpha_K)}{A^{[I,I']}(\alpha_{K+1})}\right\}. \tag{3.15}$$

The following theorem can be obtained. We omit the straightforward proof.

Theorem 3.3. *If there are distinct $\alpha_1, \alpha_2, \ldots, \alpha_{2k+2} \in F_q$ and $v_1, v_2, \ldots, v_{2k+2} \in F_q^*$, such that properties (L1) and (L2) are satisfied, then any $GRS_{k+1}(q, \ell, \alpha, \mathbf{v})$ code of length $\ell = 2k+3$ with parameters satisfying*

$$\alpha_{2k+3} \in F_q \setminus \left(\{\alpha_1, \ldots, \alpha_{2k+2}\} \bigcup_{I,I'} A(I,I')\right)$$

$$v_{2k+3} \in F_q \setminus \left(\bigcup_{I,I'} V(I,I')\right)$$

is capable of correcting a single deletion.

The following shows that the choice of parameters depends on the value of k only. So for sufficient large q, all α_i and v_i required in (3.14) and (3.15) exist.

Lemma 3.1. *Suppose $\ell = K+1$. There are $(k+1)(k+2)/2$ pairs I, I' of sets as defined in* (3.1) *with $i_{j_0} < i'_{j_0}$ where $j_0 = \min\{j : i_j \neq i'_j\}$.*

Proof. Let $\ell = K+1$ and I, I' be two sets satisfying (3.1). Suppose $|\{j : i_j = i'_j\}| = m$. Then

$$\{j : i_j = i'_j\} =$$
$$\begin{cases} \{1, 2, \ldots, m\}, \\ \{1, 2, \ldots, m_1\} \cup \{K - m + m_1 + 2, K - m + m_1 + 3, \ldots, K+1\}, 0 < m_1 < m \\ \{K - m + 1, K - m + 2, \ldots, K\}, \end{cases}$$

I and I' are uniquely determined by $\{j : i_j = i'_j\}$. If $\{j : i_j = i'_j\} = \{1, 2, \ldots, m\}$, $I = \{1, \ldots, m, m+1, \ldots, K\}$, $I' = \{1, \ldots, m, m+2, \ldots, K+1\}$; if $\{j : i_j = i'_j\} = \{K-m+1, K-m+2, \ldots, K\}$, $I = \{1, \ldots, m, m+2, \ldots, K+1\}$, $I' = \{2, \ldots, m+1, m+2, \ldots, K+1\}$; if $\{j : i_j = i'_j\} = \{1, 2, \ldots, m_1\} \cup \{K-m+m_1+2, \ldots, K+1\}$, $I = \{1, \ldots, m_1, m_1+2, \ldots, K-m+m_1+1, K-m+m_1+2, \ldots, K+1\}$, $I' = \{1, \ldots, m_1, m_1+1, \ldots, K-m+m_1, K-m+m_1+2, \ldots, K+1\}$. So

$$|\{(I, I') : i_j = i'_j \text{ for } m \text{ values of } j\}| = m + 1$$

and hence the total number of required pairs (I, I') is

$$\sum_{m=0}^{k} (m+1) = \frac{(k+1)(k+2)}{2} .$$

□

Comments on Theorem 3.2

The most restricted parameter is α_5. The conditions in Theorem 3.2 imply that there may be at most seven values that α_5 cannot take. We therefore note that for $q > 8$ there will always be sets of α and $\mathbf{v}$ satisfying Theorem 3.2, and therefore always $\mathrm{GRS}_{k+1}(q, \ell, \alpha, \mathbf{v})$ codes capable of correcting at least a single deletion. For $q \leq 8$, however, it is possible elements in the set V_4, or in the set A_5, are all distinct implying no available choice for v_4 or α_5, respectively. Elements of V_4 or A_5 may coincide however, in particular for A_5 they may equal one of $\alpha_1, \alpha_2, \alpha_3$ or α_4. Should enough elements coincide it is possible for the theorem to generate codes capable of correcting single deletions even if $q \leq 8$.

We include below a small sample of codes generated using the structure in Theorem 3.2. We give ones with $q < 8$ to illustrate the point made above.

q	v_1	v_2	v_3	v_4	v_5	α_1	α_2	α_3	α_4	α_5
5	4	3	2	3	4	4	1	3	0	2
5	3	2	3	3	3	0	3	2	4	1
7	2	4	2	2	1	4	1	5	6	2
7	1	6	3	1	1	2	1	6	0	3

4. Experiments on Deletion Correcting in GRS Codes

In this section we present the results of exhaustive searches where possible, and selective searches otherwise, on a range of q and ℓ to find the codes with the best deletion rates. To interpret the results we consider three categories of codes based on the value of q.

Prime GRS codes: q is prime.

Prime extension GRS codes: $q = p^m$, p prime and integer $m > 1$. The field elements are not integers, rather they may be represented as m-tuples or as powers of a primitive polynomial.

GRS–like codes: The parameter q is not expressible in the form $q = p^m$, for p prime and m integer.

The definitions in Section 2 hold explicitly for the first two cases, the third are not strictly GRS codes. They are included in this paper since deletion correction properties have been observed. Let us formally define them.

Let Z_q be the ring of integers $\{0, \ldots, q-1\}$, with addition and multiplication modulo q, where q is not a prime power. Let k be an integer. Let

$$Z_q[x]_k = \{f(x) : f(x) \text{ is a polynomial over } Z_q \text{ with } \deg(f) \leq k\}.$$

Let $\alpha_1, \alpha_2, \ldots, \alpha_\ell \in Z_q, \ell \leq q$ be distinct elements of F_q and let $v_1, v_2, \ldots, v_\ell \in Z_q$ be non–zero elements. Write $\alpha = (\alpha_1, \alpha_2, \ldots, \alpha_\ell)$ and $\mathbf{v} = (v_1, v_2, \ldots, v_\ell)$. A $(k+1)$–dimensional GRS–like code of length ℓ is the set of all vectors

$$(v_1 f(\alpha_1),\, v_2 f(\alpha_2),\, \ldots,\, v_\ell f(\alpha_\ell)),\ \forall f \in Z_q[x]_k$$

We shall denote this code by $\mathrm{GRS}_{k+1}(q, \ell, \alpha, \mathbf{v})$ also, and again refer to α as the *selector* and $\mathbf{v}$ as the *multiplier.* The results we give later consider the GRS–like codes explicitly generated as having q^{k+1} codewords, even if some coincide. Thus when there are repeated codewords no deletion correcting is possible. The treatment of GRS–like codes with such redundancy removed will be the subject of later work.

We note that, unlike for the error–correcting capability of codes which is unaffected by permuting columns of the code, deletion correcting capability depends on the order of the columns. Codes with reordered α are thus necessarily treated as distinct codes.

Definition 4.1. *For fixed ℓ, q, k, let $G(\ell, q, k)$ be the collection of codes obtained by taking all $GRS_{k+1}(q, \ell, \alpha, \mathbf{v})$ codes with all possible α and $\mathbf{v}$.*

Recall α are the selectors denoting the field element values used to build the codewords from the field polynomials, while $\mathbf{v}$ are multipliers. Since α and $\mathbf{v}$ are both ℓ elements long, the former distinct, the latter non-zero, $G(\ell, q, k)$ has a cardinality of at most

$$|G(\ell, q, k)| \leq (q-1)^\ell \frac{q!}{(q-\ell)!} \approx e^{-q} q^{2q}\,. \tag{4.1}$$

This increases rapidly; for $q = \ell = 7$ $|G(\ell, q, k)| \approx 2^{30}$ while by $q = 11, \ell = 7$, $|G(\ell, q, k)| \approx 2^{44}$ and by $q = \ell = 11$ $|G(\ell, q, k)| \approx 2^{62}$.

For a code $\mathrm{GRS}_{k+1}(q, \ell, \alpha, \mathbf{v})$ in $G(\ell, q, k)$ we define $s(\ell, q, k, \alpha, \mathbf{v}) = \rho + 1$, where ρ was defined in Section 2. We then define

$$\sigma(\ell, q, k) = \min_{\alpha} \min_{\mathbf{v}} s(\ell, q, k, \alpha, \mathbf{v}) \tag{4.2}$$

to be the minimum $s(\ell, q, k, \alpha, \mathbf{v})$ for $G(\ell, q, k)$. This gives the highest deletion capability for given parameters ℓ, q, k.

The efficiency of a code is measured with the *code rate*, which for a linear code of length ℓ and dimension k is given by k/ℓ. For $\mathrm{GRS}_{k+1}(q, \ell, \alpha, \mathbf{v})$ this equals

$(k+1)/\ell$. We define the *deletion rate* of a code to be s/ℓ. This is the proportion of symbols in a codeword that need to be received in order to recover the word.

4.1. Methods and Results

Let us briefly consider $k = 0$, where the codewords are constants. For the multipliers being all 1 each codeword is a string of ℓ copies of the same fixed element. This implies a unique substring length of 1.

We sketch our algorithm for finding $s(\ell, q, k, \alpha, \mathbf{v})$ using pre–generated codewords of the code $\mathrm{GRS}_{k+1}(q, \ell, \alpha, \mathbf{v})$.

```
Generate the codebook $\Gamma$ for $\mathrm{GRS}_{k+1}(q,\ell,\alpha,\mathbf{v})$
    Set $c=\ell, s(\ell,q,k,\alpha,\mathbf{v})=0$.
    While $s(\ell,q,k,\alpha,\mathbf{v})=0$
        Generate all length $c$ subwords of $\Gamma$
        Sort subwords and compare adjacent subwords
        If some subword is from more than one codeword
            $s(\ell,q,k,\alpha,\mathbf{v})=c+1$.
        else
            Set $c=c-1$.
        End If
        If $c=0$ set $s(\ell,q,k,\alpha,\mathbf{v})=1$.
    End While
```

If the algorithm has been used for α and $\mathbf{v}$, one applies equation (4.2) to find $\sigma(\ell, q, k)$. There are $q^{k+1}\frac{\ell!}{c!(\ell-c)!}$, possibly indistinct, subwords of length c for each code. By $\ell = 7, k = 1, q = 7$ we have over 1000 subwords of length 5 for each code. Our script requires approximately 2^{37} floating point operations to exhaustively find $\sigma(7,5,1)$, increasing to 2^{45} by $\sigma(7,7,1)$, and as such we have limited our exhaustive searches to $k = 1$ and (ℓ, q)=(5,3-5),(6,3-6),(7,3-5),(8-12,3). The results are given in Table 1.

TABLE 1. The exhaustive searches provide the percentage of codes in $G(\ell, q, k)$ with particular $s = s(\ell, q, k, \alpha, \mathbf{v})$. For *GRS*–like codes, some percentages are tabulated for $s = \ell + 1$. In these cases there are repeated codewords, so no substring uniquely specifies a single polynomial.

ℓ	3						4			5			6
q	5	6	7	10	11	12	5	6	7	5	6	7	6
s													
3	100	34	100	33	100	22		0.05	1				
4		66		67		78	100	60	99	42	2		
5								40		58	75	60	12
6											23	40	75
7													13

For large values of ℓ, q and k we have performed selective searches. The selections were based on previous searches. As suggested in Section 5 some sets of parameters appear more likely to give better deletion correcting properties. The results for the selective searches are given in Tables 2, 3 and 4. These results provide upper bounds on $\sigma(\ell, q, k)$ for prime and prime extension GRS codes, and GRS–like codes respectively. For example, although we have located a $G(9, 31, 5)$ code with $s = 5$, there may exist codes in $G(9, 31, 5)$ with $s = 4$. Testing some cases by our method, even randomly, for example $q = 23, \ell = 10$, requires extensive resources. We include in Table 5 some specific examples of codes found for different $G(\ell, q, k)$ classes, with the highest deletion correcting properties known for that class.

TABLE 2. A tabulation of experimental upper bounds on the value of the $\sigma(\ell, q, k)$ for q a prime. Where entries have triple entries they correspond to $k = 1, k = 2$ and $k = 3$, while double entries correspond to $k = 1, k = 2$ and single entries correspond to $k = 1$. This notation is also used in subsequent tables.

						ℓ				
q	4	5	6	7	8	9	10	11	12	13
5	4,4,4	4,5,5								
7	3,4,4	4,5,5	4,5,6	5,6,7						
11	3,4,4	4,5,5	4,5,6	4,6,7	5,6,7	6,7,8	6,8,9	7,8,10		
13	3,4,4	3,5,5	4,5,6	4,6,7	5,6,7	5,7,8	6,7,8	6,8,9	7,9	8,9
17	3,4	3,5	4,5	4,5	4,6	5,7	5,7	6,8	7	7
19	3,4	3,5	4,5	4,5	4,6	5,6	5,7	6	6	7
23	3,4	3,5	4,5	4,5	4,6	5	5	6	6	7
29	3,4	3,5	3,5	4,5	4,6	5	5	5	6	
31	3	3	3	4	4	4	5	5	6	

TABLE 3. A tabulation of experimental upper bounds on the value of $\sigma(\ell, q, k)$, when q is a prime power.

						ℓ				
q	4	5	6	7	8	9	10	11	12	13
8	3,4	4,5	4,5	5,6	6,7					
9	3,4	4,5	4,5	5,6	5,7	6,7				
16	3,4	3,5	4,5	4,6	5,6	5,7	6,7	6,8	7	8
25	3,4	3,5	4,5	4,6	4,6	5,7	5	6	6	
27	3	3	4	4	4	5	5	5	6	
32	3	3	3	4	4	5	5	5	6	

TABLE 4. A tabulation of experimental upper bounds on the value of $\sigma(\ell, q, k)$ for q not a prime power (GRS–like codes).

q	ℓ 4	5	6	7	8	9	10	11	12	13
6	3	4	5							
10	3	4	4	5	5	6	7			
12	3	4	5	5	6	6	7	8	9	
14	3	4	4	5	5	6	7	7	8	9
15	3,4	4,5	4,6	5,6	5,7	6,7	7,8	7,9	8,10	9,11
18	3	4	4	5	6	6	7	7	8	9
20	3	4	4	5	6	6	7	7	8	9
21	3	4	4	5	5	6	7	7	8	9
22	3	4	4	5	5	6	6	7	8	8

4.2. Observations

Recall the number of deletions correctable by a code is given by $r = \ell - s$. We have obtained r as high as 6. For example, $q \geq 17, \ell = 13$ and $k = 1$. More significantly, the deletion rate, s/ℓ, is as low as $3/7$ for $k = 1$ and $q = 67$. Even the GRS–like codes allow for up to four deletions and provide codes with deletion rates of 0.6.

Increasing k decreases the number of correctable deletions. For $k = 2$ we can obtain up to three corrections, for example for $q = 11, \ell = 11$ and $q = 13, \ell = 10$. For $k = 3$ we have codes correcting two deletions. We see certain trends in Tables 2–4, which we summarize in two propositions.

Proposition 4.2 (Code length variance). *If* $\ell_1 < \ell_2$ *then*

$$\sigma(\ell_1, q, k) \leq \sigma(\ell_2, q, k) \ . \tag{4.3}$$

Proof. (Sketch) Any α' and $\mathbf{v}'$ of length ℓ_2 can be expressed as the concatenations of α and β, and of $\mathbf{v}$ and $\mathbf{w}$, respectively, for some $\alpha, \mathbf{v}$ of length ℓ_1. It follows that $s(\ell_1, q, k, \alpha, \mathbf{v}) \leq s(\ell_2, q, k, \alpha', \mathbf{v}')$. □

Proposition 4.3 (Polynomial order variance). *If* $k_1 < k_2$ *then*

$$\sigma(\ell, q, k_1) \leq \sigma(\ell, q, k_2) \ . \tag{4.4}$$

Proof. (Sketch) By definition $\mathrm{GRS}_{k_2+1}(q, \ell, \alpha, \mathbf{v})$ contains $\mathrm{GRS}_{k_1+1}(q, \ell, \alpha, \mathbf{v})$. For $\alpha, \mathbf{v}$ then, the set of substrings in the code of dimension $k_2 + 1$ contains the set of substrings from the code of dimension k_1+1. Thus $s(\ell, q, k_1, \alpha, \mathbf{v}) \leq s(\ell, q, k_2, \alpha, \mathbf{v})$. □

5. Conclusions and Discussion

We have given an explicit construction of single deletion correcting codes by using $\mathrm{GRS}_{k+1}(q, \ell, \alpha, \mathbf{v})$. We have summarized our exhaustive and selective searches for optimal deletion correcting properties for small GRS codes. Codes correcting deletions of over half the length of the code have been obtained.

This is a novel use for GRS codes, which are well known for their error-correcting properties. One advantage is their well-established efficient decoding algorithms [4]. Further to the previous propositions, the results in Tables 2 and 3 suggest that

If $q_1 < q_2$ *are prime powers then* $\sigma(\ell, q_1, k) \geq \sigma(\ell, q_2, k)$.

That is, the deletion correction capability of codes does not decrease as q increases. This remains a conjecture however.

We note the lowest s/ℓ observed, but not tabulated, is $3/7$, for $q = 67, k = 1$. The exhaustive searches and selective trials suggest ways of selecting parameters to increase the chance of obtaining codes with the best deletion correction capability.

- The brief $k = 0$ analysis at the start of Section 4.1 shows $\mathbf{v} = \mathbf{1}$ is likely to give the best substring properties, and this is supported by our exhaustive and selective searches. We see examples illustrating this in Table 5.
- For GRS-like codes the lowest values of $s(\ell, q, k, \alpha, \mathbf{v})$ are obtained (in most observed cases) when the selectors are a mix of even and odd values, each type equally used. For $\ell = 4, q = 6, k = 1$, for example, the only codes allowing one deletion correction have α with two even and two odd values.
- A related observation on GRS-like codes is those with *all* odd or even values in α give the worst deletion correcting properties. For $\ell = 3, q = 6, k = 1$, for example, having α all even or all odd always gives a code with no deletion correcting properties, independent of $\mathbf{v}$.

References

[1] P.A.H. Bours, 'On the construction of perfect deletion-correcting codes using design theory'. *Designs, Codes and Cryptography* **6** (1995) 5–20.

[2] L. Calabi and W.E. Hartnett, 'Some general results of coding theory with applications to the study of codes for the correction of synchronisation errors'. *Information and Control* **15** (1969) 235–249.

[3] M.C. Davey and D.J.C. MacKay, 'Reliable communication over channels with insertions, deletions and substitutions'. *IEEE Transactions on Information Theory* **47**(2) (2001) 687–696.

[4] V. Guruswami and M. Sudan, 'Improved decoding of Reed-Solomon and algebraic-geometry codes'. *IEEE Transactions on Information Theory* **45**(6) (1999) 1757–1767.

[5] T. Kløve, 'Codes correcting a single insertion/deletion of a zero or a single peak-shift'. *IEEE Transactions on Information Theory* **41** (1995) 279–283.

[6] V.I. Levenshtein, 'Binary codes capable of correcting deletions, insertions and reversals'. *Soviet Physics – Doklady* **10**(8) (1966) 707–710.

[7] V.I. Levenshtein, 'One method of constructing quasilinear codes providing synchronisation in the presence of errors.' *Problems of Information Transmission* **7**(3) (1971) 215–222.

[8] A. Mahmoodi, 'Existence of perfect 3-deletion-correcting codes'. *Designs, Codes and Cryptography* **14** (1998) 81–87.

[9] R. Safavi-Naini and Y. Wang, 'Traitor tracing for shortened and corrupted fingerprints'. *ACM-DRM'02, LNCS* **2696** (2003) 81–100.

[10] N. Shalaby, J. Wang and J. Yin, 'Existence of perfect 4-deletion-correcting codes with length six'. *Designs, Codes and Cryptography* **27** (2002) 145–156.

[11] N.J.A. Sloane, 'On single-deletion-correcting codes'. *Codes and Designs*. Mathematical Research Institute Publications **10** (Ohio University, 2002) 273–292.

[12] E. Tanaka and T. Kasai, 'Synchronisation and substitution error-correcting codes for the Levenshtein metric'. *IEEE Transactions on Information Theory* **22** (1976) 156–162.

[13] J. Yin, 'A combinatorial construction for perfect deletion-correcting codes'. *Designs, Codes and Cryptography* **23** (2001) 99–110.

Appendix

TABLE 5. A tabulation of a selector α and multiplier $\mathbf{v}$ pair which give the best value for $\sigma(\ell, q, k)$ known. Although multipliers which are all 1, the identity element, are given here, those are not the only solutions.

(ℓ, q, k)	s	α	$\mathbf{v}$
(5,5,1)	4	(4 0 3 1 2)	(1 1 1 1 1)
(4,7,1)	3	(5 1 2 4)	(1 1 1 1)
(5,7,1)	4	(1 6 0 2 5)	(1 1 1 1 1)
(6,7,1)	4	(1 2 0 6 4 5)	(1 1 1 1 1 1)
(7,7,1)	5	(5 3 0 6 4 2 1)	(1 1 1 1 1 1 1)
(4,11,1)	3	(8 1 5 6)	(1 1 1 1)
(5,11,1)	4	(9 5 3 6 1)	(1 1 1 1 1)
(6,11,1)	4	(10 6 5 8 4 7)	(1 1 1 1 1 1)
(7,11,1)	4	(6 1 0 7 8 3 4)	(1 1 1 1 1 1 1)
(8,11,1)	5	(9 0 1 4 2 6 7 3)	(1 1 1 1 1 1 1 1)
(9,11,1)	6	(6 9 10 7 3 5 2 8 0)	(1 1 1 1 1 1 1 1 1)
(10,11,1)	6	(7 2 5 9 8 3 10 6 4 0)	(1 1 1 1 1 1 1 1 1 1)
(11,11,1)	7	(1 6 7 9 10 8 3 4 0 2 5)	(1 1 1 1 1 1 1 1 1 1 1)

Yejing Wang, Luke McAven and Reihaneh Safavi-Naini
School of Information Technology and Computer Science
University of Wollongong
Wollongong, NSW 2522, Australia
e-mail: `yejing@uow.edu.au`
e-mail: `lukemc@uow.edu.au`
e-mail: `rei@uow.edu.au`

Progress in Computer Science and Applied Logic, Vol. 23, 359–367

A Note on the Linear Complexity Profile of the Discrete Logarithm in Finite Fields

Arne Winterhof

Abstract. We essentially improve lower bounds on the linear complexity of a sequence representing the residues of the discrete logarithm in a finite field modulo a divisor of the order of the multiplicative group. More generally, we present the result as a bound on the linear complexity profile. The proof is based on character sum bounds.

Mathematics Subject Classification (2000). Primary 11T71; Secondary 11T24 65C10 68Q25 94A60.

Keywords. Discrete logarithm, linear complexity profile, finite fields, cyclotomic generator.

1. Introduction

Let p be a prime, r a positive integer, $q = p^r$, $\mathbb{F}_q$ the finite field of order q, and γ a primitive element of $\mathbb{F}_q$. The *discrete logarithm* (or *index*) of a nonzero element $\xi \in \mathbb{F}_q$ to the base γ, denoted $\text{ind}_\gamma(\xi)$, is the unique integer l with $0 \le l \le q-2$ such that $\xi = \gamma^l$. The *discrete logarithm problem* is to find a computationally feasible method for determining the discrete logarithm. The security of many public-key cryptosystems depends on the presumed intractability of the discrete logarithm problem (see e.g. [14]). This paper provides some theoretical support to this assumption of hardness of the discrete logarithm problem.

For an integer $N \ge 2$ the *linear complexity profile* $L(a_i, N)$ at N of a sequence (a_i) over a commutative ring R with 1 is the least L such that there are constants $c_1, \dots, c_L \in R$ satisfying

$$-a_i = c_1 a_{i-1} + c_2 a_{i-2} + \dots + c_L a_{i-L} \qquad \text{for all } L \le i \le N-1, \tag{1}$$

with the convention that $L(a_i, N) = 0$ if the first N terms of (a_i) are all 0 and $L(a_i, N) = N$ if $a_0 = a_1 = \dots = a_{N-2} = 0$ and $a_{N-1} \neq 0$. The *linear complexity* of (a_i) is defined by

$$L(a_i) = \sup_{N \ge 2} L(a_i, N).$$

Linear complexity and linear complexity profile are important cryptographic characteristics of sequences and provide information on the predictability and thus unsuitability for cryptography. Hence, a low linear complexity (profile) has turned out to be undesirable in cryptography.

Consider the following representation of the discrete logarithm via linear recurring sequences. Let $\{\beta_0, \beta_1, \ldots, \beta_{r-1}\}$ be a basis of $\mathbb{F}_q$ over $\mathbb{F}_p$. For $0 \le i \le q-1$ we define $\xi_i \in \mathbb{F}_q$ by

$$\xi_i = i_0\beta_0 + i_1\beta_1 + \ldots + i_{r-1}\beta_{r-1} \tag{2}$$

if

$$i = i_0 + i_1 p + \ldots + i_{r-1}p^{r-1} \quad \text{with } 0 \le i_0, i_1, \ldots, i_{r-1} \le p-1.$$

Let $d > 1$ be a divisor of $(q-1)$ and (s_i) a sequence over $\mathbb{Z}_d$ with the property

$$s_i = \mathrm{ind}_{\gamma,d}(\xi_i) \quad \text{for } 1 \le i \le q-1, \tag{3}$$

where $\mathrm{ind}_{\gamma,d}(\xi)$ denotes the residue class of $\mathrm{ind}_\gamma(\xi)$ modulo d for $\xi \in \mathbb{F}_q^*$. We prove the following lower bound on the linear complexity profile of (s_i).

Theorem 1. *Let $d > 1$ be a divisor of $q-1$ and (s_i) a sequence with the property (3). Then for $2 \le N \le q$ we have*

$$L(s_i, N) > \frac{1}{4}\frac{N^{1/2}}{p^{1/2}q^{1/4}}.$$

If N is a power of p then we can slightly improve Theorem 1.

Corollary 1. *If $N = p^n$ with an integer $1 \le n \le r$, then we have*

$$L(s_i, p^n) > \frac{1}{4}\frac{N^{1/2}}{q^{1/4}}.$$

In particular for the linear complexity we have the lower bound

$$L(s_i) > \frac{1}{4}q^{1/4}.$$

This result improves earlier results of [12] if $r \ge 5$ (order of magnitude $p^{r/4}$ vs. rp). We prove Theorem 1 and Corollary 1 in the next section. The proofs are based on a bound of incomplete character sums.

2. Proofs

Before we prove the results we recall some facts on character sums. With the method of Polya and Vinogradov [2, 19, 20, 21, 22] and Weil's Theorems [15, Theorems 2C and 2G] we get the following bound on incomplete character sums.

Lemma 1. *Let χ be a multiplicative character of $\mathbb{F}_q$ of order $d > 1$ and $f(X) \in \mathbb{F}_q[X]$ a monic polynomial which is not a dth power and has m distinct zeros in its splitting field over $\mathbb{F}_q$. For $0 \le k_0, k_1, \ldots, k_{r-1} \le p-1$ let $B \subseteq \{0, 1, \ldots, q-1\}$ be a box of the form*

$$B = \{i_0 + i_1 p + \ldots + i_{r-1}p^{r-1} : 0 \le i_j \le k_j \textit{ for } 0 \le j \le r-1\}.$$

Let s be the number of $0 \le j \le r-1$ with $1 \le k_j \le p-2$. Then we have for each $a \in \mathbb{F}_q^*$,

$$\left|\sum_{x\in B} \chi(af(x))\right| \le mq^{1/2}\left(\left(1-\frac{1}{p}+\log p\right)^s - \frac{|B|}{mq}\right).$$

For the convenience of the reader we add the most important steps for the deduction of Lemma 1. For more details we refer to [2, 19, 20, 21, 22].

We may restrict ourselves to the case $a = 1$. From

$$\chi(f(x)) = \frac{1}{q}\sum_{y\in\mathbb{F}_q}\chi(f(y))\sum_{z\in\mathbb{F}_q}\psi((x-y)z)$$

we get

$$\begin{aligned}
\left|\sum_{x\in B}\chi(f(x))\right| &\le \frac{1}{q}\sum_{z\in\mathbb{F}_q}\left|\sum_{y\in\mathbb{F}_q}\chi(f(y))\psi(-yz)\right|\left|\sum_{x\in B}\psi(xz)\right| \\
&\le \frac{m}{q^{1/2}}\sum_{z\in\mathbb{F}_q}\left|\sum_{x\in B}\psi(xz)\right| - \frac{|B|}{q^{1/2}}
\end{aligned}$$

by Weil's Theorems, where ψ denotes the additive canonical character of $\mathbb{F}_q$. Now we have

$$\left|\sum_{x\in B}\psi(xz)\right| = \prod_{j=0}^{r-1}\left|\sum_{i_j=0}^{k_j}\psi(\beta_j z)^{i_j}\right|.$$

The right hand side is zero if $\psi(\beta_j z) \ne 1$ for some j with $k_j = p-1$. For z with $\psi(\beta_j z) = 1$ for all j with $k_j = p-1$ we get

$$\begin{aligned}
\left|\sum_{x\in B}\psi(xz)\right| &= p^{r-s}\prod_{\substack{j=0\\k_j<p-1}}^{r-1}\left|\sum_{i_j=0}^{k_j}\psi(\beta_j z)^{i_j}\right| \\
&\le p^{r-s}\prod_{\substack{j=0\\k_j<p-1}}^{r-1}\min\left(k_j+1, \frac{1}{\sin(\mathrm{Tr}(\beta_j z)\pi/p)}\right).
\end{aligned}$$

Since the mapping $z \mapsto (\mathrm{Tr}(\beta_0 z), \ldots, \mathrm{Tr}(\beta_{r-1}z))$ is a bijection we get

$$\begin{aligned}
\sum_{z\in\mathbb{F}_q}\left|\sum_{x\in B}\psi(xz)\right| &\le p^{r-s}\prod_{\substack{j=0\\k_j<p-1}}\left(k_j+1+\sum_{u=1}^{p-1}\frac{1}{\sin(u\pi/p)}\right) \\
&\le p^{r-s}\prod_{\substack{j=0\\k_j<p-1}}^{r-1}(k_j+1+p\log p) \le q\left(1-\frac{1}{p}+\log p\right)^s,
\end{aligned}$$

which yields the desired upper bound. □

Now we prove the main results.

Proof of Theorem 1. Put $L := L(s_i, N)$ and let l, L_l, and n be the integers defined by

$$p^l \le L < p^{l+1}, \quad L_l p^l \le L < (L_l + 1)p^l,$$

and

$$p^n \le N < p^{n+1}.$$

Since otherwise the result is trivial we may assume $r \ge 2$ and

$$n \ge \frac{r+1}{2}. \tag{4}$$

Put $c_0 = 1$. By (1) there exist $c_1, c_2, \ldots, c_L \in \mathbb{Z}_d$ such that

$$\sum_{j=0}^{L} c_j \mathrm{ind}_{\gamma,d}(\xi_{i-j}) = 0 \quad \text{for all } L \le i \le N-1. \tag{5}$$

By (2) we have

$$\xi_{i-j} = \xi_i - \xi_j \quad \text{for all } 0 \le j \le L$$

if

$$i = i_0 + i_1 p + \ldots + i_{r-1} p^{r-1}$$

with

$$i_0 = i_1 = \ldots = i_{l-1} = p-1, \quad L_l \le i_l \le p-1,$$

and

$$0 \le i_{l+1}, \ldots, i_{n-1} \le p-1.$$

For $i \ne L$ of this form, (5) is equivalent to

$$\chi\left(\prod_{j=0}^{L} (\xi_i - \xi_j)^{c_j}\right) = 1,$$

where χ denotes a multiplicative character of $\mathbb{F}_q$ of order d. Put

$$S_1 := \left| \sum_{i_{l+1},\ldots,i_{n-1}=0}^{p-1} \chi\left(\prod_{j=0}^{L} (\xi_i - \xi_j)^{c_j}\right) \right| \ge p^{n-l-1} - 1$$

and

$$S_2 := \left| \sum_{i_l=L_l}^{p-1} \sum_{i_{l+1},\ldots,i_{n-1}=0}^{p-1} \chi\left(\prod_{j=0}^{L} (\xi_i - \xi_j)^{c_j}\right) \right| \ge (p - L_l)p^{n-l-1} - 1.$$

For $l \le n - r/2 - 1$ we have

$$\begin{aligned} p^{n-l-1} - 1 &\le S_1 \le (L+1)q^{1/2}\left(1 - \frac{p^{n-l-1}}{(L+1)q}\right) \\ &\le (L+1)q^{1/2} - 1 \le p^{r/2+l+1} - 1 \end{aligned}$$

by Lemma 1. This yields $n - l - 1 \le r/2 + l + 1$ or equivalently since $l \in \mathbb{Z}$

$$l \ge \left\lceil \frac{2n-r}{4} \right\rceil - 1 = \frac{2n-r+s}{4} - 1, \tag{6}$$

where s denotes the least residue of $r - 2n$ modulo 4. For $l \ge n - r/2 - 1/2$ the lower bound (6) is also valid by (4). If $l \ge (2n-r)/4$ then we have

$$L \ge \frac{p^{n/2}}{q^{1/4}} > \frac{N^{1/2}}{p^{1/2}q^{1/4}}. \tag{7}$$

Hence we may assume $l = (2n - r + s)/4 - 1$ and thus

$$L \ge L_l \frac{p^{n/2+s/4-1}}{q^{1/4}} > L_l \frac{N^{1/2}p^{s/4-1}}{p^{1/2}q^{1/4}}. \tag{8}$$

For $p \le 3$ we get $L > \frac{1}{3}\frac{N^{1/2}}{p^{1/2}q^{1/4}}$ and for $s = 3$ and $p \le 251$ we obtain $L > \frac{1}{4}\frac{N^{1/2}}{p^{1/2}q^{1/4}}$. For $p \ge 5$ we use Lemma 1 again and get

$$\begin{aligned} (p - L_l)p^{n-l-1} &\le S_2 + 1 < (L+1)q^{1/2}\left(1 - \frac{1}{p} + \log p\right) + 1 \\ &\le (L_l+1)p^{r/2+l}\left(1 - \frac{1}{p} + \log p\right) + 1 \\ &< 1.05(L_l+1)p^{r/2+l}\left(1 - \frac{1}{p} + \log p\right), \end{aligned}$$

where we used $r \ge 2$ in the last step. With $l = (2n - r + s)/4 - 1$ we get

$$(p - L_l)p^{1-s/2} < 1.05(L_l + 1)\left(1 - \frac{1}{p} + \log p\right)$$

and thus

$$L_l > \frac{p^{2-s/2} - 1.05(1 + \log p)}{p^{1-s/2} + 1.05(1 + \log p)}.$$

We get

$$L_l > \begin{cases} p/2, & s = 0,\ p \ge 5, \\ p/3 > p^{3/4}/3, & s = 1,\ p \ge 5, \\ 0.19p/\log p > p^{1/2}/4, & s = 2,\ p \ge 5, \\ 0.45p^{1/2}/\log p > p^{1/4}/4 & s = 3,\ p \ge 257, \end{cases} \tag{9}$$

which completes the proof. □

Proof of Corollary 1. We proceed as in the proof of Theorem 1 and use the same notation. For $q \leq 256$ the result is trivial and we may assume $q > 256$. For $r = 1$ we have

$$L(s_i, p) > \frac{p^{1/2}}{2\log p} > \frac{p^{1/4}}{4}$$

by (10) below and we may assume $r \geq 2$. Analogously to (7) and (8) we get

$$L \geq \frac{p^{n/2}}{q^{1/4}} = \frac{N^{1/2}}{q^{1/4}} \quad \text{if } l \geq \frac{2n-r}{4}$$

and

$$L \geq L_l \frac{p^{n/2+s/4}}{pq^{1/4}} = L_l \frac{N^{1/2}p^{s/4}}{pq^{1/4}} \quad \text{if } l = \frac{2n-r+s}{4} - 1$$

and the result follows with (9). □

3. Final Remarks

The results of this paper can be extended to some extent to sequences (s_i) over $\mathbb{Z}_m$ defined by

$$s_i := \mathrm{ind}_{\gamma,m}(\xi_i), \quad 1 \leq i \leq q-1,$$

where m is not a divisor of $q-1$ in the following way. We restrict ourselves to the case that $r = 1$ and $\xi_i = i$. We proceed as in the proof of Theorem 1 and start with an analog of (5),

$$\sum_{j=0}^{L} c_j \mathrm{ind}_{\gamma,m}(i-j) \quad \text{for all } L \leq i \leq N-1$$

or equivalently

$$\prod_{j=0}^{L} \mathrm{e}_m(c_j \mathrm{ind}_\gamma(i-j)) = 1,$$

where $\mathrm{e}_k(x) = \exp(2\pi\sqrt{-1}x/k)$. Now for x with $|x| \leq p-1$ we have

$$0 \leq \left|\frac{x}{d} - \frac{x}{m}\right| \leq \frac{|m-d|(p-1)}{dm}.$$

Since $|\mathrm{e}(u) - 1| \leq 2\pi|u|$ for real u, we have for any positive divisor of $q-1$,

$$|\mathrm{e}_m(x) - \mathrm{e}_d(x)| \leq \frac{2\pi|m-d|(p-1)}{dm}.$$

Hence, we get

$$\begin{aligned} N-L-1 &= \left|\sum_{i=L+1}^{N-1}\prod_{j=0}^{L} \mathrm{e}_m(c_j \mathrm{ind}_\gamma(i-j))\right| \\ &\le \frac{2\pi(N-L-1)|m-d|(p-1)}{dm} + \left|\sum_{i=L+1}^{N-1}\prod_{j=0}^{L} \mathrm{e}_d(c_j \mathrm{ind}_\gamma(i-j))\right| \\ &< \frac{2\pi(N-L-1)|m-d|(p-1)}{dm} + (L+1)p^{1/2}\log p, \end{aligned}$$

where we used the fact that $\mathrm{e}_d(\mathrm{ind}_\gamma(x))$ is a multiplicative character. In particular, for the most interesting case $m = p$ and $d = p-1$ we get

$$L(s_i, N) > \left(1 - \frac{2\pi}{p}\right)\frac{N}{1 - 2\pi/p + p^{1/2}\log p} - 1.$$

For $r \le 4$ Theorem 1 and Corollary 1 can be improved with a slight modification of the proof. In particular we have

$$L(s_i, N) > \frac{N}{1 + p^{1/2}\log p} - 1 \quad \text{if } r = 1 \tag{10}$$

by [17, Theorem 9.2] (see also [16]). For large d the lower bound of [8] (see also [17, Theorem 9.1]) of the order of magnitude $Nd/(p\log d)$ is better than (10). The method of [8] can be extended to $r > 1$ but yields a somewhat weaker lower bound (cf. [12, Theorem 3]). This method can also be extended to the case that d is not a divisor of $q-1$.

If we define $s_0 = 0$ and continue the sequence with period q then for d with $d\not|(q-1)/(p-1)$ we know the following lower bound on the linear complexity [12, Theorem 1],

$$L(s_i) \ge \begin{cases} r(p-1)(d-1)/d + \varepsilon & \text{if } d \text{ is a } d\text{th power in } \mathbb{F}_q, \\ r(p-1) + \varepsilon & \text{otherwise,} \end{cases}$$

where

$$\varepsilon = \begin{cases} 0 & \text{if } d > 2 \text{ or } d = 2 \text{ and } q \equiv 1 \bmod 4, \\ 1 & \text{otherwise,} \end{cases}$$

which improves Corollary 1 if $r \le 4$. For $r = 1$ we have equality (see [1, 3, 4]). In this case the sequence (s_i) is also called *dth-order cyclotomic generator.*

We can identify the discrete logarithm via p-adic expansion with representations of the elements of $\mathbb{F}_q$ with respect to some fixed basis. Then we can also prove lower bounds on the linear complexity of sequences over $\mathbb{F}_q$ related to the discrete logarithm [12, Theorem 5]. Similarly one can investigate sequences over $\mathbb{F}_d$, where d is a prime power divisor of $q-1$ (see [5]).

Besides the linear complexity (profile) the autocorrelation is an important measure for the randomness of sequences. For results on the autocorrelation of the sequences (s_i) investigated in this paper see [3, 6, 13].

Generalized Sidelńikov sequences (σ_i) are the $q-1$ periodic sequences defined by

$$\sigma_i = \chi_d(\gamma^i - 1), \quad i = 0, 1, \ldots, q-2,$$

where χ_d is a multiplicative character of $\mathbb{F}_q$ of order d. Equivalently, we can define $q-1$ periodic sequences (s_i) over $\mathbb{Z}_d$ by

$$s_i = \mathrm{ind}_{\gamma,d}(\gamma^i - 1), \quad i = 0, 1, \ldots, q-2.$$

In the following we use the term (generalized) Sidelńikov sequences for the sequences over $\mathbb{Z}_d$. The autocorrelation of (generalized) Sidelńikov sequences was determined in [10, 11, 18]. The method of this paper combined with the method of Polya and Vinogradov provides a lower bound on the linear complexity profile of the order of magnitude $N/(q^{1/2} \log q)$. (See e. g. [17, Lemma 3.3] for the used character sum bound.) For first results on the linear complexity in the binary case see [7, 9]. Although Sidelńikov sequences and the sequences investigated in this paper look very similar at first glance, the latter sequences have some advantages. For large p and $r > 1$ we have two different nice autocorrelation properties [13, Theorems 1 and 2], and the ordering ξ_i, $i = 0, 1, \ldots, q-1$, can be faster generated than the ordering $\gamma^i - 1$, $i = 0, 1, \ldots, q-2$.

Acknowledgment
The author is supported by DSTA research grant R-394-000-011-422.

References

[1] T.W. Cusick, C. Ding, and A. Renvall, Stream Ciphers and Number Theory. Amsterdam: North-Holland, 1998.

[2] H. Davenport and D.J. Lewis, Character sums and primitive roots in finite fields. Rend. Circ. Mat. Palermo (2) **12** (1963), 129–136.

[3] C. Ding and T. Helleseth, On cyclotomic generator of order r, Inform. Process. Lett. **66** (1998), 21–25.

[4] C. Ding, T. Helleseth, and W. Shan, On the linear complexity of Legendre sequences, IEEE Trans. Inform. Th. **44** (1998), 1276–1278.

[5] Z. Dai, J. Yang, G. Gong, and P. Wang, On the linear complexity of generalised Legendre sequence. Sequences and their applications (Bergen, 2001), 145–153, Discrete Math. Theor. Comput. Sci. (Lond.), Springer, London, 2002.

[6] T. Helleseth, On the crosscorrelation of m-sequences and related sequences with ideal autocorrelation. Sequences and their applications (Bergen, 2001), 34–45, Discrete Math. Theor. Comput. Sci. (Lond.), Springer, London, 2002.

[7] T. Helleseth and K. Yang, On binary sequences of period $n = p^m - 1$ with optimal autocorrelation. Sequences and their applications (Bergen, 2001), 209–217, Discrete Math. Theor. Comput. Sci. (Lond.), Springer, London, 2002.

[8] S. Konyagin, T. Lange, and I. Shparlinski, Linear complexity of the discrete logarithm, Designs, Codes, and Cryptography **28** (2003), no. 2, 135–146.

[9] G. M. Kyureghyan and A. Pott, On the linear complexity of the Sidelnikov-Lempel-Cohn-Eastman sequences, Designs, Codes, and Cryptography **29** (2003), 149–164.

[10] A. Lempel, M. Cohn, and W. L. Eastman, A class of balanced binary sequences with optimal autocorrelation properties. IEEE Trans. Information Th. **23** (1977), no. 1, 38–42.

[11] H.D. Lüke, H.D. Schotten, and H. Hadinejad-Mahram, Generalized Sidelnikov sequences with optimal autocorrelation properties. Electronic Letters **36** (2000), no. 6, 525–527.

[12] W. Meidl and A. Winterhof, Lower bounds on the linear complexity of the discrete logarithm in finite fields, IEEE Transactions on Information Theory **47** (2001), 2807–2811.

[13] W. Meidl and A. Winterhof, On the autocorrelation of cyclotomic generators. Finite fields and applications (Toulouse, 2003), to appear.

[14] A.J. Menezes, P.C. van Oorschot, and S.A. Vanstone, Handbook of applied cryptography. CRC Press Series on Discrete Mathematics and its Applications. CRC Press, Boca Raton, FL, 1997.

[15] W.M. Schmidt, Equations over finite fields. An elementary approach. Lecture Notes in Mathematics, Vol. 536. Springer-Verlag, Berlin-New York, 1976.

[16] I.E. Shparlinski, Number Theoretic Methods in Cryptography. Basel: Birkhäuser, 1999.

[17] I.E. Shparlinski, Cryptographic Applications of Analytic Number Theory. Basel: Birkhäuser, 2003.

[18] V.M. Sidelńikov, Some k-valued pseudo-random sequences and nearly equidistant codes. Problems of Information Transmission **5** (1969), no. 1, 12–16.; translated from Problemy Peredači Informacii **5** (1969), no. 1, 16–22 (Russian).

[19] A. Tietäväinen, Vinogradov's method and some applications. Number theory and its applications (Ankara, 1996), 261–282, Lecture Notes in Pure and Appl. Math., 204, Dekker, New York, 1999.

[20] A. Winterhof, On the distribution of powers in finite fields. Finite Fields Appl. **4** (1998), no. 1, 43–54.

[21] A. Winterhof, Some estimates for character sums and applications. Des. Codes Cryptogr. **22** (2001), no. 2, 123–131.

[22] A. Winterhof, Incomplete additive character sums and applications. Finite fields and applications (Augsburg, 1999), 462–474, Springer, Berlin, 2001.

Arne Winterhof
Temasek Laboratories, National University of Singapore
10 Kent Ridge Crescent, Singapore 119260, Republic of Singapore
and

Johann Radon Institute for Computational and Applied Mathematics
Austrian Academy of Sciences
c/o Johannes Kepler University Linz, Altenbergerstraße 69
A-4040 Linz, Austria
e-mail: `tslwa@nus.edu.sg` and `arne.winterhof@oeaw.ac.at`

Progress in Computer Science and Applied Logic, Vol. 23, 369–383

Speeding Up RSA and Elliptic Curve Systems by Choosing Suitable Moduli

Huapeng Wu, M. Anwar Hasan and Ian F. Blake

Abstract. In this paper we propose a method to speed up the modular operation by choosing suitable moduli. When the modulus N can be represented as a sum of a few positive or negative powers of 2, we show that a modular operation ($X \bmod N$), where X is not greater than the square of the modulus N, can be computed with a few addition/subtraction operations with the operands of about the same size as the modulus. No evidence has been shown that use of such moduli in RSA and elliptic curve cryptosystems can compromise the security of the systems.

Mathematics Subject Classification (2000). Primary 68W99; Secondary 94A60.

Keywords. Modular operation, number system, efficient computation.

1. Introduction

Modular reduction is essential in computation over an integer ring $\mathbb{Z}_n$. Consequently it has important applications in residue number system and number theory transforms. There are also many public-key cryptography systems that involve extensive computation of integer modular reduction operation, for example, RSA, elliptic curve system defined over field of odd characteristic, XTR and GH systems.

It is well known that integers of certain forms are more suited for modular reduction The best examples are the Mersenne numbers and the Fermat numbers [4]. Families of the generalized Mersenne numbers and generalized Fermat numbers have also been discussed in many publications, for example, [6, 3, 2]. Other proposals for efficient integer modular reduction include [5, 1]. The main difference between our proposal and the above methods is that the former provides us with only an *incomplete solution.* Subsequently, other methods can still be applied to this incomplete solution to further speedup the operation.

We organize the paper as follows. Main results are shown in Section 2. In Section 3, an algorithm which can partially evaluate modular reduction operation is presented. A proof of the correctness of the algorithm is also provided in the same section. When the modulus N has a binary SD form of Hamming weight 3 the incomplete solutions to the modular reduction are derived in Section 4. Finally, concluding remarks are made in the Section 5.

2. Main Results

We first define certain notations.

1. Let a signed-digit binary representation of the modulus N be given by:[1]
$$N = 2^n + s'_{e_{w-2}} 2^{e_{w-2}} + s'_{e_{w-3}} 2^{e_{w-3}} + \cdots + s'_{e_1} 2^{e_1} + s'_{e_0} 2^{e_0}, \tag{2.1}$$
where $n > e_{w-2} > e_{w-3} > \cdots > e_1 > e_0 = 0$ and $s'_{e_i} \in \{-1, 1\}$, $i = 0, 1, \ldots, w-2$. Clearly the Hamming weight of the signed-digit representation of N is w. We write expression (2.1) as follows
$$2^n \bmod N \equiv s_{e_{w-2}} 2^{e_{w-2}} + s_{e_{w-3}} 2^{e_{w-3}} + \cdots + s_{e_0} 2^{e_0}, \tag{2.2}$$
where $s_j = -s'_j$ for $j = e_i$ and $i = 0, 1, \ldots, w-2$.
2. Let $d_i = e_i - e_{i-1}$ for $i = 1, 2, \ldots, w-1$, where $e_{w-1} = n$. Then the binary form of X can be partitioned and given by
$$\begin{aligned} X &= A_w \mid\mid A_{w-1} \mid\mid \cdots \mid\mid A_4 \mid\mid A_3 \mid\mid A_2 \mid\mid A_1 \\ &= (((\cdots(A_w \times 2^{d_2} + A_{w-1}) \times 2^{d_3} + \cdots + A_4) \\ &\quad \times 2^{d_{w-2}} + A_3) \times 2^{d_{w-1}} + A_2) \times 2^n + A_1, \end{aligned} \tag{2.3}$$
where the symbol "$\mid\mid$" denotes a cascade operation.

Then the main results obtained in this paper can be summarized into the following two lemmas.

Lemma 2.1. *Let N be given in (2.2) and X be given in (2.3), respectively. Let*
$$B_i = A_w \mid\mid A_{w-1} \mid\mid \cdots \mid\mid A_{i+1} \quad \text{for } i = 1, 2, \ldots, w-1. \tag{2.4}$$
If $e_{w-2} < \frac{n+1}{2}$, then the modular operation $X \pmod N$ can be solved with the following congruence.
$$X \pmod N \equiv A_1 + \sum_{i=0}^{w-2} s_{e_i} 2^{e_i} [(A_{w-i} \mid\mid A_{w-(i+1)} \mid\mid \cdots \mid\mid A_2) + \sum_{i=1}^{w-2} s_{e_i} B_{w-i}]. \tag{2.5}$$

Lemma 2.2. *Let $N = 2^n - s_m 2^m - s_0$, where $0 < m < n$ and $s_m, s_0 \in \{-1, 1\}$. Let k be an integer determined by n and m such that $\frac{n}{n-m} \leqslant k < \frac{n}{n-m} + 1$. We partition the binary number X, $0 \leqslant X < N^2$ as follows*
$$\begin{aligned} X &= A_{k+1} \mid\mid A_k \mid\mid \cdots \mid\mid A_3 \mid\mid A_2 \mid\mid A_1 \\ &= ((\cdots(A_{k+1} \times 2^{n-m} + A_k) \times 2^{n-m} + \cdots + A_3) \times 2^{n-m} + A_2) \times 2^n + A_1. \end{aligned}$$

[1]If N is given in the non-adjacent form (NAF), the maximum efficiency using this method can be achieved.

Then the following congruence holds

$$X \pmod N \equiv A_1 + 2^m \sum_{i=2}^{k+1} s_m^{i-1} A_i + s_0 \sum_{i=2}^{k+1} s_m^{i-2}(A_{k+1}||A_k|| \cdots ||A_i). \qquad (2.6)$$

In the following we give a simple example to show how these results work for efficient computation of the modular operation.

Example. Let $N = 2^n + 2^m - 1$ and $m < \frac{n+1}{2}$. It can be seen that this case is covered by both lemmas. Let us use the incomplete solution obtained in Lemma 2.1.

From (2.3) a partition of binary number X is given by

$$X = (A_3 \times 2^{n-m} + A_2) \times 2^n + A_1 = A_3 \mid\mid A_2 \mid\mid A_1.$$

Then it follows from (2.4) $B_1 = A_3 \mid\mid A_2$ and $B_2 = A_3$. Since $w = 3$, $s_{e_1} = s_m = -1$ and $s_{e_0} = 1$, it follows from Lemma 2.1,

$$\begin{aligned} X \pmod N &\equiv A_1 + \sum_{i=0}^{1} s_{e_i} 2^{e_i} [(A_{w-i} \mid\mid A_{w-(i+1)} \mid\mid \cdots \mid\mid A_2) + \sum_{i=1}^{1} s_{e_i} B_{w-i}] \\ &\equiv A_1 + (A_3 \mid\mid A_2) - 2^m A_2 + (2^m - 1) B_2 \\ &\equiv A_1 + (A_3 \mid\mid A_2) - 2^m A_2 + (2^m - 1) A_3 \\ &\equiv A_1 + (A_3 \mid\mid A_2) - 2^m (A_2 - A_3) - A_3. \end{aligned} \qquad (2.7)$$

We can estimate the complexity of solving this modular operation using (2.7). Clearly, we have $0 \le A_1 < 2^n < N$. To decide the range of $(A_3 \mid\mid A_2)$, let us consider

$$N^2 = (2^n + 2^m - 1)^2 = 2^{2n} + 2^{n+m+1} + 2^{2m} - 2^{n+1} - 2^{m+1} + 1,$$

from $X = (A_3||A_2||A_1) < N^2$ and note $m < \frac{n+1}{2}$, we have $(A_3||A_2) < 2^n + 2^{m+1} - 1$. It follows that $0 \le A_3 \le 2^m + 1$, $0 \le A_2 < 2^{n-m}$ and $|2^m(A_2 - A_3)| < 2^n$. Then it can be seen that not more than five modular additions (subtractions) are required to solve the modular reduction.

3. An Algorithm for Modular Reduction

A proof of the two lemmas in Section 2 can be given in three steps: Firstly we present an algorithm which can partially solve the modular operation. Then a proof of the correctness of the algorithm is given. Finally, we obtain the incomplete solutions (2.5) and (2.6) by analyzing the output of the proposed algorithm. In this section we describe the first two steps and the final step is discussed in the following sections.

3.1. A New Algorithm for Partial Evaluation of Modular Operation

Consider the modular operation $Y \equiv X \bmod N$, where the NAF of N is given in (2.2). Let X be an $n + t$-bit integer. The basic idea of the algorithm is as follows.

For ith bit of $X = a_{n-1}a_{n-2}\dots a_0$, $i = 0, 1, \dots, n+t-1$, we introduce a *weight list* ℓ_i. First, the weight lists are initialized as $\ell_i = \langle a_i \rangle$.

From (2.2), it follows

$$2^{n+j} \bmod N \equiv s_{e_{w-2}} 2^{e_{w-2}+j} + s_{e_{w-3}} 2^{e_{w-3}+j} + \dots + s_{e_0} 2^{e_0+j}, \ j = 0, 1, \dots, t-1. \tag{3.1}$$

Then the weight lists ℓ_i, $i = 0, 1, \dots, n+t-1$, are updated based on (3.1) in a manner that the final form of the weight lists has the following properties:

1. Each ℓ_i contains a_i and several other a_j's or their negatives for $i=0,1,\dots,n+t-1$.
2. Let S_{ℓ_i} be the sum of all the numbers in the finally updated ℓ_i and be obtained one by one in the order that i decreases. Then the value of $X \bmod N$ is decided only by $S_{\ell_0}, S_{\ell_2}, \dots, S_{\ell_{n-1}}$. In fact, an incomplete solution for the modular reduction is

$$X \bmod N \equiv \sum_{i=0}^{n-1} S_{\ell_i} 2^i.$$

The process of updating a weight list depends on the distribution of nonzero bits in $2^n \bmod N$. When the jth bit of $2^{n+i} \bmod N$ is not a zero, we append the number a_{n+i} to the weight list ℓ_j. When we finish the updating process with all the nonzero bits in $2^{n+i} \bmod N$, and for $i = t-1, t-2, \dots, 1, 0$, the weight lists are referred to as the *prepared weight lists.*

The updating of weight lists can be considered as the *precomputation* part of the algorithm, where the value of X is not required. In the second part of the algorithm (Main Program), with the prepared weight lists and value of X available, we can obtain an incomplete solution to $X \pmod N$ by summing up the terms in the weight list ℓ_i sequentially from $i = n+t-1$ to 0 and viewing the sum as the binary coefficient of B at weight position 2^i. The algorithm is given below:

Algorithm 3.1. Partial Evaluation of Modular Reduction

Part 1. *Precomputation: Setup a prepared weight list*
Input: N,
Output: Prepared weight lists $\ell_0, \ell_1, \dots, \ell_{n+t-1}$.
1. *Initialization of weight lists:*
$2^j : \ \ell_j := \langle a_j \rangle, \ j = 0, 1, \dots, n+t-1.$
/* *Note that* a_j *is a variable here.* */
2. *Obtain the minimal weight SD representation of* $2^n \bmod N$:

$$2^n \bmod N \equiv s_{e_{w-2}} 2^{e_{w-2}} + s_{e_{w-3}} 2^{e_{w-3}} + \dots + s_{e_0} 2^{e_0},$$

where $n > e_{w-2} > e_{w-3} > \dots > e_0 \geqslant 0$ *and* $s_{e_i} \in \{-1, 1\}$.
3. *Obtain the prepared weight lists:*
For $i := t-1$ *To* 0, *Step* -1
For $j := 0$ *To* $w-2$
Append $s_{e_j} a_{n+i}$ *as one element to the weight list* ℓ_{i+e_j} .

Part 2. *Main Program*

Input: X and the prepared weight lists $\ell_0, \ell_1, \ldots, \ell_{n+t-1}$.

Output: Incomplete solution $Y \equiv X \pmod N$.

1. *For* $i := n+t-1$ *To* 0, *Step* -1

   $a_i :=$ *the sum of all the elements in* ℓ_i;

2. $Y := a_0; d := 1;$

   For $i := 1$ *To* $n-1$,

   $d := d \times 2;$

   $Y := Y + d \times a_i;$

3.2. A Proof for the Algorithm

A proof of the correctness of Algorithm 3.1 is given as follows.

Proof. Let

$$X = \sum_{i=0}^{n+t-1} a_i 2^i.$$

From (2.2) it follows

$$\begin{aligned}
\sum_{i=0}^{n-1} a_i 2^i &= a_{n-1}2^{n-1} + a_{n-2}2^{n-2} + \cdots + a_1 2 + a_0 \\
a_n 2^n \bmod N &\equiv a_n s_{e_{w-2}} 2^{e_{w-2}} + a_n s_{e_{w-3}} 2^{e_{w-3}} + \cdots + a_n s_{e_0} 2^{e_0} \\
a_{n+1} 2^{n+1} \bmod N &\equiv a_{n+1} s_{e_{w-2}} 2^{e_{w-2}+1} + a_{n+1} s_{e_{w-3}} 2^{e_{w-3}+1} + \cdots \\
&\quad + a_{n+1} s_{e_0} 2^{e_0+1} \\
a_{n+2} 2^{n+2} \bmod N &\equiv a_{n+2} s_{e_{w-2}} 2^{e_{w-2}+2} + a_{n+2} s_{e_{w-3}} 2^{e_{w-3}+2} + \cdots \\
&\quad + a_{n+2} s_{e_0} 2^{e_0+2} \\
&\vdots \quad \vdots \quad \vdots \\
a_{n+t-2} 2^{n+t-2} \bmod N &\equiv a_{n+t-2} s_{e_{w-2}} 2^{e_{w-2}+t-2} + a_{n+t-2} s_{e_{w-3}} 2^{e_{w-3}+t-2} + \cdots \\
&\quad + a_{n+t-2} s_{e_0} 2^{e_0+t-2} \\
a_{n+t-1} 2^{n+t-1} \bmod N &\equiv a_{n+t-1} s_{e_{w-2}} 2^{e_{w-2}+t-1} + a_{n+t-1} s_{e_{w-3}} 2^{e_{w-3}+t-1} + \cdots \\
&\quad + a_{n+t-1} s_{e_0} 2^{e_0+t-1}
\end{aligned} \tag{3.2}$$

Clearly, if we sum up the above $t+1$ congruences together, the left-hand side of the congruence is $X \bmod N$. On the other hand, it can be seen that there are totally $tw+n$ terms on the right-hand side of the above $t+1$ congruences (3.2). Let us rewrite these $tw+n$ terms in the following array of $t+1$ rows with the column on the far left being the indices. Clearly the sum of the $tw+n$ terms is congruent to X modulo N. Now we want to reduce the weight of all $tw+n$ terms to 2^{n-1} or less. Let us start with the term of the highest weight.

$$\begin{array}{llllll}
(-1): & a_{n-1}2^{n-1} & a_{n-2}2^{n-2} & \cdots & a_0 \\
(0): & a_n s_{e_{w-2}} 2^{e_{w-2}} & a_n s_{e_{w-3}} 2^{e_{w-3}} & \cdots & a_n s_{e_0} 2^{e_0} \\
(1): & a_{n+1} s_{e_{w-2}} 2^{e_{w-2}+1} & a_{n+1} s_{e_{w-3}} 2^{e_{w-3}+1} & \cdots & a_{n+1} s_{e_0} 2^{e_0+1} \\
\vdots & \vdots & \vdots & \vdots & \vdots \\
(i): & a_{n+i} s_{e_{w-2}} 2^{e_{w-2}+i} & a_{n+i} s_{e_{w-3}} 2^{e_{w-3}+i} & \cdots & a_{n+i} s_{e_0} 2^{e_0+i} \\
\vdots & \vdots & \vdots & \vdots & \vdots \\
(t-2): & a_{n+t-2} s_{e_{w-2}} 2^{e_{w-2}+t-2} & a_{n+t-2} s_{e_{w-3}} 2^{e_{w-3}+t-2} & \cdots & a_{n+t-2} s_{e_0} 2^{e_0+t-2} \\
(t-1): & a_{n+t-1} s_{e_{w-2}} 2^{e_{w-2}+t-1} & a_{n+t-1} s_{e_{w-3}} 2^{e_{w-3}+t-1} & \cdots & a_{n+t-1} 2^{e_0} 2^{e_0+t-1}
\end{array} \quad (3.3)$$

Obviously, the term at the bottom-left corner $a_{n+t-1}s_{e_{w-2}}2^{e_{w-2}+t-1}$ is of the highest weight among the $tw+n$ terms. From Step 3, Part 1 of Algorithm 3.1, it can be seen that the prepared weight lists $\ell_{e_{w-2}+t-1} = \langle a_{e_{w-2}+t-1}, s_{e_{w-2}}a_{n+t-1}\rangle$, and $\ell_{e_{w-2}+t+i} = \langle a_{e_{w-2}+t+i}\rangle, i \geqslant 0$. Subsequently, in Step 1 of Main Program, the first addition operation is performed and then the value of $a_{e_{w-2}+t-1}$ is set to be the sum.

$$a_{e_{w-2}+t-1} := a_{e_{w-2}+t-1} + s_{e_{w-2}}a_{n+t-1}. \quad (3.4)$$

For the purpose of avoiding ambiguousness, we put a prime on the updated coefficient $a_{e_{w-2}+t-1}$ and rewrite (3.4) as follows

$$a'_{e_{w-2}+t-1} := a_{e_{w-2}+t-1} + s_{e_{w-2}}a_{n+t-1}.$$

This step is equivalent to removing the term $a_{n+t-1}2^{e_{w-2}+t-1}$ from the array (3.3), and updating all the terms with coefficient $a_{e_{w-2}+t-1}$. That is, to change the $(e_{w-2}+t-n-1)^{\text{st}}$ row in the array into

$$a'_{e_{w-2}+t-1}2^{2e_{w-2}-n+t-1} + a'_{e_{w-2}+t-1}2^{e_{w-2}+e_{w-3}-n+t-1} + \cdots +$$
$$a'_{e_{w-2}+t-1}2^{e_{w-2}+e_1-n+t-1} + a'_{e_{w-2}+t-1}2^{e_{w-2}+e_0-n+t-1},$$

where $a'_{e_{w-2}+t-1} = a_{e_{w-2}+t-1} + s_{e_{w-2}}a_{n+t-1}$. Thus, it can be seen that the highest weight in the array is now reduced to $e_{w-2}+t-2$. Again, in Step 3, Part 1, it can be seen that the prepared weight list

$$\ell_{e_{w-2}+t-2} = \begin{cases} \langle a_{e_{w-2}+t-2}, s_{e_{w-2}}a_{n+t-2}\rangle & \text{if } e_{w-2} > e_{w-3}+1, \\ \langle a_{e_{w-2}+t-2}, s_{e_{w-2}}a_{n+t-2}, s_{e_{w-3}}a_{n+t-1}\rangle & \text{if } e_{w-2} = e_{w-3}+1, \end{cases}$$

We first discuss the case of $e_{w-2} > e_{w-3}+1$.

1. If $e_{w-2} > e_{w-3}+1$, then, in Step 1, Part 2, it performs

$$a'_{e_{w-2}+t-2} := a_{e_{w-2}+t-2} + s_{e_{w-2}}a_{n+t-2}.$$

   It is equivalent to removing $a_{n+t-2}2^{e_{w-2}+t-2}$ from the array (3.3) and updating all the terms with coefficient $a_{e_{w-2}+t-2}$ by changing the coefficient into $a'_{e_{w-2}+t-2} = a_{e_{w-2}+t-2} + a_{n+t-2}$. When this is done, the highest weight in the array (3.3) is reduced to $e_{w-2}+t-3$.

2. If $e_{w-2} = e_{w-3} + 1$, in Step 1, Part 2, it performs

$$a'_{e_{w-2}+t-2} := a_{e_{w-2}+t-2} + s_{e_{w-2}} a_{n+t-2} + s_{e_{w-3}} a_{n+t-1}.$$

It is equivalent to removing two terms $a_{n+t-2} s_{e_{w-2}} 2^{e_{w-2}+t-2}$ and $a_{n+t-1} s_{e_{w-3}} 2^{e_{w-3}+t-1}$ from the array (3.3), and updating all the terms with coefficient $a_{e_{w-2}+t-2}$ by changing the coefficient into $a'_{e_{w-2}+t-2} = a_{e_{w-2}+t-2} + s_{e_{w-2}} a_{n+t-2} + s_{e_{w-3}} a_{n+t-1}$. When it is done the highest weight in the array (3.3) is $e_{w-2} + t - 3$.

Clearly, this removing and updating process does not alter the fact that the sum of all the terms in the updating array is congruent to Y modulo N. Then above removing and updating process continues until all the terms of weight higher than $n - 1$ have been removed from the array (3.3). Let us divide Step 1, Part 2, into two sub-steps as follows:

1. For $i := n + t - 1$ To n, Step -1
   $a_i :=$ the sum of all the elements in ℓ_i;
2. For $i := n - 1$ To 0, Step -1
   $a_i :=$ the sum of all the elements in ℓ_i.

Then it can be seen that the first sub-step corresponds to the removing job, while the second sub-step is responsible for updating the terms of weight lower than 2^n and summing them up to yield the output Y. Therefore, we have Y is congruent to X modulo N, or $Y \equiv X \bmod N$.

Since there may be multiple terms with value of either 1 or -1 at the weight position 2^i, the finally updated coefficient a_i can be out of the proper range of $\{0, 1\}$, where $i = n - 1, n - 1, \ldots, 0$. Consequently, the output Y can exceed the modulus N or have a negative value. □

4. Analysis of Algorithm's Output

We first consider the special case $w = 3$.

4.1. Family of Moduli of Form $N = 2^n \pm 2^m \pm 1$

Let N be given as in (2.2) for $w = 3$ and $e_1 = m$, and X be given in (2.3). Then we proceed with Algorithm 3.1 when it takes inputs of N and X.

Part 1. Precomputation: preparing the weight lists.

1. Initialize the weight lists:

$$2^i : \ell_i := \langle a_i \rangle, \quad i = 0, 1, \ldots, 2n - 1.$$

2. Obtain the binary expansion of $2^n \bmod N$:

$$2^n \bmod N \equiv s_m 2^m + s_0, \; s_0, s_m \in \{-1, 1\}.$$

3. Compute the prepared weight lists:
   It is easy to see that the prepared weight lists are as follows:

$$
\begin{array}{lclll}
\ell_{2n-1} & = & \langle a_{2n-1}\rangle & & \\
\vdots & \vdots & \vdots & & \\
\ell_{n+m} & = & \langle a_{n+m}\rangle & & \\
\ell_{n+m-1} & = & \langle a_{n+m-1}, & s_m a_{2n-1}\rangle & \\
\ell_{n+m-2} & = & \langle a_{n+m-2}, & s_m a_{2n-2}\rangle & \\
\vdots & \vdots & \vdots & \vdots & \\
\ell_{n} & = & \langle a_{n}, & s_m a_{2n-m}\rangle & \\
\ell_{n-1} & = & \langle a_{n-1}, & s_m a_{2n-m-1}, & s_0 a_{2n-1}\rangle \\
\ell_{n-2} & = & \langle a_{n-2}, & s_m a_{2n-m-2}, & s_0 a_{2n-2}\rangle \\
\vdots & \vdots & \vdots & \vdots & \vdots \\
\ell_{m} & = & \langle a_{m}, & s_m a_{n}, & s_0 a_{n+m}\rangle \\
\ell_{m-1} & = & \langle a_{m-1}, & & s_0 a_{n+m-1}\rangle \\
\vdots & \vdots & \vdots & & \vdots \\
\ell_{0} & = & \langle a_{0}, & & s_0 a_{n}\rangle
\end{array}
$$

Part 2. Main Program

We can divide the first step of Main Program into three sub-steps as follows:

1. For $i := n+m-1$ To n, Step -1,
   $a_i := a_i + s_m a_{n-m+i}$;
2. For $i := n-1$ To m, Step -1,
   $a_i := a_i + s_m a_{n-m+i}$;
3. For $i := n-1$ To 0, Step -1,
   $a_i := a_i + s_0 a_{n+i}$.

Let

$$
\begin{aligned}
X &= A_3 \,||\, A_2 \,||\, A_1 \\
&= (A_3 \times 2^{n-m} + A_2) \times 2^n + A_1,
\end{aligned}
$$

and $B_1 = A_3 \,||\, A_2$ and $B_2 = A_3$. Then it can be seen that Step 1 of the Main Program performs $B_1 := B_1 + s_m A_3$ if $m < \frac{n+1}{2}$. This is because, otherwise, say that $m = \frac{n+1}{2}$, the value of a_{2n-m} would not be an original bit of X when performing $a_n := a_n + s_m a_{2n-m}$, since it has been changed when $a_{n+m-1} := a_{n+m-1} + s_m a_{2n-1}$ is performed. Similar argument applies to the case that $m > \frac{n+1}{2}$. Step 2 performs $A_1 := A_1 + s_m 2^m (A_2 + s_m A_3)$ while Step 3 finishes with $A_1 := A_1 + s_0(B_1 + s_m A_3)$.

Note the order that each sub-step is performed. Then we have

$$\begin{aligned} X \pmod N &\equiv A_1 + s_m 2^m A_2 + 2^m A_3 + s_0(A_2 + A_3 \times 2^{n-m}) + s_0 s_m A_3 \\ &\equiv A_1 + (s_m 2^m + s_0)A_2 + (2^m + s_0 2^{n-m} + s_0 s_m)A_3 \\ &\equiv A_1 + s_0 B_1 + s_m 2^m A_2 + (2^m + s_0 s_m)A_3. \end{aligned}$$

We summarize the results in the following lemma.

Lemma 4.1. *Let N and X be given in (2.2) and (2.3), respectively. Let*

$$w = 3 \quad \textit{and} \quad e_1 = m < \frac{n+1}{2}.$$

Then the modular operation $X \pmod N$ can be partially solved as follows:

1. *When $N = 2^n - 2^m - 1$,*

$$X \bmod N \equiv A_1 + (A_3 || A_2) + 2^m A_2 + (2^m + 1)A_3.$$

2. *When $N = 2^n - 2^m + 1$,*

$$X \bmod N \equiv A_1 - (A_3 || A_2) + 2^m A_2 + (2^m - 1)A_3.$$

3. *When $N = 2^n + 2^m - 1$,*

$$X \bmod N \equiv A_1 + (A_3 || A_2) - 2^m A_2 + (2^m - 1)A_3.$$

4. *When $N = 2^n + 2^m + 1$,*

$$X \bmod N \equiv A_1 - (A_3 || A_2) - 2^m A_2 + (2^m - 1)A_3.$$

4.2. Family of Moduli of Form $\mathbf{2^n \pm 2^m \pm 1, \frac{n+1}{2} \leqslant m \leqslant \frac{2}{3}n}$

Let

$$N = 2^n \pm 2^m \pm 1, \frac{n+1}{2} \leqslant m \leqslant \frac{2}{3}n, \quad \text{and} \quad X, 0 \leqslant X < N^2,$$

be given in its binary form $X = (a_{2n-1}a_{2n-2}\dots a_1 a_0)_2$. Then we proceed with Algorithm 3.1 step by step when it takes inputs N and X.

1. *Initialize the weight lists:*

$$\langle a_i \rangle, \quad i = 0, 1, \dots, 2n-1.$$

2. *Obtain the prepared weight list:*
   It is easy to see that the prepared weight lists are as follows:

$$\begin{array}{lllll}
\langle a_{2n-1}\rangle & & & & \\
\vdots & & & & \\
\langle a_{n+m}\rangle & & & & \\
\langle a_{n+m-1}\rangle & := & \langle a_{n+m-1} & +s_m a_{2n-1}\rangle & \\
\langle a_{n+m-2}\rangle & := & \langle a_{n+m-2} & +s_m a_{2n-2}\rangle & \\
\vdots & \vdots & \vdots & & \\
\langle a_{n}\rangle & := & \langle a_{n} & +s_m a_{2n-m}\rangle & \\
\langle a_{n-1}\rangle & := & \langle a_{n-1} & +s_m a_{2n-m-1} & +s_0 a_{2n-1}\rangle \\
\langle a_{n-2}\rangle & := & \langle a_{n-2} & +s_m a_{2n-m-2} & +s_0 a_{2n-2}\rangle \\
\vdots & \vdots & \vdots & & \\
\langle a_{m}\rangle & := & \langle a_{m} & +s_m a_{n} & +s_0 a_{n+m}\rangle \\
\langle a_{m-1}\rangle & := & \langle a_{m-1} & & +s_0 a_{n+m-1}\rangle \\
\vdots & \vdots & \vdots & & \\
\langle a_{0}\rangle & := & \langle a_{0} & & +s_0 a_{n}\rangle
\end{array}$$

3. *Update and evaluate the list obtained from the previous step:*
   The updated weight list is shown in Fig. 1.
4. *Obtain incomplete solution:*
   Let

$$\begin{aligned} X &= A_4 \mid\mid A_3 \mid\mid A_2 \mid\mid A_1 \\ &= ((A_4 \times 2^{n-m} + A_3) \times 2^{n-m} + A_2) \times 2^n + A_1. \end{aligned} \tag{4.1}$$

   Then from the updated weight list it follows

$$\begin{aligned} X \bmod N \equiv\; & A_1 + 2^m s_m[A_2 + s_m(A_3 + s_m A_4)] \\ & + s_0[A_4||A_3||A_2 + s_m(A_4||A_3 + s_m A_4)]. \end{aligned} \tag{4.2}$$

   We summarize the results in the following lemma.

Lemma 4.2. *Let $N = 2^n - s_m 2^m - s_0$, where $s_0, s_m \in \{-1, 1\}$ and $\frac{n+1}{2} \leqslant m \leqslant \frac{2}{3}n$. Let the binary number X, $0 \leqslant X < N^2$, be partitioned and given in (4.1). Then the modular operation $X \pmod N$, $0 \leqslant X < N^2$, can be partially solved as follows:*

1. *When $N = 2^n - 2^m - 1$,*

$$X \bmod N \equiv A_1 + 2^m(A_2 + A_3 + A_4) + A_4||A_3||A_2 + A_4||A_3 + A_4.$$

2. *When $N = 2^n - 2^m + 1$,*

$$X \bmod N \equiv A_1 + 2^m(A_2 + A_3 + A_4) - A_4||A_3||A_2 - A_4||A_3 - A_4.$$

$$
\begin{array}{lllllllll}
\langle a_{2n-1}\rangle & & & & & & & & \\
\vdots & & & & & & & & \\
\langle a_{n+m}\rangle & & & & & & & & \\
\langle a_{n+m-1}\rangle & := & \langle a_{n+m-1} & +s_m a_{2n-1}\rangle & & & & & \\
\langle a_{n+m-2}\rangle & := & \langle a_{n+m-2} & +s_m a_{2n-2}\rangle & & & & & \\
\vdots & \vdots & \vdots & \vdots & & & & & \\
\langle a_{2n-m}\rangle & := & \langle a_{2n-m} & +s_m a_{3n-2m}\rangle & & & & & \\
\vdots & \vdots & \vdots & \vdots & & & & & \\
\langle a_{2m}\rangle & := & \langle a_{2m} & +s_m a_{n+m}\rangle & & & & & \\
\langle a_{2m-1}\rangle & := & \langle a_{2m-1} & +s_m(a_{n+m-1} & +s_m a_{2n-1})\rangle & & & & \\
\vdots & \vdots & \vdots & \vdots & \vdots & & & & \\
\langle a_{n}\rangle & := & \langle a_{n} & +s_m(a_{2n-m} & +s_m a_{3n-2m})\rangle & & & & \\
\langle a_{n-1}\rangle & := & \langle a_{n-1} & +s_m(a_{2n-m-1} & +s_m a_{3n-2m-1}) & & +s_0 a_{2n-1}\rangle & & \\
\vdots & \vdots & \vdots & \vdots & \vdots & & \vdots & & \\
\langle a_{3m-n}\rangle & := & \langle a_{3m-n} & +s_m(a_{2m} & +s_m a_{n+m}) & & +s_0 a_{3m}\rangle & & \\
\langle a_{3m-n-1}\rangle & := & \langle a_{3m-n-1} & +s_m[a_{2m-1} & +s_m(a_{n+m-1} & +s_m a_{2n-1})] & +s_0 a_{3m-1}\rangle & & \\
\vdots & \vdots & \vdots & \vdots & \vdots & \vdots & \vdots & & \\
\langle a_{m}\rangle & := & \langle a_{m} & +s_m[a_{n} & +s_m(a_{2n-m} & +s_m a_{3n-2m})] & +s_0 a_{n+m}\rangle & & \\
\langle a_{m-1}\rangle & := & \langle a_{m-1} & & & & +s_0(a_{n+m-1} & +s_m a_{2n-1})\rangle & \\
\vdots & \vdots & \vdots & & & & \vdots & \vdots & \\
\langle a_{2m-n}\rangle & := & \langle a_{2m-n} & & & & +s_0(a_{2m} & +s_m a_{n+m})\rangle & \\
\langle a_{2m-n-1}\rangle & := & \langle a_{2m-n-1} & & & & +s_0[a_{2m-1} & +s_m(a_{n+m-1} & +s_m a_{2n-1})]\rangle \\
\vdots & \vdots & \vdots & & & & \vdots & \vdots & \vdots \\
\langle a_{0}\rangle & := & \langle a_{0} & & & & +s_0[a_{n} & +s_m(a_{2n-m} & +s_m a_{3n-2m})]\rangle
\end{array}
$$

FIGURE 1. The updated weight list for the case $N = 2^n \pm 2^m \pm 1, \frac{n+1}{2} \leqslant m \leqslant \frac{2}{3}n$.

3. *When* $N = 2^n + 2^m - 1$,

$$X \bmod N \equiv A_1 - 2^m(A_2 - A_3 + A_4) + A_4||A_3||A_2 - A_4||A_3 + A_4.$$

4. *When* $N = 2^n + 2^m + 1$,

$$X \bmod N \equiv A_1 - 2^m(A_2 - A_3 + A_4) - A_4||A_3||A_2 + A_4||A_3 - A_4.$$

4.3. Family of Moduli of Form $\mathbf{2^n \pm 2^m \pm 1, 0 < m < n}$

Given n and m, we can determine an integer k, $2 \leqslant k \leqslant n$, such that $(k-1)(n-m) < n \leqslant k(n-m)$, or

$$\frac{n}{n-m} \leqslant k < \frac{n}{n-m} + 1. \tag{4.3}$$

When $k = 2$, it follows $0 < m \leqslant \frac{n}{2}$, which is the case discussed in Section 4.1. When $k = 3$, we have $\frac{n}{2} < m \leqslant \frac{2n}{3}$ and this case has been discussed in Section 4.2.

For the general case that $2 \leqslant k \leqslant n$, we proceed with Algorithm 3.1 step by step when it takes inputs of N and X.

1. *Step 1. Initialize the weight lists:*

$$\langle a_i \rangle, \quad i = 0, 1, \ldots, 2n-1.$$

2. *Step 2. Obtain the prepared weight list:*

$$\begin{array}{llllll}
\langle a_{2n-1}\rangle & & & & & \\
\vdots & & & & & \\
\langle a_{n+m}\rangle & & & & & \\
\langle a_{n+m-1}\rangle & := & \langle a_{n+m-1} & +s_m a_{2n-1}\rangle & & \\
\langle a_{n+m-2}\rangle & := & \langle a_{n+m-2} & +s_m a_{2n-2}\rangle & & \\
\vdots & \vdots & \vdots & & & \\
\langle a_{n}\rangle & := & \langle a_{n} & +s_m a_{2n-m}\rangle & & \\
\langle a_{n-1}\rangle & := & \langle a_{n-1} & +s_m a_{2n-m-1} & +s_0 a_{2n-1}\rangle & \\
\langle a_{n-2}\rangle & := & \langle a_{n-2} & +s_m a_{2n-m-2} & +s_0 a_{2n-2}\rangle & \\
\vdots & \vdots & \vdots & & & \\
\langle a_{m}\rangle & := & \langle a_{m} & +s_m a_{n} & +s_0 a_{n+m}\rangle & \\
\langle a_{m-1}\rangle & := & \langle a_{m-1} & & +s_0 a_{n+m-1}\rangle & \\
\vdots & \vdots & \vdots & & & \\
\langle a_{0}\rangle & := & \langle a_{0} & & +s_0 a_{n}\rangle &
\end{array}$$

3. *Step 3. Update and evaluate the prepared weight list:*
   It is shown in Fig. 2.

		1	2	3	$\cdots$	$k-1$	k	$k+1$	$k+2$	$k+3$	$k+4$	$\cdots$	$2k+1$
$\langle a_{2n-1}\rangle$													
$\vdots$													
$\langle a_{n+m}\rangle$													
$\langle a_{n+m-1}\rangle$	$:=$	$\langle a_{n+m-1}$	$+s_m a_{2n-1}\rangle$										
$\vdots$	$\vdots$	$\vdots$	$\vdots$										
$\langle a_{2m}\rangle$	$:=$	$\langle a_{2m}$	$+s_m a_{n+m}\rangle$										
$\langle a_{2m-1}\rangle$	$:=$	$\langle a_{2m-1}$	$+s_m(a_{n+m-1}$	$+s_m a_{2n-1})\rangle$									
$\vdots$	$\vdots$	$\vdots$	$\vdots$	$\vdots$									
$\langle a_{(k-1)m-(k-3)n}\rangle$	$:=$	$\langle a_{(k-1)m-(k-3)n}$	$+s_m(\cdots(a_{(k-2)m-(k-4)n}$	$+s_m(a_{(k-3)m-(k-5)n}$	$+\cdots$	$+s_m a_{m+n})\cdots)\rangle$							
$\langle a_{(k-1)m-(k-3)n-1}\rangle$	$:=$	$\langle a_{(k-1)m-(k-3)n-1}$	$+s_m(\cdots(a_{(k-2)m-(k-4)n-1}$	$+s_m(a_{(k-3)m-(k-5)n-1}$	$+\cdots$	$+s_m(a_{n+m-1}$	$+s_m a_{2n-1})\cdots)\rangle$						
$\vdots$	$\vdots$	$\vdots$	$\vdots$	$\vdots$	$\vdots$	$\vdots$	$\vdots$						
$\langle a_n\rangle$	$:=$	$\langle a_n$	$+s_m(\cdots(a_{2n-m}$	$+s_m a_{3n-2m}$	$+\cdots$	$+s_m(a_{(k-1)n-(k-2)m}$	$+s_m a_{kn-(k-1)m})\cdots)\rangle$						
$\langle a_{n-1}\rangle$	$:=$	$\langle a_{n-1}$	$+s_m(\cdots(a_{2n-m-1}$	$+s_m(a_{3n-2m-1}$	$+\cdots$	$+s_m(a_{(k-1)n-(k-2)m-1}$	$+s_m a_{kn-(k-1)m-1})\cdots)$		$+s_0 a_{2n-1}\rangle$				
$\vdots$	$\vdots$	$\vdots$	$\vdots$	$\vdots$	$\vdots$	$\vdots$							
$\langle a_{km-(k-2)n}\rangle$	$:=$	$\langle a_{km-(k-2)n}$	$+s_m(\cdots(a_{(k-1)m-(k-3)n}$	$+s_m(a_{(k-2)m-(k-4)n}$	$+\cdots$	$+s_m a_{2m}$	$+s_m a_{n+m})\cdots)$		$+s_0 a_{km-(k-3)n}\rangle$				
$\langle a_{km-(k-2)n-1}\rangle$	$:=$	$\langle a_{km-(k-2)n-1}$	$+s_m(\cdots(a_{(k-1)m-(k-3)n-1}$	$+s_m(a_{(k-2)m-(k-4)n-1}$	$+\cdots$	$+s_m a_{2m-1}$	$+s_m(a_{n+m-1}$	$+a_{2n-1}))\cdots)$	$+s_0 a_{km-(k-3)n-1}\rangle$				
$\vdots$	$\vdots$	$\vdots$	$\vdots$	$\vdots$	$\vdots$	$\vdots$	$\vdots$	$\vdots$	$\vdots$				
$\langle a_m\rangle$	$:=$	$\langle a_m$	$+s_m(a_n$	$+s_m(a_{2n-m}$	$+\cdots$	$+s_m(a_{(k-2)n-(k-3)m}$	$+s_m(a_{(k-1)n-(k-2)m}$	$+s_m a_{kn-(k-1)m})\cdots)$	$+s_0 a_{n+m}\rangle$				
$\langle a_{m-1}\rangle$	$:=$	$\langle a_{m-1}$							$+s_0(a_{n+m-1}$	$+s_m a_{2n-1})\rangle$			
$\vdots$	$\vdots$	$\vdots$							$\vdots$	$\vdots$			
$\langle a_{2m-n}\rangle$	$:=$	$\langle a_{2m-n}$							$+s_0(a_{2m}$	$+s_m a_{n+m})\rangle$			
$\langle a_{2m-n-1}\rangle$	$:=$	$\langle a_{2m-n-1}$							$+s_0(a_{2m-1}$	$+s_m(a_{n+m-1}$	$+s_m a_{2n-1}))\rangle$		
$\vdots$	$\vdots$	$\vdots$							$\vdots$	$\vdots$	$\vdots$	$\vdots$	
$\langle a_0\rangle$	$:=$	$\langle a_0$							$+s_0(a_n$	$+s_m(\cdots(a_{2n-m}$	$+s_m(a_{3n-2m}$	$+\cdots$	$+s_m a_{kn-(k-1)m})\cdots)\rangle$

FIGURE 2. The updated weight list for the case $N = 2^n \pm 2^m \pm 1, 0 < m < n$.

4. *Step 4. Obtain the incomplete solution:*
   We partition the binary number X, $0 \leqslant X < N^2$, as follows:

$$\begin{aligned} X &= A_{k+1} \mid\mid A_k \mid\mid \cdots \mid\mid A_3 \mid\mid A_2 \mid\mid A_1 \\ &= ((\cdots(A_{k+1} \times 2^{n-m} + A_k) \times 2^{n-m} + \cdots \\ &\qquad \cdots + A_3) \times 2^{n-m} + A_2) \times 2^n + A_1, \end{aligned}$$

   then it can be seen that the top n rows in Fig. 2 are equivalent to performing

$$B_1 = \sum_{i=2}^{k+1} s_m^{i-2}(A_{k+1}||A_k||\cdots||A_i).$$

   And the bottom n rows perform

$$X \pmod N \equiv A_1 + 2^m \sum_{i=2}^{k+1} s_m^{i-1} A_i + s_0 B_1$$

   Then we can summarize the results in Lemma 2.2.

When the modulus N is given in the NAF, then we have $m \leqslant n-2$ and it follows that $k \leqslant \frac{n+1}{2}$. It can be seen that the first $k+1$ terms on the right-hand side of (2.6) are of a binary length no longer than the modulus N, while $k-1$ out of the last k terms are shorter than N with exception of the term $(A_{k+1}||\cdots||A_2)$.

The results in Lemma 2.1 can be obtained in a similar way by analyzing the output of Algorithm 3.1.

5. Discussions

In this paper we have proposed closed-form incomplete solutions to modular reduction for a class of moduli. The results can be used to speed up RSA, elliptic curve cryptosystems defined over finite fields with odd characters, XTR and GH cryptosystems. No evidence has been shown that use of such moduli can compromise the security of the systems.

References

[1] A. Bosselaers, R. Govaerts, and J. Vandewalle. Comparison of three modular reduction functions. In *Crypto 1993*, pp. 175–186. Springer-Verlag, Berlin, 1993.

[2] R.E. Crandall. Method and apparatus for public key exchange in a cryptographic system. U.S. Patent No. 5159632, 1992.

[3] S.W. Golomb. Properties of the sequences $3 \cdot 2^n + 1$. *Math. Comp.*, 30:657–663, 1976.

[4] D.E. Knuth. *The Art of Computer Programming: Seminumerical Algorithms.* Addison-Wesley Publ. Co., Reading, MA, 1981.

[5] C.K. Koc and C.Y. Hung. A fast algorithm for modular reduction. *IEE Proc.: Computers and Digital Techniques*, 145(4):265–271, 1998.

[6] J.M. Pollard. Implementation of number-theoretic transforms. *Electronics Letters*, 15(12):378–379, 1976.

Huapeng Wu
Dept. of Electrical and Computer Eng.
University of Windsor
Windsor, ON, Canada
e-mail: hwu@uwindsor.ca

M. Anwar Hasan
Dept. of Electrical and Computer Eng.
University of Waterloo
Waterloo, ON, Canada
e-mail: ahasan@ece.uwaterloo.ca

Ian F. Blake
Dept. of Electrical and Computer Eng.
University of Toronto
Toronto, ON, Canada
e-mail: ifblake@comm.utoronto.ca

Progress in Computer Science and Applied Logic, Vol. 23, 385–396

Security Analysis of Three Oblivious Transfer Protocols

Gang Yao, Feng Bao and Robert H. Deng

Abstract. An m out of n oblivious transfer (OT) protocol is a cryptographic protocol for a sender to transfer m out of n messages to a receiver such that the sender has no idea which m messages are obtained by the receiver (*receiver security*) and at the same time the receiver cannot obtain more than m messages (*sender security*). Three such protocols are proposed in [1], which have the advantage that the communication overhead of the protocols is much smaller than that of m implementations of a 1 out of n OT protocol. In this paper we give a security analysis of the three protocols. First we show that the first protocol cannot guarantee both the sender security and the receiver security simultaneously. Next, we point out an obvious security flaw in the second protocol which allows the receiver to obtain all the n messages. The third protocol is nicely designed to be non-interactive. However, we show that the security of the protocol is based on a sort of parallel discrete logarithm problem, instead of the discrete logarithm problem as claimed in the paper. Using the technique of "generalized birthday attack", the former problem can be solved with a computation complexity much smaller than that required to solve the discrete logarithm problem.

1. Introduction

Oblivious Transfer (OT) protocol is an important primitive in modern cryptography. Since its introduction by Rabin in 1981 [2], the subject has attracted a lot of attentions and has become the basis for realizing a broad class of cryptographic protocols [3, 4, 5, 6, 7]. Stadler and Piveteau [3] present a fair blind signature scheme using the concept of an OT protocol. Even, et al., [4] describe a protocol for signing contracts which also uses the OT protocol as the basic building block. Aiello, et al., [5] propose a priced oblivious transfer protocol that can protect the privacy of customers buying digital goods. Goldreich and Vainish [6] reduce two problems of general secure multiparty computation to a variant of the OT problem, and then solve the problems. Crépeau, et al., [7] exploit "committed oblivious transfer" to obtain a protocol for private multi-party computation, without making the assumption that a specific number or fraction of participants are honest.

Rabin's original proposal can be considered as a protocol between two parties, Alice and Bob. Alice sends a message to Bob in such a way that: (1) Bob has the probability of 1/2 of actually receiving the message and has the probability of 1/2 of receiving nothing. (2) Alice does not know whether Bob received the message but Bob does.

Various OT protocols have been proposed during the last 20 years, such as 1 out of 2 oblivious transfer, denoted as $\binom{2}{1}$-OT, 1 out of n oblivious transfer, denoted as $\binom{n}{1}$-OT, and m out of n ($n > m$) oblivious transfer, denoted as $\binom{n}{m}$-OT. Informally, in an $\binom{n}{m}$-OT, Alice sends n messages to Bob such that Bob can receive only m messages while Alice has no knowledge which messages Bob has received.

Till now, OT has been studied in various flavors [7, 8, 9, 10, 11, 12, 13]. Crépeau, et al., [7] present an efficient protocol for "committed oblivious transfer" to perform OT on committed bits. The protocol, based on the properties of error correcting codes, uses bit commitment and $\binom{2}{1}$-OT as black boxes. Boer [8] presents a protocol for oblivious transfer in which the secrecy is protected (almost) unconditionally and a bit-commitment scheme based on factoring where the secrecy is unconditional. Brassard, et al., [9] give formal definitions of oblivious transfer based on information-theoretic consideration. Stern [10] presents a new and efficient all-or-nothing disclosure of secrets protocol. Cachin, et al., [11] propose a protocol for OT that is unconditionally secure under the sole assumption that the memory size of the receiver is bounded. Naor and Pinkas [12] propose a two-round $\binom{n}{1}$-OT protocol that is computationally efficient in amortized analysis, that is, one modular exponentiation per invocation. Tzeng [13] presents an efficient (string) $\binom{n}{1}$-OT protocol for any $n \geq 2$. He builds his $\binom{n}{1}$-OT protocol from fundamental cryptographic techniques directly. The privacy of the receiver's choice is unconditionally secure and the secrecy of the un-chosen secrets is based on hardness of the decisional Diffie-Hellman problem.

Distributed OT protocols distribute the task of the sender between several servers. Security is ensured as long as a limited number of these servers collude. Naor and Pinkas [14] identify the important attributes of distributed oblivious transfer. Their work describes distributed protocols for oblivious transfer, in which the role of the sender is distributed between several servers, and a chooser (e. g., receiver) must contact a threshold of these servers in order to run the OT protocol. These distributed OT protocols provide information theoretic security, and do not require the parties to compute exponentiations or any other kind of public key operations. Tzeng [13] also extends his $\binom{n}{1}$-OT protocol to distributed OT protocol.

Very recently, Mu, Zhang and Varadharajan [1] present three novel $\binom{n}{m}$-OT protocols and claim that their protocols demonstrate significant improvement over the existing protocols in terms of completeness, robustness and flexibility. In this paper, we will give the security analysis to the three protocols. The first protocol in [1], called the first MZV $\binom{n}{m}$-OT protocol in this paper, is an interactive $\binom{n}{m}$-OT protocol. Unfortunately, this protocol cannot guarantee both the sender security and the receiver security simultaneously. The second protocol in [1], denoted as the second MZV $\binom{n}{m}$-OT protocol here, is an efficient interactive $\binom{n}{m}$-OT protocol but has an obvious security flaw which can be exploited by the receiver to obtain all the n messages. The third protocol in [1], called the third MZV $\binom{n}{m}$-OT protocol in this paper, is a nicely designed non-interactive $\binom{n}{m}$-OT protocol. Both the second and the third MZV $\binom{n}{m}$-OT protocols are based on the same key setup scheme which is claimed in [1] to be based on the discrete logarithm problem. However, we will show in this paper that the security of the key setup scheme is actually based on the parallel discrete logarithm problem which is subject to the "generalized birthday attack" [17].

2. Analysis of the First MZV $\binom{n}{m}$-OT Protocol

The first MZV $\binom{n}{m}$-OT protocol is an interactive protocol making use of generic symmetric and asymmetric key cryptographic algorithms. Suppose that Alice has n messages M_i, $i = 1, 2, \ldots, n$, and she wants Bob to receive m of them. Denote the symmetric encryption and decryption with respect to a secret key k by $E_k(\cdot)$ and $D_k(\cdot)$, respectively, and denote the asymmetric encryption using a public key e and the corresponding asymmetric decryption using a private key d by $\langle\cdot\rangle_e$ and $\langle\cdot\rangle_d$, respectively. The first MZV $\binom{n}{m}$-OT protocol is carried out as follows:

1. Alice chooses n private/public key pairs (d_i, e_i), $i = 1, 2, \ldots, n$, and sends all the n public keys to Bob.
2. Bob chooses m symmetric keys k_j, $j = 1, 2, \ldots, m$. He also chooses m public keys (e_{s_j}) from the n public keys received from Alice, where the indices (s_j) are known to Bob only. Bob encrypts the symmetric key k_j with the public key (e_{s_j}), $j = 1, 2, \ldots, m$, and sends, $\langle k_j\rangle_{e_{s_j}}$, $j = 1, 2, \ldots, m$, to Alice.
3. Alice decrypts each of $k'_j = \langle k_j\rangle_{e_{s_j}}$ using all n private keys d_i to get $k_{ij} = \langle k'_j\rangle_{d_i}$, and encrypts all messages M_i using k_{ij}, $i = 1, 2, \ldots, n$, $j = 1, 2, \ldots, m$, respectively. The encrypted messages $C_{ij} = E_{k_{ij}}(M_i)$ are then sent to Bob.
4. Bob decrypts the m received messages using k_j, $j = 1, 2, \ldots, m$, based on the indices of the public key known to him only.

The description of this protocol as presented in [1] is generic in nature. It does not provide the details such as how Bob encrypts the symmetric keys k_j and how Alice chooses n private/public key pairs (d_i, e_i). In the following subsection, we will give the security analysis to this protocol.

2.1. Discussion on Encrypting the Symmetric Keys k_j

Suppose that Alice and Bob agree on a public key algorithm, say RSA, and a symmetric key algorithm, say the Advanced Encryption Standard (AES).

The lengths of the symmetric key k_j, $j = 1, 2, \ldots, m$, of the AES are 128 bits, 192 bits or 256 bits, which are much smaller compared with the modulus size of the RSA algorithm. Therefore, when we encrypt this symmetric key using the private/public key pair (d_i, e_i), $i = 1, 2, \ldots, n$, of the RSA, we need to pad some bits to the symmetric key k_j. There are several standards on the padding formats for RSA signature and encryption, such as, ANSI X9.31 [15], PKCS #1 V2.0 [16], etc. In the protocol, if Bob uses one of these standard to pad the symmetric key, Alice will know which message Bob has received at the end of protocol.

Let N be the modulus in the RSA algorithm and (d_i, e_i), $i = 1, 2, \ldots, n$, be the RSA private/public key pairs that Alice has chosen. Suppose that Bob uses ANSI X9.31 [15] to pad the symmetric key k. We use $\mu(\cdot)$ to denote the padding function and $a \cdot b$ to denote the concatenation of a and b. Now, let us go through the steps of the first MZV $\binom{n}{m}$-OT protocol outlined in the last subsection.

In step 2, Bob chooses m symmetric keys k_j, $j = 1, 2, \ldots, m$, and encrypts them using m public keys (e_{s_j}) selected from the n public keys e_i, $i = 1, 2, \ldots, n$. When Bob encrypts k_j, he uses the ANSI X9.31 standard to pad the symmetric key k_j:

$$\mu(k_j) = 6B_{16} \cdot BBBB_{16} \cdot \ldots \cdot BBBB_{16} \cdot BA_{16} \cdot k_j \cdot 33CC_{16}$$

where $\cdot_{16}$ denotes a hexadecimal digit. Then, Bob uses the public key e_{s_j} to encrypt $\mu(k_j)$ and gets k_j':

$$k_j' = (6B_{16} \cdot BBBB_{16} \cdot \ldots \cdot BBBB_{16} \cdot BA_{16} \cdot k_j \cdot 33CC_{16})^{e_{s_j}} \mod N$$

After he encrypts all the m keys, he sends the encrypted keys to Alice. In step 3, Alice decrypts each encrypted key using all the n private keys d_i to obtain k_{ij}, and encrypts all messages using k_{ij}, to get the encrypted messages C_{ij}.

In the above decryption process, when the encrypted key k_j' is decrypted using the right private key d_{s_j}, Alice gets a well-formatted message

$$\begin{aligned} \mu(k_j) &= (6B_{16} \cdot BBBB_{16} \cdot \ldots \cdot BBBB_{16} \cdot BA_{16} \cdot k_j \cdot 33CC_{16})^{e_{s_j} d_{s_j}} \mod N \\ &= 6B_{16} \cdot BBBB_{16} \cdot \ldots \cdot BBBB_{16} \cdot BA_{16} \cdot k_j \cdot 33CC_{16} \mod N \end{aligned}$$

and then gets the correct key k_j; however, when k_j' is decrypted by the other $n-1$ private keys than d_{s_j}, Alice obtains not well-formatted messages with very high probabilities. So, Alice knows which key is the correct key by observing whether the decrypted message is well formatted or not. As a result, she knows which messages Bob will receive.

If Bob uses other standards to pad the symmetric key k, we can obtain similar results as above. Therefore, when Bob encrypts the symmetric key k, it is not secure for him to pad the symmetric key k using any well-defined padding format such as ANSI X9.31, for Alice can guess which key is the correct key that Bob chooses by checking the padding format in the decrypted message. For this reason, Bob needs to add random bits for padding when he encrypts the symmetric key such that every message that Bob encrypts appears as a random string when decrypted by Alice.

2.2. Discussion on Choosing the RSA Key Pairs (d_i, e_i)

Again suppose that Alice and Bob agree on using the RSA algorithm. The protocol does not describe how Alice chooses n private/public key pairs (d_i, e_i). From the above subsection, every message that Bob encrypts can be regarded as a random string. If the key pairs that Alice chooses have the same modulus, then there is a way for Bob to get more than m messages from Alice.

In the first MZV $\binom{n}{m}$-OT protocol, Alice chooses N, the RSA modulus and (d_i, e_i), $i = 1, 2, \ldots, n$, the RSA private/public key pairs. She sends the public keys to Bob. Bob chooses a random number k and computes

$$t = k^{e_1 e_2} \mod N$$

Then, Bob chooses other $m-1$ keys and encrypts them with e_{s_j} which is different to e_1 and e_2. Bob sends t and the other $m-1$ encrypted keys k'_j, $j = 2, 3, \ldots, m$ to Alice. After Alice obtains the m encrypted keys, she carries out the protocol and sends C_{ij} to Bob. From C_{ij}, Bob can get $m+1$ messages.

In fact, when Alice decrypts the encrypted keys, she gets

$$t^{d_1} = k^{e_1 e_2 d_1} = k^{e_2} \mod N$$

$$t^{d_2} = k^{e_1 e_2 d_2} = k^{e_1} \mod N$$

Then she forms k_{11} and k_{12} from k^{e_2} and k^{e_1}, respectively. So, $C_{11} = E_{k_{11}}(M_1)$ and $C_{12} = E_{k_{12}}(M_2)$. For Bob also knows k, e_1 and e_2, he can compute $k^{e_2} \mod N$ and $k^{e_1} \mod N$ and get k_{11} and k_{12} using the same method as Alice used. Then Bob has $M_1 = D_{k_{11}}(C_{11})$ and $M_2 = D_{k_{12}}(C_{12})$. That is, he gets two messages using one "encrypted key". So, Bob gets $m+1$ instead of m messages.

Generalized the above technique, Bob can get all n messages by sending one "encrypted" key to Alice. He chooses a random number k, computes

$$t = k^{e_1 e_2 \ldots e_n} \mod N$$

and sends t to Alice. Alice decrypts this "encrypted" key and gets

$$t^{d_1} = k^{e_1 \ldots e_n d_1} = k^{e_2 e_3 \ldots e_n} \mod N$$
$$t^{d_2} = k^{e_1 \ldots e_n d_2} = k^{e_1 e_3 \ldots e_n} \mod N$$
$$\ldots$$
$$t^{d_n} = k^{e_1 \ldots e_n d_n} = k^{e_1 e_2 \ldots e_{n-1}} \mod N$$

She then forms n keys $k_{11}, k_{12}, \ldots, k_{1n}$ from the n decryptions. Alice uses these n keys to encrypt n messages, respectively, and sends the n encrypted messages to Bob. Since Bob also knows k and $e_1, e_2, \ldots, e_n$, which are the public keys, he can get $k_{11}, k_{12}, \ldots, k_{1n}$. He decrypts all the encrypted messages from Alice to obtain all the n messages. For this reason, it is not secure for Alice to choose the RSA public key pairs using the same modulus.

Now, let us consider the situation where Alice chooses the RSA public key pairs with different moduli. Without loss of generality, denote the n public keys chosen by Alice as (N_i, e_i), $i = 1, 2, \ldots, n$, where

$$N_1 > N_2 > \cdots > N_n$$

If Bob selects a random key k and encrypts it using the first public key (N_1, e_1), then the encrypted key is given by

$$k' = k^{e_1} \mod N_1$$

If k' is bigger than N_2, then Alice knows that this encrypted key is encrypted using the first public pair, and she knows one message Bob will receive.

Further, there is a method for Alice to know which messages Bob is interested in, but she cannot accomplish the protocol. Alice chooses n "public keys" (N_i, e_i), $i = 1, 2, \ldots, n$, as follows: First Alice chooses n different primes $p'_1, p'_2, \ldots, p'_n$. Then she chooses n primes $p_1, p_2, \ldots, p_n$ such that p'_i divides $p_i - 1$, $i = 1, 2, \ldots, n$, and n primes $q_1, q_2, \ldots, q_n$. For $i = 1, 2, \ldots, n$, Alice computes

$$\lambda_i = \operatorname{lcm}(p_i - 1, q_i - 1)$$

which is the least common multiple of $p_i - 1$ and $q_i - 1$. At last, Alice computes

$$N_i = p_i q_i \qquad e_i = \lambda_i / (2^{u_i} p'_i)$$

where u_i is the greatest integer such that 2^{u_i} divides λ_i. For e_i has no inverse modulus $(p-1)(q-1)$, the "public keys" (N_i, e_i) are pseudo keys.

If Bob encrypted a key k_j using the i-th public key (N_i, e_i), he computes

$$k' = \langle k_j \rangle_{e_i} = k_j^{e_i} \mod N_i$$

It is easy to see that the order of k' is $2^{u_i} p'_i$. And we can see that the order of the encrypted key k' is distinct if Bob encrypts the key k using different public key. So, by testing the order of the encrypted key, Alice will know the encrypted key is encrypted by which public key, i.e., Alice will know which message Bob is interested in. If Alice generates public keys using this method, she can not accomplish the protocol for there is no decryption key corresponding to e_i.

2.3. Discussion on Choosing the ElGamal Key Pairs (a_i, b_i)

Suppose that Alice and Bob agree on using the ElGamal algorithm. From the above subsection, every message that Bob encrypts can be regarded as a random string. In this situation, if the key pairs that Alice chooses have the same system parameters, then there is a way for Bob to get more than m messages from Alice.

Let p and g be the system parameters of the ElGamal algorithm where p is a large prime and g is a generator of $\mathbb{Z}_p^*$. In the first MZV $\binom{n}{m}$-OT protocol, Alice chooses a_i, the private keys, and $b_i = g^{a_i} \mod p$, the corresponding public key, $i = 1, 2, \ldots, n$. Alice sends the public keys to Bob. Bob chooses a symmetric key k and random number w and constructs an "encrypted" message as follows:

$$(y_1, y_2) = (g^w, k(b_1 b_2)^w) \mod p$$

Bob sends (y_1, y_2) to Alice. Alice decrypts the encrypted key and gets:

$$y_2 y_1^{-a_1} = k(b_1 b_2)^w g^{-a_1 w} = k b_2^w \mod p$$
$$y_2 y_1^{-a_2} = k(b_1 b_2)^w g^{-a_2 w} = k b_1^w \mod p$$

Then she forms k_{11} and k_{12} from kb_2^w and kb_1^w, respectively. Alice encrypts all messages using k_{ij}. Then she sends C_{ij} to Bob. Since Bob knows k, w, b_1 and b_2, he also can compute $kb_2^w \mod p$ and $kb_1^w \mod p$ and get k_{11} and k_{12}. Then Bob can get two messages, M_1 and M_2, using one "encrypted" key (y_1, y_2). So, Bob obtains one extra message from Alice.

Similar to using the RSA algorithm, Bob can get all n messages by sending one "encrypted" key to Alice. Bob chooses random numbers k and w, and computes

$$(y_1, y_2) = (g^w, k(b_1 \ldots b_n)^w) \mod p$$

and sends them to Alice. Alice decrypts this "encrypted" key and gets

$$y_2 y_1^{-a_1} = k(b_1 \ldots b_n)^w g^{-a_1 w} = k(b_2 \ldots b_n)^w \mod p$$
$$y_2 y_1^{-a_2} = k(b_1 \ldots b_n)^w g^{-a_2 w} = k(b_1 b_3 \ldots b_n)^w \mod p$$
$$\ldots$$
$$y_2 y_1^{-a_n} = k(b_1 \ldots b_n)^w g^{-a_n w} = k(b_1 \ldots b_{n-1})^w \mod p$$

She then forms n keys $k_{11}, k_{12}, \ldots, k_{1n}$ from these n numbers. Alice uses these n keys to encrypt n messages, respectively, and sends them to Bob. For Bob also knows k, m, b_1, b_2, ..., b_n, he can get $k_{11}, k_{12}, \ldots, k_{1n}$. So Bob can decrypt all the encrypted messages from Alice to obtain all the n messages.

On the other hand, if Alice chooses the public key pairs using different system parameters, Alice has a method to know which message Bob is interested in. Alice chooses n public keys (p_i, g_i, b_i), $i = 1, 2, \ldots, n$, as follows: First Alice chooses n different primes $p'_1, p'_2, \ldots, p'_n$. Then she chooses n primes $p_1, p_2, \ldots, p_n$, such that p'_i divides $p_i - 1$, $i = 1, 2, \ldots, n$. At last, she chooses g_i in order p'_i.

If Bob encrypts a key k using the ith public key pair (p_i, g_i, b_i), he computes

$$y_1 = g_i^w \mod p_i \qquad y_2 = kb_i^w \mod p_i$$

where w is a random number. It is easy to see that the order of g_i^w is p'_i. And we can see that the order of the encrypted key y_1 is different if Bob encrypts the key k by different public key pair. So, Alice will know which message Bob is interested in by testing the order of the encrypted key. If Alice generates public key pairs using this method, she can accomplish the protocol. This is not like the situation when we use the RSA algorithm.

The above analysis shows that neither RSA nor ElGamal can make the protocol secure. When the system parameters are the same, to prevent Bob get one extra message, the encryption scheme needs to satisfy the following condition: Bob cannot find two numbers k_1 and k_2, such that $\langle k_1 \rangle_{e_1} = \langle k_2 \rangle_{e_2}$. This is a very different problem from the conventional one and hard to fulfill.

3. Analysis of the Setup Protocol

The second and the third MZV $\binom{n}{m}$-OT protocol are based on a same key setup system. Here, we analyze the setup protocol and show that the security of the setup protocol is based on a sort of parallel discrete logarithm, instead of the discrete logarithm as claimed in [1]. The setup protocol is built as follows.

Let p be a large prime number, $\mathbb{Z}_p^*$ be a multiplicative group, $g \in \mathbb{Z}_p^*$ be the generator of order $q = p - 1$, and $x_i \in \mathbb{Z}_p^*$, $i = 1, 2, \ldots, n$, be a set of integers. For simplicity, we omit modulus p in the rest of the presentation in this section.

The public key setup is done by Bob who selects m private keys $s_i \in \mathbb{Z}_q$ and then computes $y_i = g^{s_i}$, $i = 1, 2, \ldots, m$, $(m < n)$. Given x_i, the n public keys are constructed by using a set of m linear equations with respect to m unknown numbers $a_1, a_2, \ldots, a_m$,

$$a_1 x_i + a_2 x_i^2 + \cdots + a_m x_i^m = y_i, \quad i = 1, 2, \ldots, m.$$

The corresponding linear equations in a matrix form are as follows:

$$\begin{pmatrix} x_1 & x_1^2 & \cdots & x_1^m \\ x_2 & x_2^2 & \cdots & x_2^m \\ \vdots & \vdots & & \vdots \\ x_m & x_m^2 & \cdots & x_m^m \end{pmatrix} \begin{pmatrix} a_1 \\ a_2 \\ \vdots \\ a_m \end{pmatrix} = \begin{pmatrix} y_1 \\ y_2 \\ \vdots \\ y_m \end{pmatrix}$$

Since the determinant of the coefficient matrix is non-zero, the equations have a unique solution over the field $\mathbb{Z}_p$.

After Bob gets the unique solution $a_1, \ldots, a_m$, he can calculate other $n - m$ "public keys" (their discrete logs are unknown), using the following formula:

$$y_j = a_1 x_j + \cdots + a_m x_j^m, \quad m < j \leq n.$$

As a result, he has n public keys $\{x_i, y_i\}_{i=1}^n$.

Bob shuffles his public keys such that the order is known to himself only and publishes the public keys. For convenience, we denote by $\mathbb{U}$ the subset of public key indices whose associated public key discrete logs are unknown to Bob and by $\mathbb{K}$ those known. Denote the shuffled public key set by $\{x_i, y_i\}_{i=1}^n$.

Let A' be a $n \times (m+1)$ matrix where

$$A' = \begin{pmatrix} x_1 & x_1^2 & \cdots & x_1^m & y_1 \\ x_2 & x_2^2 & \cdots & x_2^m & y_2 \\ \vdots & \vdots & & \vdots & \vdots \\ x_n & x_n^2 & \cdots & x_n^m & y_n \end{pmatrix}$$

The authors give a lemma that to prevent Bob from cheating by pre-selecting all $\{y_i\}_{i=m+1}^n$, the rank of matrix A' must be $m + 1$. But the lemma is not correct.

In fact, from the construction of $\{x_i, y_i\}_{i=1}^n$, we can get that x_i and y_i, $i = 1, 2, \ldots, n$, satisfy the equation

$$y_j = a_1 x_j + \cdots + a_m x_j^m$$

That is to say that the last column of matrix A' is the linear combination of the former m columns of matrix A', so the rank of matrix A' can not be $m+1$.

In the following, we discuss how to attack this setup scheme using birthday attack. The result of the attack is that: at the end of the setup scheme, Bob gets n public keys $\{x_i, y_i\}_{i=1}^{n}$ and knows $m+1$ private keys $\{s_i\}_{i=1}^{m+1}$, not m private keys.

If $m = 1$, Bob can construct $\{x_i, y_i\}_{i=1}^{n}$ and $\{s_i\}_{i=1}^{n}$ as follows: First, he chooses a random number x_1 and $s_1, \ldots, s_n$. Then he computes $a_1 = g^{s_1} x_1^{-1}$ and $x_i = g^{s_i} a_1^{-1}$. So he gets n public keys $\{x_i, y_i\}_{i=1}^{n}$ and knows n private keys $\{s_i\}_{i=1}^{n}$, not one private key.

Assuming that $m \geq 2$, Bob chooses $m+1$ random numbers $x_i \in \mathbb{Z}_p^*$, i =1, 2, $\ldots, m+1$. If Bob wants to get $m+1$ private keys, he needs to find $m+1$ numbers $s_j \in \mathbb{Z}_p^*$, such that equations

$$x_j a_1 + \cdots + x_j^m a_m - g^{s_j} = 0 \qquad j = 1, 2, \ldots, m+1$$

have a non-zero solution $(a_1, a_2, \ldots, a_m, -1)$. For this reason, the determinant of coefficients of the equations equals zero, i.e.,

$$\det B = \det \begin{pmatrix} x_1 & x_1^2 & \cdots & x_1^m & g^{s_1} \\ x_2 & x_2^2 & \cdots & x_2^m & g^{s_2} \\ \vdots & \vdots & & \vdots & \vdots \\ x_{m+1} & x_{m+1}^2 & \cdots & x_{m+1}^m & g^{s_{m+1}} \end{pmatrix} = 0$$

Computing this determinant, we can get that $s_j \in \mathbb{Z}_p^*$, j = 1, 2, $\ldots, m+1$ satisfy an equation

$$c_1 g^{s_1} + c_2 g^{s_2} + \cdots + c_{m+1} g^{s_{m+1}} = 0 \tag{3.1}$$

where

$$c_i = (-1)^{i+m+1} \det B_{i,m+1} \qquad i = 1, 2, \ldots, m+1$$

here $B_{i,m+1}$ be the matrix obtained by omitting the ith row and the $(m+1)$th column of B.

To find a solution of equation (3.1) is still difficult. Wagner [17] studies a k-dimensional generalization of the birthday problem: given k lists of l-bit values, find some way to choose one element from each list so that the resulting k values XOR to zero. He shows a cube-root time algorithm for the case of $k = 4$ lists, and gives an algorithm with sub-exponential running time when k is unrestricted.

The algorithm of Wagner can also be applied to find a solution of equation (3.1). Let the length of prime p be l. We define the list L_i as follows:

$$L_i = \{c_i g^{s_i} | 0 \leq s_i \leq p-1\}$$

For c_i is a non-zero element and g is a generator of $\mathbb{Z}_p^*$, list L_i is equal to the set $\mathbb{Z}_p^*$. Then, to find a solution of equation (3.1) can transfer to the following problem:

Given $m+1$ lists $L_1, \ldots, L_{m+1}$ of elements drawn uniformly and independently at random from $\{1, \ldots, p-1\}$, find $z_1 \in L_1$, $\ldots$, $z_{m+1} \in L_{m+1}$ such that $z_1 + \cdots + z_{m+1} = 0 \mod p$.

When $m = 3$, using the algorithm of Wagner, we can find a solution of equation (3.1) with $O(2^{l/3})$ time and space. In general case, for $m > 3$, using the algorithm of Wagner, we can find a solution of equation (3.1) with $O((m+1)2^{l/(1+\lg(m+1))})$ time and space. Especially, when $k = 2^{\sqrt{l}-1}$, we can find a solution of equation (3.1) with $O(2^{2\sqrt{l}})$ time and $O(\sqrt{l}\, 2^{\sqrt{l}})$ space.

4. Analysis of the Second MZV $\binom{n}{m}$-OT Protocol

Given the setup protocol, the second MZV $\binom{n}{m}$-OT Protocol is carried out as follows. In the protocol, the public keys $\{x_i, y_i\}_{i=1}^n$ are generated by Bob, and assume that the public keys are shared by Alice and Bob only.

1. Bob:
   - Chooses m session keys, k_j.
   - Encrypts k_j: Bob selects a random number $w_j \in_R \mathbb{Z}_q$, $j \in \mathbb{K}$, and computes

   $$r_j = k_j g^{w_j}, \quad k'_j = s_j r_j + w_j \mod q$$

   The ciphertext is the doublet (r_j, k'_j), which can be decrypted with y_j, $j \in \mathbb{K}$. Here, we treat the signature as encryption since the public keys are shared by Alice and Bob only.
   - Extends the k'_j, $j \in \mathbb{K}$, set by adding $n - m$ random numbers (as dummies) to those positions whose indices are in $\mathbb{U}$ to form k''_i, $i = 1, 2, \ldots, n$, which are placed in order. Similarly, extends r_j to r'_i.
   - Sends k''_i and r'_i to Alice.
2. Alice:
   - Decrypts k''_i using all y_i in order, $k_i = g^{-k''_i} y_i^{r'_i} r'_i$, to get n "keys". Only m of them are correct.
   - Encrypts n messages using k_i in order, $C_i = E_{k_i}(M_i)$.
   - Sends C_i to Bob.
3. Bob decrypts m of C_i, and gets $M_j = D_{k_j}(C_j)$, $j \in \mathbb{K}$.

The authors consider this protocol is a $\binom{n}{m}$-OT protocol. *Completeness* of this protocol is obvious, for if Bob has encrypted the m keys using his correct private keys, Alice can correctly decrypt m out of n keys in Step 2, and if Alice correctly follows the process, Bob can then correctly decrypt m messages in Step 3. But *Soundness* of this protocol is NOT correct, i.e., the conclusion that Bob can obtain at most m correct messages is not correct.

At the end of this protocol, Bob can obtain more than m messages from Alice. In fact, he can obtain all the messages from Alice. This is because Bob can compute all the keys k_i used in Step 2. k_i is given by

$$k_i = g^{-k''_i} y_i^{r'_i} r'_i \qquad i = 1, 2, \ldots, n$$

In this formula, all the parameters k_i'', r_i' and y_i, $i = 1, \ldots, n$, are chosen by Bob. So Bob can compute k_i, $i = 1, \ldots, n$, and then he can compute

$$D_{k_i}(C_i) = D_{k_i}(E_{k_i}(M_i)) = M_i \qquad i = 1, 2, \ldots, n$$

That is to say Bob can obtain all the messages from Alice, i.e., this protocol is not a secure $\binom{n}{m}$-OT protocol.

5. Conclusion

Oblivious Transfer protocol is an important cryptographic primitive in modern cryptography. $\binom{n}{m}$-OT protocols have practical applications in the privacy protection of e-commerce of digital goods. Mu, Zhang and Varadharajan present three $\binom{n}{m}$-OT protocols, which have the advantage that the communication of the protocols is much smaller than that of m implementations of a $\binom{n}{1}$-OT protocol. In this paper, we presented our security analysis to the three protocols. Their first protocol employs public key encryption schemes without mentioning the specific public key schemes being used. Our analysis showed that neither RSA nor ElGamal can make the protocol secure. Their second protocol has an obvious security flaw such that the receiver can obtain all the n messages. The third protocol is nicely designed to be non-interactive. However, we showed that the security of the protocol is based on a sort of parallel discrete logarithm problem. By the technique of "generalized birthday attack", the parallel discrete logarithm problem can be solved with a computational complexity much smaller than that for solving the discrete logarithm problem.

Acknowledgements

This work is supported by the National Natural Science Foundation of China (Grant No.60073021) and the research collaboration program between Institute for InfoComm Research, Singapore and Institute of Software, Chinese Academy of Sciences.

References

[1] Y. Mu, J. Zhang and V. Varadharajan, "m out of n Oblivious Transfer", *Australasian Conference, Information Security and Privacy - ACISP 2002, LNCS 2384*, pp. 395–405, Springer-Verlag, 2002.

[2] M. Rabin, "How to exchange secrets by oblivious transfer", Technical Report TR-81, Aiken Computation Laboratory, Harvard University, 1981.

[3] M. Stadler, J.-M. Piveteau and Jan L. Camenisch "Fair blind signature", *Advances in Cryptology - EUROCRYPT'95, LNCS 921*, pp. 209–219, Springer-Verlag, 1995.

[4] S. Even, O. Goldreich, A. Lempel, "A randomized protocol for signing contracts", *Communications of the ACM 28*, pp. 637–647, 1985.

[5] B. Aiello, Y. Ishai, O. Reingold, "Priced oblivious transfer: how to sell digital goods", *Advances in cryptology - EUROCRYPT 2001, LNCS 2045*, pp. 119–135, Springer-Verlag, 2001.

[6] O. Goldreich, R. Vainish, "How to solve any protocol problem: an efficient improvement", *Advances in Cryptology - CRYPTO'87, LNCS 293*, pp. 73–86, Springer-Verlag, 1988.

[7] C. Crépeau, J. van de Graff, A. Tapp, "Committed oblivious transfer and private multi-party computations", *Advances in Cryptology - CRYPTO'95, LNCS 963*, pp. 110–123, Springer-Verlag, 1995.

[8] B. den Boer, "Oblivious transfer protecting secrecy", *Advances in Cryptology - EUROCRYPT'90, LNCS 473*, pp. 31–45, Springer-Verlag, 1991.

[9] G. Brassard, C. Crépeau, M. Santha, "Oblivious transfer and intersecting codes", *IEEE Transactions on Information Theory*, 42(6), pp. 1769–1780, 1996.

[10] J.P. Stern, "A new and efficient all-or-nothing disclosure of secrets protocol", *Advances in Cryptology - ASIACRYPT'98, LNCS 1514*, pp. 357–371, Springer-Verlag, 1998.

[11] C. Cachin, C. Crépeau, J. Marcil, "Oblivious transfer with a memory-bounded receiver", *the 39th IEEE Symposium on Foundations of Computer Science*, pp. 493–502, 1998.

[12] M. Naor, B. Pinkas, "Efficient oblivious transfer protocols", *12th Annual Symposium on Discrete Algorithms (SODA)*, pp. 448–457, 2001.

[13] W.G. Tzeng, "Efficient 1-Out-n oblivious transfer schemes", *Public Key Cryptography - PKC 2002, LNCS 2274*, pp. 159–171, Springer-Verlag, 2002.

[14] M. Naor, B. Pinkas, "Distributed oblivious transfer", *Advances in Cryptology - ASIACRYPT 2000, LNCS 1976*, pp. 205–219, Springer-Verlag, 2000.

[15] ANSI X9.31, "Digital signatures using reversible public-key cryptography for the financial services industry (rDSA)", 1998.

[16] RSA Laboratories, "PKCS #1 v2.0: RSA Encryption Standard", 1998.

[17] D. Wagner, "A generalized birthday problem", *Advances in Cryptology - CRYPTO 2002, LNCS 2442*, pp. 288–304, Springer-Verlag, 2002.

Gang Yao, Feng Bao, Robert H. Deng
Infocomm Security Department
Institute for InfoComm Research, Singapore 119613
e-mail: `yaogang@i2r.a-star.edu.sg`
e-mail: `baofeng@i2r.a-star.edu.sg`
e-mail: `deng@i2r.a-star.edu.sg`

Gang Yao
Laboratory of Computer Science, Institute of Software
Chinese Academy of Sciences, Beijing 100080, P.R. China
e-mail: `g.yao@is.iscas.ac.cn`

Progress in Computer Science and Applied Logic, Vol. 23, 397–405

An Improved Identification Scheme

Gang Yao, Guilin Wang and Yong Wang

Abstract. Kim and Kim recently proposed a new identification scheme based on the Gap Diffie-Hellman problem, and proved that their scheme is secure against active attacks if the Gap Diffie-Hellman problem is intractable. However, their identification scheme is NOT secure in fact. In this paper, we first point out the reason why their scheme is not secure, and then improve their scheme such that the modified scheme is secure against active attacks if the Gap Diffie-Hellman problem is intractable.

Keywords. Identification scheme, Weil pairing, Gap-problem.

1. Introduction

It is well known that an identification scheme is a very important and useful cryptographic tool. It is a method by which a user may prove his or her identity to somebody else, without revealing essential knowledge that may be used by either an eavesdropper or the recipient to impersonate the user. In such a scheme, a prover, $\mathcal{P}$, tries to convince a verifier, $\mathcal{V}$, of his identity. The objectives of an identification scheme include the following [6]:

1. In the case of honest parties $\mathcal{P}$ and $\mathcal{V}$, $\mathcal{P}$ is able to successfully authenticate itself to $\mathcal{V}$, i.e., $\mathcal{V}$ will complete the protocol having accepted $\mathcal{P}$'s identity.
2. (*transferability*) $\mathcal{V}$ cannot reuse an identification exchange with $\mathcal{P}$ so as to successfully impersonate $\mathcal{P}$ to a third party $\mathcal{C}$.
3. (*impersonation*) The probability is negligible that any party $\mathcal{C}$ distinct from $\mathcal{P}$, carrying out the protocol and playing the role of $\mathcal{P}$, can cause $\mathcal{V}$ to complete and accept $\mathcal{P}$'s identity.

These objectives allow a person to identify himself to another person without giving any information to the other person.

In general, an identification scheme is said to be broken if an adversary succeeds in an impersonation attempt (making the verifier accept with non-negligible probability). We can classify the type of attacks according to the interaction allowed to the adversary before an impersonation attempt [10].

The weakest form of attack is a passive attack, where the adversary is not allowed to interact with the system at all before attempting an impersonation; the only available information to the adversary is the public key of the prover.

The strongest form of attack is an active attack, in which the adversary is allowed to interact with $\mathcal{P}$ several times, posing as $\mathcal{V}$. We may consider active attacks as adaptive chosen-cipher text attacks. We should note that active attacks are quite feasible in practice.

Fiat and Shamir [3] proposed an identification scheme based on the factorization problem in 1986. Their scheme is secure against active attacks if factorization is hard. After Fiat-Shamir Identification Scheme is proposed, several practical identification schemes have been proposed, such as Feige-Fiat-Shamir Identification Scheme [2], Schnorr Identification Scheme [9], Guillou-Quisquater Identification Scheme [4] and Okamoto Identification Scheme [7].

Recently, Kim and Kim proposed a new identification scheme based on Gap Diffie-Hellman problem in [5]. They also provided security proofs to show that their scheme is secure against active attacks if the Gap Diffie-Hellman problem is intractable. However, this identification scheme is NOT secure at all. In this paper, we first point out the reason why their scheme is not secure, and then propose a modified scheme which is secure against active attacks if the Gap Diffie-Hellman problem is intractable.

2. Notions

In this section, we formally state the definition of security and the basic notions used in the paper. We use the same notions used in [5].

2.1. Notions of Security

If S is a probability space, then $[S]$ denotes the set of elements in this space that occur with non-zero probability, and $\Pr_S[e]$ denotes the probability that S associates with the element e. If S is any probability space, then $x \leftarrow S$ denotes the algorithm which assigns to x an element randomly selected according to S. The notation $\Pr[p(x_1, x_2, \ldots)|x_1 \leftarrow S_1; x_2 \leftarrow S_2; \ldots]$ denotes the probability that the predicate $p(x_1, x_2, \ldots)$ will be true after the ordered execution of the algorithms $x_1 \leftarrow S_1, x_2 \leftarrow S_2, \ldots$.

In addition, we use the same conventions in [2]: $\bar{\mathcal{P}}$ represents an honest prover who follows its designated protocol, $\tilde{\mathcal{P}}$ does a polynomial-time cheater. $\bar{\mathcal{V}}$ represents a valid verifier who follows the designated protocol, $\tilde{\mathcal{V}}$ does an arbitrary polynomial-time algorithm which may try to extract additional information from $\mathcal{P}$. $(\mathcal{P}, \mathcal{V})$ represents the execution of the two party protocol where $\mathcal{P}$ is the prover and $\mathcal{V}$ is the verifier.

In general, an identification scheme $(\mathcal{G}, \mathcal{P}, \mathcal{V})$ consists of a probabilistic polynomial time algorithm $\mathcal{G}$, and two probabilistic polynomial-time interactive algorithms $\mathcal{P}$ and $\mathcal{V}$ with the following properties [10, 2]:

1. The algorithm $\mathcal{G}$ is a *key generation algorithm.* It takes a string of the form 1^k as input, and outputs a pair of string (I, S). k is called a *security parameter*, I is called a *public key*, and S is called a *secret key.*
2. $\mathcal{P}$ receives as input the pair (I, S) and $\mathcal{V}$ receives as input I. After an interactive execution of $\mathcal{P}$ and $\mathcal{V}$, $\mathcal{V}$ outputs either a 1 (indicating "accept") or a 0 (indicating "reject"). For a given I and S, the output of $\mathcal{V}$ at the end of this interaction is a probability space which is denoted by $\langle \mathcal{P}(I, S), \mathcal{V}(I)\rangle$.
3. A valid prover should always be able to succeed in convincing the verifier. Formally speaking, for all k and for all $(I, S) \in [\mathcal{G}(1^k)]$, $\langle \mathcal{P}(I, S), \mathcal{V}(I)\rangle = 1$ with probability 1.

An *adversary* $\langle \tilde{\mathcal{P}}, \tilde{\mathcal{V}}\rangle$ is a pair of probabilistic polynomial-time interactive algorithms. For given key pair (I, S), we denote by $\langle \bar{\mathcal{P}}(I, S), \tilde{\mathcal{V}}(I)\rangle$ the string h output by $\tilde{\mathcal{V}}$ after interacting with $\bar{\mathcal{P}}$ several times. The string h (called a "help string") is used as input to $\tilde{\mathcal{P}}$ who attempts to convince $\bar{\mathcal{V}}$. We denote by $\langle \tilde{\mathcal{P}}(h), \bar{\mathcal{V}}(I)\rangle$ the output of $\bar{\mathcal{V}}$ after interacting with $\tilde{\mathcal{P}}(h)$.

An identification scheme $(\mathcal{G}, \mathcal{P}, \mathcal{V})$ is *secure against active attacks* if for all adversaries $\langle \tilde{\mathcal{P}}, \tilde{\mathcal{V}}\rangle$, for all constants $c > 0$, and for all sufficiently large k,

$$\Pr\left[\sigma = 1 \,\middle|\, \begin{array}{l} (I, S) \leftarrow \mathcal{G}(1^k); \\ h \leftarrow \langle \bar{\mathcal{P}}(I, S), \tilde{\mathcal{V}}(I)\rangle; \\ \sigma \leftarrow \langle \tilde{\mathcal{P}}(h), \bar{\mathcal{V}}(I)\rangle \end{array}\right] < k^{-c}.$$

2.2. The Gap Diffie-Hellman Problem

Okamoto and Pointcheval [8] define a new class of problems, called the *Gap-problem.* Let $f : \{0,1\}^* \times \{0,1\}^* \to \{0,1\}$ and $R : \{0,1\}^* \times \{0,1\}^* \times \{0,1\}^* \to \{0,1\}$ be any relation.

- The *inverting problem* of f is, given x, to compute any y such as $f(x, y) = 1$ if it exists, or to answer Fail.
- The *R-decision problem* of f is, given (x, y), to decide whether $R(f, x, y) = 1$ or not. Here y may be the null string, $\perp$.

The R-gap problem of f is to solve the inverting problem of f with the help of the oracle of the R-decision problem of f.

Okamoto and Pointcheval [8] claim that the DH problems are the typical instance of the Gap problem. Let $\mathbf{G}$ be any group of prime order q.

- The C-DH problem: given a triple of $\mathbf{G}$ elements (g, g^a, g^b), find the element $C = g^{ab}$.
- The D-DH problem: given a quadruple of $\mathbf{G}$ elements (g, g^a, g^b, g^c), decide whether $c = ab \mod q$ or not.
- The G-DH problem: given a triple of $\mathbf{G}$ elements (g, g^a, g^b), find the element $C = g^{ab}$ with the help of a D-DH oracle (which answers whether a given quadruple is a DH quadruple or not).

Now we formally define groups in which the Weil pairing works using notions defined above. Let $\mathbf{G}$ be a cyclic group of a prime order with an arbitrary generator. For any polynomial-time probabilistic algorithm $\mathcal{A}$:

- $\mathbf{G}$ is said to be a τ-breakable D-DH group if the D-DH problem can be computed on $\mathbf{G}$ by $\mathcal{A}$ whose running time is bounded by τ.
- $\mathcal{A}$ is said to (t, ϵ)-break C-DH problem in $\mathbf{G}$ if the C-DH problem can be solved by $\mathcal{A}$ whose running time is bounded by t, the success probability $\mathsf{Succ}^{\mathbf{G}}(\mathcal{A}) \geq \epsilon$.
- $\mathbf{G}$ is said to be a (τ, t, ϵ)-G-DH group if it is a τ-breakable D-DH group and no algorithm (t, ϵ)-breaks C-DH on it.

2.3. Weil Pairing

We can make use of any bilinear map on an elliptic curve to construct a group $\mathbf{G}$ in which the C-DH problem is intractable, but the D-DH problem is tractable. In particular, we make use of the Weil pairing among bilinear maps.

Let E be a elliptic curve over a base field K and let $\mathbf{G}_1$ and $\mathbf{G}_2$ be two cyclic groups of order q for some large prime p. The Weil pairing [11] is defined by a bilinear map e,

$$e : \mathbf{G}_1 \times \mathbf{G}_1 \rightarrow \mathbf{G}_2,$$

where $\mathbf{G}_1$ corresponds to the additive group of points of E/K, and $\mathbf{G}_2$ corresponds to the multiplicative group of an extension field $\bar{K}$ of K.

Let $P, Q \in \mathbf{G}_1$. The Weil pairing e has the following properties:

1. *Identity*: For all $P \in \mathbf{G}_1$, $e(P, P) = 1$.
2. *Alternation*: For all $P, Q \in \mathbf{G}_1$, $e(P, Q) = e(Q, P)^{-1}$.
3. *Bilinearity*: For all $P, Q, R \in \mathbf{G}_1$, $e(P+Q, R) = e(P, R) \cdot e(Q, R)$ and $e(P, Q+R) = e(P, Q) \cdot e(P, R)$.
4. *Non-degeneracy*: If $e(P, Q) = 1$ for all $Q \in \mathbf{G}_1$, then $P = \mathcal{O}$, where $\mathcal{O}$ is a point at infinity.

In addition to these properties, we have an efficient algorithm to compute $e(P, Q)$ for all $P, Q \in \mathbf{G}_1$.

3. Original Scheme and Our Attack

Let the modified Weil pairing be $\hat{e}(P, Q) = e(P, \phi(Q))$, where ϕ is an automorphism on the group of points of E. As noted in [1], the existence of the bilinear map e implies (1) DLP in $\mathbf{G}_1$ can be reduced to DLP in $\mathbf{G}_2$, (2) C-DH problem in $\mathbf{G}_1$ is still hard even though D-DH in $\mathbf{G}_1$ is easy.

In [5], for a security parameter k, the generation of a pair of secret and public parameters and the actions of the scheme are described as follows.

Key generation. On input k, the key generation algorithm $\mathcal{G}$ works as follows:

1. Generates two cyclic groups $\mathbf{G}_1$ and $\mathbf{G}_2$ of order m for some large prime p and a bilinear map $\hat{e} : \mathbf{G}_1 \times \mathbf{G}_1 \rightarrow \mathbf{G}_2$.
2. Generates an arbitrary generator $P \in \mathbf{G}_1$.

3. Chooses randomly $a, b, c \in \mathbb{Z}_m^*$ and computes $v = \hat{e}(P, P)^{abc}$.
4. The public parameter is $\mathsf{Pub} = \langle \mathbf{G}_1, \mathbf{G}_2, P, aP, bP, cP, \hat{e}, v \rangle$, and the secret parameter is $\mathsf{Sec} = \langle a, b, c \rangle$. And then publishes them.

Protocol actions between $\mathcal{P}$ and $\mathcal{V}$. The identification scheme includes several rounds, and each of these rounds is performed as follows:

1. $\mathcal{P}$ chooses $r_1, r_2, r_3 \in \mathbb{Z}_m^*$ at random, computes $x = \hat{e}(P, P)^{r_1 r_2 r_3}$, $Q_1 = r_1 P$, $Q_2 = r_2 P$, and $Q_3 = r_3 P$, and sends $\langle x, Q_1, Q_2, Q_3 \rangle$ to $\mathcal{V}$.
2. $\mathcal{V}$ picks $\omega \in \mathbb{Z}_m^*$ at random, and sends ω to $\mathcal{P}$.
3. $\mathcal{P}$ computes $y = \hat{e}(\omega P, P)^{abc} \cdot \hat{e}(P, P)^{r_1 r_2 r_3}$ and sends y to $\mathcal{V}$; $\mathcal{V}$ accepts if $y = v^{\omega} \cdot x$, and rejects otherwise.

The Base of Security. The proof of security is based on the intractability of the G-DH problem. Let $\mathbf{Z}$ be a probability space consisting of the uniform distribution over all integers in $\mathbb{Z}_m^*$. Let $\mathbf{G}$ be a probability space consisting of the uniform distribution over all elements of the form $nP \neq \mathcal{O} \in \mathbf{G}_1$, where $n \in \mathbf{Z}$. *The Gap Diffie-Hellman Intractability Assumption* is defined as the following: Given $C = \hat{e}(P, P)^{abc} \in \mathbf{G}_2$, for all polynomial-time probabilistic algorithm $\mathcal{A}$, for all constant $c > 0$, and for all sufficiently large k,

$$\Pr\left[C = C' \,\middle|\, \begin{array}{l} x \leftarrow \mathbf{Z}, xP \in \mathbf{G};\ \ y \leftarrow \mathbf{Z}, yP \in \mathbf{G};\ \ z \leftarrow \mathbf{Z}, zP \in \mathbf{G}; \\ C' \leftarrow \mathcal{A}(\hat{e}, xP, yP, zP) \end{array}\right] < k^{-c}.$$

This identification scheme is NOT secure at all. In fact, any adversary can impersonate a prover $\mathcal{P}$, provided he knows $\mathcal{P}$'s public key.

Attacking the Scheme. Suppose that an adversary $\mathcal{C}$ knows the public key $\mathsf{Pub} = \langle \mathbf{G}_1, \mathbf{G}_2, P, aP, bP, cP, \hat{e}, v \rangle$ of a prover $\mathcal{P}$. Then $\mathcal{C}$ can impersonate $\mathcal{P}$ as follows:

1. $\mathcal{C}$ chooses $r_1, r_2, r_3 \in \mathbb{Z}_m^*$ at random, computes $x = \hat{e}(P, P)^{r_1 r_2 r_3}$, $Q_1 = r_1 P$, $Q_2 = r_2 P$, and $Q_3 = r_3 P$, and sends $\langle x, Q_1, Q_2, Q_3 \rangle$ to $\mathcal{V}$.
2. $\mathcal{V}$ picks $\omega \in \mathbb{Z}_m^*$ at random, and sends ω to $\mathcal{C}$.
3. For $\mathcal{C}$ knows v, a parameter in the public key, x, a value chosen by himself, ω, a value get from $\mathcal{V}$, $\mathcal{C}$ can computes $y = v^{\omega} \cdot x$. Then he sends y to $\mathcal{V}$; $\mathcal{V}$ accepts for he can verify that $y = v^{\omega} \cdot x$.

4. Modified Scheme

For the identification scheme proposed in [5] is not secure at all, we modified the scheme as follows.

Key generation. The key generation algorithm $\mathcal{G}$ is same to the algorithm in the original scheme.

Protocol actions between $\mathcal{P}$ and $\mathcal{V}$. Identification scheme includes several rounds, each of these is performed as follows:

1. $\mathcal{P}$ chooses $r_1, r_2, r_3 \in \mathbb{Z}_m^*$ at random, computes $x = \hat{e}(P, P)^{r_1 r_2 r_3}$, $Q_1 = r_1 P$, $Q_2 = r_2 P$, and $Q_3 = r_3 P$, and sends $\langle x, Q_1, Q_2, Q_3 \rangle$ to $\mathcal{V}$.
2. $\mathcal{V}$ picks $\omega \in \mathbb{Z}_m^*$ at random and computes $R = \omega P$, and sends R to $\mathcal{P}$.

3. $\mathcal{P}$ computes $Q = abcR + r_1r_2r_3P$ and sends to $\mathcal{V}$; $\mathcal{V}$ accepts if $\hat{e}(Q,P) = v^{\omega} \cdot x$, and rejects otherwise.

Security Proof. Now, we first state the security of the modified scheme as follows:

Theorem 4.1. *Under the Gap Diffie-Hellman intractability assumption, the modified identification scheme on (τ, t, ϵ)-G-DH groups is secure against active attacks.*

To prove Theorem 4.1, it is good enough to show that any adversary $\mathcal{I}$ who succeeds in impersonating with non-negligible probability can be reduced into a polynomial-time probabilistic algorithm $\mathcal{A}$ that (τ, t, ϵ)-breaks C-DH problem with non-negligible probability. This is proved in Lemma 4.2.

Firstly, to construct such an adversary $I = (\tilde{\mathcal{P}}, \tilde{\mathcal{V}})$, we associate the adversary with the following polynomials:

- $T_{\mathcal{V}}(k)$: a time bound required for $\mathcal{V}$ to run the protocol once with $\bar{\mathcal{P}}$ including $\bar{\mathcal{P}}$'s computing time.
- $N_{\mathcal{V}}(k)$: an iteration bound for $\mathcal{V}$ to run the protocol with $\bar{\mathcal{P}}$.
- $T_{\text{off}}(k)$: an off-line time bound for $\mathcal{V}$ to spend other than running the protocol with $\bar{\mathcal{P}}$.
- $T_{\mathcal{P}}(k)$: a time bound for $\tilde{\mathcal{P}}$ to run the protocol with $\bar{\mathcal{V}}$.

Then, for a given public parameter Pub and "help string" h, let

$$\Pr[(\tilde{\mathcal{P}}(h), \bar{\mathcal{V}}(Pub)) = 1] = \varepsilon(h, \mathsf{Pub}),$$

where the probability is taken over the coin tosses of $\tilde{\mathcal{P}}$ and $\bar{\mathcal{V}}$. Since we assume that the adversary succeeds in breaking the protocol, there must exist polynomial Π_1 and Π_2 such that, for sufficiently large k,

$$\Pr\left[\varepsilon(h, \mathsf{Pub}) \geq \frac{1}{\Pi_2(k)} \,\middle|\, \begin{array}{l} (\mathsf{Sec}, \mathsf{Pub}) \leftarrow \mathcal{G}(1^k); \\ h \leftarrow (\bar{\mathcal{P}}(\mathsf{Sec}, \mathsf{Pub}), \tilde{\mathcal{V}}(\mathsf{Pub})) \end{array}\right] \geq \frac{1}{\Pi_1(k)}.$$

Lemma 4.2. *Assume that there exists an adversary $\mathcal{I}$ as above. Then there exists a polynomial-time probabilistic algorithm $\mathcal{A}$ that (t, ϵ)-breaks C-DH problem, whose running time τ is defined by*

$$O((N_{\mathcal{V}}(k)T_{\mathcal{V}}(k) + T_{\mathcal{P}}(k))\Pi_2(k) + T_{\text{off}}(k))$$

and for a valid C-DH value C, the success probability ϵ is bounded by

$$\Pr\left[C = C' \,\middle|\, \begin{array}{l} x \leftarrow \mathbf{Z}, xP \in \mathbf{G};\ y \leftarrow \mathbf{Z}, yP \in \mathbf{G};\ z \leftarrow \mathbf{Z}, zP \in \mathbf{G}; \\ C' \leftarrow \mathcal{A}(\hat{e}, xP, yP, zP) \end{array}\right] \geq \Pi_1(k)^{-1}/16.$$

Proof (sketch): First Let E denote an elliptic curve over a field K, with $E[m]$ its group of m-torsion points. Let Φ be a natural map in the modified Weil pairing. Throughout this paper, the underlying probability space consists of the random choice of input $x, y, z \in \mathbb{Z}_m^*$, and $P \in_R E(K)$ and the coin tosses of the algorithm. As a proving method, rather than constructing the algorithm $\mathcal{A}$ in toto, we will increasingly construct $\mathcal{A}$ in series of "phases".

Phase 1. In the first phase, we generate a public parameter $\mathsf{Pub} = \langle P, aP, bP \rangle$ with the corresponding secret parameter $\mathsf{Sec} = \langle a, b \rangle$. This phase takes as input

P, aP, bP, runs in the expected time $O(N_{\mathcal{V}}(k)T_{\mathcal{V}}(k)\Pi_2(k) + T_{\text{off}}(k))$; and outputs $\hat{X}_i$, where $\hat{X}_i \equiv a\gamma_i^f \mod m$, $f \not\equiv (m-1) \mod m$, and $\gamma_i \in \mathbb{Z}_m^*$ is picked randomly by $\mathcal{A}$, and h is a "help string". In addition, we know

1. $\Pr[\varepsilon(h, \mathsf{Pub}) \geq \Pi_2(k)^{-1}] \geq \Pi_1(k)^{-1}$,
2. the distribution of $\Phi(\hat{X}_i)$ is uniform and independent of that of (h, Pub).

This stage runs as follows: We choose $\gamma_i \in \mathbb{Z}_m^*$, $1 \leq i \leq |\mathbb{Z}_m^*| - 1$ at random and compute $\hat{X}_i \equiv a\gamma_i^f \mod m$. With the help of D-DH oracle, we can easily verify that $(aP, \gamma_i^f P)$ is a valid DH value. We then simulate the interaction $(\bar{\mathcal{P}}(\cdot, \mathsf{Pub}), \tilde{\mathcal{V}}(\mathsf{Pub}))$.

To simulate the interaction, we employ a zero-knowledge simulation technique [10]. We then modify the identification protocol as the following:

I. $\bar{\mathcal{P}}$ chooses $\omega', r_1, r_2 \in \mathbb{Z}_m^*$ at random, computes $x = \hat{e}(aP, \gamma^f P)^{\omega'} \cdot \hat{e}(P,P)^{r_1 r_2}$, $Q_1 = r_1 P$, $Q_2 = r_2 P$, and sends $\langle x, Q_1, Q_2 \rangle$ to $\tilde{\mathcal{V}}$.

II. $\tilde{\mathcal{V}}$ chooses $\omega \in \mathbb{Z}_m^*$ at random and computes $R = \omega P$, then sends R to $\bar{\mathcal{P}}$.

III. On receiving R, $\bar{\mathcal{P}}$ checks $\hat{e}(R, P) = \hat{e}((\hat{X}_i\omega_1 + \omega_0)P, P)$. If $\omega' \not\equiv \omega_0$, we go back to step I. Otherwise, $\bar{\mathcal{P}}$ computes $Q = a\gamma_i^f \hat{X}_i\omega_1 P + r_1 r_2 P$ and sends to $\tilde{\mathcal{V}}$.

When the adversary completes the protocol, we outputs the "help string" h that $\tilde{\mathcal{V}}$ outputs, along with $\hat{X}_i$.

In this step, the distribution of C is uniformly distributed in $\mathbf{G}_2$, and its distribution is independent of every variable other than in the adversary's view up to that point, and is also independent of the hidden variable ω'. Therefore, up to this point, this simulation is perfect, and furthermore, the probability that $\omega_0 = \omega'$ is $1/|\mathbb{Z}_m^*|$. If $\omega_0 = \omega'$, then

$$v^{\omega} \cdot x = \hat{e}(aP, \gamma^f P)^{\omega} \cdot \hat{e}(aP, \gamma^f P)^{\omega'} \cdot \hat{e}(P,P)^{r_1 r_2} = \hat{e}(P,P)^{a\gamma_i^f(\omega+\omega')+r_1 r_2},$$

$$\hat{e}(Q, P) = \hat{e}(P,P)^{a\gamma_i^f \hat{X}_i\omega_1 + r_1 r_2}.$$

For $\hat{e}(R, P) = \hat{e}((\hat{X}_i\omega_1 + \omega_0)P, P)$, we have $\hat{e}(Q,P) = v^{\omega} \cdot x = C$.

Moreover, C reveals no information of $\Phi(Q_1)$, $\Phi(Q_2)$, and $\Phi(\mathsf{Sec})$, and the distribution of $\Phi(y)$ is uniform and independent of $\Phi(\mathsf{Sec})$. From the above result, the expected value of the total number of iteration rounds is $|\mathbb{Z}_m^*| \cdot N_{\mathcal{V}}(k)$.

Phase 2. This phase takes as input h, Pub, and output from Phase 1, and runs in time $O(T_{\mathcal{P}}(k)\Pi_2(k))$. It outputs Fail or Success according to success outputs Z such that $Z \equiv a\gamma_i^f \equiv ab \mod m$, since $\hat{e}(P,P)^Z = \hat{e}(P,P)^{a\gamma_i^f} = \hat{e}(P,P)^{ab}$, where $f \not\equiv (m-1) \mod m$. The probability of success, given that $\varepsilon(h, \mathsf{Pub}) \geq \Pi_2(k)^{-1}$, is at least $1/2$. For the sake of convenience, let $\varepsilon = \varepsilon(h, \mathsf{Pub})$, and assume $\varepsilon \geq \Pi_2(k)^{-1}$.

This stage runs as follows: First run $(\tilde{\mathcal{P}}(h), \bar{\mathcal{V}}(\mathsf{Pub}))$ up to $\lceil \Pi_2(k) \rceil$ times, or until $\bar{\mathcal{V}}$ accepts. If $\bar{\mathcal{V}}$ accepts, let

$$\begin{aligned} \hat{e}(Q,P) &= \hat{e}(Z\omega P + r_1 r_2 P, P) = \hat{e}(\omega P, P)^Z \hat{e}(P,P)^{r_1 r_2} \\ &= \hat{e}(\omega P, P)^{ab} \hat{e}(P,P)^{r_1 r_2} = v^{\omega} \cdot x \end{aligned}$$

be the accepting conversation. Fixing the coin tosses of $\tilde{\mathcal{P}}$, run the interaction again up to $\lceil 3\Pi_2(k)\rceil$, or until $\bar{\mathcal{V}}$ accepts again with a challenge $\omega'' \not\equiv \omega \mod m$. In this case, let $\hat{X}_j \equiv a\gamma_j^f \mod m$. If $\bar{\mathcal{V}}$ accepts this challenge, then we have another accepting conversation

$$\begin{aligned}\hat{e}(Q', P) &= \hat{e}(Z\omega'' P + r_1 r_2 P, P) = \hat{e}(\omega'' P, P)^Z \hat{e}(P, P)^{r_1 r_2}\\ &= \hat{e}(\omega'' P, P)^{ab} \hat{e}(P, P)^{r_1 r_2} = v^{\omega} \cdot x,\end{aligned}$$

where $Z \equiv a\gamma_i^f \mod m$, $Z \equiv a\gamma_j^f \mod m$, and $\omega a\gamma_i^f \equiv \omega'' a\gamma_j^f \mod m$. Therefore, we can easily calculate $f = \log_{\frac{\gamma_j}{\gamma_i}} \omega - \log_{\frac{\gamma_j}{\gamma_i}} \omega''$.

We analyze this phase using a variant of a truncated execution tree as employed in [10, 2]. The analysis is similar to the proof in [5]. We can get that: for two accepting conversations, the probability that the above procedure succeeds is at least $(1 - \exp^{-1})^2/4$. Thus, the probability that one of seven experiments succeeds is at least $1/2$.

Phase 3. This phase takes as input, the output $\hat{X}_i$ from Phase 1, and the value Z from Phase 2. When Phase 2 succeeded, the probability that it solves the C-DH problem is $1/2$. Similar to the proof in [5], we can get that: for sufficiently large k, the overall success probability of the algorithm $\mathcal{A}$ is at least $\Pi_1(k)^{-1}/4$.

Phase 4. This phase repeatedly executes Phase 1 to Phase 3 to solve the C-DH problem, $\hat{e}(P, P)^{xc}$, where $x \equiv ab \mod m$. If phases from 1 to 3 succeed, this phase must succeed with the above probability.

Phase 5. If Phase 4 succeeds with given probability, it is equivalent to solving the C-DH problem $\hat{e}(P, P)^{xc} = \hat{e}(P, P)^{abc}$ with probability

$$\Pr[C = C'] = \Pi_1(k)^{-1}/16.$$

This completes the proof of Lemma 4.2. □

Therefore, we conclude that the modified identification scheme satisfies the requirement of security.

5. Conclusion

In this paper, we first pointed out that an identification scheme proposed by Kim and Kim [5] is not secure indeed. In fact, any adversary can impersonate a prover $\mathcal{P}$, provided he knows $\mathcal{P}$'s public key. After that, we proposed an improved scheme which is also based on the G-DH problem using the Weil pairing, and proved that the improved scheme is secure against active attacks.

Acknowledgements

The authors would like to thank Dr. Feng Bao for valuable comments and helpful discussion. This work is supported by the National Natural Science Foundation of China (Grant No.60073021) and the research collaboration program between

Institute for InfoComm Research, Singapore and Institute of Software, Chinese Academy of Sciences.

References

[1] D. Boneh and M. Franklin, "ID-based encryption from the Weil-pairing", *Advances in Cryptology - Crypto 2001, LNCS 2139*, Springer-Verlag, pp. 213–229, 2001.

[2] U. Feige, A. Fiat, and A. Shamir, "Zero-knowledge proofs of identity", *Journal of Cryptology*, Vol. 1, No. 3, pp. 77–94, 1988.

[3] A. Fiat and A. Shamir, "How to prove yourself: pratical solutions to identification and signature problems", *Advances in Cryptology - Crypto'86, LNCS 263*, Springer-Verlag, pp. 186–194, 1987.

[4] L. Guillou and J. Quisquater, "A practical zero-knowledge protocol fitted to security microprocessors minimizing both transmission and memory", *Advances in Cryptology - Eurocrypt'88, LNCS 330*, Springer-Verlag, pp. 123–128, 1989.

[5] M. Kim and K. Kim, "A new identification scheme based on Gap Diffie-Hellman problem", *The 2002 Symposium on Cryptography and Information Security*, 2002.

[6] A. Menezes, P. van Oorschot and S. Vanstone, *Handbook of Applied Cryptography*, CRC Press, 1996.

[7] T. Okamoto, "Provably secure and practical identification schemes and corresponding signature schemes", *Advances in Cryptology - Crypto'92, LNCS 740*, Springer-Verlag, pp. 31–53, 1993.

[8] T. Okamoto and D. Pointcheval, "The gap-problem: a new class of problems for the security of cryptographic schemes", *PKC 2001, LNCS 1992*, Springer-Verlag, pp. 104–118, 2001.

[9] C.P. Schnorr, "Efficient signature generation for smart cards", *Journal of Cryptology*, Vol. 4, No. 3, pp. 161–174, 1991.

[10] V. Shoup, "On the security of a practical identification scheme", *Journal of Cryptology*, Vol. 12, No. 4, pp. 247–260, 1999.

[11] J. H. Silverman, *The arithmetic of elliptic curves*, GTM 106, Springer-Verlag, 1986.

Gang Yao and Guilin Wang
Cryptography Lab, Infocomm Security Department
Institute for Infocomm Research (I^{2}R), Singapore 119613
e-mail: `yaogang@i2r.a-star.edu.sg`
e-mail: `glwang@i2r.a-star.edu.sg`

Gang Yao and Yong Wang
Laboratory of Computer Science, Institute of Software
Chinese Academy of Sciences, Beijing 100080, P.R. China
e-mail: `yaogcrab@ios.ac.cn`
e-mail: `wangyong@ios.ac.cn`

MIX
Papier aus verantwortungsvollen Quellen
Paper from responsible sources
FSC® C105338

If you have any concerns about our products,
you can contact us on
ProductSafety@springernature.com

In case Publisher is established outside the EU,
the EU authorized representative is:
Springer Nature Customer Service Center GmbH
Europaplatz 3, 69115 Heidelberg, Germany

Printed by Libri Plureos GmbH
in Hamburg, Germany